Absolute Allergy and Immunology Board Review

Massoud Mahmoudi
Editor

Absolute Allergy and Immunology Board Review

Springer

Editor
Massoud Mahmoudi
Department of Medicine
University of California San Francisco
San Francisco, CA, USA

ISBN 978-3-031-12866-0 ISBN 978-3-031-12867-7 (eBook)
https://doi.org/10.1007/978-3-031-12867-7

This Springer imprint is published by the registered company Springer Nature Switzerland AG
The registered company address is: Gewerbestrasse 11, 6330 Cham, Switzerland

To the memory of my father, Mohammad H, Mahmoudi, and to my mother, Zohreh, my wife, Lily, and my sons, Sam and Sina, for their continuous support and encouragement.

Preface

Every year hundreds of allergy and immunology fellows in training and practicing allergists take board examination. My personal search for finding prep books led to disappointment as there were limited resources available for such an important task. This prompted me to search for a distinguished team of academicians to help in preparing the present collection.

I have been fortunate to gather over 70 national experts in the field which comprises allergy and immunology fellowship training directors and academicians from various regional academic institutes and hospitals.

This book consists of 31 chapters. Each chapter presents two cases and each case is followed by five questions/answers to stimulate the reader's thought process and discussion about the topic.

We understand that preparation for the board examination requires years of didactic preparation and hands-on clinical practice. We, the authors, collectively believe that the present book will help to serve as a brief review of common allergy and immunology topics and to serve as a medium to inspire the thought process for preparation of the board examination.

Preparation of this book would not have been possible without the help and support of the publisher. I would like to express my thanks to Michelle Tam for overseeing this project and the editorial and production team at Springer who helped me at all stages of the preparation. Finally, I would like to thank Richard Lansing, Editorial Director of Clinical Medicine at Springer, who has been supportive in all my projects throughout the years with the publisher.

I appreciate hearing your comments and suggestions for use in what I hope will be the second edition of this title. Please contact me at allergycure@sbcglobal.net.

San Francisco, CA Massoud Mahmoudi

Contents

Innate and Adaptive Immunity

Kranthi Nomula, Hanadys Ale, and Christopher Chang

1 Introduction

The human immune system is a complex system of cells and molecules that resist and protect the body from infections, toxins, invasion by other foreign bodies, and tumors. Conceptually, the reason why organisms need an immune system is to ward off dangerous elements that pose a threat to their health and well-being. These dangerous elements may be exogenous or may originate from the host. The immune system must be programmed to be able to recognize what is dangerous and what is benign, otherwise our bodies will reject essential elements in our body and the surrounding environment that are necessary for our survival as a species. Examples of dangerous and non-dangerous exposures can be seen in Table 1.

The immune system can be broadly divided into innate and adaptive immunity. It is important to recognize that this is an artificial distinction. In fact, there are many parts of the immune system that span both innate and adaptive immunity. Other elements may bridge innate and adaptive immunity in order to allow the immune system to function to its most efficient capability. The major differences between the innate and adaptive immunity are mentioned in Table 2.

Table 1 Dangerous and non-dangerous entities

	Endogenous (self, etc.)	Exogeneous (non-self, external)
Non-dangerous	Fetuses, commensal bacteria	Food, pets, clothes, medications
Dangerous	Tumor cells	Pathogens (viruses, bacteria, fungal, mycobacterium)

Table 2 Differences between innate and adaptive immunity

	Innate immunity	Adaptive immunity
Antigen dependency	Antigen-independent	Antigen-dependent
Immunologic memory	Not present	Present
Components	Skin, epithelium, complement system, phagocytes, NK cells	Lymphocytes
Specificity	Nonspecific. Mounts the same response for molecules shared by related microbes and molecules	Specific. Mounts response specific to the antigens
Diversity	Limited diversity	Large diversity
Response time after exposure	Immediate. Develops within minutes to few hours	Delayed. Develops within hours to days

2 Innate Immunity

Case 1
A 3-month-old baby develops recurrent skin infections. Culture of the lesions shows *Burkholderia cepacia*.

Question 1
The test that would most likely reveal the diagnosis is

A. Lymphocyte subset panel
B. Natural Killer cell function
C. Neutrophil oxidative burst assay
D. Toll-like receptors

K. Nomula
Department of Pediatrics, Joe DiMaggio Children's Hospital, Memorial Healthcare System, Hollywood, FL, USA
e-mail: knomula@mhs.net

H. Ale · C. Chang (✉)
Division of Immunology, Allergy and Pediatric Rheumatology, Joe DiMaggio Children's Hospital, Memorial Healthcare System, Hollywood, FL, USA
e-mail: chrchang@mhs.net

M. Mahmoudi (ed.), *Absolute Allergy and Immunology Board Review*, https://doi.org/10.1007/978-3-031-12867-7_1

Answer and Explanation

Answer: B

While answers A, C, and D are all components of the innate immune system, unusual infections such as *Burkholderia* and *Nocardia* are commonly seen when phagocytes are unable to kill microorganisms. This is generally due to a defect in nicotinamide adenine dinucleotide phosphate (NADPH) oxidase, which generates reactive oxygen species (ROS) to kill pathogens. Chronic granulomatous disease is a severe primary immunodeficiency in which variants in genes coding for NADPH oxidase render the enzyme ineffective, leading to infections that would normally not be seen in healthy individuals.

Question 2

Which of the following is a feature of innate immunity?

A. Diversity
B. Rapidity
C. Specificity
D. Strength

Answer and Explanation

Answer: B

The advantage of the innate immune system is that it relies on the presence of highly conserved moieties in pathogens or other dangerous external threats to mount an initial rapid response. What it does not have is a finely tuned specific response to a particular antigen. It also does not have the ability to remember previous exposures so that a subsequent exposure will result in an amplified response. One way to think of the difference between innate and adaptive immunity is that innate immunity is akin to a shotgun, whereas adaptive immunity can be represented by a precision scoped rifle operated by a skilled marksman to kill its target.

Innate immunity is an immediate and first line of defense against microorganisms and other dangerous entities. This form of immunity is genetically determined since it uses germline-encoded receptors that recognize microbial products and other molecules. It is dependent on the immune system's "innate" ability to recognize molecular characteristics that are present in a wide variety of organisms or other dangerous entities. The advantage of this response is that it can be generated quickly, usually within minutes to a few hours after exposure to a pathogen. However, speed of response comes at a price, and innate immune responses are weaker and more temporary than those of the adaptive immune system. The innate immune system consists of multiple components that will be covered in this chapter.

Question 3

Which TLR is paired correctly with its ligand?

A. TLR-1/2 and CpG
B. TLR-3 and Poly I:C
C. TLR-5 and Lipopolysaccharide
D. TLR-8 and bacterial flagellin

Answer and Explanation

Answer: B

Lipopolysaccharide was the PAMP that Bruce Beutler studied which eventually won him the Nobel Prize in Physiology and Medicine in 2011 for his work on the activation of innate immunity. Lipopolysaccharide is the ligand for TLR-4. The ligand for TLR-5 is bacterial flagellin while the ligand for TLR-9 is CpG. The TLRs and their ligands are shown in Table 4. Polyinosinic:polycytidylic acid (Poly I:C) is an immunostimulant that can simulate viral infections. It is structurally similar to double-stranded RNA, the natural stimulant of TLR3.

The response is mounted after recognition of pathogen-associated molecular patterns (PAMPs) or danger (or damage)-associated molecular patterns (DAMPs) by corresponding molecular entities on the immune cell known as pathogen recognition receptors (PRR). PAMPs are conserved molecular sequences shared by microbes that mount an innate immune response. Examples of PAMPs include lipopolysaccharides, teichoic acid, double-stranded RNA, and flagellin. Examples of PRRs include toll-like receptors (TLRs), nucleotide oligomerization domain (NOD)-like receptors, retinoic acid-inducible gene (RIG)-like receptors, C-type lectin receptors (CLRs), and scavenger receptors (SRs). The structure and function of these receptor classes are shown in Table 3. The most well-known group of PRRs are the toll-like receptors which bind to specific molecules that are part of the PAMP. Some of these entities cross-react with synthetic molecules which allow for measurement of the function of TLRs in individual patients. Some of these molecules along with recognized ligands of TLRs are illustrated in Table 4.

The mechanisms of innate immunity can be categorized under anatomic, physiologic, phagocytic, and inflammatory responses. Table 5 shows the various barriers and their role in the immune system. PRRs upon binding to PAMPs lead to the production of pro-inflammatory cytokines which contribute to the removal or destruction of the pathogens. These pro-inflammatory cytokines are often responsible for the clinical symptoms or signs that occur when the body is mounting a defense against a pathogen. The reaction can be

Table 3 Pattern recognition receptors

Class of pattern recognition receptor	Structural characteristics	Function
C-type lectin receptors (CLR)	Contains a highly conserved domain for the recognition of carbohydrates in a calcium ion-dependent manner	Binds sugar molecules and plays a role in recognizing carbohydrate structures on the surface of pathogens
Nucleotide oligomerization domain (NOD)-like receptors	Cytoplasmic PRRs consist of a NOD domain, a domain for initiation of signaling (e.g., caspase activation and recruitment domain (CARD)), pyrin domains, or baculovirus inhibitor of apoptosis repeat (BIR) domains	Detection of intracytoplasmic bacterial products, such as bacterial peptidoglycans including mesodiaminopimelic acid (NOD1) and muramyl dipeptide (NOD2), also plays a role in the activation of the inflammasome
Retinoic acid-inducible gene (RIG)-like receptors	Members of DEAD-box (SF2) helicases. Binds viral RNA through the C-terminal domain. Catalytic helicase core consists of nine highly conserved sequence motifs which function to unwind RNA	Key sensors of viral infection detect atypical nucleic acids and plays a role in host antiviral responses by mediating transcriptional induction of type 1 interferons
Toll-like receptors	Type I integral transmembrane proteins usually with an N-terminal domain external to the membrane, a middle single helix transmembrane domain within the membrane, and a C-terminal domain in the cytoplasm. The N-terminal domain contains the ligand recognition site, whereas the C-terminal domain contains effector moieties for downstream signal transduction	Binds various ligands on the surface of viral and bacterial pathogens (see Table 4)
Scavenger receptors	Consist of an N-terminal cytoplasmic tail, a C-terminal scavenger receptor cysteine-rich (SRCR) domain, as well as a spacer region, collagenous domain, and an α-helical coiled-coil domain	Removal of foreign substances and waste in the body, plays a role in homeostasis, apoptosis, inflammation, and clearance of pathogens

Table 4 Toll-like receptors, NOD-like receptors, and their ligands

Pattern recognition receptor	Ligand(s)	Adapter protein
TLR-1	Triacyl lipopeptides, PAM3CSK4[a]	MyD88/MAL
TLR-2 (CD282)	Glycolipids, proteolipids, lipopeptides, lipoteichoic acid, HSP70, Zymosan[b], PAM3CSK4[a]	MyD88/MAL
TLR-3	Double-stranded RNA, poly I:C	TRIF
TLR-4	Lipopolysaccharide (LPS), heat shock proteins, fibrinogen, heparin sulfate, hyaluronic acid, nickel, opioid drugs	MyD88/ MAL,TRIF, TRAM
TLR-5	Bacterial flagellin, profiling	MyD88
TLR-6	Diacyl lipopeptides, Zymosan[b]	MyD88/MAL
TLR-7	Single-stranded RNA, imidazoquinolone (CLO97)[c], resiquimod, bropiramine, loxoribine	MyD88
TLR-8	Single-stranded viral RNA, small synthetic compounds, imidazoquinolone (CLO97)[c]	MyD88
TLR-9	Unmethylated CpG oligodeoxynucleotide DNA	MyD88
NOD-1 (CARD4)	Peptidoglycan iE-DAP dipeptide	
NOD-2 (CARD15)	Peptidoglycan—Muramyl dipeptide (MDP)	

Compounds used to test for TLR function are in red
[a]TLR1/TLR2 ligand
[b]TLR2/TLR6 ligand
[c]CLO is a water-soluble derivative of the imidazoquinolone R848; (Resiquimod) and is a TLR7/TLR8 ligand

Table 5 Mechanism of innate immunity

	Barrier	Function
Anatomic	Skin and mucous membranes	Protect against invasion of foreign organisms
Physiologic	Temperature	High temperature destroys infectious organisms. Also, produces interferons and helps recovery from infections
	Low pH	High acidity of stomach kills microorganisms
	Complement system	Bactericidal action against pathogenic bacteria
	Chemical mediators	
	Lysozyme	Breaks down the bacterial cell wall
	Beta lysin	Antibacterial action
	Lactic acid	Antibacterial action
	Lactoperoxidase	Antibacterial action
Phagocytic	Macrophages, dendritic cells, mast cells	Destroy microorganisms by phagocytic action
Inflammation		Tissue injury leads to mobilization of phagocytic cells to the damaged area by means of inflammation

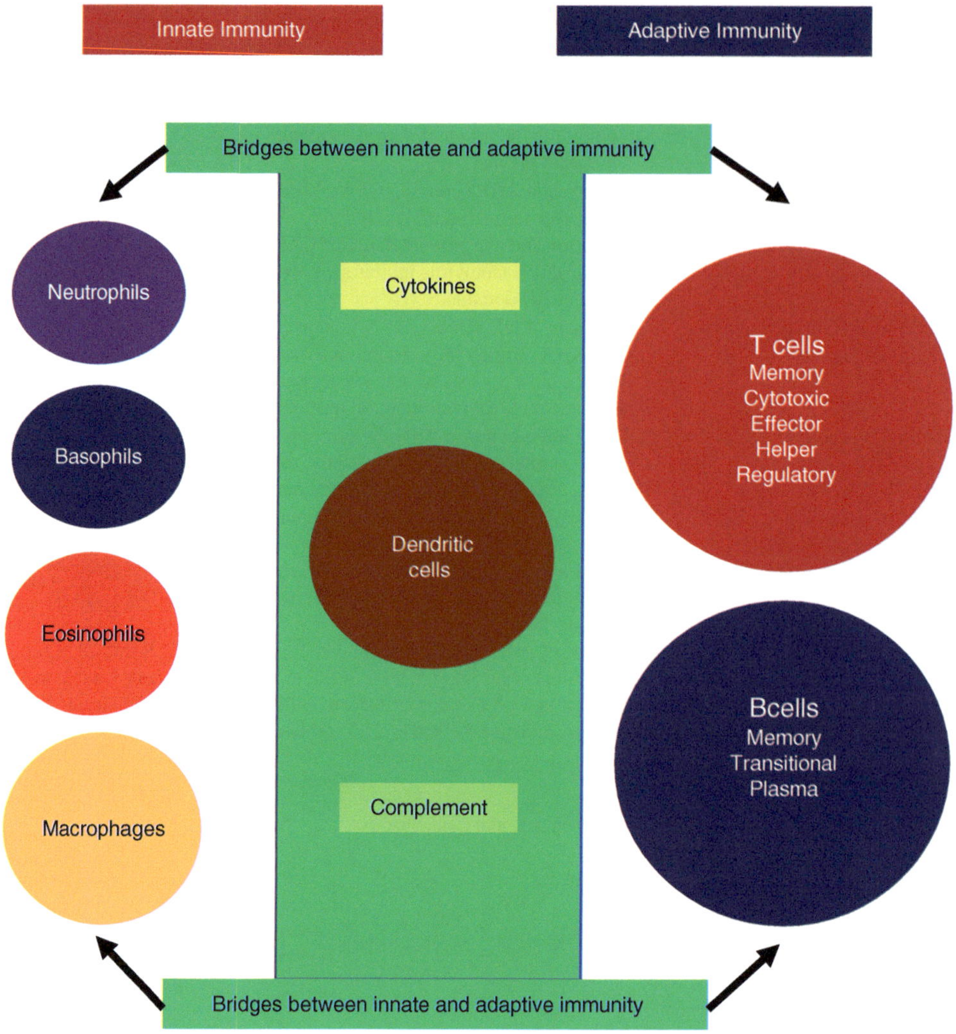

Fig. 1 Components of the innate and adaptive immune systems. (Source: Bioactive Nanoparticles for Cancer Immunotherapy - Scientific Figure on ResearchGate. Available from: https://www.researchgate.net/figure/Cells-and-components-involved-in-the-innate-and-adaptive-immune-system_fig1_329405685)

local or systemic. The hallmarks of this inflammation include heat, erythema, swelling, and pain, but when pro-inflammatory cytokines are produced unchecked, the reaction could be systemic, which can result in destruction and failure of host organ systems, leading to significant morbidity or even death.

The innate immune response also primes the adaptive immune system, which allows for a much stronger, longer lasting, and more specific response. Components of the innate and adaptive immunity are shown in Fig. 1.

3 Components of the Innate Immune System

3.1 Phagocytes

Cells that have a phagocytic function are known as phagocytes. Phagocytic cells can be broadly classified as immune cells and nonimmune cells. Neutrophils, macrophages, and monocytes form the immune component of phagocytic cells. Epithelial cells, dendritic cells, fibroblasts, and complements are some of the nonimmune phagocytic cells. The functions and their role in the immune system of some of these cells will be discussed below.

3.2 Neutrophils and Neutrophil Extracellular Traps (NETs)

Neutrophils are one of the most efficient cells of the innate immune system. Neutrophils migrate to the site of infection or inflammation by the process of chemotaxis. Neutrophils recognize pathogen patterns unique to microorganisms known as PAMPs, because of the expression of toll-like receptors (TLRs) and begin the process of phagocytosis. Neutrophils exhibit their phagocytic action by both, engulfment and secretion of neutrophil extracellular traps (NETs). Neutrophils have another interesting function by which they eradicate the microbes from the body, they extrude their nuclear contents, such as their DNA and histones by undergoing cell death in a process called NETosis. These contents are then referred to as neutrophil extracellular traps. As the name indicates, they trap and eliminate the pathogen.

Question 4

Which cytokine released by macrophages drives interferon-γ production by Th1 cells?

A. IL-2
B. IL-4
C. IL-12
D. IL-17

Answer and Explanation

Answer: C

IL-12 is a heterodimer that is secreted by macrophages, dendritic cells, and neutrophils upon stimulation by antigens. IL-12 stimulates the priming of CD4+ T cells and induces the production of interferon-γ, leading to the proliferation of Th1 cells. IL-2 is a T cell activator, and it will activate all subsets of T cells, but is obligatory for T regulatory cell proliferation. IL-4 is a Th2 cytokine which stimulates allergic reactions, while IL-17 is secreted by Th17 cells. A Th17/Treg cell imbalance that is skewed towards Th17 plays a role in the development of autoimmune diseases.

3.3 Monocytes and Macrophages

Monocytes are large white blood cells when migrated to the tissues are called macrophages. The major function of macrophages is phagocytosis and antigen presentation. Macrophages express Major Histocompatibility Complex (MHC) via which antigens are presented for recognition by lymphocytes. Another function of macrophages that plays an important role in the immune system is the production of cytokines. The production and resolution of cytokines are very intricately woven and any disruption in this leads to hyperinflammatory syndromes, such as macrophage-activating syndrome (MAS) or hemophagocytic lymphohistiocytosis (HLH).

3.4 Innate Lymphoid Cells

Innate lymphoid cells are bone marrow-derived cells with lymphocyte morphology that are predominantly residential to the lymphoid and nonlymphoid tissues and rarely found in circulation. They are present in large quantities at the mucosal surfaces. They do not have antigen-specific receptors, but they express receptors that are able to recognize cytokines. Although their overall numbers are not massive, their role in immunity is still significant. Three major types of innate lymphoid cells-ILC1, ILC2, and ILC3 have been identified. Imbalance in the function of innate lymphoid cells has been shown to cause respiratory infections, asthma, allergic rhinitis, chronic obstructive pulmonary disease, cancer, and gastrointestinal infections among others.

3.5 NK and NK-T Cells

Natural killer cells or NK cells are large granular lymphocytes that have protective action against virally infected cells and tumor cells. NK cells recognize antigens presented by MHC I and eliminate the infected cells by apoptosis. NK cells activity is measured by granzyme and perforin activity and CD107a expression. NK cells are not to be confused with NK-T cells, which possess T cell receptors but lack cytoplasmic granules. However, both NK cells and NK-T cells have a role in autoimmunity.

3.6 Epithelial Cells

Epithelial cells have an important role in the instigation and maintenance of both innate and adaptive immunity. The function of epithelial cells is of utmost importance in the innate immune system in the maintenance of protective barriers such as the skin. Like innate lymphoid cells, disruption of the normal functioning of epithelial cells has been shown to cause respiratory infections, asthma, allergic rhinitis, chronic obstructive pulmonary disease, cancer, and gastrointestinal infections among others.

Question 5

Deficiencies in which complement factor can lead to *Neisseria meningitides* infection

A. C2b
B. C4
C. Factor B
D. Properdin

Answer and Explanation

Answer: D

In general, complement deficiencies are rare. The same can be said of properdin deficiency. Properdin is the only positive regulator of the alternative complement pathway. Properdin is released by activated neutrophils, macrophages, and T cells and functions to stabilize the C3 convertase by binding with C3b thereby preventing cleavage by Factor H and Factor I. Properdin deficiency has been associated with severe *Neisseria meningitidis* infection. Deficiency of C4 is associated with autoimmune disease, especially systemic lupus erythematosus (SLE), while Factor B deficiency has even more rarely been reported. C2b is a cleavage product of C2 and is a component of the C3 convertase of the classical complement pathway.

3.7 Complement

The complement system consists of several heat-labile proteins that represent an important branch of innate immunity. These unique set of plasma proteins act as a bridge between innate and adaptive immunity. The Complement pathways as the name suggests complement the innate immune system and help in the destruction of pathological microorganisms via opsonization, phagocyte recruitment, and direct killing of microbes. Classical, alternative, and lectin pathways are the major pathways of the complement system. Although the names and the initiating mechanisms of the three pathways are different, they lead to the production of similar effector molecules. Opsonization, the process by which cells are rendered favorable for phagocytosis is mediated by the complement system. Byproducts of the complement pathways act as chemoattractants and amplify the immune response. The terminal complement is formed by a complex of proteins (C6, C7, C8, and C9) that are assembled into a membrane pore called the membrane attack complex (MAC) and damage bacterial cells by cell lysis.

Question 6
Delayed separation of the umbilical cord should raise suspicions of a defect in

A. CD18
B. CD25
C. CD40
D. CD146

Answer and Explanation
Answer: A
Delayed separation of the umbilical cord has long been accepted as a sign of leukocyte adhesion defect (LAD). There are two main forms of LAD, LAD1 and LAD2. LAD1 results from defects in the *ITGB2* gene, which encodes for CD18. CD18, along with CD11b is a component of complement receptor 3 (CR3).

4 Adaptive Immunity

Case 2
A 4-year-old female has a history of recurrent upper respiratory infections since age 6 months. She also has had a history of two pneumonias per year for the past 2 years and a history of recurrent ear infections which required pressure equalizer (PE) tube surgery. Ear infections resolved after the surgery. On immune evaluation, she was noted to have sub-protective pneumococcal antibody titers for 18 of 23 serotypes. Her immunizations are up-to-date.

Question 1
What is the most appropriate next step?

A. Administer pneumococcal polysaccharide vaccine and repeat titers after 4–6 weeks
B. Obtain genetic testing
C. Tonsillectomy/adenoidectomy
D. Treat with prophylactic antibiotics

Answer and Explanation
Answer: A
If a 4-year-old patient has a history of recurrent infections and pneumococcal titers are low, the first step would be to test humoral immunity function. We can do this by testing the patient's response to a vaccine. One would expect that a patient with a functioning humoral immune system would be able to produce specific antibodies to the majority of pneumococcal serotypes following vaccination. The generally accepted threshold for protection for most laboratories is 1.3 μg/mL, although this is somewhat of an arbitrary standard. In addition, there is some disagreement as to how many of the serotypes would be required to show a response. Most immunologists would agree that at least 60% of the serotypes should respond. Another way to interpret the results is that a response is acceptable when there is a fourfold rise in titers, but this may not always be present especially if titers of some of the serotypes are high to begin with. For this patient, it is probably too early in the evaluation to treat with prophylactic antibiotics or to obtain genetic testing. Surgical intervention is also not indicated in this patient.

Adaptive immunity is a more sophisticated type of immune response. It takes over from the innate immune system to generate a stronger and more specific response which takes longer to develop. Adaptive immunity is antigen-dependent and is mediated by lymphocytes. Lymphocytes are primarily divided into B lymphocytes and T lymphocytes. Other subsets of lymphocytes include Natural Killer (NK)-T lymphocytes which are T lymphocytes that have innate effector functions. These lymphocytes recognize glycolipid molecules in the context of MHC class I-related glycoprotein CD1d.

Lymphocytes help the body eliminate pathogens by developing immunologic memory to mount rapid response if reinfected, and recognize self-antigens from nonself-antigens. The two facets of adaptive immunity are humoral

Table 6 Differences between humoral and cell-mediated immunity

	Humoral Immunity	Cell-mediated Immunity
Major mediator	Antibodies	T lymphocytes
Other cells involved	B lymphocytes, macrophages	Macrophages and natural killer cells
Intracellular vs extracellular microbe	Works against extracellular microbes and the toxins produced by them	Works against intracellular microbes
Processed vs unprocessed antigens	Unprocessed antigens	Antigens are processed and presented by MHC complex

immunity and cell-mediated immunity. Like innate and adaptive immunity, humoral and cell-mediated immunity are also codependent and function in unison to mount appropriate immune response. Table 6 shows the major differences between humoral and cell-mediated immunity.

5 Humoral Immunity

Question 2

Which receptor facilitates active transport of immunoglobulin across the placenta

A. FcγR1
B. FcγRIII
C. FcRn
D. FcεR1

Answer and Explanation

Answer: C

IgG is the only immunoglobulin class with the ability to cross the placenta. It does so through an active transport mechanism facilitated by the neonatal Fc receptor (FcRn).

Humoral immunity or antibody-mediated immunity is mediated by B lymphocytes. Receptors on B lymphocytes, B cell receptors (BCRs) recognize antigens on antigen-presenting cells (APCs). Cytokines help in proliferation of the B cells and help in the transformation of B cells into plasma cells. Plasma cells are antibody powerhouses that bind to the extracellular pathogens or toxins produced by the pathogens. The binding of antigen and antibody enhances the process of phagocytosis. This is called opsonization.

B cells produce five types of immunoglobulins: IgG, IgA, IgM, IgD, and IgE. Table 7 provides an overview of the properties and functions of these immunoglobulins. In the latter parts of pregnancy, IgG is actively transported across the placenta to provide passive immunity to the fetus and newborn while the infant begins to produce his or her own immunoglobulin G around 6 months of age.

5.1 How Does the Adaptive Immune System Achieve Its Goals of Strength, Specificity, and Diversification?

The immune repertoire can refer to both humoral and cellular immunity. This term refers to the diversity of the immune system in recognizing and distinguishing millions of different epitopes on various molecules and organisms. The immune repertoire develops over time. Indeed, it is estimated that over 1600 genes are involved in the development of the human immune system, and even before birth, our immune systems have the ability to recognize innumerable antigenic epitopes by virtue of various physiologic functions involved in the development of tolerance. Through positive and negative selection, our immune system gradually learns to repel dangerous antigens while accepting safe ones, including self-antigens and maternal and environmental exposures. The ability to enhance immune responses to certain antigens through positive and negative selection, and the mechanisms of central and peripheral tolerance eventually shape an individual's immune repertoire.

As our immune system matures and we are gradually exposed to new antigens, our cells differentiate into specialized cells that can recall previous exposures (as in memory cells), and cells that possess effector function. A broad range of antigens can ultimately be recognized by our immune system through biochemical processes such as V(D)J recombination and somatic hypermutation. V(D)J recombination contributes to the assembly of the variable domain of immunoglobulin and T cell receptor genes. This process is regulated by precise DNA cleavage at short conserved sequences under the control of Recombination-activating gene (RAG) proteins. Further modification of immunoglobulin genes takes place through somatic hypermutation which generates mutations in variable gene segments leading to affinity maturation. This satisfies the property of amplification, generating a stronger response, in adaptive immunity.

Isotype class switching, on the other hand, alters the heavy chain immunoglobulin constant region to produce antibodies other than IgM. This process dictates the effector function of the antibody molecule. Somatic hypermutation and heavy chain isotype (class) switching are under the control of activation-induced cytidine deaminase (AID) which causes DNA deamination of transcribed target DNA. All of these processes contribute to generate the characteristics of adaptive immunity—a stronger response with tremendous diversity and specificity. The molecular details of how this happens are beyond the scope of this chapter.

Table 7 Properties and functions of the immunoglobulins

Property	IgG	IgM	IgA	IgE	IgD
Molecular weight (KDa)	153	950	162	190	185
Subunits	Monomeric	Pentameric	Monomeric in serum, dimer in secretions	Monomeric	Monomeric
Concentration in serum (varies with age and lab)	560–1800 mg/dl	39–283 mg/dl	82–470 mg/dl	~100kU/L, but varies with age	Very small amounts (0.25% of immunoglobulins)
Subclasses	IgG1, IgG2, IgG3, IgG4	IgM	IgA1, IgA2	IgE	IgD
Complement binding	Yes	Yes	No	No	No
Function	Sustained immune response	Early response to pathogens	Mucosal immunity	Allergic diseases, parasitic infections	Signal for B cell activation, activator of mast cells and basophils
Serum half-life (days)	~23	5–6	5–6	2–3	2.8 days (secreted IgD)

6 Cell-Mediated Immunity

Cell-mediated immunity is mediated by T lymphocytes. T lymphocytes do not produce antibodies. Instead, the receptors on T cells, or T cell receptors (TCRs), recognize the proteins presented by the major histocompatibility complex (MHC) and modulate the immune response by stimulating B cells to produce antibodies. The interaction between antigen-presenting cells and T lymphocytes is a critical step in the completion of an immune response. In addition to the effects on B cells, T lymphocytes also drive the production of cytokines for an amplified immune response and can lead to inflammatory responses in the same way that the innate immune system drives these responses. These inflammatory responses when not regulated appropriately can lead to a hyperinflammatory state that can result in multiorgan system damage and failure. In addition, cytotoxic T lymphocytes may also induce apoptosis of pathogen-infected cells or tumor cells. Other functions of T lymphocytes include macrophage activation which can lead to phagocytosis and oxygen-radical induced killing of microbes. Cell-mediated immunity is effective in the destruction of not just intracellular pathogens, but also of virus-infected cells and tumor cells.

7 Components of the Adaptive Immune System

Question 3

Which mitogen stimulates proliferation of both B and T lymphocytes

A. Concanavalin A
B. Lipopolysaccharide
C. Phytohemagglutinin A
D. Pokeweed Mitogen

Answer and Explanation

Answer: D

Concanavalin A and Phytohemagglutinin A are T cell mitogens. Lipopolysaccharide is a non-lectin mitogen for human B cells. Pokeweed Mitogen can stimulate both human T and B cells to proliferate.

7.1 B Lymphocytes

B lymphocytes are generated and mature in the bone marrow. After B lymphocytes are released into the peripheral circulation, the process of maturation does not begin until they encounter an antigen. The functions of B cells are antigen presentation, cytokine secretion, and antibody production. The common lymphoid progenitor progresses to form the pro-B cells which further progress to form pre-B cells. Pre-B cells express "μ" heavy chain and "κ" or "λ" light chains. Pre-B cells mature to form immature B cells and ultimately mature B cells, which express both IgM and IgD receptors. The B cell receptor is a transmembrane immunoglobulin complex that recognizes antigens.

7.2 T Lymphocytes

T cells originate in the bone and mature in the thymus. There are two types of lymphocytes—helper T cells and cytotoxic T lymphocytes (CTL). The T cell receptor is incapable of recognizing antigens unless processed and presented by the antigen-presenting cells (APCs) via the major histocompatibility complex (MHC). Cytotoxic T cells recognize antigen patterns when presented by the MHC class I. Once activated, cytotoxic T cells induce apoptosis of the infected cells. Helper T cells do not have cytotoxic action. Helper T cells recognize antigen patterns presented by the MHC class II. This leads to the activation of the helper T cells and production of cytokines that synchronize the activity of other cells such as APCs, B cells, and CTLs. The function of T and B lymphocytes can be evaluated using mitogens to stimulate their growth and proliferation.

7.3 Antigen-Presenting Cells

Antigen-presenting cells (APCs) play a mediator role in the adaptive immune system. APCs possess the MHC which enables antigen presentation to T lymphocytes. Although many cell lines have the capacity to present antigens, dendritic cells, B cells, and macrophages are designated as "professional" antigen-presenting cells. Professional APCs are efficient in internalizing, processing, and expressing antigens on the MHC. Dendritic cells are present at the interfaces between the body and the environment, such as the skin, mucosal lining of nose, stomach, and intestines.

Question 4

An inhibitory molecule in the T cell receptor which helps to regulate activation of a T cell response is

A. CD28
B. CD80
C. CTLA-4
D. LAT

Answer and Explanation

Answer: C

CD28 is a costimulatory molecule within the T cell receptor. CD80, also known as B7–1, is on B cells and interacts with CD28. It also reacts with CTLA-4, but in an inhibitory manner. LAT, or Linker for Activation of T cells, is an adaptor protein within the TCR signaling pathway.

7.4 The Function of CD4+ and CD8+ T Cells

When the body encountered a dangerous external antigen, the epithelial barrier offers the first line of protection to prevent entry of the antigen into the body. If the antigen escapes this initial protection, they may be taken up by antigen-presenting cells and presented to the T cell receptor on the surface of T lymphocytes by the MHC on the surface of the APC. The interaction of the TCR and the MHC induces the release of cytokines which modulates the immune response. For invading organisms, this contributes to the "amplification" property of the adaptive immune system, although an opposite effect may also be possible if the antigen is deemed not to be dangerous. T cells can differentiate into helper CD4+ T cells or cytotoxic CD8+ T cells. CD4+ T cells are activated after antigen presentation by the MHC Class II complex. These cells have no inherent phagocytic activity, but act as mediators and produce cytokines such as interferon-γ that assist in the elimination of the pathogen via macrophage activation. Activated CD4+ T cells help B cells to undergo clonal expansion which leads to the production of large amounts of antibodies. These are known as effector B cells or plasma cells. These antibodies can bind to dangerous antigens to induce their inactivation or removal from the body.

Cytotoxic CD8+ T cells are activated after presentation of the antigen by MHC Class I. They also have no phagocytic activity but after interacting with MHC, clonal expansion of CD8+ Th cells ensues. As the name suggests, activated CD8+ T cells can initiate direct killing of cells that have been infected by a virus. As in the case of CD4+ cells, cytokines are also released which can lead to amplification of the response against viral pathogens that have been taken up by antigen-presenting cells. The transporter associated with antigen processing 1 (TAP1) mediates antigen transfer from the cytosol to the endoplasmic reticulum within the cell. Deletion of the genes encoding TAP1 impairs presentation of antigen by MHC Class I. Another component of the MHC Class I complex is β-2-microglobulin, which is encoded by the *B2M* gene. β-2-microglobulin is required for transport of MHC Class I to the plasma membrane.

There are many other types of T cells. T regulatory cells help to temper the activity of activated CD4+ and CD8+ T cells. They are stimulated by the action of IL-2 and generally carry the surface marker FOXP3. These cells are important because they prevent the immune system from reacting to non-harmful antigens or even self-antigens. For that reason, they play a significant role in preventing autoimmune diseases (Fig. 2).

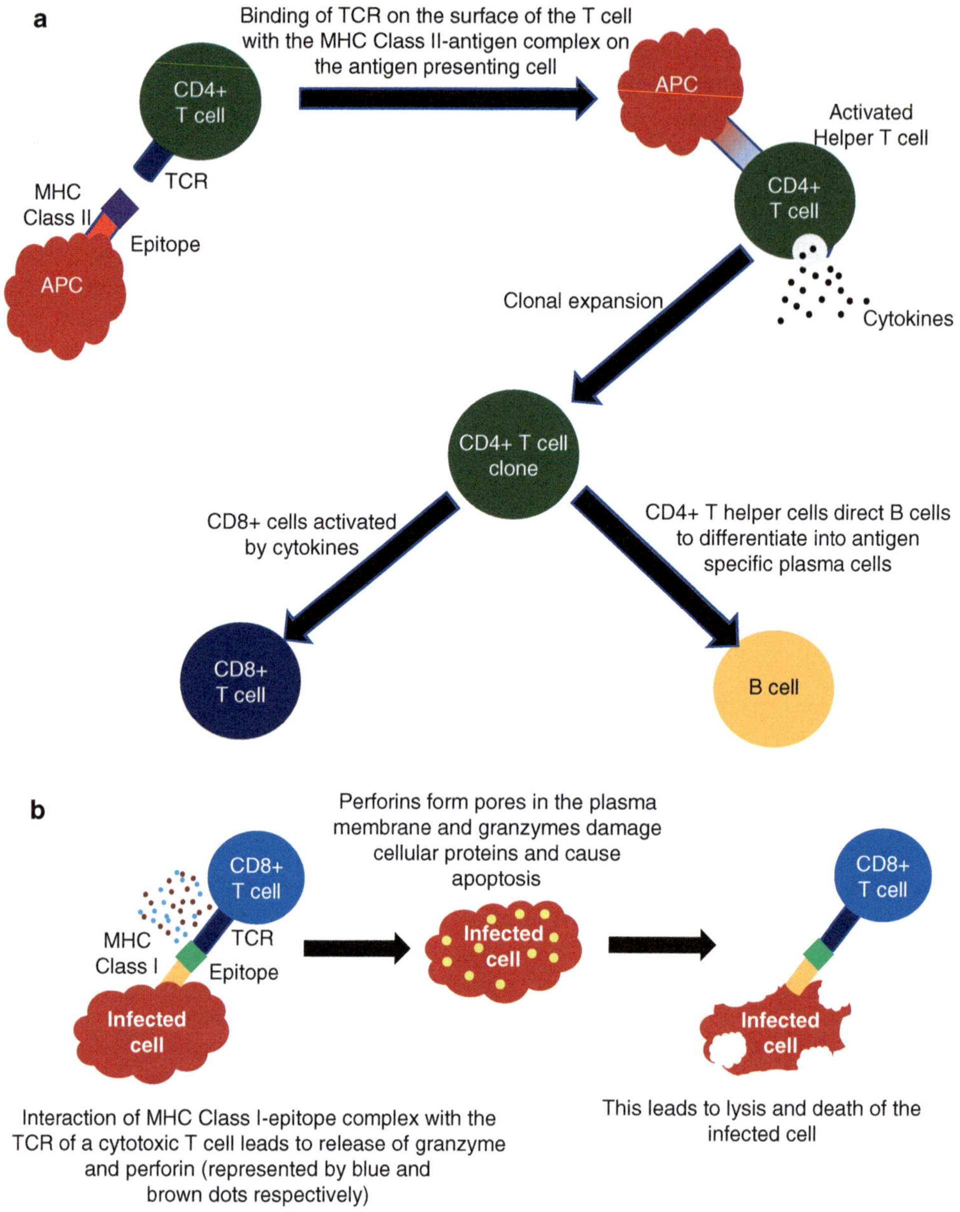

Fig. 2 CD4+ T cells and CD8+ T cells illustarted. (**a**) The function of CD4+ helper T cells (**b**) The function CD8+ cytotoxic T cells. (Source: https://opentextbc.ca/biology/chapter/23-2-adaptive-immune-response/)

Question 5
Which is a T cell independent vaccine?

A. DTaP
B. Hepatitis B vaccine
C. Pneumovax-23
D. SARS-CoV-2 mRNA vaccine

Answer and Explanation
Answer: C

When our immune system detects protein antigens, these antigens are taken up by antigen-presenting cells and presented via MHC Class II complexes to the T cell receptor on T helper cells, which then trigger antigen-specific B cells to produce antibody. This process occurs for not only natural epitopes but vaccines. Thus, a protein conjugate vaccine such as DTaP or a recombinant protein vaccine such as Hepatitis B would require T cell cooperation with B cells to mount a humoral response. However, other nonprotein antigens, such as polysaccharides, as is the case in the Pneumovax or PPSV23 vaccine, directly stimulate production of antibody by B cells without the help of T helper cells. These are referred to as T cell independent vaccines. In the case of SARS-CoV-2 mRNA vaccines, the mRNA is packaged along with the translation apparatus necessary to generate spike protein epitopes, and theoretically this would then go through the same mechanism as other protein antigens through T cell receptor recognition of MHC Class II presentation of these antigens to stimulate antibody production by B cells.

8 Summary and Conclusions

With each year, more and more information regarding the immune system is being revealed. There are millions of molecules, cells, networks, pathways, and interactions that must function in concert to maintain good health. For the purpose of organizing these complex interactions, concepts such as innate and adaptive immunity have been introduced in recent years. These concepts help build a foundation for understanding the immune system. It would be ludicrous to try and cover all the knowledge that has been accumulated over the years on human immunity and it would be impossible for any textbook to achieve this feat, let alone a single chapter. Therefore, it was not the intent of this chapter to discuss everything about innate and adaptive immunity, but to introduce the foundations on which to build the student's understanding and organization of this knowledge. It is important to remember that these classifications are artificial and not perfect, as innate immunity and adaptive immunity are not two completely separate entities but are interconnected by multiple cells and molecules that support a functioning and effective operation of our immune system to fend off dangerous elements and maintain human health.

Bibliography

1. Alberts B, Johnson A, Lewis J, et al. Molecular biology of the cell. 4th ed. New York: Garland Science; 2002. Innate Immunity. https://www.ncbi.nlm.nih.gov/books/NBK26846/
2. Alberts B, Johnson A, Lewis J, et al. Molecular biology of the cell. 4th ed. New York: Garland Science; 2002. Chapter 24, The Adaptive Immune System. https://www.ncbi.nlm.nih.gov/books/NBK21070/
3. Schleimer RP, Kato A, Kern R, Kuperman D, Avila PC. Epithelium: at the interface of innate and adaptive immune responses. J Allergy Clin Immunol. 2007;120(6):1279–84. https://doi.org/10.1016/j.jaci.2007.08.046.
4. Merle NS, Noe R, Halbwachs-Mecarelli L, Fremeaux-Bacchi V, Roumenina LT. Complement system part II: role in immunity. Front Immunol. 2015;6:257. https://doi.org/10.3389/fimmu.2015.00257.
5. Kaplan MJ, Radic M. Neutrophil extracellular traps: double-edged swords of innate immunity. J Immunol (Baltimore, Md: 1950). 2012;189(6):2689–95. https://doi.org/10.4049/jimmunol.1201719.
6. Lim JJ, et al. Diversity and versatility of phagocytosis: roles in innate immunity, tissue remodeling, and homeostasis. Front Cell Infect Microbiol. 2017;7:191. https://doi.org/10.3389/fcimb.2017.00191.
7. Panda SK, Colonna M. Innate lymphoid cells in mucosal immunity. Front Immunol. 2019;10:861. https://doi.org/10.3389/fimmu.2019.00861.
8. Vivier E, Tomasello E, Baratin M, et al. Functions of natural killer cells. Nat Immunol. 2008;9:503–10. https://doi.org/10.1038/ni1582.
9. Rudolph MG, Stanfield RL, Wilson IA. How TCRs bind MHCs, peptides, and coreceptors. Annu Rev Immunol. 2006;24:419–66.

Immune Cells and Functions

Kenneth Paris and Andrew M. Abreo

Case 1

A 7-month-old boy with failure to thrive and recurrent pneumonia was admitted to the hospital in respiratory distress. Past medical history was notable for two previous episodes of pneumonia and suppurative lymphadenitis requiring incision and drainage with wound cultures growing normal flora. There is no family history of primary immunodeficiency. His birth history was unremarkable.

The patient's symptoms included productive cough and tachypnea. Clinical symptoms did not improve despite antimicrobial therapy for community-acquired pneumonia. Computed tomography (CT) scan of the chest was obtained which revealed mediastinal lymphadenopathy and right lower lobe consolidation with scattered abscesses (Fig. 1). Bacterial and fungal blood cultures were negative throughout the admission. IR-guided drainage of a lung abscess revealed growth of *Phaeoacremonium parasiticum* (Fig. 2). The patient was treated with broad-spectrum antibiotic and antifungal therapy. The neutrophil oxidative burst was abnormal compared to normal control. The remainder of the immunologic evaluation was normal. Genetic testing identified a pathogenic variant in CYBB confirming the diagnosis of X-linked chronic granulomatous disease (CGD).

Question 1

Which cytokine is critical for neutrophil recruitment?

A. Interleukin 4 (IL-4)
B. Interleukin 5 (IL-5)
C. Interleukin 6 (IL-6)
D. Interleukin 8 (IL-8)
E. Interleukin 13 (IL-13)

Answer: The correct answer is D

Phagocytes, including neutrophils and macrophages, are a major component of the innate immune system. These cells are recruited to sites of infection to eliminate microbes and damaged cells. Neutrophils are the most abundant phagocyte in the bloodstream. They are also known as polymorphonuclear leukocytes (PMNs) because the nucleus is composed of multiple, connected lobules. Neutrophil production in the bone marrow is stimulated by granulocyte colony-stimulating factor (G-CSF) and granulocyte-macrophage colony-stimulating factor (GM-CSF). G-CSF and GM-CSF are produced by cells at the site of infection to replenish short-lived and depleted neutrophils. G-CSF interacts with the G-CSF receptor (CD114) on myeloid progenitor cells to drive the differentiation of myeloblasts and neutrophils. Neutrophils are recruited from the bone marrow to sites of infection by various inflammatory cytokines, such as interleukin 8 (IL-8, CXCL8) and leukotriene B4 (LTB4). Chemotaxis of neutrophils from the blood into tissues depends on choreographed interactions between selectins, cytokines, integrins, and cell-cell junction proteins. IL-8 (CXCL8) interacts with CXCR1 and CXCR2 on neutrophils to increase the affinity of the LFA-1 (CD11aCD18) and Mac-1 (CD11bCD18) integrins to their ligands ICAM-1 and ICAM-2 on the endothelial surface. Stable integrin-mediated arrest of neutrophils to the endothelium is required prior to transmigration into the tissue site.

Question 2

Which neutrophil granule contains human alpha-defensins?

A. Specific granules
B. Azurophilic granules
C. Secretory granules
D. Gelatinase granules
E. Cytotoxic granules

Answer: The correct answer is B

Neutrophils contain membrane-bound granules filled with preformed proteins. The most abundant are azurophilic

K. Paris (✉) · A. M. Abreo
Division of Allergy and Immunology, Department of Pediatrics, Louisiana State University Health Sciences Center, Children's Hospital New Orleans, New Orleans, LA, USA
e-mail: kparis@lsuhsc.edu; aabre1@lsuhsc.edu

M. Mahmoudi (ed.), *Absolute Allergy and Immunology Board Review*, https://doi.org/10.1007/978-3-031-12867-7_2

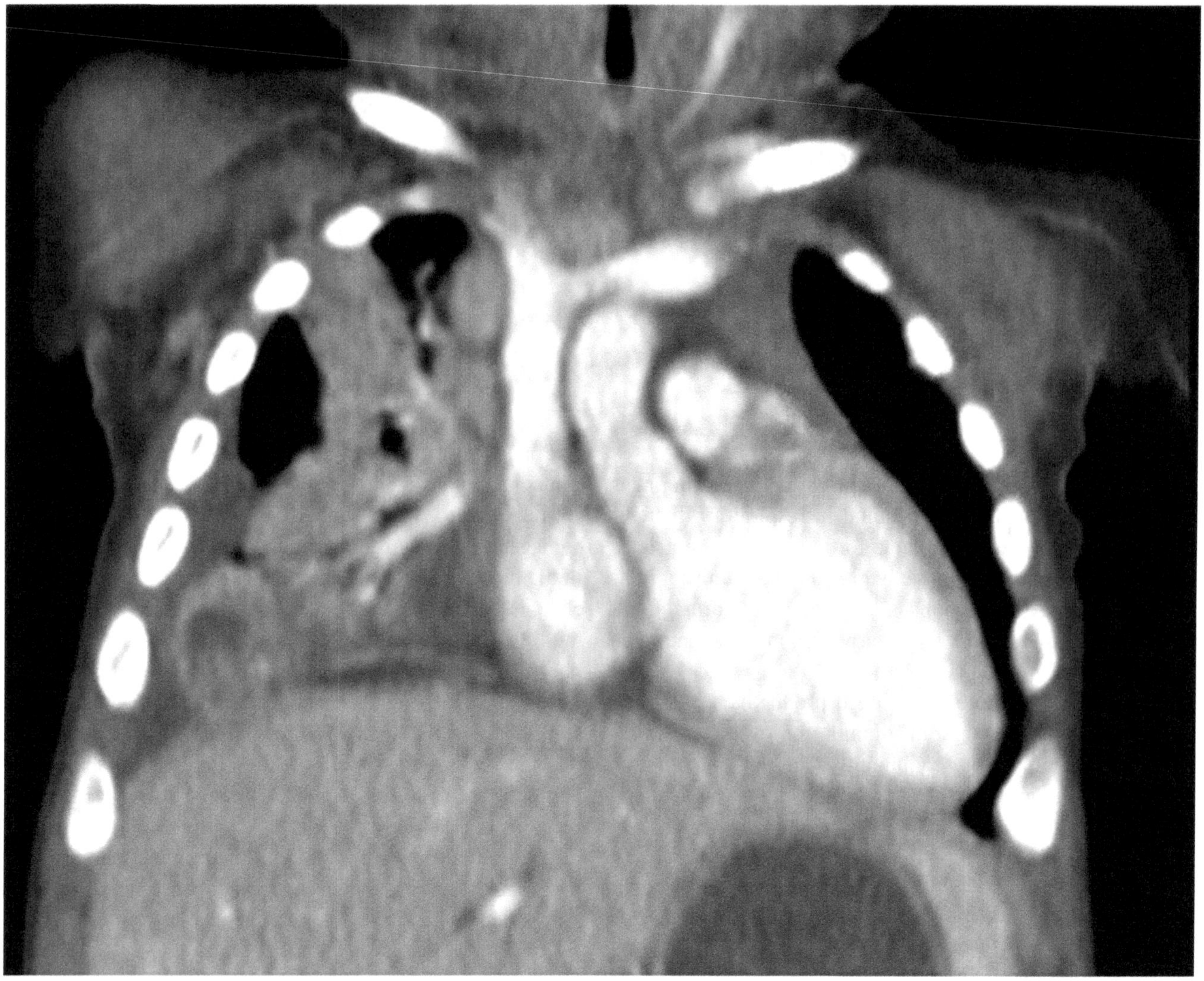

Fig. 1 CT scan of the lung showing right lung pneumonia with parenchymal abscesses

(primary), specific (secondary), and gelatinase granules. Neutrophilic granule contents stain neutral pink with hematoxylin and eosin staining, unlike eosinophils (bright red) and mast cells (dark blue). The granules contain a variety of preformed antimicrobial peptides and proteases that contribute to the degradation of the extracellular matrix and microbicidal activity. Specific granules contain a variety of proteins including lactoferrin, cathelicidin, and matrix metalloprotease 8. Azurophilic granules contain human alpha-defensins, myeloperoxidase, proteinase 3, elastase, and cathepsin G/C. Defensins are cationic peptides that insert into the microbial membrane leading to disruption and death. Defensins and other granule contents also interact with chromatin extruded from neutrophils to form neutrophil extracellular traps (NETs). NETs capture microbes in the extracellular scaffold leading to microbial death but also neutrophil cell death. Activated neutrophils are also capable of synthesizing cytokines and other inflammatory mediators.

Question 3

Which helper T cell produces cytokines that contribute to neutrophil production and recruitment?

A. T helper 1 cell (Th1)
B. T helper 2 cell (Th2)
C. T helper 9 cell (Th9)
D. T helper 17 cell (Th17)
E. T follicular helper cell (Tfh)

Answer: The correct answer is D

CD4+ effector T cells are divided into four subsets: T helper 1 cell (Th1), T helper 2 cell (Th2), T helper 17 cell

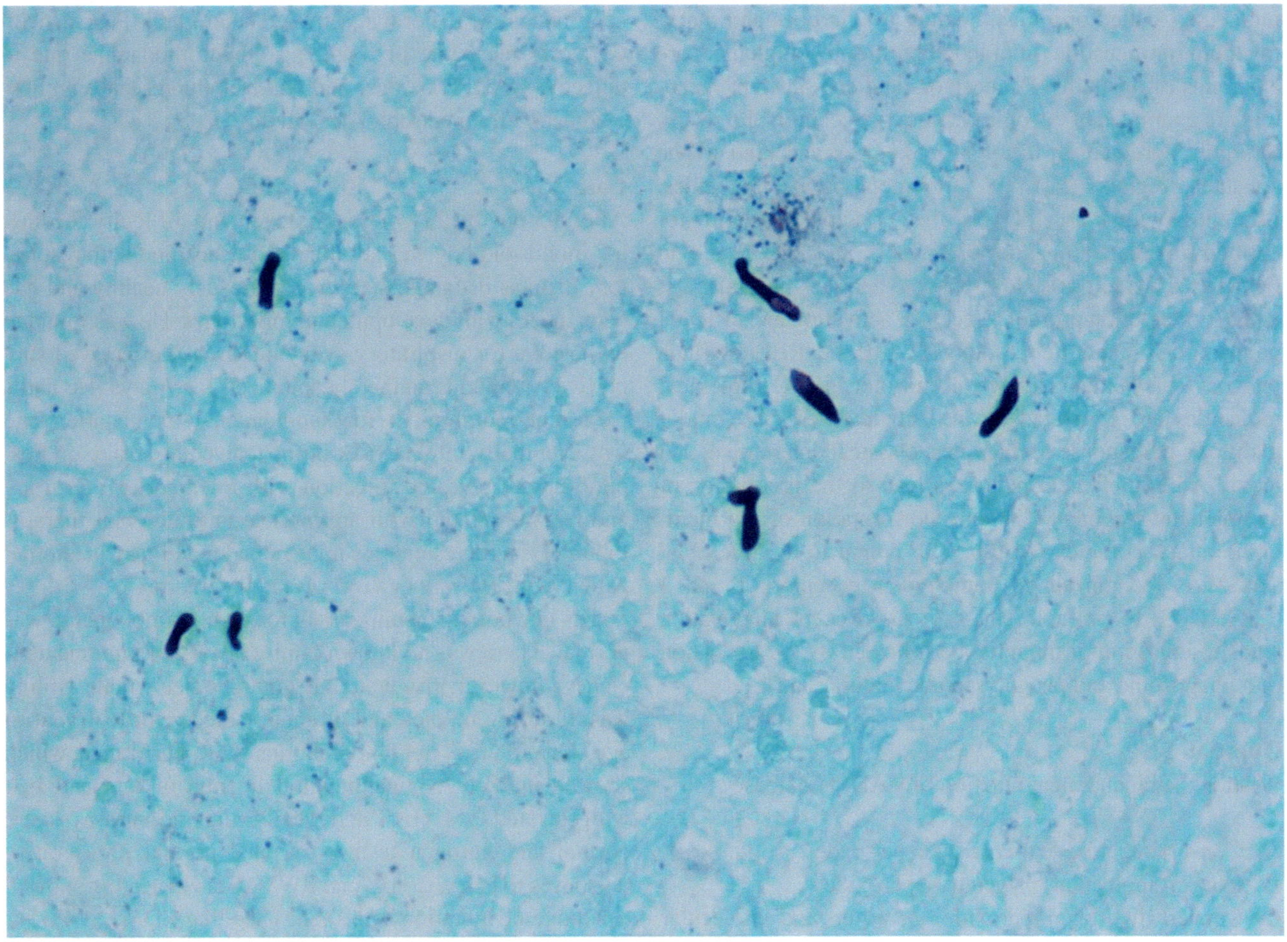

Fig. 2 Left hilar lymph node biopsy, silver stain highlights *Phaeoacremonium parasiticum* fungal hyphae (GMS, final magnification 600×). (Image courtesy of Randall Craver, MD. Children's Hospital New Orleans, LA)

(Th17), and T follicular helper cell (Tfh). Each subset produces distinct cytokines that contribute to different functions in host defense. Th17 cells develop under the influence of interleukin 1 (IL-1), interleukin 6 (IL-6), interleukin 23 (IL-23), and TGF-β. Th17 cells produce interleukin 17A (IL-17A) and interleukin 22 (IL-22) in response to activation by bacteria and fungi. IL-17A induces the epithelium to produce G-CSF, IL-8 (CXCL8), and TNF-α. G-CSF stimulates the bone marrow to produce neutrophils. IL-8 (CXCL8) is a chemokine that recruits neutrophils to tissue. Therefore, IL-17A produced by Th17 cells indirectly leads to the production and recruitment of neutrophils. IL-22 is involved in the maintenance of the epithelial barrier but does not play a role in neutrophil chemotaxis. Abnormalities in the IL-17 pathway result in chronic mucocutaneous candidiasis due to defective recruitment of neutrophils to the site of *Candida* infections. Th1 cells produce IFN-γ to activate macrophages. Th2 cells produce IL-4, IL-5, and IL-13 to activate eosinophils and mast cells in defense against parasitic infections and allergic disease. Tfh cells produce IL-21 to enhance B cell antibody production in germinal centers.

Question 4

Which immunoglobulin is the principal opsonin for neutrophil phagocytosis of microbes?

A. Immunoglobulin G (IgG)
B. Immunoglobulin A (IgA)
C. Immunoglobulin M (IgM)
D. Immunoglobulin E (IgE)
E. Immunoglobulin D (IgD)

Answer: The correct answer is A

Phagocytes engulf microbes through a process called phagocytosis. Neutrophils recognize infected and damaged cells through pattern recognition receptors, Fc receptors, and complement receptors. Phagocytosis is enhanced by host molecules, called opsonins, binding to microbes and marking them for ingestion. Immunoglobulin and complement are examples of opsonins. Fc receptors and complement receptors are examples of opsonic phagocytic receptors. Fc receptors are expressed on different leukocytes and bind to the constant region (Fc) of immunoglobulin molecules. Immunoglobulin G (IgG) is the principal opsonin for antibody-mediated phagocytosis. IgG1 and IgG3 bind to FcγRI (CD64) expressed on phagocytes with high affinity, and the interaction promotes receptor activation and efficient phagocytosis. Other Fc receptors (FcγRII, FcγRIII) facilitate phagocytosis but have a lower affinity for IgG. Complement molecules are generated by the activation of three major pathways in response to microbes: classical, alternative, and lectin. All complement pathways lead to the production of complement fragment C3b and the formation of the membrane attack complex (MAC). C3b bound to the microbial surface is recognized by complement receptors 1 and 3 (CR1, CR3) to promote phagocytosis. Receptor activation leads to ingestion of the microbe followed by the formation of a phagosome and phagolysosome maturation.

Question 5

What is the biochemical function of NADPH oxidase?

A. Convert superoxide to hydrogen peroxide
B. Convert molecular oxygen to superoxide
C. Convert hydrogen peroxide to water and molecular oxygen
D. Convert hydrogen peroxide to hypochlorous acid
E. Convert molecular oxygen to hydrogen peroxide

Answer: The correct answer is B

Phagocytosed microbes are killed in phagolysosomes by hypohalous acids generated by a process called the respiratory burst. NADPH oxidase is a complex of six proteins that converts molecular oxygen molecules to superoxide anions. Superoxide anions are converted to hydrogen peroxidase by superoxide dismutase. Myeloperoxidase combines hydrogen peroxide and chloride anions to produce hypochlorous acid. Hypochlorous acid is, directly and indirectly, toxic to a variety of organisms, including *Aspergillus* species, *Staphylococcus aureus, Burkholderia cepacia* complex, *Serratia marcescens,* and *Nocardia* species. Patients with chronic granulomatous disease (CGD) have a defect in the NADPH oxidase complex leading to an inability to generate the respiratory burst. The most common genetic defect is a mutation in cytochrome b-245 beta subunit (CYBB) which encodes gp91phox and causes X-linked CGD. Patients with CGD have recurrent and deep-seated bacterial and fungal infections due to an inability to generate hypohalous acids. CGD is diagnosed by measuring superoxide production via the dihydrorhodamine (DHR) test. Nonfluorescent dihydrorhodamine is oxidized to fluorescent rhodamine in normal, activated neutrophils. Patients with CGD are incapable of oxidizing dihydrorhodamine due to defective NADPH oxidase and will lack fluorescence on flow cytometry after neutrophil activation.

Case 2

After a 5-day history of cough, fever, and difficulty feeding, a 7-month-old boy was admitted to the hospital with respiratory distress in January. He was previously healthy and had been meeting all developmental milestones. He was found to be positive for respiratory syncytial virus. He received supportive care and remained hospitalized for 7 days. He had a persistent requirement for supplemental oxygenation and was eventually intubated. A bronchoalveolar lavage on hospital day #9 revealed infection with *Pneumocystis jirovecii* and he was treated with appropriate antimicrobial therapy. Figure 3 is a representative image of a lung biopsy in another patient with *Pneumocystis jirovecii* pneumonia.

His newborn screening assay for T cell lymphopenia was normal. There was no known family history of immune deficiency; however, his maternal grandmother's male sibling was ill during infancy and ultimately died prior to 1 year of age. Quantitative immunoglobulins were performed which showed marked hypogammaglobulinemia of IgG and IgA with elevated IgM. His T and B cell enumeration assays were normal and mitogen testing revealed normal proliferative responses to PHA, ConA, and PWM. Genetic testing showed a pathogenic mutation in the gene encoding CD40L (CD154) confirming the suspected diagnosis of X-linked hyper-IgM syndrome.

Question 1

Which cytokine is a major growth factor for dendritic cells?

A. FLT3LG
B. T-bet
C. GATA3
D. RORγt
E. PAX5

Answer: The correct answer is A

Dendritic cells are found in the blood, epithelium, and tissue compartments. They are bone marrow-derived and develop under the influence of fms-related tyrosine kinase 3 ligand (FLT3LG). Dendritic cells are divided into two major subsets: classical (conventional) and plasmacytoid. Classical dendritic cells are the most efficient professional antigen-presenting cell (APC) due to their ability to constantly sample the environment and migrate to the lymph nodes to present to CD4+ T cells. Other professional APCs include macrophages and B cells due to their ability to capture and present antigen in the context of class 2 major histocompatibility complex (MHC). Dendritic cells recognize microbial

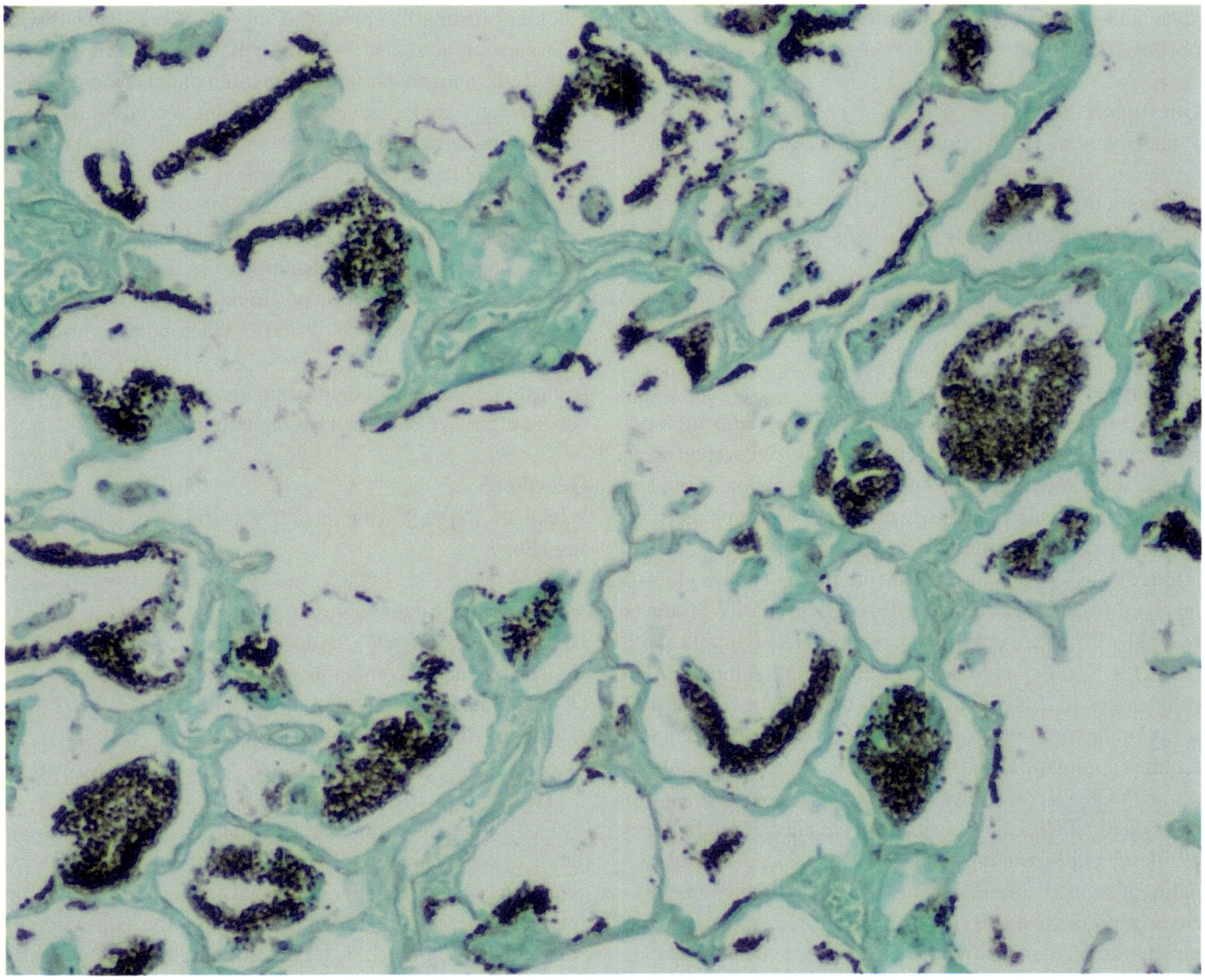

Fig. 3 Medium power view of lung biopsy, silver stain highlights the alveoli filled with fungus *Pneumocystis jirovecii* (GMS, final magnification 100×). Image courtesy of Matthew Stark, MD. Children's Hospital New Orleans, LA

antigens using a variety of cell surface and cytoplasmic pattern recognition receptors. Plasmacytoid dendritic cells are primarily found in the blood. They are a major source of type I interferons (IFN-α, IFN-β) in response to viral infections. The development of CD4+ effector T cell subsets is driven by other transcription factors: Th1 (T-bet), Th2 (GATA3), and Th17 (RORγt). PAX5 is a transcription factor involved in B cell differentiation.

Question 2

Which interaction increases the expression of CD80 and CD86 on antigen-presenting cells?

A. PD-L1 and PD-1
B. CD86 and CTLA-4
C. ICOS-L and ICOS
D. CD40L and CD40
E. OX40L and OX40

Answer: The correct answer is D

Dendritic cells, macrophages, and B cells are professional antigen-presenting cells. Professional APCs express class 2 MHC molecules and can present to CD4+ T cells. T cell activation by APCs requires not only recognition of the antigen-MHC complex but also costimulation. The interaction between CD28 on T cells and CD80 and CD86 on activated APCs is a critical costimulatory pathway. Expression of CD80 and CD86 by APCs is dependent on the activation of the APC by other molecules. CD40 ligand (CD40L, CD154) is expressed on T cells after recognition of the antigen-MHC complex. CD40L engages CD40 on APCs to enhance the expression of CD80 and CD86. Increased expression of CD80 and CD86 augments costimulation, T cell activation, and T cell proliferation. PD-1 and CTLA-4 are inhibitory receptors on T cells. ICOS is a T cell receptor required for the development of T follicular helper cells (Tfh). OX40 is a T cell receptor that promotes T cell survival and polarization of the immune response to a Th2 phenotype.

Question 3
Which cytokine is the primary activator of macrophages?

A. Interleukin 2 (IL-2)
B. Interleukin 10 (IL-10)
C. Interleukin 12 (IL-12)
D. Interferon-β (IFN-β)
E. Interferon-γ (IFN-γ)

Answer: The correct answer is E.

Macrophages are a type of phagocyte derived from blood monocytes. Monocytes are recruited to tissue compartments during infection or injury. The primary function of macrophages is to engulf and destroy microbes. Macrophage function is enhanced by interferon-γ (IFN-γ). IFN-γ is a type II interferon as opposed to type I interferons (IFN-α, IFN-β) involved in anti-viral defense. IFN-γ is primarily produced by CD4+ T helper 1 cells (Th1), type 1 innate lymphoid cells (ILC1), NK cells, and CD8+ T cells. IL-12 stimulates Th1 cells to secrete IFN-γ which signals through the STAT1 transcription factor in macrophages. Macrophage activation also requires stimulation from CD40 on the macrophage interacting with CD40L on activated Th1 cells. The combination of IFN-γ and CD40-CD40L signaling enhances the efficiency of microbe killing in the phagolysosome. CD40-CD40L signaling also increases the expression of costimulatory molecules on macrophages to enhance T cell activation and proliferation. Activated classical macrophages also synthesize and secrete cytokines involved in the acute inflammatory response to microbes.

Question 4
What is the function of alternatively activated macrophages?

A. Secretion of IL-1 and TNF-α
B. Phagocytosis and killing of microbes
C. Tissue repair and fibrosis
D. Antigen presentation to T cells
E. Removal of apoptotic cells

Answer: The correct answer is C

Alternatively activated macrophages are a subset involved in tissue repair and suppression of inflammation. Alternative activation is driven by IL-4 and IL-13, whereas IFN-γ leads to classically activated macrophages. IL-4 and IL-13 are Th2 cytokines primarily produced by Th2 cells and ILC2. Alternative macrophages produce IL-10 and TGF-β. IL-10 inhibits the antigen-presenting function of dendritic cells and macrophages and prevents them from secreting IL-12, a potent stimulator of IFN-γ production by Th1 cells. The lack of IFN-γ inhibits the development of classically activated macrophages and impairs the function of macrophages in cell-mediated immunity. TGF-β inhibits the production of classical macrophages but also drives tissue repair and fibrosis by enhancing fibroblast proliferation and collagen synthesis. Classically activated macrophages secrete pro-inflammatory cytokines, present antigen to T cells in the context of MHC, phagocytose and kill microbes, and participate in efferocytosis to remove apoptotic cells.

Question 5
Which of the following is not a result of CD40-CD40L signaling?

A. Macrophage activation
B. Inhibition of T cell activation
C. Class switch recombination
D. Germinal center reaction
E. Increased expression of CD80/CD86

Answer: The correct answer is B

T cell activation increases the expression of CD40L on the T cell surface. CD40 is expressed on macrophages, dendritic cells, and B cells. CD40-CD40L signaling enhances the APC function by increasing the expression of costimulatory molecules. The interaction promotes the killing function of macrophages in combination with IFN-γ. CD40 on B cells interact with CD40L on T follicular helper cells (Tfh). CD40 signaling on B cells not only stimulates B cell proliferation but leads to the development of the extrafollicular and germinal center response. CD40-mediated expression of activation-induced cytidine deaminase enables class switch recombination from IgM to IgG, IgA, or IgE. Abnormalities in CD40-CD40L signaling are illustrated by X-linked hyper-IgM syndrome (HIGM). HIGM is caused by a mutation in CD40L (CD154). Patients have decreased levels of IgG and IgA due to absent class switching. Immune response to intracellular microbes is defective due to an abnormal interaction between T cells and professional APCs, and many patients suffer from severe *Pneumocystis jirovecii* infection. CD40-CD40L is not involved in T cell inhibition. The best described inhibitory T cell receptors are PD-1 and CTLA-4.

Bibliography

1. Abbas AK, Lichtman AH, Pillai S. Cellular and molecular immunology. 10th ed. Philadelphia, PA: Elsevier; 2022.
2. de la Morena MT. Clinical phenotypes of hyper-IgM syndromes. J Allergy Clin Immunol Pract. 2016;4(6):1023–36.
3. Lehman HK, Segal BH. The role of neutrophils in host defense and disease. J Allergy Clin Immunol. 2020;145(6):1535–44.
4. Thomsen IP, Smith MA, Holland SM, Creech CB. A comprehensive approach to the management of children and adults with chronic granulomatous disease. J Allergy Clin Immunol Pract. 2016;4(6):1082–8.

Histocompatlbility Antigens (HLA) and Transplantation

Michell M. Lozano Chinga, David Buchbinder, and Jolan E. Walter

1 Introduction

In this chapter, we will use a case-based approach to review topics pertinent to the use of hematopoietic stem cell transplantation (HSCT) in the setting of inborn errors of immunity (IEI). IEI are disorders that result from pathogenic variants of genes involved in the immune response which place the host at increased risk of infections, autoimmunity, hyperinflammation, and other manifestations of immune dysregulation. We will review the most common indications for HSCT in IEI as well as important considerations including donor selection and conditioning which affect the outcomes of patients with IEI undergoing HSCT. We will also discuss the complications associated with HSCT including infectious and noninfectious complications such as graft-versus-host disease (GVHD), graft failure, and autoimmunity among others. Finally, we will discuss late effects and survivorship in this unique population. Despite the improvement in outcomes, complications could occur during transplant as well as in the short-term and long-term post-HSCT period. Comprehensive evaluation includes screening for complications (toxicities), subsequent chronic health conditions, and immune reconstitution along with careful attention to psychosocial health and well-being. The evaluation for HSCT and surveillance posttransplant is best achieved by a collaborative relationship between immunologists and transplant specialists.

M. M. Lozano Chinga (✉)
Division of Immunology, Rheumatology and Allergy, University of Iowa Hospitals and Clinics, Iowa City, IA, USA
e-mail: michell-lozanochinga@uiowa.edu

D. Buchbinder
Division of Hematology, Children's Hospital of Orange County, Orange, CA, USA

Department of Pediatrics, University of California at Irvine, Orange, CA, USA
e-mail: DBuchbinder@choc.org

J. E. Walter
Division of Allergy and Immunology, Department of Pediatrics, Morsani College of Medicine, University of South Florida, Tampa, FL, USA

Division of Allergy/Immunology, Department of Pediatrics, Johns Hopkins All Children's Hospital, St. Petersburg, FL, USA

Division of Allergy and Immunology, Massachusetts General Hospital for Children, Boston, MA, USA
e-mail: jolanwalter@usf.edu

Case 1

An 11-day-old Caucasian male was born at 39 weeks of gestation to a nulliparous 26-year-old mother with unremarkable prenatal history and ongoing breastfeeding. The family history was noncontributory and there was no reported consanguinity. The neonate was discharged home on the day of life two. His newborn screen was presumptive positive for SCID with undetectable T cell receptor excision circles (TRECs). The mother was instructed to hold breastfeeding and store breastmilk until maternal cytomegalovirus (CMV) status is clear. The neonate was brought to the Immunology outpatient clinic for further clinical and laboratory evaluation. Complete blood count with differential documented absolute lymphopenia: 1700 cells/mm^3 (normal range 2000–11,000 cells/mm^3). Flow cytometry evidenced severe T cell lymphopenia (CD3+ = 5 cells/mm^3, CD3+CD4+ = 2 cells/mm^3, CD3+CD8+ = 5 cells/mm^3) with poor thymic production and full absence of naive of T cells (CD3+ CD45RA+); however, normal number of B cells (995 cell/mm^3) and natural killer (NK) cells (175 cell/mm^3). The patient was admitted due to high suspicion for SCID and treatment with prophylactic antimicrobials including acyclovir, fluconazole, and atovaquone was initiated immediately. The following day, second tier testing confirmed normal IgG (912 mg/dL, presumably maternal), low IgM (18 mg/dL), and undetectable IgA (<7 mg/dL). Within 3 days, lymphocyte proliferation to mitogen phytohemagglutinin revealed low lymphocyte function (<10% of normal, radioactive thymidine incorporation assay). After 3 weeks, genetic testing revealed a pathogenic missense mutation in the interleukin 2 receptor γ chain (*IL2RG)* (c.676C>T, p.Arg226Cys). This same genetic variation was

M. Mahmoudi (ed.), *Absolute Allergy and Immunology Board Review*, https://doi.org/10.1007/978-3-031-12867-7_3

confirmed in the mother. The family and the bone marrow registry were queried to find an appropriate donor. In absence of a HLA-matched sibling and 10/10 unrelated donor, the family agreed to pursue unconditioned maternal haploindentical HSCT at 3 months of age. The patient demonstrated immune reconstitution with 99% donor short tandem repeat (STR) in T cell compartment; however, B cells remained of recipient origin and the patient required immunoglobulin replacement therapy. Around 4 years of age, the patient presented with acute CMV infection in the lungs, oral candidiasis, and norovirus enteritis. Further evaluation evidenced a declining number of T cells and function. After treatment of the infections and initiation of antimicrobial prophylaxis, the patient was enrolled for a research protocol with *IL2RG* gene therapy and low dose busulfan conditioning. At 6 years of age, the patient remains with normal T cell count and function and no need for immunoglobulin replacement therapy.

1.1 Indications for Hematopoietic Stem Cell Transplant in Inborn Errors of Immunity

Factors that determine the indication of HSCT includes the patient's primary IEI diagnosis as well as age, sex, and clinical status prior to HSCT. For example, HSCT candidates may include an asymptomatic and healthy appearing newborn with X-linked SCID requiring urgent HSCT or a young adult with severe clinical manifestations of RAG deficiency including granulomatous interstitial lung disease and autoimmune enteropathy.

Question 1

As for indications, in which adaptive IEI has HSCT been most commonly used as definitive therapy in the United States?

A. Severe combined immunodeficiency (SCID)
B. Wiskott-Aldrich syndrome (WAS)
C. X-linked agammaglobulinemia (XLA)
D. A and B

Answer and Explanation

Answer: A

Among the adaptive immune deficiencies, those with absence or impaired function of T cell, such as SCID and WAS are commonly considered for HSCT. With newborn screening for SCID available in all 50 states in the United States, an estimated 75–80 new patients diagnosed yearly require HSCT (1 in every 58,000 live birth). In contrast, there is no newborn screening for WAS and currently incidence is estimated to be 1:100,000. The patients with WAS are diagnosed later and not all are transplanted. XLA is an antibody deficiency syndrome with preserved T cell compartment that is primarily managed with immunoglobulin replacement therapy. HSCT has been reported mainly in the developing world, where access to immunoglobulin replacement therapy is limited.

Question 2

As for indications, which is the most common IEI for which HSCT has been used as definitive therapy?

A. Chronic granulomatous disease (CGD)
B. Leukocyte adhesion defect (LAD)
C. Warts, hypogammglobulinemia, infections, myelokathexis (WHIM) syndrome
D. Chediak-Higashi (CHS) syndrome

Answer and Brief Explanation

Answer: A

Among the innate immune deficiencies, those with profound neutrophil dysfunction, such as CGD, are commonly considered for HSCT. The worldwide birth prevalence is 1 in 200,000 and most patients are diagnosed with infections and/or family history. Depending on the region, HSCT is increasingly used with success among patients even with advanced disease and noninfectious complications. LAD, WHIM, and CHS syndromes are less common (>1 in million) and only selected cases are transplanted for severe disease.

1.1.1 Indications for Hematopoietic Stem Cell Transplant in Inborn Errors of Immunity

The first successful HSCTs for IEI were reported in one patient with SCID and one with WAS. These events took place in 1968, 1 year after the characterization of the human major histocompatibility complex (MHC). Since that time, the understanding of IEI and HSCT has greatly improved. HSCT has been proven to be curative for multiple genetic causes of IEI including adaptive immune defects (e.g., SCID, DOCK8 deficiency, DOCK2 deficiency, IPEX [immune dysregulation, polyendocrinopathy, enteropathy, X-linked], CD40 ligand deficiency, CD40 deficiency, X-linked lymphoproliferative disease, activated PI3K delta syndrome, familial hemophagocytic lymphohistiocytosis, GATA2 deficiency, MHC class II deficiency, autosomal dominant IgE syndrome, CTLA4 haploinsufficiency, LRBA deficiency), and innate immune defects (e.g., CGD, RAB27A deficiency, reticular dysgenesis, and LAD type I). Table 1 lists all IEI, based on the International Union of Immunological Societies (IUIS) criteria, where HSCT has been used.

The heterogeneity of IEI is enormous and it is vital to understand that HSCT is not indicated for the treatment of all IEI, especially antibody deficiency syndromes where immunoglobulin replacement therapy is sufficient to maintain

Table 1 Inborn errors of immunity reported with treatment approach of HSCT

IUIS I. Immunodeficiencies affecting cellular and humoral immunity	
ADA deficiency, AR[a]	IL2RG deficiency, XL[a]
AK2 deficiency (reticular dysgenesis), AR	IL-7Ra deficiency, AR
B2M (MHC class I deficiency), AR	IL-21 deficiency, AR
BCL10 deficiency, AR	IL-21R deficiency, AR
CARD11 deficiency, AR	ITK deficiency, AR
CD27 deficiency, AR	JAK3 deficiency, AR
CD3d, AR	LCK deficiency, AR
CD3e, AR	LRBA deficiency, AR
CD3g, AR	MAGT1 deficiency, XL
CD3z deficiency, AR	MALT1 deficiency, AR
CD8 deficiency, AR	MST1 deficiency, AR
CD40 deficiency, AR	NIK deficiency, AR
CD40 ligand deficiency, XL	Omenn syndrome, AR
CD45 deficiency, AR	OX40 deficiency, AR
Cernunnos/XLF deficiency, AR	RAG1 deficiency, AR[a]
CIITA (MHC class II deficiency), AR	RAG2 deficiency, AR
Coronin-1A deficiency, AR	RFX5 (MHC class II deficiency), AR
CTPS1 deficiency, AR	RFXAP (MHC class II deficiency), AR
DCLRE1C deficiency (Artemis), AR[a]	RFXANK (MHC class II deficiency), AR
DNA ligase IV deficiency, AR	RhoH deficiency, AR
DNA PKcs deficiency, AR	TAP1 (MHC class I deficiency), AR
DOCK2 deficiency, AR	TAP2 (MHC class I deficiency), AR
DOCK8 deficiency, AR	TAPBP (MHC class I deficiency), AR
ICOS deficiency, AR	TCRa deficiency, AR
IKBKB deficiency, AR	ZAP-70 deficiency, AR
IUIS II. Combined immunodeficiencies with associated syndromic feature	
ATM deficiency (ataxia-telangiectasia-AT), AR	RMRP deficiency (cartilage hair hypoplasia, AR)
Complete DiGeorge syndrome[b]	RECQL3 deficiency (bloom syndrome, AR)[c]
Dyskeratosis congenita (TERC, TINF2, AD)[c]	STAT3 deficiency (AD-HIES)
Dyskeratosis congenita (NOLA2, NOLA3, DCLRE1B, PARN, AR)[c]	STAT5b deficiency, AR
Dyskeratosis congenita (TERT, RTEL1, TPP1, AD/AR)[c]	STIM-1 deficiency, AR
Dyskeratosis congenita (DKC1, XL)[c]	TBX1 deficiency, HS[b]
NBS1 deficiency (Nijmegen breakage syndrome), AR[d]	TTC7A deficiency (ID with multiple intestinal atresias), AR
NEMO/IKBKG deficiency (ectodermal dysplasia with ID/EDA-ID, XL)	WAS deficiency (Wiskott-Aldrich syndrome), XL[a]
ORAI-I deficiency, AR	
IUIS III. Predominantly antibody deficiencies	
BTK deficiency, XL[e]	NFKB2 deficiency (AD)
CARD11 deficiency (AD, GOF)	PI3KR1 deficiency (AD, LOF)
Common variable immunodeficiency (CVID), AR	PI3KR1 deficiency (AR)
	PI3K-d activated (AD, GOF)
IUIS IV. Diseases of immune dysregulation	
ADA2 deficiency, AR	RNASEH2A deficiency (Aicardi-Goutieres syndrome, AGS4), AR
ADAR1 deficiency (Aicardi-Goutieres syndrome, AGS6), AD	RNASEH2B deficiency (Aicardi-Goutieres syndrome, AGS2), AR
CTLA4 deficiency, AD	RNASEH2C deficiency (Aicardi-Goutieres syndrome, AGS3), AR
IL-10 deficiency, AR	SAMHD1 deficiency (Aicardi-Goutieres syndrome, AGS5), AR
FOXP3 deficiency (IPEX syndrome, XL)	SH2D1A deficiency (XLP1)
STXBP2 deficiency (FHL5), AR/AD	STAT3 deficiency (AD, GOF)
PRKCd deficiency, AR	XIAP deficiency (XLP2)
IUIS V. congenital defects of phagocyte number, function, or both	
CGD, AR (p22phox deficiency)	GATA-2 deficiency, AD
CGD, AR (p40phox deficiency)	ITGB2 deficiency (leukocyte adhesion deficiency type 1, LAD1), AR[a]
CGD, AR (p47phox deficiency)	Rac 2 deficiency, AD
CGD, AR (p67phox deficiency)	HAX1 deficiency (Kostmann disease, SCN3), AR
CGD, XL (gp91 phox deficiency)[a]	SBDS deficiency (Shwachman-diamond syndrome), AR
	SLC35C1 deficiency (leukocyte adhesion deficiency type 2, LAD2), AR
	WAS GOF (X-linked neutropenia/myelodysplasia)
IUIS VI. Defects in intrinsic and innate immunity	
CXCR4 gain-of-function (WHIM syndrome, AD)	IL12/IL23RB1 deficiency (MSMD, AR)
IFNγR1 deficiency (MSMD, AD)	IL-12p40 (MSMD, AR)
IFNγR1 deficiency (MSMD, AR)	STAT1 deficiency (MSMD, AD)
IFNγR2 deficiency (MSMD, AR)	STAT 1 deficiency (predisposition to severe viral infection, AR)
	STAT 1 GOF (chronic mucocutaneous candidiasis/CMC, AD)

In gray the most common defects are highlighted
[a] Gene therapy **available**
[b] Thymus transplant is preferred
[c] HSCT is mainly indicated with bone marrow failure
[d] HSCT is mainly indicated in case of cancer
[e] HSCT in the developing world

patient infection-free and with good quality of life. Here we describe the most common IEI for which patients receive HSCT.

Severe Combined Immunodeficiency

SCID encompasses a group of heterogeneous genetic disorders that have in common profound T cell impairment secondary to absent or diminished T cells or T cell function. Clinical manifestations are usually present early in life, after the loss of maternal antibodies. Features include failure to thrive, recurrent infections by common and opportunistic pathogens as well as infections after live-virus vaccines (e.g., Bacille Calmette-Guerin [BCG]) and/or signs of inflammatory or graft-versus-host disease due to expansion of oligoclonal lymphocytes (Omenn syndrome) or the engraftment of maternal T cells. If not treated there is almost 100% mortality during the first year of life, unless the patient spontaneously develop revertant T cell lines that partially correct the underlying severe immunodeficiency.

The first successful HSCT was reported in 1968, over 50 years ago; however, the treatment was not available in a timely manner for most infants for decades. The importance of early HSCT was established in the early 2000s and summarized in a milestone paper by Dr. Rebecca Buckey in 2011. This study described the long-term outcomes of 166 infants with SCID that received a nonconditioned related donor HSCT in a single institution and compared it to HSCT done at other institutions where chemotherapy as conditioning prior HSCT was used. The study concluded that if a HSCT from a relative or matched unrelated control could be done in the first few months of life (<3.5 months) with or without conditioning the survival would be excellent (>90%) in comparison with older patients. The best transplant approach was examined by Sung-Yun Pai et al. based on a 25 multicenter study among 240 infants with SCID patients in 2014. This study concluded that the survival rate was high regardless of donor type among infants who received transplants at 3.5 months of age or younger and among older infants without prior infection or with an infection that had resolved. The only exception is adenosine deaminase (ADA) deficiency that may evolve slower. In patients with ADA deficiency SCID, enzyme replacement therapy can be used as a bridge during a prolonged period until the conditions/requirements for the best definitive therapy are available.

The most important risk factor for poor outcome in patients with SCID undergoing HSCT is an active infection. The quality of CD3+ T cell recovery, but not survival, is affected by the type of conditioning and genetic subtype of SCID. In fact, the genetic, phenotypic, and functional diversity of SCID adds further complexity to the best approach.

Wiskott-Aldrich Syndrome

WAS is an X-linked disorder with an incidence of 1 in 100,000 live male births. This disorder is caused by hemizygous mutations in the *WAS* gene and characterized by microthrombocytopenia, eczema, and immunodeficiency affecting B and T cells. Patients with WAS are also at increased risk of autoimmunity and malignancy. Management of WAS includes immunoglobulin supplementation, prevention and timely treatment of infections, and control of bleeding and autoimmune complications. These supportive care efforts have increased the survival of these patients; however, successful HSCT is the only definitive treatment and corrects immunodeficiency and thrombocytopenia. A scoring system was implemented to describe the severity of the phenotype of patients with WAS gene variations. This scoring system is based on clinical and laboratory findings including thrombocytopenia, small platelets, eczema, immunodeficiency, infections, autoimmunity, malignancy, congenital neutropenia, and myelodysplasia. A score of 0 is given to patients with X-linked neutropenia and <1 to intermittent X-linked thrombocytopenia (XLT). A score of 1 or 2 defines patients with XLT, a score of 3 and 4 is reserved for those with a classic WAS phenotype, and a score of 5 identifies patients who have developed autoimmunity and/or malignancies. In 2020, Burroughs et al. reported 129 patients with WAS that underwent HSCT. Fifty-three patients (41%) had a WAS score of 1–2, the majority of whom were <2 years of age at the time of HSCT. Forty-three patients (33%) had a WAS score of 3–4, and 33 (26%) had a score of 5 consistent with the most severe phenotype. Overall survival (OS) at 5 years was better in patients younger than 5 years (94% vs. 66%). Moreover, OS was excellent regardless of the donor source. Over half of the patients had received a myeloablative conditioning regimen. Patients who received reduced intensity regimens had a lower level of myeloid donor chimerism after HSCT which is associated with decreased platelet counts following transplant.

Chronic Granulomatous Disease

CGD is an IEI that results from pathogenic variants in the genes encoding subunits of the nicotinamide adenine dinucleotide phosphate (NADPH) oxidase complex with subsequent impaired production of reactive oxygen intermediates. This prevents effective microbe killing by neutrophils, monocytes, and macrophages. The incidence of CGD is 1 in 200,000 live births in the United States. Clinical manifestations include infections by bacterial and fungal organisms and organ-specific inflammation including inflammatory lung and bowel disease as examples.

HSCT is a definitive treatment for CGD which can abrogate the risk for infectious and inflammatory complications. In 2020 Chiesa et al. reported a study of 635 children and 77 adults with CGD that underwent HSCT. OS at 3 years was higher in younger patients (87% vs. 76%). The use of a matched family donor as well as the use of well-matched unrelated donors was associated with superior survival. There is no consensus about the role of inflammatory complications in patients survival (the US vs. European experience).

1.2 Donor Selection for Hematopoietic Stem Cell Transplant in Inborn Errors of Immunity

Donor selection constitutes a critical step in ensuring a successful HSCT for patients with IEI. Importantly, transplant center expertise and preferences may shape donor selection, including consideration of autologous (gene therapy) beyond allogeneic (conventional HSCT).

Question 3
What are the most important molecular markers used to identify an adequate donor for HSCT in IEI?

A. Blood-Group-Specific Polysaccharides
B. Human Leukocyte Antigens (HLA)
C. CD34
D. Alkaline phosphatase

Answer and Explanation
Answer: B

Donor selection is dependent on many factors, with the most important being the degree of HLA disparity. The HLA genes are part of the MHC in humans, and they are located in the short arm of chromosome 6 (6p21.3). HLA genes encode cell surface markers and antigen-presenting molecules which allow the immune system to discern self from non-self. The MHC is divided into three regions: class I, class II, and class III regions. Class I HLAs are present in almost every cell except for erythroblasts and trophoblasts. Class II HLAs are expressed solely on antigen-presenting cells as dendritic cells, macrophages, and B cells. The region between class I and class II MHC is known as class III and no HLA genes are encoded there; however, it contains other genes that are important for the immune response as complement, tumor necrosis factor, and heat shock protein. HLA genes are highly polymorphic though some alleles are more conserved within ethnic groups. This polymorphism constitutes a barrier to identify a match, especially in underrepresented minorities.

Molecular typing of HLA Class I (−A, -B, -C) and class II (-DRB1, -DQB1, and -DPB1) genes allows for the identification of a potential allogeneic donor. The definition of HLA matching depends on the number of loci tested and the molecular testing resolution employed. The match is generally tested for 10–12 HLA antigens. HLA typing is now done using DNA sequencing (high resolution). HLA Class I and Class II are tightly linked and a child coinherits a haplotype from each parent, explaining why a parent is usually a potential haploidentical donor.

Question 4
In regards to donor selection, which forms of SCID can receive autologous gene-modified-HSCT, termed gene therapy?

A. X-linked SCID (IL2RG)
B. Adenosine deaminase deficiency SCID
C. ARTEMIS
D. All of the above

Answer and Explanation
Answer: D

Beyond the clinical use of ADA-gene therapy in Europe, there are clinical trials worldwide on gene therapy for other genetic SCIDs, such as X-SCID (IL2RG), ARTEMIS, and RAG1 deficiencies.

Question 5
In regards to donor selection, which is the best and least favorable donor selection?

A. Best: haploidentical donor; Least: 10/10 HLA-matched unrelated donor
B. Best: matched sibling donor from bone marrow; Least: cord blood from unmatched unrelated donor
C. Best: 10/10 matched unrelated donor; Least: haploidentical donor
D. Best: matched sibling donor from peripheral PBMC; Least: haploidentical from bone marrow

Answer and Brief Explanation
Answer: B

Matched sibling donor is the best possible option for donor selection. Unmatched selection (unrelated or haploidentical from parents) are inferior; however, recent conditioning regimens have shown to improve outcomes when haploidentical donors are used.

Donors may be categorized as autologous or allogeneic where autologous donation focuses on the collection of the patient's own hematopoietic stem cells (HSC). Few IEI so far may be subject to the utilization of autologous donation in the setting of autologous gene-modified HSCT such as X-linked, ADA deficiency, RAG1, and ARTEMIS deficiency. This list continues to expand, even though most are in clinical trials. The only approved gene therapy is ADA-Gene therapy (Strimvelis) which is approved in Europe.

In contrast, allogeneic HSCT is most frequently utilized in the care of patients with IEI and involves the collection of HSC from related or unrelated donors. Related donor options typically focus on siblings and other matched family members although advances in HSCT have supported the utilization of mismatched family members including haploidentical donors (parents with > = 50% match). As many IEI are autosomal recessive disorders, evaluating extended family members as potential donors may be considered especially if there is consanguinity.

1.2.1 Matched Related Donors

Historically, the gold standard for donors has been a matched sibling donor (MSD). A matched related donor is the preferred source for HSCT; however, only less than 30% of patients that require HSCT have a matched-related donor (MRD) available. The possibility of having an adequate MSD or other MRD in patients with IEI is usually diminished due to the possibility that family members may be carriers of the genetic abnormality and in some, but not all IEI, the use of a carrier family member may not be optimal. IEI are monogenic disorders which can be inherited as X-linked, autosomal recessive or autosomal dominant diseases. The expression and penetrance of the genetic abnormality can be variable, and some possible donors may be asymptomatic which requires careful genetic and immune evaluation of potential family donors. Genetic testing of asymptomatic family members, including minors, may also generate significant stress for the patient undergoing HSCT and the family. Furthermore, there are patients whose genetic abnormalities are not identified after an extensive evaluation which often precludes the use of a family donor. Similar issues arise when considering other related donor possibilities such as haploidentical donors. Specific benefits associated with the use of a MRD donor include a shorter time to HSCT and less expense as procurement of an unrelated donor is often costly. Importantly, additional HSC can be obtained for a boost or donor lymphocyte infusion (DLI) later to address issues such as graft failure or graft dysfunction.

1.2.2 Haploidentical Donors

A haploidentical donor HSCT uses stem cells from "half-matched" donors. Potential haploidentical donors include siblings as well as parents ensuring that most patients with IEI requiring HSCT have an adequate HSC donor available. As above, the possibility of having an adequate haploidentical donor in patients with IEI is usually diminished due to the possibility that family members may be carriers of the genetic abnormality and in some, but not all IEI the use of a carrier family member may not be optimal. Advances in haploidentical HSCT using posttransplant cyclophosphamide (PTCy) for in-vivo T cell depletion has supported the adoption of haploidentical HSCT in recent years across the globe, especially in multiracial setting or in the low- to middle-income countries where access to large donor banks is less readily available. The use of haploidentical HSCT has increased the pool of available donors for patients requiring HSCT and is within acceptable risk of graft failure and rejection as well as acute and chronic graft-versus-host disease (GVHD). Moreover, specific benefits associated with the use of haploidentical donors include a shorter time to HSCT and less expense as procurement of an unrelated donor is often costly. Importantly, additional HSCs can be obtained for a boost or DLI at a later time to address issues such as graft failure or graft dysfunction. When considering a haploidentical HSCT, an assessment for anti-HLA antibodies should be completed. Anti-HLA antibodies directed to mismatched HLA antigens can be associated with a risk of primary graft failure and rejection. To improve outcome, haploidentical HSCT can be coupled with in vivo T cell depletion using PTCy or (50 mg/kg) on days +3 and +4 after HSCT or a variety of ex vivo T cell depletion techniques such as selected αβ T cell depletion combined with CD19+ B cell depletion. This process preserves the γδ T cells and NK cells, which contribute to immune reconstitution and antiviral function while eliminating αβ T cells responsible for GVHD. Unfortunately, these techniques are costly and available only at a limited number of transplant centers.

1.2.3 Unrelated Donors

Unrelated donors including matched unrelated donors and mismatched unrelated donors are an excellent option when faced with a desire to avoid using a related donor that may be a carrier of the same monogenic disorder, especially in case of autosomal dominant inheritance. This is important for specific IEI where the use of a carrier donor is not optimal and may also be important when the genetic etiology causing the IEI is unknown. Traditionally a 10/10 match has been considered a suitable match for an unrelated donor. This includes a donor-recipient match for HLA-A, HLA-B, HLA-C, HLA-DRB1, and HLA-DQB1. For a 12/12 HLA match, there is additional matching for HLA-DP1. A mismatched unrelated donor (9/10, 8,10), which is more common, can be acceptable for HSCT. However, single mismatches for HLA-A, -B, -C, or -DRB1 have been associated to increase risk of complications, while mismatches for HLA-DQB1 may be better tolerated. Despite, the World Marrow Donor Association and its associated 40 million registered donors registered worldwide there are a host of formidable barriers when considering the use of an unrelated donor. There are significant barriers when considering the use of an unrelated donor including the high costs associated with donor procurement and often longer waiting times. Furthermore, the availability of the unrelated donor may be limited, and additional HSCs can be challenging to obtain for a boost or DLI at a later time to address issues such as graft failure and graft dysfunction.

Table 2 Human stem cell sources

Source	Method of obtention	CD34+ cell dose obtained	Mean time to engraftment (mean)	Risk of GVHD (related to T cell number)	Limitations
BM	BM harvest—20 ml/kg max	3–5 × 10^6 CD34+/kg	14 days	++	Cell dose may not be able to obtain in a single harvest
PBSC	Requires mobilizations of stem cells with G-CSF and/or plerixafor, via apheresis	5–10 × 10^6 CD34+/kg Or 300–500 × 10^6 CD3+/kg	10 days	+++	G-CSF: Myalgias, pain, headache Apheresis: Vascular access, risk of thrombosis, hypocalcemia (due to citrate)
UCB	Collected at birth	0.2 × 10^6 CD34+/kg 2.5 × 10^6 CD34+/kg	21 days	+	Dose is limited—The patient may require double cord or cord blood expansion

Abbreviations: *BM* bone marrow, *PBSC* peripheral blood stem cells, *UCB* umbilical cord blood, *G-CSF* granulocyte colony-stimulating factor

Question 6

What is the most common graft source used to obtain human stem cells?

A. Bone marrow
B. Peripheral blood
C. Umbilical cord blood
D. A and C

Answer and Explanation

Answer: B

HSCs can be obtained from bone marrow, peripheral blood, or cord blood. Multiple factors are considered when selecting a graft source including recipient and donor (e.g., size), IEI-specific, HSCT-specific (e.g., conditioning intensity), and center-specific factors (e.g., center expertise) as just a few examples. Graft source selection remains a critical factor in the success of HSCT. The gold standard graft source for HSCT is bone marrow; however, the most commonly utilized graft source is peripheral blood stem cells (PBSC). Table 2 shows an overview of graft sources.

1.2.4 Bone Marrow

Bone marrow is the gold standard HSC source for HSCT. A bone marrow harvest is done under general anesthesia. The procedure consists of direct aspiration of bone marrow with bone marrow needles targeting different areas typically from the posterior iliac crest. Bone marrow then is transferred to a collection bag which contains an anticoagulant. It is filtered carefully in order to remove any debris. The maximum allowed volume to be collected is 20 ml/kg of the donor body weight. This donation maximum must be compared with the traditional target for cell collection which is 2–5 × 10^8 total nucleated cells (TNC) per kg of recipient body weight with the average marrow possessing 0.2–0.3 × 10^8 TNC per mL. Bone marrow collections will typically yield a CD34+ cell count of 3–5 × 10^6 CD34+ cells per kg of recipient body weight and a CD3+ cell count of 30–50 × 10^6 CD3+ per kg of recipient body weight. Most commonly, side effects of the procedure include pain and discomfort in the area of the collection. Other side effects include fatigue, tiredness, and lower back pain. Bone marrow engraftment occurs usually by day +14 following HSCT, which is faster than cord blood engraftment and slower than PBSC engraftment. Bone marrow as graft has a lower number of T cells than the PBSC; therefore, it is associated with a lower rate of GVHD than PBSC, but higher than umbilical cord blood (UCB).

1.2.5 Peripheral Blood Stem Cells

In the recent decade, PBSCs have become the most common source of stem cell graft due to reduced costs, safety of the collection procedure, and more rapid hematopoietic recovery. The normal concentration of HSCs in circulation is 10–100 times lower than the concentration of HSCs in the BM which requires that PBSC donors receive mobilization agents as growth factors. These growth factors are administered subcutaneously and include granulocyte colony-stimulating factor (G-CSF) or plerixafor, a C-X-C Motif Chemokine Receptor 4 (CXCR4) antagonist, that disrupts the interaction between CXCR4 and Stromal Cell-Derived Factor 1 (SDF1) which allows the release of stem cells into the peripheral blood. HSCs are then collected utilizing apheresis. Apheresis collection of HSCs include risks for a variety of side effects. The growth factors used for mobilization can cause myalgias, bone pain, and fatigue whereas the vascular access required for apheresis can cause discomfort, infection, and thrombosis. Citrate-based anticoagulation which is required during the apheresis procedure can also cause bleeding as well as symptomatic hypocalcemia. Moreover, volume changes and fluid shifts can be associated with changes in blood pressure. PBSC collections approximate 350 mL and typically will yield a CD34+ cell count of 5–10 × 10^6 CD34+ cells per kg of recipient body weight and a CD3+ cell count of 300–500 × 10^6 CD3+ per kg of recipient body weight. PBSCs are associated with predictable engraftment kinetics with engraftment typically occurring on day +10 following HSCT. Collection of lymphocyte numbers tends to be higher in contrast to bone marrow harvests with subsequent increased risk for GVHD.

1.2.6 Cord Blood

Umbilical cord blood can be used as a source of stem cells in the setting of allogeneic transplant, especially for patients of minorities that are underrepresented in donor registries. The World Marrow Donor Association has >700,000 cord blood units available worldwide. Given the unique ability of cord blood to be used successfully despite greater HLA disparity, single or double HLA mismatches can be considered. Importantly, cord blood unit donors are therefore available for over 80% of the possible recipients of any ethnic group. Cord blood is collected at the time of birth, after which it is processed and cryopreserved in the cord blood bank. The time of cryopreservation has shown no effect on post-thaw cell viability and neutrophil engraftment probability. Benefits associated with the use of UCB as a potential donor source include immediate availability and lack of risks to the donor. On the other hand, important barriers to the use of umbilical cord blood must be acknowledged. The limited cell dose often precludes the use of cord blood in many older children and adults. The use of two cord blood units and ex vivo expansion of cord blood stem cells can increase the number of available stem cells allowing adults to be suitable recipients. High costs compared to the acquisition of a haploidentical donor are notable. Cord blood units typically approximate 100 mL or less and will yield a CD34+ cell count of 0.2×10^6 CD34+ cells per kg of recipient body weight and a CD3+ cell count of 2.5×10^6 CD3+ per kg of recipient body weight. Regarding GVHD risk, cord blood as a graft source is associated with the lowest risk of GVHD in comparison to other graft sources.

Case 2

A 23-year-old male with diagnosis of X-linked chronic granulomatous disease (pathogenic truncating nonsense mutation: CYBB c.469C > T; protein mutation: p.Arg157X) is being evaluated post-HSCT. The patient was diagnosed with CGD at 6 months of age after presenting with a liver abscess. During childhood, the patient had recurrent painful oral ulcers and later in life developed recurrent perianal abscesses and fistulas. Colonoscopy revealed granulomatous duodenitis and pancolitis. The patient was receiving prophylactic trimethoprim-sulfamethoxazole, posaconazole, and terbinafine for prior history of fungal pneumonia with chronic pulmonary infiltrates. A decision was made to avoid the use of biologic therapies and steroids due to concerns of worsening immunosuppression. The patient did not have an active infection at the time of HSCT and received a non-myeloablative conditioning regimen with busulfan (5 mg/kg), total body irradiation (TBI) (300 cGy), and alemtuzumab and GVHD prophylaxis consisting of sirolimus. He underwent a 10/10 HLA-matched unrelated donor PBSC transplant. The patient had neutrophil engraftment (first of three consecutive days of an absolute neutrophil count >500 cells/mm3) on Day +19. At day +100 posttransplant, lineage-specific (myeloid, CD3, and NK) chimerism was 100%, 80%, and 95%, respectively. Posttransplant complications included Grade I GVHD involving the skin which was successfully treated with corticosteroids and Klebsiella pneumoniae bacteremia treated with meropenem. Six months posttransplant the patient developed cervical lymphadenopathy and hepatosplenomegaly. A biopsy confirmed Epstein-Barr virus (EBV)-associated monomorphic PTLD treated successfully with rituximab.

Question 1

What type of conditioning regimen is associated with excellent outcomes and stable mixed myeloid chimerism in patients with CGD?

A. Not conditioning
B. Reduced intensity conditioning
C. Myeloablative conditioning
D. Antibody-based conditioning

Answer and Explanation

Answer: B

Consideration of the choice of conditioning is important to the success of HSCT in IEI. Success equates to the establishment of adequate donor-derived T cell immunity (naïve T cell production), and B cell immunity (no need for immunoglobulin supplementation, and specific antibody production) in case of adaptive immune defects, and myeloid lineage for innate IEI. The purpose of conditioning with myeloablative agents such as alkylating agents (e.g., busulfan, treosulfan) is to "make space" in the bone marrow niche to allow engraftment of donor HSCs. Conditioning with immunosuppressive agents (e.g., fludarabine, cyclophosphamide) serves to ensure ablation of the host immune system to prevent rejection of the donor HSCs. In some forms of IEI, there is an inability to reject the donor graft, such as in SCID, making conditioning less needed with subsequent lower risk of morbidity and mortality. However, pitfalls of unconditioned HSCT procedures include incomplete immune reconstitution and waning immunity. Novel approaches to conditioning are being developed to limit toxicity (e.g., JSP191, humanized monoclonal anti-CD117 c-kit in clinical trials, which clears hematopoietic stem cells targeted) and the increased risk for chronic health conditions or so-called late effects while supporting robust T and B cell immunity among survivors of HSCT for IEI.

1.2.7 Conditioning Intensity and Donor Chimerism

Donor Chimerism

The establishment of adequate donor chimerism is impacted by the intensity of conditioning and remains a critical element to the success of HSCT in IEI. Donor chimerism refers to a state which exists after engraftment in which cell populations (myeloid, T cells, B cells) may be derived from different individuals. The measurement of donor chimerism relies on differences in short tandem repeat (STR) of the DNA of the donor versus the recipient (patient). This can be assessed at lymphocyte subset level.

Complete donor chimerism is associated with the absence of recipient cell populations whereas mixed donor chimerism is associated with the presence of both recipient and donor cell populations. Mixed myeloid chimerism and mixed lymphoid chimerism have different implications. Cells derived from the myeloid lineage (mainly neutrophils) have a short lifespan and require bone marrow production; therefore, they are reflective of engraftment of HSCs and associated donor chimerism of the HSCs. Cells derived from the lymphoid lineage (mainly T cells) have a longer lifespan and are not reflective of the engraftment of HSCs, but stable donor T cell chimerism is necessary for adequate engraftment of HSCs.

In SCID, there is a selective advantage for the donor-derived T cells to survive and proliferate in the periphery. If full B cell chimerism is also achieved, the patient may not need immunoglobulin replacement therapy. However, in most circumstances only mixed chimerism is achieved, and thus patients require long-term monitoring for changes in their disease phenotype. On the contrary, in patients with CGD, the correction of myeloid compartment is of high importance. Even low neutrophil chimerism with residual NADPH activity as low as 10% may be sufficient to prevent clinical disease. Therefore, data on donor chimerism should be analyzed in light of the underlying IEI.

Conditioning Intensity

Table 3 provides an overview of aspects of conditioning in HCT for IEI. Myeloablative conditioning (MAC) regimens include a variety of agents or combinations of agents that are associated with the destruction of myeloid cells and opening of niches for HSC engraftment. These regimens also possess the capacity to destroy lymphoid cells leading to immunoablation as well. MAC regimens that are utilized in allogeneic HCT for IEI include cyclophosphamide in combination with alkylating agents such as busulfan or treosulfan. MAC regimens allow for donor T cells to engraft and are typically associated with the establishment of complete donor chimerism. This is due to complete or near complete clearance of recipient T cells facilitating complete donor T cell chimerism. Reduced intensity conditioning (RIC) regimens are associated with diminished toxicity and have allowed for the extension of HSCT to older patients with IEI or patients with significant comorbid conditions in association with their IEI. RIC regimens are also associated with a greater risk of mixed chimerism and graft failure or graft dysfunction following HSCT. RIC regimens that are frequently utilized in allogeneic HSCT include fludarabine in combination with alkylating agents such as busulfan or melphalan. Low dose TBI is also utilized in RIC regimens (e.g., 2 Gy TBI) for IEI. Conditioning regimens may also be non-myeloablative (NMA) and are associated with minimal toxicity. There is no need for stem cell support for autologous recovery. Large numbers of donor T cells and HSCs as well as adequate immunoablation are required for the success of these regimens to ensure donor engraftment. NMA conditioning regimens may include fludarabine in combination with low dose TBI or fludarabine in combination with cyclophosphamide as examples. The use of non-myeloablative conditioning is also associated with the presence of mixed chimerism.

Examples of Disease-Specific Considerations

SCID is an IEI in which unconditioned HSCT may be considered; however, outcomes are variable. OS is excellent (>80%) in the setting of unconditioned matched sibling donor HSCT for severe combined immunodeficiency; however, unconditioned HSCT with unrelated donors or cord blood may be less favorable. Although classically, many HSCT procedures for SCID were unconditioned, a greater interest in the use of conditioning has occurred in recent decades as limited engraftment of donor HSC which fail to support adequate T and B cell immunity long-term. Conditioning prior to HSCT for SCID is associated with

Table 3 Aspects of conditioning in inborn errors of immunity

Intensity	Myeloablation	Examples
Non-myeloablative	NA	Fludarabine (150–160 mg/m^2) + Total body irradiation (2 Gy) Fludarabine (150–160 mg/m^2) + Cyclophosphamide (20–40 mg/kg)
Reduced intensity	+	Fludarabine (150–160 mg/m^2) + Melphalan (140 mg/m^2) Fludarabine (150–160 mg/m^2) + Busulfan (AUC 60–70 mg*h/L)
Myeloablative	++	Fludarabine (150–160 mg/m^2) + Busulfan (AUC 85–95 mg*h/L) Busulfan (AUC 85–95 mg*h/L) + Cyclophosphamide (120–200 mg/kg)

Abbreviations: *AUC* area under the curve, *Gy* gray, *NA* not applicable

improved T cell and B cell immunity, and a diminished need for a second HSCT procedure. This is particularly salient for specific forms of SCID with a tendency for inflammation (e.g., RAG deficiency). As anticipated, conditioning leads to a greater risk for tissue injury and complications (e.g., acute GVHD and late effects). Specific forms of SCID with DNA repair defects require special consideration (e.g., ARTEMIS deficiency) due to sensitivity to alkylating agents and greater risk for devastating late effects (e.g., growth failure, dental issues) with conditioning. Similar to SCID, conditioning intensity may be tailored to each specific IEI or category of IEI and recipient, incorporating other factors such as donor and stem cell source. As an example, WAS is an IEI that requires greater consideration of MAC regimens. RIC regimen use in WAS is associated with mixed myeloid and lymphoid chimerism and risk for persistent thrombocytopenia and autoimmunity, respectively. CGD is an IEI that requires greater consideration of RIC regimens. RIC regimen use in CGD is associated with excellent outcomes (OS >90%) and stable mixed myeloid chimerism (>90%) which is curative for patients with CGD including those with intractable and often invasive infections and inflammatory complications.

Concluding Thoughts and Future Directions

Ongoing studies are attempting to tailor conditioning exposures to ensure adequate T cell and B cell engraftment while limiting the risk for late effects. Conditioning SCID infants diagnosed early (C-SIDE) represents the first randomized trial to assess if lower doses of busulfan can support robust establishment of T and B cell immunity while mitigating toxicity and late effects in newborns with SCID. Antibody-based conditioning is also under investigation. Recent clinic data demonstrates that anti-CD117 (c-Kit) antibody is able to clear the niches of HSCs facilitating donor HSC engraftment without toxicity in patients with SCID. These approaches and others are currently being evaluated to optimize the health and well-being of patients with IEI.

Question 2

What organs/system is not included in the evaluation of acute graft-versus-host disease?

A. Skin
B. Liver
C. Blood
D. Gut

Answer and Explanation

Answer: C

GVHD occurs frequently following allogeneic HSCT for IEI and offers no benefit for patients with IEI as it may in patients with selected malignant disorders. An absence of GVHD is associated with robust immune reconstitution. On the other hand, treatment of GVHD with immunosuppressive agents is associated with impaired immune reconstitution, infectious risk, and an increased risk of morbidity and mortality. A variety of factors shape the risk of GVHD such as PBSCs as a graft source (greater numbers of T cells), MAC conditioning, and greater HLA disparity. DLI used to boost antiviral immunity or engraftment following HSCT for IEI is associated with a greater risk for GVHD. Other risk factors for GVHD specific to IEI include the presence of transplacental maternal engraftment often detected in SCID patients or early mixed donor chimerism with a gradual increase in donor T cell chimerism which may be protective.

Classically, acute GVHD was described by the presence of signs and symptoms of GVHD occurring prior to Day +100 post-HCT and chronic GVHD after Day +100 post-HCT. Currently, it is recognized that acute GVHD and its manifestations may occur after Day +100. Moreover, acute GVHD can be classified as persistent, recurrent, or de-novo. There may also be a so-called overlap syndrome in which features of both acute and chronic GVHD coexist.

Table 4 provides a brief description of pharmacologic and non-pharmacologic approaches frequently utilized for GVHD prophylaxis as well as their mechanisms and toxicities. Prevention of GVHD is paramount and requires the use of pharmacologic agents which interfere with T cell activation, T cell function, as well as T cell proliferation. Decreasing the number of T cells using T cell depletion techniques including both in vivo and ex vivo approaches are also frequently utilized.

Pathophysiology of Acute and Chronic GVHD

Tissue damage which may occur in relation to the underlying IEI, and its treatment including conditioning regimens (even when RIC or NMA regimens are utilized) supports the development of GVHD. Tissue damage directly results in the elaboration of proinflammatory cytokines such as tumor necrosis factor (TNF)-alpha and IL-1. The elaboration of proinflammatory cytokines also occurs indirectly. For example, damage to the gut results in the translocation of microbial products such as lipopolysaccharide (LPS) which activates host antigen-presenting cells including monocytes and macrophages leading to the elaboration of proinflammatory cytokines. LPS may also lead to the augmentation of the inflammatory response through activation of the innate immune system (e.g., Toll-Like Receptors [TLRs], Nod-Like Receptors [NLRs]). Donor T cells are then activated by this inflammatory milieu. Target cell death ensues via a variety of mechanisms (e.g., FAS-mediated apoptosis, etc.) in association with the destruction of epithelial cells in specific target organs (e.g., skin, liver, gut).

Chronic GVHD also involves a breakdown in central and peripheral tolerance to self-antigens with resultant autoimmunity. This occurs in an analogous fashion to the break-

Table 4 Aspects of graft-versus-host disease prophylaxis

Category	Examples	Mechanism of action	Toxicities
Calcineurin inhibitors	Cyclosporine tacrolimus	Bind immunophilins which interact with calcineurin, a calmodulin-dependent phosphatase involved in the activation of nuclear factor of activated T cells (NFAT) proteins. The NFAT proteins translocate to the nucleus and are associated with the transcription of various cytokine genes that are vital for T cell proliferation, activation, and function such as interleukin (IL)-2	Increased risk of infection, hypertension, nephrotoxicity, encephalopathy, and microangiopathy
Antimetabolites	Methotrexate	Methotrexate is an anti-metabolite which inhibits dihydrofolate reductase (DHFR). DHFR reduces folic acid to tetrahydrofolate (THF) which is necessary for thymidylate and purine synthesis	Nausea, vomiting, myelosuppression, hepatotoxicity, nephrotoxicity, and infection risk
	Mycophenolate mofetil	A prodrug of mycophenolic acid, an inhibitor of inosine 5′ monophosphate dehydrogenase which results in depletion of guanosine nucleotides. This is associated with the inhibition of T cell and B cell proliferation	Gastrointestinal toxicity including nausea, vomiting, and myelosuppression, and infection risk
mTOR inhibitors	Sirolimus	Bind FKBP12 and complexes with mTOR resulting in the inhibition of IL-2 mediated T cell proliferation	Risk of infection, hyperlipidemia, and microangiopathy
In vivo T cell depletion	Antithymocyte globulin	T cell depletion in blood and peripheral lymphoid tissue via complement-dependent lysis and T cell activation associated apoptosis	Risk of infection, infusion reactions, flu-like illness, serum sickness
	Alemtuzumab	Anti-CD52 which results in T cell and B cell depletion in the peripheral blood and peripheral lymphoid tissues via complement-dependent lysis, antibody-dependent cytotoxicity, and direct apoptosis	Risk of infection, infusion reactions, flu-like illness
	Posttransplant cyclophosphamide	Depletion of host T cells which proliferate in response to donor antigens and the depletion of donor T cells which proliferate in response to host antigens. Specific subsets of T cells are spared which are important for viral immunity and immune reconstitution	Risk of infection, nausea, vomiting, and hemorrhagic cystitis
Ex vivo T cell depletion	TCR alpha/beta + CD19 depletion	Depletion of alloreactive T and B cells which are implicated in the pathogenesis of GVHD and Epstein-Barr virus (EBV)-related disease while ensuring that T cells that support engraftment and surveillance against malignant cells and viral pathogens remain	Increased risk of graft rejection, delayed immune reconstitution, and risk of infectious complications

Table 5 Aspects of acute graft-versus-host disease staging and grading

Organ	**Stage**	**1**	**2**	**3**	**4**
Skin (rash)[a]		<25% of BSA	25–50% BSA	>50% BSA	Desquamation and/or bullae
Liver (bilirubin)[b]		2–3 mg/dL	4–6 mg/dL	7–15 mg/dL	> 15 mg/dL
Gut (diarrhea)[c]		> 500 mL/day	> 1000 ml/day	> 1500 ml/day	Severe abdominal pain and/or ileus
	Grade	**I**	**II**	**III**	**IV**
Skin		1 or 2	3 or	–	4 or
Liver		–	1 or	2–3 or	4
Gut		–	1	2–4	–

[a]Exclude other causes: drug reaction, viral exanthem, engraftment syndrome, chemotherapy-related, radiation-related
[b]Exclude other causes: sinusoidal obstruction syndrome, viral infection, drug toxicity, sepsis
[c]Exclude other causes: chemotherapy-related, drug-related, infection

down of central and peripheral tolerance which occurs in selected IEI due to defects in thymic stroma or HSC leading to disordered thymopoiesis and diminished thymic output. Tissue injury from GVHD or conditioning exposures extends to thymic epithelial cells which is associated with a breakdown in central tolerance as the cortical thymic epithelial cells and medullary thymic epithelial cells are essential to the process of positive and negative selection. During positive and negative selection, thymocytes whose T cell receptors do not recognize self-antigen or thymocytes that recognize self-antigen intensively in the context of peptides presented by MHC are deleted. Regulatory T cells which are vital to the maintenance of peripheral tolerance may also become impaired in chronic GVHD. B cell dysregulation may also contribute to chronic GVHD as developing B cells also engage with antigens in the bone marrow with a process similar to the positive and negative selection of T cells by which the deletion of autoreactive B cells occurs.

Staging and Grading of Acute and Chronic GVHD

Table 5 provides a brief overview of the organ staging and grading of acute GVHD as well as potential diagnoses which have substantial clinical overlap with the features of acute GVHD. Organ staging of acute GVHD is based on criteria by Glucksberg. This staging requires the assignment of a stage for each organ system involved including skin, liver, and gut. Stages are assigned for each organ using a score of 0–4. The Seattle grading system is then used which allows for an overall grade to be assigned from 0 to 4 using the organ-specific staging results. The diagnosis of acute GVHD is challenging due to clinical diagnoses which may present with features that overlap with acute GVHD. Serum biomarkers are being utilized to a greater extent in the diagnosis of acute GVHD and its prognosis. As an example, markers of intestinal crypt damage including regenerating islet derived 3 alpha (REG3alpha) as well as Suppressor of Tumorigenesis 2 (ST2) have been shown to be predictive of mortality in association with acute GVHD.

Features of chronic GVHD including organ-specific manifestations as well as a brief overview of the staging and grading of chronic GVHD are depicted in Table 6. The most frequently involved organs including the skin, mouth, eye, gut, and liver are those mentioned above; however, joints, muscles, genitalia, esophagus, nails, and lungs may also be involved in the setting of chronic GVHD. The diagnosis of chronic GVHD requires the confirmation of diagnostic (e.g., lichen planus or lichen sclerosis) and distinctive manifestations (e.g., papulosquamous lesions). Signs and symptoms are utilized to define organ-specific scores from 0 to 3 (no involvement or symptoms to severe functional compromise). A global severity score is assigned (mild, moderate, severe) based on the number of organs involved and their organ-specific scores. Serum biomarkers are also under intense investigation as tools in the diagnosis of chronic GVHD and its prognosis.

Table 6 Aspects of chronic graft-versus-host disease assessment[a]

Evaluation	**Organ System**	**Symptoms**
History	Skin	Tight, raw, sores, itchy, rash, shiny scars, scaly, flaky, dark or light spots, hair thin or loss, nail changes
	Eyes	Dry, use of artificial tears, gritty, excess tearing, discomfort, trouble opening, impacts daily activity
	Mouth	Sores, ulcers, sensitivity to temperature/toothpaste/spicy foods/drinks, dry mouth, trouble opening mouth, issues impact daily intake
	Sinus	Drainage, congestion, pain, sinus infection, surgery
	Gastrointestinal	Nausea, vomiting, diarrhea, appetite changes, weight changes, and swallowing difficulties impacts daily intake
	Musculoskeletal	Tightness, tightness muscle cramps, muscle weakness, swollen joints, limits daily activity
	Genitals	Penile/vaginal dryness/discomfort
	Lungs	Short of breath, need for supplemental oxygen, cough or wheezing
Exams	**Organ system**	**Assessment**
Physical exam	Skin	Extent of cGVHD within skin surface surveying for diagnostic or distinctive signs of cGVHD and quantify extent of erythema, dyspigmentation, and sclerosis
	Mouth	Extent of cGVHD within oral cavity surface surveying for erythema, lichenoid lesions, ulcers, mucoceles
	Musculoskeletal	Extent of cGVHD-associated sclerosis or fasciitis by surveying for diminished joint range of motion
Laboratory	Gastrointestinal	AST, ALT, alkaline phosphatase, bilirubin
	Hematologic	Platelet count
Adjunctive	Lungs	Forced expiratory volume (1 s)
Overall assessment	**Global severity**	**Definition**
	Mild	1–2 organs involved, severity = mild (excluding lung)
	Moderate	3 or 3+ organs involved, severity = mild to moderate (lung only mild)
	Severe	3 or 3+ organs involved, severity = severe (lung moderate to severe)

[a]Adapted from Carpenter PA. How I conduct a comprehensive chronic graft-versus-host disease assessment. Blood 2011

Treatment of Acute and Chronic GVHD

Acute and chronic GVHD requires prompt therapy. The first-line agent which is used for the treatment of acute and chronic GVHD is corticosteroids. For acute GVHD, standard therapy includes methylprednisolone 2 mg/kg/day divided into two or three equal doses followed by a short taper. Steroid refractory acute GVHD is diagnosed if there is a lack of clinical improvement or worsening in the signs or symptoms of acute GVHD after 3–5 days of treatment. Second-line therapy is considered if a diagnosis of steroid-dependent or refractory GVHD is present. Examples of second-line agents utilized in acute GVHD are depicted in Table 7.

For chronic GVHD, standard therapy includes methylprednisolone 1 mg/kg/day followed by a prolonged taper. Steroid refractory chronic GVHD is often considered if there is a worsening in the signs or symptoms of chronic GVHD after 1–2 weeks of treatment or a lack of improvement after 1–2 months of treatment. Second-line therapy is considered if a diagnosis of steroid-dependent or refractory chronic GVHD is present. A variety of approaches now exist for the treatment of chronic GVHD. Various examples of agents are depicted in Table 8. Importantly, patients with acute or chronic GVHD also require aggressive methods of infection prophylaxis.

Table 7 Examples of second-line acute graft-versus-host disease treatment

	Examples	Mechanism of action	Toxicities
Anti-cytokine antibodies	Basiliximab	Monoclonal antibody directed at interleukin (IL)-2 receptor preventing formation of the IL-2 binding site abrogating activation of T cells	Risk of infection and infusion reactions
	Infliximab	Monoclonal antibody directed at tumor necrosis factor (TNF) alpha which play an important role in the development of GVHD	Risk of infection and infusion reactions
Antimetabolites	Methotrexate	An anti-metabolite which inhibits dihydrofolate reductase (DHFR). DHFR reduces folic acid to tetrahydrofolate (THF) which is necessary for thymidylate and purine synthesis	Nausea, vomiting, myelosuppression, hepatoxicity, nephrotoxicity, and infection risk
	Mycophenolate mofetil	A prodrug of mycophenolic acid, a inhibitor of inosine 5′ monophosphate dehydrogenase which results in depletion of guanosine nucleotides. This is associated with the inhibition of T cell and B cell proliferation	Nausea, vomiting, myelosuppression, and infection risk
JAK inhibitor	Ruxolitinib	A selective Janus kinase (JAK) 1/2 inhibitor with a impact on interferon (IFN) gamma and IFN gamma receptor signaling and subsequent reductions in T cell infiltration	Risk of infection, cytopenias
T cell depletion	Antithymocyte globulin	T cell depletion in blood and peripheral lymphoid tissue via complement dependent lysis and T cell activation associated apoptosis.	Risk of infection, infusion reactions, flu-like illness, serum sickness
	Alemtuzumab	Anti-CD52 which results in T cell and B cell depletion in the peripheral blood and peripheral lymphoid tissues via complement-dependent lysis, antibody-dependent cytotoxicity, and direct apoptosis	Risk of infection, infusion reactions, flu-like illness
Other	Extracorporeal photopheresis	Extracorporeal exposure of peripheral blood mononuclear cells (PBMCs) to photoactivated 8-methoxypsoralen and reinfusion of the treated cells. When exposed to ultraviolet (UV)-A light it is activated and cross-links DNA resulting in apoptosis with changes in lymphocytes, monocytes, and dendritic cells	Requires vascular access, high cost, and specialized center

Table 8 Examples of second line chronic graft-versus-host disease treatment

Category	Examples	Mechanism of action	Toxicities
B cell directed	Rituximab	Monoclonal antibody directed at CD20 on B cells leading to selective depletion of CD20+ B cells	Hypogammaglobulinemia, risk of infection, and infusion reactions
	Ibrutinib	Inhibits Bruton tyrosine kinase (BTK) in B cells and Interleukin-2-inducible T cell kinase (ITK) in T cells leading to diminished B cell survival and inhibition of selected subsets of T cells; respectively	Risk of infection, fatigue, nausea, vomiting, diarrhea, muscle spasms, and bruising
Antimetabolites	Methotrexate	An anti-metabolite which inhibits dihydrofolate reductase (DHFR). DHFR reduces folic acid to tetrahydrofolate (THF) which is necessary for thymidylate and purine synthesis	Nausea, vomiting, myelosuppression, hepatoxicity, nephrotoxicity, and infection risk
	Mycophenolate mofetil	A prodrug of mycophenolic acid, an inhibitor of inosine 5′ monophosphate dehydrogenase which results in depletion of guanosine nucleotides. This is associated with the inhibition of T cell and B cell proliferation	Nausea, vomiting, myelosuppression, and infection risk
mTOR inhibitors	Sirolimus	Binds FK binding protein-12 and complexes with mTOR resulting in the inhibition of IL-2 mediated T cell proliferation	Risk of infection, hyperlipidemia, and microangiopathy
JAK inhibitor	Ruxolitinib	A selective Janus kinase (JAK) 1/2 inhibitor with an impact on interferon (IFN) gamma and IFN gamma receptor signaling and subsequent reductions in T cell infiltration	Risk of infection, cytopenias
Tyrosine kinase inhibitor	Imatinib	Inhibits both platelet-derived growth factor (PDGF) and transforming growth factor (TGF)-beta signaling intracellularly which reduces the development of fibrosis	Risk of infection, cytopenias, fluid retention
Other	Extracorporeal photopheresis	Extracorporeal exposure of peripheral blood mononuclear cells (PBMCs) to photoactivated 8-methoxypsoralen and reinfusion of the treated cells. When exposed to ultraviolet (UV)-A light it is activated and cross-links DNA resulting in apoptosis with changes in lymphocytes, monocytes, and dendritic cells	Requires vascular access, high cost, and specialized center

Question 3
During the post-engraftment period, what predisposes bacterial infections?

A. Neutropenia
B. Hypogammaglobulinemia
C. Mucositis
D. Intravascular devices

Answer and Explanation
Answer: B

Infections which occur prior to HSCT and those that occur following HSCT are an important cause of early and late morbidity and mortality when considering HSCT for IEI. There are multiple factors which shape the impact of infection in the setting of a specific diagnosis of an IEI including pathogens, prior treatment of IEI, time from diagnosis of the IEI to HSCT, as well as the degree of HLA disparity, donor and graft source, intensity of conditioning, and complications including GVHD.

Severe Combined Immunodeficiency
SCID is one particular example of an IEI in which infections are a vitally important aspect of pre-HSCT planning and post-HSCT outcomes. Pre-HSCT infections are associated with adverse HSCT outcomes. The 5-year survival for patients with SCID post-HSCT is 80–95% when transplanted prior to the onset of infections. With the successful application of newborn screening, the number of newborns with SCID whose diagnosis was triggered by infection has continued to diminish. Despite newborn screening efforts, many newborns with SCID still experience infections prior to HSCT which can potentially cause organ damage. Judicious use of supportive care therapy including early prophylaxis with Bactrim and azoles has diminished the burden of pre-HSCT infections with Pneumocystis Jirovecci pneumonia (PJP) and Candida. Suppression of double-stranded DNA viruses such as cytomegalovirus (CMV) remain paramount prior to HSCT. Importantly, the use of conditioning in SCID patients with active infections is also associated with inferior OS which requires thoughtful consideration. Post-HSCT immune reconstitution is also an important consideration in the setting of SCID. As an example, for those patients that do not achieve robust B cell reconstitution, hypogammaglobulinemia and specific antibody production remain poor and the risk of severe bacterial infections remain elevated in the absence of immunoglobulin supplementation.

Other Inborn Errors of Immunity
Pre-HSCT infections are also important when considering other IEIs as many patients present to HSCT with a significant burden of infections which requires careful identification and treatment. Fungal infections, for example, are an important cause of morbidity in patients with phagocytic defects. Herpesviruses as CMV and bacterial infections can be problematic in patients with T cell deficiencies. To reduce the risk of complications, treatments targeted at colonizing bacteria can be helpful prior to and during HSCT, for example, the use of antibiotics for patients with eczema and Staphylococcus aureus or those with significant pulmonary disease and colonizing organisms such as non-tuberculous mycobacteria. Those at high risk of CMV may benefit from prophylaxis with novel agents such as letermovir. For patients with phagocytic defects, such as CGD, patients are able to undergo HSCT with active infections including fungal infections although specific considerations may include the use of granulocyte transfusions during the period of neutropenia.

Post-HSCT infections also remain of particular importance following HSCT for IEI. As an example, hypogammaglobulinemia and risk of severe bacterial infections may be abrogated with immunoglobulin supplementation. Viral reactivation with double-stranded DNA viruses may be abrogated with antiviral agents such as ganciclovir and foscarnet. For other viral infections such as Epstein-Barr virus (EBV) and associated lymphoproliferative disease post-HSCT, adoptive cellular therapies may be required such as EBV cytotoxic T lymphocytes.

Bacterial Infections
Bacterial infections can appear anytime following HSCT; however, they are more common in the pre-engraftment period secondary to severe neutropenia, presence of intravascular devices, and mucositis. Infections can be caused by gram-negative and gram-positive bacteria resulting in bacteremia, pneumonia, colitis as well as central nervous system, urinary, skin, and soft tissue infections. Patients are not routinely placed on prophylactic antibiotics; however, if there is a concern for a bacterial infection the recommendation is to start broad-spectrum antibiotics following guidelines for neutropenic fever. During the post-engraftment period, bacterial infections are secondary to hypogammaglobulinemia and hyposplenism. The most common bacterial infections during this period are secondary to encapsulated bacteria as Haemophilus influenzae, Streptococcus pneumonia, and Neisseria meningitidis. Chronic GVHD can be associated with further immunosuppression in these patients with a subsequent increased risk of infections. Figure 1 shows the infection risk in relationship with the specific period after HSCT.

Fungal Infections
Fungal infections are an important cause of morbidity and mortality in patients following HSCT. The diagnosis of fungal infections can be challenging since the fungal pathogen is not always identified with a culture and diagnosis may be

Fig. 1 Infection risk during HSCT

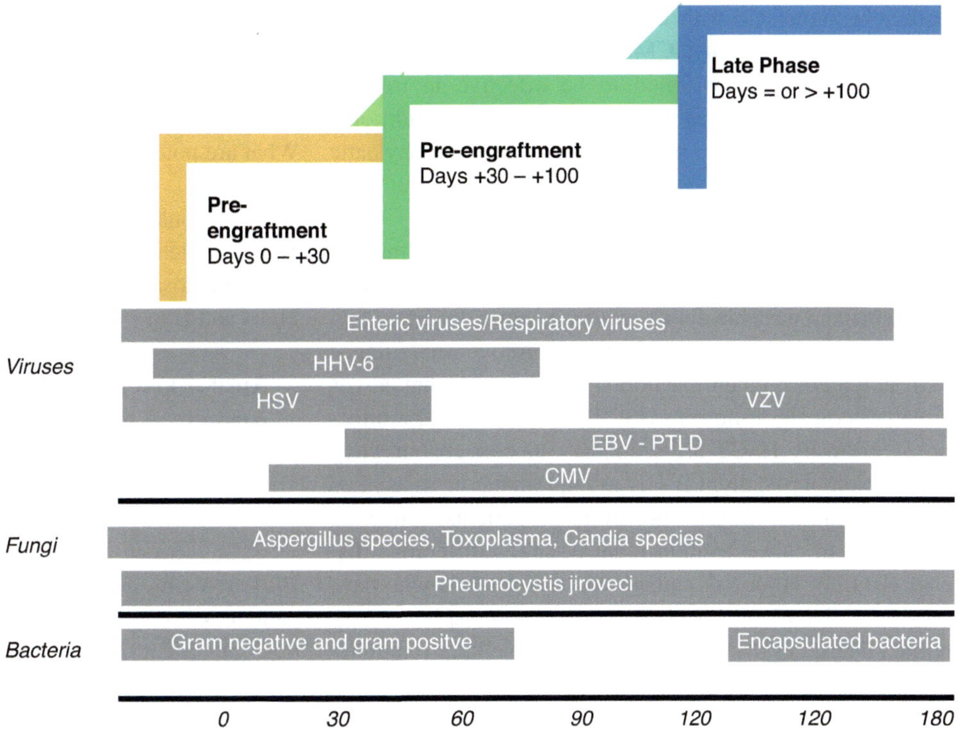

suggested by symptoms, imaging, and indirect evidence of fungal infection such as a positive galactomannan or beta D-glucan assay. The risk of invasive candidiasis and mold infections is higher during the pre-engraftment period due to the associated severe neutropenia, mucositis, and central venous access devices. After engraftment, the presence of a central venous access device in conjunction with severe gastrointestinal GVHD can also be associated with a greater risk of invasive candidiasis. The risk of opportunistic mold infections such as Aspergillosis in the post-engraftment period is also related to prolonged and severe cell-mediated immunodeficiency. Aspergillus is the most common cause of fungal pneumonia in patients following HSCT. Prophylaxis against fungal infections is routinely used during the pre-engraftment period. Azoles affect cytochrome P450 metabolism; therefore, attention must be paid to possible interactions with other medications often utilized during HSCT. Alternative medications include echinocandins such as micafungin. The risk of PJP is approximately 5–15% among patients not receiving prophylaxis; however, effective prophylaxis decreases the risk of PJP significantly. The preferred regimen is treatment with trimethoprim-sulfamethoxazole (TMP-SMX). Other medications used for PJP prophylaxis include dapsone, pentamidine, and atovaquone.

Viral Infections

Viral infections are easily transmissible and thus are quite common. Following HSCT patients can acquire a variety of viral infections including respiratory and gastrointestinal viruses such as metapneumovirus, respiratory syncytial virus, parainfluenza, influenza, and adenovirus. Viral pathogens that establish latency such as CMV, EBV, varicella (VZV), BK, and huma herpesvirus 6 (HHV6) can also reactivate and can be associated with significant morbidity and mortality. Herpes virus prophylaxis with acyclovir is usually given to seropositive patients following HSCT. Periodic laboratory surveillance of certain viruses such as CMV and EBV can help detect the viruses before patients become symptomatic allowing for preemptive therapy. If a viral pathogen is identified, then antivirals and other supportive care agents are given as clinically indicated.

Cytomegalovirus is one of the most common viruses causing infections in patients following HSCT. CMV is a herpesvirus typically acquired early in life. Patients undergoing HSCT and their possible donors have their CMV IgG positivity assessed to identify CMV serostatus. If the recipient is seronegative, the preferred donor is seronegative as well. Similarly, if the recipient is positive, the preferred donor is seropositive. A high risk of CMV reactivation is seen with a negative donor and a positive recipient. Leukoreduction and CMV testing of blood products has reduced the likelihood of patients acquiring CMV through blood transfusions. CMV is monitored periodically with PCR-based techniques in patients following HSCT in order to treat the infection as soon as its diagnosed. CMV can affect multiple organs with the most common infections being pneumonitis and colitis. Ganciclovir is the first-line treatment for CMV. Foscarnet is similarly effective; however, it is associated with a greater burden of side effects including nephrotoxicity. CMV-specific cytotoxic T cells have also been demonstrated to be effective in treating patients with CMV and an inadequate response to traditional therapies. Letermovir has been studied as a new prophylactic medication which has been demonstrated to be safe, well-tolerated, and effective.

EBV is another herpes virus associated with clinically significant viral infections in patients following HSCT. HSCT recipients and possible donors undergo testing to detect IgG against EBV prior to transplantation. Active surveillance with polymerase chain reaction (PCR)-based techniques is done periodically following HSCT. Most of the EBV infections following HSCT are secondary to endogenous virus reactivation. Typically, EBV is asymptomatic or associated with self-limited illness; however, in patients following HCST it can cause posttransplant lymphoproliferative disease (PTLD). High-risk patients include patients who have received a T cell-depleted HSCT or UCB transplant. EBV-related PTLD results from the inability of cellular immunity to eliminate B cells that have been infected with EBV allowing for abnormal lymphoproliferation. Management of PTLD includes decreasing doses of immunosuppressants and Rituximab, a chimeric murine antibody against CD20. Newer therapies include EBV-specific cytotoxic T lymphocytes. Prophylactic therapy against EBV is not recommended because of lack of efficacy.

Question 4
What are potential treatments for graft failure?

A. Additional graft from same donor
B. Donor leukocyte infusion
C. Immunosuppressant medication
D. A and B

Answer and Explanation
Answers: A and B

1.3 Noninfectious Complications

HCT remains a recognized curative treatment for many patients with nonmalignant disorders including IEI. Despite this, noninfectious complications involving a myriad of organ systems remain a formidable barrier to successful HSCT. Below we will outline a few examples that are particularly pertinent to patients with IEI. Table 9 provides a brief summary of these and other noninfectious complications which arise following HSCT for IEI.

Table 9 Examples of noninfectious complications following hematopoietic cell transplantation

Category	Examples	Description	Treatment
Pulmonary	Idiopathic pneumonia syndrome	IPS occurs in the first 100 days (within the first 30 days) following HCT is idiopathic pneumonia syndrome (IPS). IPS results from diffuse alveolar injury manifest radiographically as multilobar infiltrates on a chest radiograph or chest CT. Clinical manifestations of IPS following HCT include cough, difficulty breathing, and rales on clinical exam	Treatment of IPS includes the use of anti-tumor necrosis factor (TNF)-alpha agents such as etanercept
	Diffuse alveolar hemorrhage	DAH occurs early following HCT and within the first 30 days following HCT. DAH is manifest radiographically as diffuse patchy consolidations on chest radiograph or chest CT. Clinical manifestations of DAT following HCT include difficulty breathing, hypoxia, and hypoxemia. A cough may also be present in addition to fever; however, hemoptysis remains infrequent. Bronchoalveolar lavage fluid may be used to document blood and hemosiderin-laden macrophages which are supportive of a diagnosis of DAH	Treatment of DAH has focused on the use of corticosteroids
	Bronchiolitis obliterans syndrome	Destruction of small airways of the lung and manifest radiographically by often normal chest radiographs; however, on chest CT there is air trapping, bronchiectasis, and other changes. Clinically there is dyspnea, dry cough, and wheezing. Pulmonary function testing will demonstrate an obstructive pattern. Diagnostic criteria for BOS include a decrease in the forced expiratory volume (FEV1)	Treatment of BOS includes the use of corticosteroids; however, the use of inhaled corticosteroids, macrolide antibiotics, and leukotriene antagonists is considered
	Bronchiolitis obliterans organizing pneumonia/cryptogenic organizing pneumonia	BOOP/COP affects the alveoli and is manifest radiographically by what appears to be pneumonia. Chest radiographs and chest CT demonstrate infiltrates, ground glass opacities, and nodular opacities. Clinical features include fever, dyspnea, and cough. Pulmonary function testing demonstrates a restrictive pattern with a diminished diffusing capacity (DLCO) and diminished total lung capacity (TLC)	Treatment of BOOP/COP has focused on the use of corticosteroids

Table 9 (continued)

Category	Examples	Description	Treatment
Hepatic	Sinusoidal obstruction syndrome	SOS occurs during the first 30 days following HCT although late SOS is possible. Two or more of the following features must also be present: Elevated bilirubin, hepatomegaly, ascites, and weight gain. Liver biopsy is not required to make the diagnosis of SOS. Ultrasonography can be used to document hepatomegaly and/or ascites. Alterations in portal venous flow and hepatic venous outflow may also be documented but are often considered to be a late finding of SOS	Defibrotide is used for the treatment of severe SOS. Although the exact mechanism by which defibrotide leads to diminished organ-specific compromise related to SOS is unknown, it is thought that defibrotide possesses anti-inflammatory as well as anticoagulant properties
Genitourinary	Thrombotic microangiopathy	Transplant-associated thrombotic microangiopathy (TA-TMA) is an endothelial damage syndrome manifest as hypertension, thrombocytopenia, and elevated lactate dehydrogenase. TA-TMA results from the activation of endothelial cells to produce a pro-coagulant state, activation of antigen-presenting cells and lymphocytes, and activation of the complement cascade with microthrombi formation	For patients who develop signs of TA-TMA while receiving CNIs they are discontinued. Treatment with terminal complement blockade is considered. Eculizumab is a monoclonal antibody against C5 that prevents the formation of the MAC
	Hemorrhagic cystitis	Conditioning leads to damage of the urothelium and immune suppression is permissive for the reactivation of viruses (e.g., BK polyomavirus) in the urothelium. This results in inflammation and hemorrhage. The diagnosis of HC requires signs and symptoms of cystitis (dysuria and frequency). Hematuria is also required (microscopic or gross). Ultrasonography of the bladder is a tool to evaluate for bladder wall thickening and debris. Quantification of virus (e.g., BK polyomavirus) in the blood and urine is also diagnostic	Prophylactic approaches including the use of hyper-hydration and MESNA are employed in high-risk patients. Treatment involves the use of cidofovir with or without probenecid. Other agents have been utilized as well such as Levaquin or leflunomide. Adjunctive therapies have also been associated with improvements such as intravesicular therapies (e.g., cidofovir)
Other	Graft failure	Graft failure can be defined as primary or secondary based on the lack of HSC engraftment or HSC engraftment followed by a subsequent loss of HSC engraftment. Graft failure may also be defined by a lack of T cell engraftment manifest as a lack of CD3+ T cells or absence of donor T cell chimerism and/or the lack of HSC engraftment manifest as an absolute neutrophil count (ANC) < 500/μL or an ANC > 500/μL without evidence of donor myeloid chimerism. Graft failure may also be considered following HCT for IEI when persistent immune dysfunction occurs typically in association with recurrent or severe infections or autoimmunity following HCT. Graft failure may also manifest as inadequate hematopoiesis despite clinical and laboratory evidence of HSC engraftment	Options for treatment may include a so-called boost which requires an additional graft from the same donor, often CD34+ selected and administered without conditioning. Donor lymphocyte infusion (DLI) can also be utilized to stabilize mixed chimerism that is falling or to convert a state of mixed chimerism to full donor chimerism. Graft failure can also be treated with retransplantation
	Autoimmunity	Organ-specific manifestations of autoimmunity following HCT are diverse. One of the most common and frequent manifestation is an autoimmune cytopenias. Other organ-specific manifestations are infrequent, but include the thyroid (e.g., hypothyroidism), central and peripheral nervous systems (e.g., myasthenia gravis), integument (e.g., vitiligo), liver (e.g., hepatitis), and kidney as examples	Treatment typically include corticosteroids, high dose immunoglobulin, rituximab, and combinations of agents such as sirolimus, bortezomib, mycophenolate mofetil, splenectomy, and in refractory case retransplantation

1.3.1 Graft Failure

Graft failure is a rare complication following HSCT for IEI. Graft failure can be defined as primary or secondary based on the lack of HSC engraftment or HSC engraftment followed by a subsequent loss of HSC engraftment. Graft failure may also be defined by a lack of T cell engraftment manifest as a lack of CD3+ T cells or absence of donor T cell chimerism and/or the lack of HSC engraftment manifest as an absolute neutrophil count (ANC) < 500/μL or an ANC > 500/μL without evidence of donor myeloid chimerism. The interval of time elapsed following HCT required to define graft failure is a subject of debate. It is important to consider various clinical factors and HCT-related factors (e.g., specific IEI diagnosis, degree of HLA disparity, conditioning exposures, T cell depletion, stem cell source) when approaching the diagnosis of graft failure as well as the potential etiologic agents such as infections (e.g., CMV, HHV6) and medications (e.g., Bactrim, Ganciclovir). Graft failure may also be considered following HCT for IEI when persistent immune dysfunction occurs typically in association with recurrent or severe infections or autoimmunity following HCT. Graft failure may also be manifest as inadequate hematopoiesis despite clinical and laboratory evidence of HSC engraftment.

Options for treatment may include a so-called boost which requires an additional graft from the same donor, often CD34+ selected and administered without conditioning. DLI can also be utilized to stabilize mixed chimerism that is falling or to convert a state of mixed chimerism to full donor chimerism; however, the risk of GVHD in the setting of IEI must be considered. Graft failure can also be treated with retransplantation. Retransplantation may occur using the same donor; however, the procurement of the second source of HSCs can be considered. More importantly, the health of the recipient must also be considered adequate as conditioning is typically utilized. The choice of conditioning, donor, and graft source in the setting of retransplantation for graft failure is controversial. Successful HSC engraftment has been documented in association with long-term OS following retransplantation for patients with IEI.

1.3.2 Pulmonary Complications

Patients with IEI are particularly susceptible to pulmonary complications prior to HCT. Recurrent upper and lower respiratory traction infections are frequently observed leading to issues such as bronchiectasis. Other pulmonary complications are also frequently observed such interstitial lung disease. In IEI, pulmonary disease serves as a risk factor for morbidity (e.g., a need for respiratory support) following HCT as well as mortality following HCT. Importantly, pulmonary risk factors should be carefully assessed prior to HCT using pulmonary function tests, high-resolution computed tomography of the chest, bronchoalveolar lavage, and biopsy if indicated. Defining the presence of interstitial lung disease or pulmonary infection (e.g., aspergillus, atypical mycobacteria, etc.) is important to ensure optimal control prior to HCT and can also assist with the tailoring of the HCT procedure.

Noninfectious pulmonary complications which occur most commonly in the early period (the first 100 days following HCT) are idiopathic pneumonia syndrome (IPS) and diffuse alveolar hemorrhage (DAH). IPS typically occurs early following HCT and typically within the first 30 days following HCT. IPS results from diffuse alveolar injury which is manifest radiographically as multilobar infiltrates on a chest radiograph or chest CT. Clinical manifestations of IPS following HCT may include cough, difficulty breathing, and rales on clinical exam. Treatment of IPS has focused on the use of anti-tumor necrosis factor (TNF)-alpha agents such as etanercept. DAH also occurs early following HCT and typically within the first 30 days following HCT. DAH is manifest radiographically as diffuse patchy consolidations on chest radiograph or chest CT. Clinical manifestations of DAT following HCT may include difficulty breathing, hypoxia, and hypoxemia. A cough may also be present in addition to fever; however, hemoptysis remains infrequent. Blood and hemosiderin-laden macrophages on a bronchoalveolar lavage are typically documented. Treatment of DAH has focused on the use of corticosteroids.

Pulmonary complications which occur most commonly in the late period (following the first 100 days following HCT) are bronchiolitis obliterans syndrome (BOS), and bronchiolitis obliterans organizing pneumonia/cryptogenic organizing pneumonia (BOOP/COP). BOS is associated with the destruction of small airways of the lung. BOS is manifest radiographically by often normal appearing chest radiographs; however, on chest CT there is often evidence of air trapping as well as bronchiectasis, and other findings. BOS is manifest clinically by dyspnea, dry cough, and wheezing as well as an absence of fever. Pulmonary function testing will typically demonstrate an obstructive pattern. Diagnostic criteria for BOS include a decrease in the forced expiratory volume (FEV1) by 20% with a FEV1 to forced vital capacity (FVC) ratio of <0.7. Treatment of BOS has focused on the use of corticosteroids; however, a combination of inhaled corticosteroids, macrolide antibiotics, and leukotriene antagonists including fluticasone, azithromycin, and montelukast should be considered. BOOP/COP is a noninfectious pulmonary complication which typically affects the alveoli. BOOP/COP is manifest radiographically by the presence of what appears to be pneumonia. Chest imaging may demonstrate infiltrates, ground glass opacities, and nodular opacities. BOOP/COP is manifest by a spectrum of signs and symptoms including fever, dyspnea, and cough. Pulmonary function testing typically demonstrates a restrictive pattern with a diminished diffusing capacity (DLCO) and diminished total lung capacity (TLC). Treatment of BOOP/COP has focused on the use of corticosteroids.

1.3.3 Autoimmunity

Autoimmunity is a frequent complication of HCT in the setting of nonmalignant disorders including IEI, and it is more common in nonmalignant disorders than malignant disorders following HCT. Risk factors for autoimmunity following HCT are numerous including the use of an unrelated donor and T cell depletion. Autoimmunity in the setting of HCT for IEI may be associated with chronic GVHD as discussed earlier in this chapter. Autoimmunity following HCT that is not related to chronic GVHD is also complex and due to a variety of factors such as recent and donor characteristics, HCT-related factors as well as infections, and immune reconstitution-related issues. As an example, post-HCT donor T cells including regulatory T cells help to mediate peripheral immune tolerance. Following HCT, these regulatory T cells may be unable to suppress B cell expansions which occur following viral infections which may support imbalances in T and B cell-mediated immunity following HCT for IEI.

Organ-specific manifestations of autoimmunity following HCT are diverse. One of the most common and frequent manifestations are autoimmune cytopenias. Estimates vary; however, reports have suggested an incidence of approximately 10% (autoimmune cytopenias which are one of the most frequent manifestations) following HCT for IEI. They often occur early on during the post-HCT process (a median of 6.5 months following HCT). Treatment is often necessary and typically include corticosteroids, high-dose immunoglobulin, rituximab, and combinations of agents such as sirolimus, bortezomib, mycophenolate mofetil, splenectomy, and in refractory case retransplantation. Importantly, the majority (~90%) of those with autoimmune cytopenias are able to maintain a long-term remission and in recent decades mortality associated with autoimmunity following HCT is low. Other organ-specific manifestations are infrequent, but include the thyroid (e.g., hypothyroidism), central and peripheral nervous systems (e.g., myasthenia gravis), integument (e.g., vitiligo), liver (e.g., hepatitis), and kidney as examples.

Question 5
What evaluation is important in the post-HSCT period?

A. Growth and nutrition
B. Dental evaluation
C. Quality of life
D. All of the above

Answer and Explanation
Answer: D

1.4 Late Effects and Survivorship

1.4.1 General Considerations

Since the first successful HSCT in the 1960s there has been consistent progress in the practice of HSCT for IEI. In fact, patients with SCID, the most common transplantable IEI, have an OS at 3 years of >90%. This evolution has supported an increasing number of HSCT survivors with a prior diagnosis of IEI worldwide. Survivors of HSCT for IEI can be impacted by late-occurring chronic health conditions or so-called late effects which can occur following HSCT.

Routinely, patients are followed closely after HSCT. Depending on the preferences of the center, follow-up, as well as surveillance for potential late effects and management of late effects, may be carried out by the HSCT team or by their primary clinical immunologist or primary care provider or a combination of providers. The risk of late effects following HSCT is shaped by a variety of factors such as age at the time of HSCT, clinical status prior HSCT including history of infections and associated organ damage, donor source, graft manipulation, conditioning, and the presence of GVHD. Importantly, patients with specific forms of SCID remain at greater risk of late effects than others. Some forms of SCID can be associated with neurocognitive deficits as may be the case in ADA deficiency or DNA ligase IV deficiency. Conditioning exposures and the HSCT process can lead to further decrements in neurocognitive function. ARTEMIS SCID patients that receive conditioning with alkylating agents may develop a significant burden of late effects including growth retardation and endocrinologic (e.g., central growth hormone deficiency, central hypothyroidism, insulin-dependent diabetes, exocrine pancreatic deficiency), renal (e.g., tubulopathy), pulmonary (e.g., fibrosis), and dental (e.g., abnormal development of permanent teeth) issues.

For patients with other non-SCID IEI, a significant burden of late effects may also develop. Conditioning, which is often required for non-SCID IEI, can be associated with the worsening of preexisting organ damage. For example, an HSCT survivor with RAG deficiency and pre-HSCT granulomatous interstitial lung disease requires ongoing surveillance for post-HSCT pulmonary insufficiency. Moreover, necessary conditioning exposures can be associated with infertility and endocrinopathies as examples. As mentioned previously, post-HSCT autoimmunity may also be significant. As an example, WAS patients who experience mixed chimerism post-HSCT remain at risk for autoimmunity. Importantly, careful monitoring of psychosocial health is also vital as the diagnosis of IEI and the need for HSCT and its associated risk for late effects is associated with a spectrum of decrements in physical, mental, and social health culminating in decrements in quality of life.

In 2017, recommendations for screening of late effects in pediatric patients with SCID after HCST were published. Although they are specific to SCID, they can be applied non-SCID IEI. For specific organ/system evaluations post-HSCT see Table 10. Recommendations include evaluation of immune reconstitution starting at least 3 months after HSCT to identify if the patient has need of ongoing immunoglobulin replacement as well as to provide prophylactic medications. Immune reconstitution should be evaluated periodically, especially if there are concerns about recurrent infections.

In pediatric patients, continuous monitoring of height and weight allows to identify patients who may require nutritional and endocrinological evaluation. Developmental screening should also be done when patients are able to do so

Table 10 Specific organ/system evaluations post-HSCT

Organ/system	Associated late effects	Evaluation/management
Endocrine	Growth retardation/obesity	Height, weight, BMI, Tanner staging every 6 months Bone age Referral to endocrinology if clinically indicated
	Hypothyroidism	Physical examination of thyroid, TSH, and free T4 Referral to endocrinology if clinically indicated
	Gonadal insufficiency	Annual assessment of pubertal and reproductive function Males at risk: Morning testosterone, semen analysis Females at risk: LH, FSH, and estradiol. Gynecologic evaluation if at risk for vaginal fibrosis Referral to endocrinology if clinically indicated
	Diabetes/dyslipidemia	Fasting blood glucose of hemoglobin A1C and lipid profile every 2 years Referral to endocrinology if clinically indicated
	Osteoporosis/avascular necrosis	DEXA scan 1 year after transplant and as clinically indicated MRI if suspicious of avascular necrosis Recommendations about regular physical activity and adequate intake of calcium and vitamin D Referral to endocrinology if clinically indicated
Ophthalmologic	Impaired vision, cataracts, xerophthalmia	Visual acuity and fundoscopic examinations every 1–3 years by ophthalmology
Audition	Hearing loss, tinnitus, vertigo	Audiology evaluation 1 year after transplant and as clinically indicated
Dental	Teeth development	Dental hygiene education Oral assessments every 6–12 months
Neurocognitive	Developmental delay	Neuropsychological evaluation including processing speed, attention, visual motor integration, memory, comprehension, verbal fluency, executive function and planning
Cardiovascular	Coronary artery disease, heart failure, arrhythmias, peripheral arterial disease, cerebrovascular disease	Annual history regarding cardiac symptoms (dyspnea, chest pain, palpitations, syncope) Echocardiogram done based on personal risk Referral to cardiology if clinically indicated
Pulmonary	Chronic cough, interstitial lung disease, chronic GVHD	Pulmonary function tests yearly Imaging considered depending on clinical assessment Counseling about smoking and second-hand smoking
Renal	Hypertension, chronic kidney disease, tubular injury	Blood pressure with every medical visit Annual urinalyses, blood urea nitrogen, creatinine, and electrolytes 1 year after transplant and as clinically indicated Referral to nephrology if clinically indicated
Gastrointestinal	Liver dysfunction, strictures, liver GVHD	ALT, AST, bilirubin, serum ferritin 1 year posttransplant and as clinically indicated Counseling in limiting alcohol use
Skin	Skin GVHD (poikiloderma, ichthyosis, eczema, keratosis pilaris)	Annual skin check Sun protection
Immunologic	Increase risk of infections (depending on stage and use of immunosuppressant for GVHD)	Immunizations CMV monitoring in patients at risk of CMV reactivation
Psychosocial	Decrease quality of life, social withdrawal, under-employment, chronic fatigue, depression, anxiety	Psychosocial assessment including access to healthcare and insurance

to facilitate therapies if needed. Neurocognitive deficits can be related to the IEI itself as mentioned above or to conditioning, for example, the use of TBI, which is avoided in younger children.

Any patient who received alkylators as part of their conditioning may be at increased risk of hormonal imbalances such as hypothyroidism and hypogonadism with subsequent infertility. In eligible patients, egg and sperm preservation can be offered prior to HSCT.

Pulmonary dysfunction can be common, especially in patients with prior recurrent respiratory infections. Additionally, GVHD and inadequate immunoglobulin replacement can cause further lung damage. All patients should be assessed for smoking history including second-hand smoke. Pulmonary function tests are recommended yearly for patients after HSCT. Imaging such as chest CT imaging is indicated depending on the need for additional clinical assessment.

Finally, because children and adults may have significant psychosocial impairments resulting from their own disease, chronic illness, prolonged hospitalizations, and social isolation, their physical, emotional, and social health as well as their overall quality of life can be assessed through standardized surveys and measurement tools. This will allow for the identification of survivors who may benefit from specific supportive care therapies. Evaluations can be repeated as needed across the trajectory of survivorship care.

Bibliography

1. Castagnoli R, Delmonte OM, Calzoni E, Notarangelo LD. Hematopoietic stem cell transplantation in primary immunodeficiency diseases: current status and future perspectives. Front Pediatr. 2019;7:295. https://doi.org/10.3389/fped.2019.00295.
2. Buckley RH, Schiff SE, Schiff RI, et al. Hematopoietic stem-cell transplantation for the treatment of severe combined immunodeficiency. New Engl J Med. 1999;340(7):508–16. https://doi.org/10.1056/nejm199902183400703.
3. Burroughs LM, Petrovic A, Brazauskas R, et al. Excellent outcomes following hematopoietic cell transplantation for Wiskott-Aldrich syndrome: a PIDTC report. Blood. 2020;135(23):2094–105. https://doi.org/10.1182/blood.2019002939.
4. Chiesa R, Wang J, Blok HJ, et al. Hematopoietic cell transplantation in chronic granulomatous disease: a study of 712 children and adults. Blood. 2020;136(10):1201–11. https://doi.org/10.1182/blood.2020005590.
5. Copelan EA. Hematopoietic stem-cell transplantation. New Engl J Med. 2006;354(17):1813–26. https://doi.org/10.1056/nejmra052638.
6. Wachsmuth LP, Patterson MT, Eckhaus MA, Venzon DJ, Gress RE, Kanakry CG. Post-transplantation cyclophosphamide prevents graft-versus-host disease by inducing alloreactive T cell dysfunction and suppression. J Clin Invest. 2019;129(6):2357–73. https://doi.org/10.1172/jci124218.
7. Abdelhakim H, Abdel-Azim H, Saad A. Role of αβ T cell depletion in prevention of graft versus host disease. Biomed. 2017;5(3):35. https://doi.org/10.3390/biomedicines5030035.
8. Champlin RE, Schmitz N, Horowitz MM, et al. Blood stem cells compared with bone marrow as a source of hematopoietic cells for allogeneic transplantation. IBMTR histocompatibility and stem cell sources working committee and the European Group for Blood and Marrow Transplantation (EBMT). Blood. 2000;95(12):3702–9.
9. Rocha V, Cornish J, Sievers EL, et al. Comparison of outcomes of unrelated bone marrow and umbilical cord blood transplants in children with acute leukemia. Blood. 2001;97(10):2962–71. https://doi.org/10.1182/blood.v97.10.2962.
10. Shaw P, Shizuru J, Hoenig M, Veys P, IEWP-EBMT. Conditioning perspectives for primary immunodeficiency stem cell transplants. Front Pediatr. 2019;7:434. https://doi.org/10.3389/fped.2019.00434.
11. Heimall J, Puck J, Buckley R, et al. Current knowledge and priorities for future research in late effects after hematopoietic stem cell transplantation (HCT) for severe combined immunodeficiency patients: a consensus statement from the second pediatric blood and marrow transplant consortium international conference on late effects after pediatric HCT. Biol Blood Marrow Transplant. 2017;23(3):379–87. https://doi.org/10.1016/j.bbmt.2016.12.619.
12. Dvorak CC, Hassan A, Slatter MA, et al. Comparison of outcomes of hematopoietic stem cell transplantation without chemotherapy conditioning by using matched sibling and unrelated donors for treatment of severe combined immunodeficiency. J Allergy Clin Immun. 2014;134(4):935–943.e15. https://doi.org/10.1016/j.jaci.2014.06.021.
13. Thakar M, Logan B, Buckley R, et al. Transplantation outcomes for children with severe combined immune deficiency (SCID) have improved over time: a 36-year summary report by the primary immune deficiency treatment consortium (PIDTC). Biol Blood Marrow Transplant. 2020;26(3):S18–9. https://doi.org/10.1016/j.bbmt.2019.12.083.
14. Schuetz C, Neven B, Dvorak CC, et al. SCID patients with ARTEMIS vs RAG deficiencies following HCT: increased risk of late toxicity in ARTEMIS-deficient SCID. Blood. 2014;123(2):281–9. https://doi.org/10.1182/blood-2013-01-476432.
15. Gooptu M, Antin JH. GVHD prophylaxis 2020. Front Immunol. 2021;12:605726. https://doi.org/10.3389/fimmu.2021.605726.
16. Zeiser R, Blazar BR. Pathophysiology of chronic graft-versus-host disease and therapeutic targets. New Engl J Med. 2017;377(26):2565–79. https://doi.org/10.1056/nejmra1703472.
17. Flinn AM, Gennery AR. Treatment of pediatric acute graft-versus-host disease—lessons from primary immunodeficiency? Front Immunol. 2017;8:328. https://doi.org/10.3389/fimmu.2017.00328.
18. Servais S, Lengline E, Porcher R, et al. Long-term immune reconstitution and infection burden after mismatched hematopoietic stem cell transplantation. Biol Blood Marrow Transplant. 2014;20(4):507–17. https://doi.org/10.1016/j.bbmt.2014.01.001.
19. Satwani P, Baldinger L, Freedman J, et al. Incidence of viral and fungal infections following busulfan-based reduced-intensity versus myeloablative conditioning in pediatric allogeneic stem cell transplantation recipients. Biol Blood Marrow Transplant. 2009;15(12):1587–95. https://doi.org/10.1016/j.bbmt.2009.08.006.
20. Girmenia C, Bertaina A, Piciocchi A, et al. Incidence, risk factors and outcome of pre-engraftment gram-negative bacteremia after allogeneic and autologous hematopoietic stem cell transplantation: an Italian prospective multicenter survey. Clin Infect Dis. 2017;65(11):1884–96. https://doi.org/10.1093/cid/cix690.
21. Tomblyn M, Chiller T, Einsele H, et al. Guidelines for preventing infectious complications among hematopoietic cell transplantation recipients: a global perspective. Biol Blood Marrow Transplant. 2009;15(10):1143–238. https://doi.org/10.1016/j.bbmt.2009.06.019.
22. Ozdemir ZN, Bozdag SC. Graft failure after allogeneic hematopoietic stem cell transplantation. Transfus Apher Sci. 2018;57(2):163–7. https://doi.org/10.1016/j.transci.2018.04.014.

23. Smith CC, Lakin P, Lehmann LE, Pai SY, Whangbo J. Pulmonary complications in patients with primary immunodeficiency undergoing hematopoietic stem cell transplantation. Biol Blood Marrow Transplant. 2020;26(3):S16–7. https://doi.org/10.1016/j.bbmt.2019.12.081.
24. Buxbaum NP, Pavletic SZ. Autoimmunity following allogeneic hematopoietic stem cell transplantation. Front Immunol. 2020;11:2017. https://doi.org/10.3389/fimmu.2020.02017.
25. Chow EJ, Anderson L, Baker KS, et al. Late effects surveillance recommendations among survivors of childhood hematopoietic cell transplantation: a children's oncology group report. Biol Blood Marrow Transplant. 2016;22(5):782–95. https://doi.org/10.1016/j.bbmt.2016.01.023.
26. Pai SY, Logan BR, Griffith LM, et al. Transplantation outcomes for severe combined immunodeficiency, 2000–2009. New Engl J Medicine. 2014;371(5):434–46. https://doi.org/10.1056/nejmoa1401177.

Allergic Diseases of the Eye

Matthew R. Norris and Neeti Bhardwaj

Abbreviations

AKC	Atopic keratoconjunctivitis
GPC	Giant papillary conjunctivitis
IgE	Immunoglobulin E
MC_T	Mast cell (Tryptase variant)
MC_{TC}	Mast cell (Tryptase and chymase variant)
OA	Ocular allergy
PAC	Perennial allergic conjunctivitis
SAC	Seasonal allergic conjunctivitis
VKC	Vernal keratoconjunctivitis

1 Introduction

The term ocular allergy is used to describe a heterogeneous collection of immunological inflammatory processes affecting the anterior surface of the eye that affects between 20 and 40% of the United States population. It costs Americans an estimated 2–4 billion dollars annually. Despite having a significant economic and clinical burden, less than 10% of patients with allergic conjunctivitis seek medical attention. In this chapter, we use case-based discussion to review the spectrum of ocular hypersensitivity disorders, which includes ocular nonallergic hypersensitivity disorders (such as giant papillary conjunctivitis) and ocular allergic disorders (such as seasonal and perennial allergic conjunctivitis, vernal keratoconjunctivitis, and atopic keratoconjunctivitis). We will also review diagnostic criteria, key differential diagnoses, principles of evaluation, and management for each of them. We will conclude with a discussion of emerging therapies.

Case 1

A 35-year-old male presents to the office in August with complaint of red, itchy eyes. His symptoms began 6 years prior to the visit. They are intermittent, beginning around May and persisting until late October, and tend to flare while mowing the lawn. Although typically responsive to oral diphenhydramine as needed in the past, his symptoms have been worsening despite use several times per day and now include the sensation of excessive tearing and a "gritty" sensation when he closes his eyes. On examination, he has conjunctival injection with peri-limbal sparing. No conjunctival papillae are noted. Aeroallergen skin testing identifies sensitization to grass mix and a tear osmolarity of 320 mOsm/L.

Question 1

What is the most likely diagnosis?

A. Seasonal allergic conjunctivitis
B. Seasonal allergic conjunctivitis with concomitant dry eye disease
C. Atopic keratoconjunctivitis
D. Dry eye disease
E. Perineal allergic conjunctivitis

Answer and Explanation

Answer: B

The patient's history and concordant skin testing support the diagnosis of seasonal allergic conjunctivitis. However, overuse of diphenhydramine has resulted in tear film instability and development of concomitant dry eye disease. This is described in greater detail in the section below on

M. R. Norris
Department of Medicine, Division of Pulmonary, Allergy and Critical Care Medicine, Pennsylvania State College of Medicine, Penn State Health-Milton S Hershey Medical Center, Hershey, PA, USA
e-mail: mnorris3@pennstatehealth.psu.edu

N. Bhardwaj (✉)
Department of Pediatrics, Division of Pediatric Allergy-Immunology, Pennsylvania State College of Medicine, Penn State Health-Milton S Hershey Medical Center, Hershey, PA, USA
e-mail: nbhardwaj@pennstatehealth.psu.edu

M. Mahmoudi (ed.), *Absolute Allergy and Immunology Board Review*, https://doi.org/10.1007/978-3-031-12867-7_4

differential diagnoses. Furthermore, his lack of blurring of vision, photophobia, or periocular involvement suggests atopic keratoconjunctivitis.

2 Allergic Conjunctivitis: Seasonal or Perennial

Both seasonal and perennial allergic conjunctivitis are inflammatory processes of the ocular surface that occur in individuals sensitized to aeroallergens. They are defined by duration of illness, with seasonal symptoms being intermittent and perennial symptoms being year-round. The seasonal form is more prevalent and accounts for the majority of cases of allergic conjunctivitis. It is triggered by pollen (grasses, trees, weeds) that surface during specific seasons. The exact species and onset of pollen season vary depending on geographical location in the United States. Tracking of regional pollen counts is often performed by allergists, who are increasingly making their data available to the public through the American Academy of Allergy, Asthma, and Immunology's National Allergy Bureau. However, actual pollen counts are not readily available to most stakeholders as the process of counting is both laborious and not typically reimbursed. For this reason, we often rely on a pollen index (a prediction based on historical data) to see which species may be correlating with the patient's symptoms. Perennial allergic conjunctivitis, as the name implies, is associated with perennial allergens such as animal dander, dust mite, and cockroach. These allergens are present in the environment year-round. Patients may have a mixed phenotype of allergic conjunctivitis, characterized by year-long symptoms from a perennial allergen with seasonal flares during a pollen season.

Question 2

You stop the patient's diphenhydramine and initiate a combination of cetirizine 10 mg daily and artificial tears as needed up to four times daily. He returns 2 weeks later stating the intensity of his symptoms has improved dramatically; however, they are still bothersome. On examination, you observe improved but persistent conjunctival injection with peri-limbal sparing. Decreased activity of which cells is most likely responsible for the patient's symptomatic improvement?

A. Mast Cells (Tryptase variant)
B. Mast Cells (Tryptase/Chymase variant)
C. Basophils
D. Eosinophils
E. Th2 Cells

Answer and Explanation

Answer: A

MC_T is more commonly found in the conjunctival epithelium than MC_{TC} and responds best to treatment with antihistamines. Basophils, eosinophils, and Th2 cells play a role in the late-phase reaction, which does not respond well to antihistamine therapy.

3 Pathogenesis

Human mast cells are categorized based on granule-associated neutral proteases (tryptase and chymase) which are unique to mast cells. On this basis, mast cells have been divided into MC_T (tryptase) and MC_{TC} (tryptase/chymase) phenotypes. The normal human conjunctival epithelium rarely has mast cells but when they exist, they are mainly the MC_T phenotype. Individuals with seasonal and perennial allergic conjunctivitis have increased numbers of MC_T mast cells and eosinophils. In addition, they may have IgE antibodies, histamine, tryptase, eotaxin, and eosinophil cationic protein in the tear film.

The pathogenesis of allergic conjunctivitis has long been recognized as biphasic, with an early phase reaction and a late phase reaction. The early phase reaction is the product of IgE antibody-dependent interaction, where disruptions in epithelial cell tight junctions result in penetration by a pre-sensitized allergen. This exposure results in cross-linking of specific IgE molecules to the high-affinity receptors (FceRI) on mast cells (MC_T predominant) and subsequent mast cell release of preformed (histamine and tryptase) and newly synthesized mediators (prostaglandin D2 and leukotrienes C4, D4, and E4) in the substantia propria. Symptoms attributable to the release of histamine only last 20–30 min; however, the cascade of the other inflammatory mediators set up the late-phase reaction through the recruitment of inflammatory cells from the bloodstream into tissue via promotion of adhesion molecules (e.g., VCAM-1, ICAM-1/CD54, and E-selectin). T cells are also recruited simultaneously to differentiate and secrete an inflammatory cytokine milieu (Fig. 1). The late phase reaction occurs between 4 and 24 h after symptom provocation in 33–100% of patients. The overlapping biological effects of these mediators (both in the early and late phases) contribute to the characteristic symptoms of ocular allergy—namely itching, redness, and watery discharge.

Question 3

Which of the following physical exam findings is most characteristic of allergic conjunctivitis?

A. Allergic shiners
B. Peri-limbal sparing
C. Horner-Trantas dots
D. Hyperosmotic tear film
E. Herbert follicles

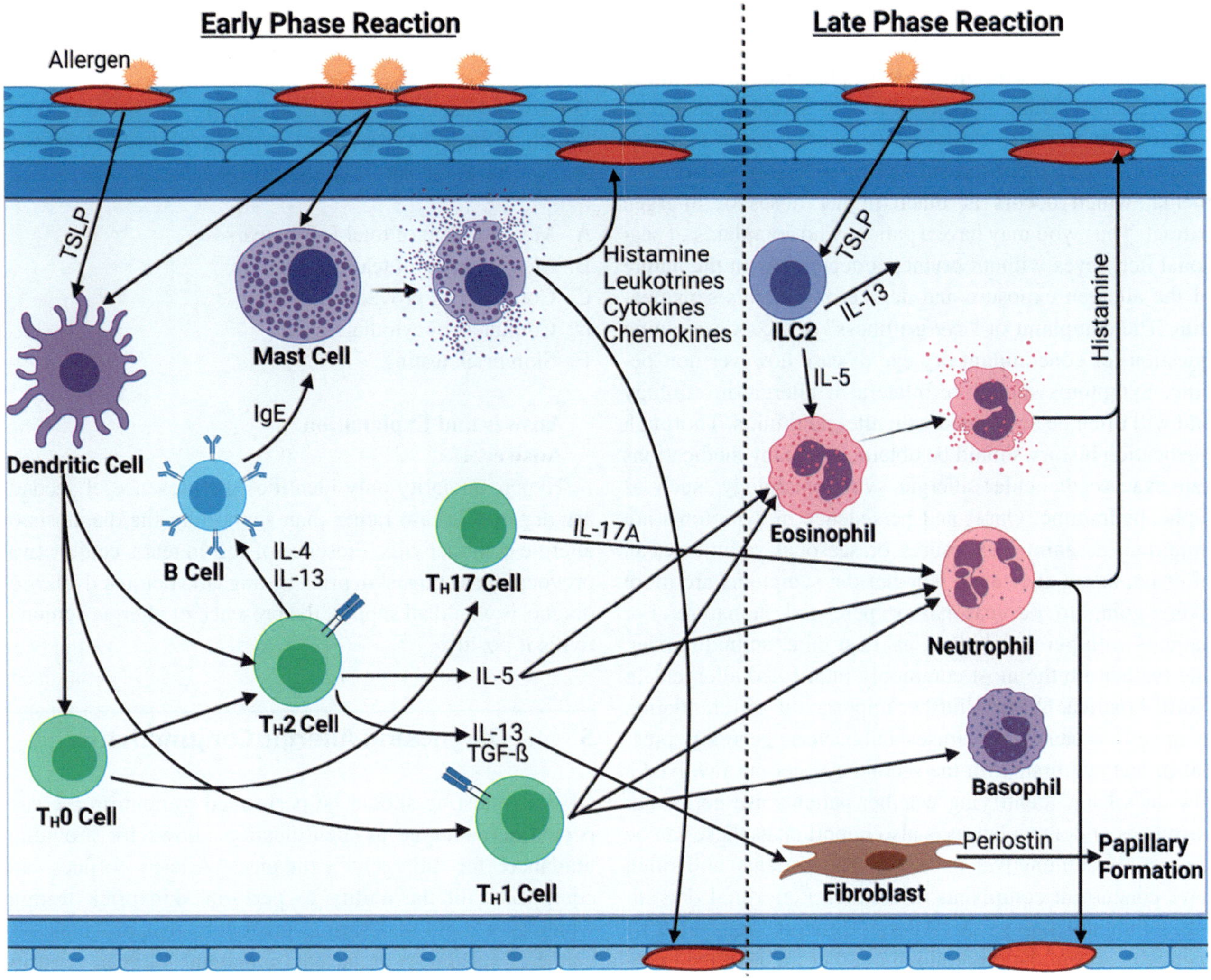

Fig. 1 Allergic Inflammation of the ocular surface. (Reproduced with permission from Valentine L, Norris MR, Bielory L. Comparison of structural components and functional mechanisms within the skin vs. the conjunctival surface. Curr Opin Allergy Clin Immunol. 2021;21(5): 472–479)

Answer and Explanation

Answer: B

In allergic conjunctivitis, the intensity of vascular engorgement decreases toward the limbus of the eye. Perilimbal redness (aka: ciliary flush) occurs secondary to dilation of the radial vessels and is better characterized as iritis. Horner-Trantas dots (gelatinous papillae on the limbal conjunctiva) and Herbert follicles (superior limbal follicles) are physical exam findings more suggestive of vernal conjunctivitis. Allergic shiners, or periorbital venous congestion, are associated with allergic rhinitis rather than allergic conjunctivitis. Although hyperosmolar tear film can be seen in cases of allergic conjunctivitis, it often signifies a concomitant dry eye syndrome.

4 Clinical Features of Allergic Conjunctivitis

In taking the history of suspected ocular allergy, it is most important to screen for ocular pain or blurred vision with photophobia. These are alarming if present and may merit further evaluation by an ophthalmologist for uveitis, keratitis, or glaucoma depending on the patient's history and rest of the physical exam. Ocular discharge is not always present however when present is either watery or "stringy/ropey." Mucoid, or purulent discharge, with morning crusting and difficulty opening the eyelids is typically seen in bacterial infection rather than SAC or PAC. Ocular pruritus and erythema are important to identify, for if neither are present you should consider an alternative condition such as dry eye disease. Ocular itching is the dominant symptom reported in allergic conjunctivitis. On review of pharmaceutical studies looking at minimum allergen dose for symptom provocation, ocular pruritus occurs at significantly lower doses of allergen extract independently of the development of erythema, which occurs at much higher doses of allergen extract. Thus, you may have a patient who complains of seasonal itchy eyes without erythema depending on the nature of the allergen exposure and degree of patient's sensitization. The complaint of "eye grittiness" is a good screening question for concomitant dry eye disease however nonspecific. Symptoms should be bilateral if allergic in etiology and will often be associated with allergic rhinitis. Thorough medication history should be obtained as many medications can exacerbate ocular allergic symptomatology, such as diphenhydramine. Onset and persistence of symptoms are important as animal exposures or seasonal variations can offer important cues as to whether the symptoms are more likely going to be seasonal or perennial in nature. For patients with perennial allergies, dust mite, animal dander, and feather are the most commonly implicated allergens in North America. Onset is further important in differentiation as infections caused by viruses and bacteria generally present in one eye first, with the second eye getting involved a few days later. Identifying whether patients use periocular cosmetics or contact lenses is also important as these can be sources of conjunctival irritation. The patients will often have concurrent complaints of rhinitis. Periorbital darkening commonly known as "Allergic Shiners" may also be seen with accumulation of blood or other fluids in the infraorbital groove.

Symptoms can be objectively measured using the Total Ocular Symptom Score (TOSS), a validated metric that asks patients whether they have experienced any of the following four ocular symptoms in the past week: itchiness, redness, tearing (or eyes watering), or swelling. For each symptom they are asked whether they experienced symptoms all of the time (4 points), most of the time (3 points), half of the time (2 points), some of the time (1 point), or none of the time (0 points). The score from all four symptoms is added together, then divided by 16 and multiplied by 100 to get the average score. Calculating a TOSS score can be of assistance to physicians caring for SAC or PAC, with utility ranging from an objective evaluation of a new therapeutic's efficacy to trending patient perception of symptom severity objectively over time.

On physical examination, symptoms may or may not be present depending on the severity of the illness or time from patient's last exposure to a suspected allergen. If conjunctival injection is present, there should be minimal erythema in the peri-limbal region of the eye. Vision, shape of the pupil, ocular movement, light reactivity, and retinal reflexes should not be affected if the etiology of the patient's complaints is secondary to allergic conjunctivitis.

Question 4

Which of the following tests, if positive, offers the least support for the diagnosis of allergic conjunctivitis?

A. Measurement of total IgE in tears
B. Measurement of tear osmolarity
C. Conjunctival provocation testing
D. Conjunctival cytodiagnosis
E. Skin prick testing

Answer and Explanation

Answer: B

Hyperosmolarity only identifies the presence of secondary dry eye disease rather than supporting the diagnosis of allergic conjunctivitis. Presence of IgE in tears, conjunctival provocation testing, skin prick testing, and conjunctival cytodiagnosis would all support the presence of allergic conjunctivitis if positive.

5 Diagnosing Allergic Conjunctivitis

Objective testing should be performed to confirm the suspected aeroallergen as identification allows for providing guidance for allergen avoidance. Allergy offices are equipped with the ability to perform skin prick testing, which is a rapid and simple procedure that provides evidence of specific external environmental allergen sensitivity. Although sensitive for allergic rhinitis and allergic asthma, skin testing may not always correlate with ocular surface allergen sensitization. In these cases, serum-specific IgE measurements should be considered when SPT results are discordant with the medical history. Other tests that can be performed but are not usually available to allergists in the clinic include measurement of total IgE in tears (which should be negative), tear osmolarity (if hyperosmotic suggestive of dry eye disease), Schirmer testing (abnormal if $\leq$5 mm wetting after 5 min without anesthesia, $\leq$3 mm with anesthesia), and conjunctival provocation testing (mainly used in drug studies). Additionally, conjunctival cytodiagnosis can be performed by an ocular specialist or ophthalmologist to better elucidate harder to diagnose ocular pathology as even the presence of one eosinophil is highly suggestive of atopic pathology. Additional discussion regarding diagnostics for ocular allergy is mentioned later in this chapter.

Question 5

As you discuss his primary diagnosis of seasonal allergic conjunctivitis, he recollects that for the previous 3 years he has observed symptoms are milder and take longer to provoke in early June than in late August. What is the most likely explanation for his observation?

A. Sensitization
B. Tachyphylaxis to diphenhydramine
C. Secondary allergen sensitization
D. Priming effect
E. Variations in pollen allergenicity

Answer and Explanation
Answer: D

The priming effect is a phenomenon by which repeated exposure to allergen results in reversible increased reactivity to the allergen. Sensitization is the process by which one becomes allergic to a particular substance. Tachyphylaxis is defined as rapidly diminishing response to successive doses of a drug. There is a widespread belief in the medical community that taking long-term antihistamines makes them less effective; to date, there is no literature to support this. Skin testing only identified monosensitization to grass pollen mix, reducing the likelihood of secondary allergen-provoking symptoms. Pollen allergenicity variation may explain differences in symptom provocation on a day-by-day basis but would not explain a general trend throughout a pollen season that resets each year.

6 Additional Factors Affecting Allergic Conjunctivitis

One key factor to understand when managing allergic conditions is the priming effect. The priming effect describes the phenomenon where chronic allergen exposure results in reversible increased reactivity to a given allergen. Although the mechanism has not been fully elucidated, it is believed that chronic stimulation of the inflammatory cascade results in increased immune cell migration to the conjunctival surface. The increased number of immune cells provides additional sites for IgE-allergen interaction, a greater quantity of preformed mediators, and greater poststimulation capacity for producing cytokines. This is important as symptom severity scores linearly rise until a threshold pollen level is reached, at which point symptom scores increase nonlinearly, and eventually level off. The inflection point where the linear increase becomes nonlinear has been hypothesized to be the point where the receptors responsible for the IgE-mediated reaction saturate. A similar relationship between symptom scores and medication scores occurs throughout a pollen season. These findings support the theory that the predominance of symptoms secondary to the late phase reaction is responsible for increased symptom scores and decreased medication efficacy as pollen counts rise.

Although sensitization to an aeroallergen such as pollen requires a combination of environmental exposure and genetic predisposition to develop allergies, the rate of pollen sensitization is often confounded by variations in regional geography, variations in CO_2 and ozone levels, seasonal and annual differences in temperature, and pollen immunogenicity (which itself can vary from plant to plant as well as varies from a single plant over the course of its pollen season). With the boom of urbanization in the past half century, allergenic plants such as ragweed are thriving near humans in areas not traditionally thought to be problematic for the given plant. A key concern to the allergist treating allergic conjunctivitis is global warming, as warmer temperatures and higher carbon dioxide levels cause plants to grow more vigorously, create more pollen, and create more allergen per pollen grain than otherwise expected. Additionally, pollen seasons are getting longer as temperatures remain warmer. With longer pollen seasons characterized by higher pollen counts of increasing allergenicity, we can predict SAC will continue to have increasing clinical and pharmacoeconomic impact.

Case 2

An 11-year-old male, who recently migrated from Algeria is referred to you by his pediatrician. His father reports that the child has been experiencing eye symptoms, especially itching and burning for the past 2 years with progressive worsening. He notes that the symptoms are more pronounced during spring. The itchiness is persistent. He experiences runny nose and watery eyes during spring. Upon exam, you notice moderate injection of conjunctiva bilaterally, with white ropy discharge in the lower fornices. Skin test is negative for seasonal and perennial allergens.

Question 1

What is the most likely diagnosis?

A. Seasonal allergic conjunctivitis
B. Atopic keratoconjunctivitis
C. Vernal keratoconjucntivitis
D. Blepharoconjunctivitis
E. Viral conjunctivitis

Answer and Explanation
Answer: C

The correct answer is C (vernal keratoconjunctivitis). This condition is seen primarily in prepubescent males, in dry climates, which fits the description of this patient. Moreover, mucoid ropey discharge is also a classic finding in this condition. Choice A is not correct as this patient experiences pruritus year-round. The nature of ocular discharge in seasonal allergic conjunctivitis is typically clear and watery, and these patients usually report nasal symptoms of allergy as well. Choice B is incorrect as this patient is not experiencing symptoms involving the eyelids or periocular skin and does not have concomitant eczema. Choice D is incorrect as blepharoconjunctivitis is characterized by inflammation of the eyelid margins, with the accumulation of crust worse in the morning. These symptoms have not been described for this patient. Choice E is incorrect because viral conjunctivitis presents with clear discharge, and is an acute condition, contrary to the chronic symptoms reported by this patient.

7 Vernal Keratoconjunctivitis

Vernal keratoconjunctivitis (VKC) is a chronic inflammatory disorder of the conjunctiva. The onset is generally before age 10 years. The duration is typically 2–10 years, with resolution in late puberty. Although VKC is considered rare in most developed countries (with an estimated mean annual incidence of 1.24 per 10,000 persons in the United States), the prevalence is significantly higher in warm/dry subtropical countries of Africa, the Middle East, Latin America, and Asia. While it affects males predominantly at younger ages, gender distribution equalizes in older age patients. Other atopic manifestations can be seen in 40–75% of patients with VKC, with 40–60% of patients having a history of atopy. Characteristic is a marked seasonal flaring of symptoms, frequently being in the spring (hence the term vernal).

Question 2
Which of the following therapies would have the least likely chance of offering symptomatic relief to the patient?

A. Allergen immunotherapy
B. Use of refrigerated artificial tears
C. Use of topical corticosteroids
D. Use of oral steroids
E. Use of topical antihistamines

Answer and Explanation
Answer: A
Not all patients with VKC have positive skin tests, nor do their flares follow a particular environmental aeroallergen stimulus. Refrigerated artificial tears would serve to lubricate the ocular surface and offer temporary analgesic relief. Topical/oral corticosteroids and topical antihistamines can potentially be therapeutic in the treatment of VKC.

8 Pathogenesis

The pathogenesis of VKC is controversial and has yet to be fully elucidated. It has historically been considered an allergic condition as there is evidence of mast cell (MC_{TC}), eosinophil, Th2, and IgE involvement. The tarsal conjunctival epithelium and the substantia propria have been shown to have large numbers of mast cells (mainly MC_{TC}) than individuals without VKC. T cells enhancing IgE and IL-4 production have been found in both layers, and major basic protein has been found throughout the conjunctival epithelium. Allergen-specific antibodies, as well as mast cell mediators, have been demonstrated in the tears of VKC patients. However, not all patients have positive skin tests or sensitization to a particular allergen. Genetic influence has been established by observations of increased prevalence of atopy in affected families as well as in studies of monozygotic versus dizygotic twins. In addition to the increasing recognition of genetic factors is the identification of possible neuroendocrine-immune interaction that may further be contributing to a non-IgE-mediated presentation of VKC.

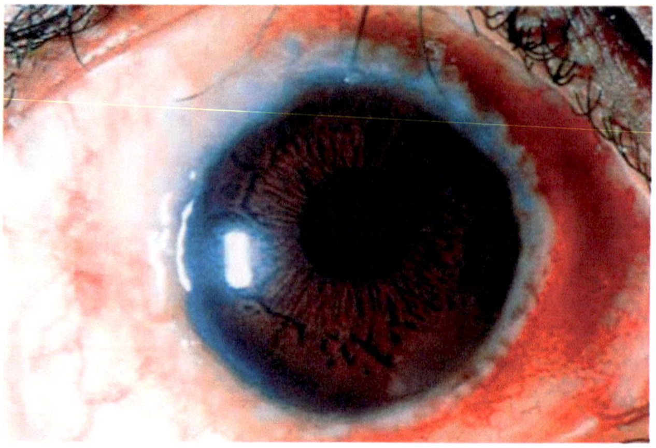

Fig. 2 Horner-Trantas dots. Reproduced with permission from Bielory L. Allergic and immunologic disorders of the eye. Part II: ocular allergy. J Allergy Clin Immunol. 2000 Dec;106(6):1019–32

Question 3
The focal white dots in Fig. 2 are primarily aggregates of which of the following?

A. T lymphocytes
B. B lymphocytes
C. Monocytes
D. Neutrophils
E. Eosinophils

Answer and Explanation
Answer: E
Horner-Trantas dots are aggregates of degenerated epithelial cells and eosinophils accumulating on a gelatinous, hyperplastic limbus, and are commonly seen in VKC and AKC.

9 Clinical Features and Diagnosis

There are three clinical forms: (1) the palpebral form with giant papillae on the tarsal conjunctiva; (2) The limbal form with Horner-Tranta dots (clumps of necrotic eosinophils and epithelial cells); (3) a mixed form of both the palpebral and limbal forms. The limbal and mixed forms are more common in central and southern Africans whereas palpebral VKC is most frequent in Europe and America. VKC is characterized by severe ocular pruritus and photophobia. These patients may also experience foreign body sensation, mucoid discharge, and bletharospasm. Most patients experience a thin, copious milk-white fibrinous secretion (composed of eosinophils, epithelial cells, and Charcot-Leyden crystals). Compared to AKC, the skin of the lids and lid margins is spared and symptoms are confined to the conjunctiva and cornea. Papillary response of the conjunctiva, primarily of the upper tarsus or limbus, is seen, with flattened papillae greater than 1 mm in diameter. These classic "cobblestone" tarsal papillae are often associated with thick, ropy mucus. With disease progression, fibrous tissue proliferation can lead to "giant papillae" reaching 7–8 mm in diameter, predominantly on the upper tarsal plate. This is described as "cobblestoning," with the cobblestones persisting during the quiescent phase, and becoming extremely swollen during the active phase, which is usually spring (Fig. 3). Horner-Trantas dots are aggregates of degenerated epithelial cells and eosinophils accumulating on the gelatinous, hyperplastic limbus. Corneal changes, especially shield ulcers and subsequent scarring can be sight-threatening. Other ocular complications reported include steroid-induced cataract, steroid-induced glaucoma, central corneal scars, keratoconus, irregular astigmatism, and dry eye syndrome. Diagnosis is primarily clinical, based on history and physical findings.

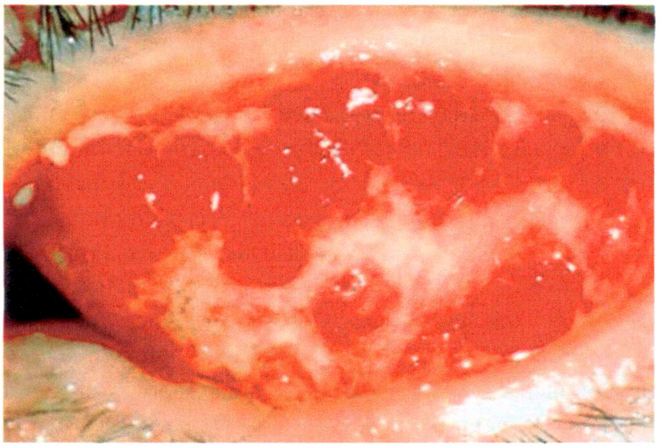

Fig. 3 Vernal keratoconjunctivitis: Note the excessive cobblestoning of conjunctival surface and mucus forming pseudomembrane over the epithelial surface. (Reproduced with permission from Bielory L. Allergic and immunologic disorders of the eye. Part II: ocular allergy. J Allergy Clin Immunol. 2000 Dec;106(6):1019–32)

The presentation is typically in young males in warm climates, who may experience intense photophobia, and ptosis and show the classic cobblestone papillae. Earlier age of presentation and lack of dermatitis of periocular skin help distinguish VKC from AKC.

Question 4
Which of the following findings is more commonly associated with atopic keratoconjunctivitis rather than vernal keratoconjunctivitis?

A. Dennie Morgan lines
B. Lack of periorbital involvement
C. Younger age of onset
D. Corneal involvement
E. Seasonal flaring of symptoms

Answer and Explanation
Answer: D

Corneal involvement is more common in AKC than in VKC. Dennie Morgan lines are nonspecific infraorbital creases that may be seen in many forms of ocular allergy. Atopic keratoconjunctivitis frequently presents with periorbital eczema. Patients with VKC are often younger when they initially present than in AKC. Seasonal flaring will be seen in SAC and VKC rather than in AKC where remission and flaring are without a seasonal correlation.

10 Atopic Keratoconjunctivitis (AKC)

AKC is characterized by bilateral, chronic inflammation involving the conjunctiva and eyelids that can be potentially sight-threatening. Despite relative rarity, AKC is thought to occur anywhere between 20 and 40% of individuals with atopic dermatitis. It is more prevalent in men than women, with peak incidence between the second and fifth decade of life. Although AKC has not been associated with a racial or geographic predilection, a family history of atopy is common. Similarly, ~95% of patients with AKC will have concomitant eczema and ~87% of patients will have concomitant asthma.

An understanding of the pathogenesis of AKC comes from histologic and immunohistochemical analysis of conjunctival biopsy specimens and from tear film analysis for mediators and cells. These studies demonstrate the role of type 1 and 4 hypersensitivity mechanisms. The conjunctival epithelium of patients with AKC has mast cells (MC_T type) and eosinophils, unlike normal individuals. Moreover, there is an increase in helper T cells, amplifying the immune response. The substantia propria in AKC has an increased number of mast cells (compared to normal) as well as eosinophils (not normally present in substantia propria). Moreover,

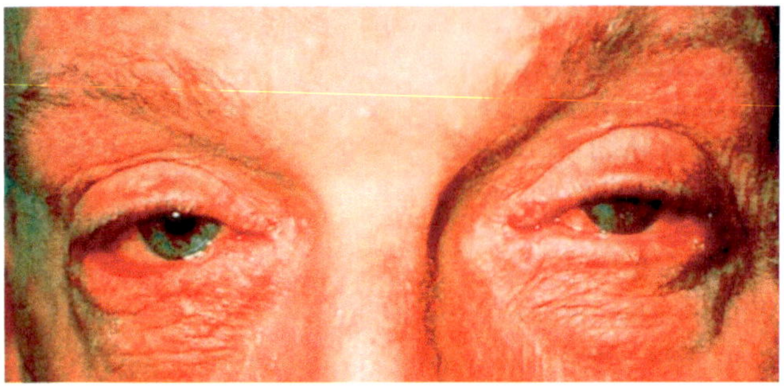

Fig. 4 Atopic keratoconjunctivitis. Note the injection and chemosis of bulbar conjunctiva. Periorbital skin is affected by the formation of infraorbital creases (Dennie Morgan Lines). (Reproduced with permission from Bielory L. Allergic and immunologic disorders of the eye. Part II: ocular allergy. J Allergy Clin Immunol. 2000 Dec;106(6):1019–32)

the substantia propria in AKC shows increased T cells, B cells, and Langerhans cells leading to a complex immune cell profile. The greater the inflammatory cell density in the conjunctiva, the poorer is the tear stability. While Th2 cytokines predominate most allergic processes, lymphocytes with Th1 cytokine profiles have been demonstrated in substantia propria of AKC patients.

The main symptom of AKC is itching, followed by discharge, redness, blurring of vision, photophobia, and pain. The periocular skin may show scaling and flaking on a reddened base. The skin of the lids may become leathery, leading to the development of cicatricial ectropion and lagophthalmos (incomplete closure of the eyelids). The infraorbital skin of the eyelid may show single or double infraorbital creases called Dennie Morgan lines caused by edema or thickening (Fig. 4). Extensive chronic eye rubbing may lead to loss of lateral eyebrows in older patients (de Hertoghe's sign). Other manifestations include lateral canthal ulceration, cracking, and loss of eyelashes (madarosis). Secondary staphylococcal blepharitis may occur due to maceration and induration of eyelids. The tarsal conjunctiva may show papillary reaction, more prominently in the inferior conjunctival fornix. Peri-limbal, gelatinous hyperplasia may occur on the bulbar conjunctiva. A reported finding is Horner-Trantas dots (aggregates of degenerated epithelial cells and eosinophils accumulating on the gelatinous, hyperplastic limbus). The most common corneal finding is punctate epithelial keratopathy. Persistent epithelial defects, scarring, microbial ulceration, and neovascularization all lead to vision loss. Some patients may experience noninflammatory progressive thinning of the cornea (keratoconus). Corneal scarring and neovascularization may result in blindness. Frequent use of steroids in treatment of this condition can lead to the development of cataracts, although the lens opacity typically associated with AKC is an anterior or subcapsular cataract. Anterior uveitis or iris abnormalities are not reported. Older age at presentation and significant lid involvement may help distinguish AKC from VKC. Near to complete resolution of symptoms when out of their season and lack of chronic conjunctival inflammation in SAC also helps to distinguish VKC from AKC.

Question 5

Which of the following is best for differentiation of vernal keratoconjunctivitis from giant papillary conjunctivitis?

A. Contact lens use preceding onset of symptoms
B. Intermittent nature of symptoms
C. Seasonal flaring
D. Photophobia
E. Periocular eczema

Answer and Explanation

Answer: A

History of contact lens use is the main differentiating factor from vernal keratoconjunctivitis which presents similarly.

11 Giant Papillary Conjunctivitis

Giant papillary conjunctivitis (GPC) is characterized by a chronic inflammatory process leading to formation on giant papillae of the upper tarsal conjunctiva. It has been reported in individuals wearing soft, hard, and rigid gas-permeable contact lenses. It is moreover seen in association with ocular prostheses and exposed sutures coming in contact with the conjunctiva. As many as 20% of soft contact lens wearers may be affected by GPC. The incidence is ten times higher in people wearing regular soft contact lenses than rigid (gas-permeable) contact lenses. The incidence among daily-wear disposable contact lens users and rigid-contact lens users is the same. Wearing disposable contact lenses during sleep increases the risk of developing GPC. Asthma and allergic rhinitis are also risk factors.

The inflammatory response seen in GPC is likely multifactorial, with mechanical trauma and chronic irritation of the conjunctiva secondary to contact lens use believed to cause the onset of GPC. The buildup of protein on the surface of the contact lens may cause an allergic reaction. Poor tear clearance in these patients may further facilitate protein buildup on the lens. Biopsy studies show increased mast cells (MC_T) in the conjunctival epithelium although no significant increase is seen in the substantia propria. Release of mediators from mast cells (as evidenced by tryptase found in the tears of these patients) causes increased capillary permeability and tissue infiltration with inflammatory cells, similar to that seen in seasonal and perennial allergic conjunctivitis.

Patients complain of ocular itching, redness, burning, increased mucus discharge in the morning, and photophobia.

Table 1 Clinical and histopathological features of allergic diseases of the eye

Feature	SAC	PAC	VKC	AKC	GPC
Clinical attributes	Intermittent, usually spring Seasonal allergens Coexistent allergic rhinitis, asthma Bilateral involvement	Persistent, year-round Perennial allergens Coexistent allergic rhinitis, asthma Bilateral involvement	Peak incidence 3–20 years Male predominance 3:1 Bilateral involvement Prevalent in warm, dry climate, worse in spring	Peak incidence 20–50 years Both males and females Persistent with exacerbations Bilateral involvement Coexistent atopic dermatitis, asthma, allergic rhinitis, chronic symptoms	Both males and females, bilateral involvement, exposure to contact lenses and prosthesis, chronic symptoms
Signs/symptoms	Ocular itching, tearing, redness, chemosis, not sight-threatening	Ocular itching, tearing, redness, chemosis, not sight-threatening	Severe ocular itching, photophobia, stringy mucoid discharge, cobblestoning and giant papillae on upper tarsal conjunctiva, corneal involvement with shield ulcers, Trantas' dots on limbus, sight-threatening	Severe itching, erythema and flaking of periocular skin, photophobia, cobblestoning and giant papillae in inferior fornix, corneal erosions, conjunctival scarring, Trantas' dots, keratoconus (rare) anterior subcapsular cataract, sight-threatening	Mild ocular itching and mucoid discharge, giant papillae, contact lens intolerance, protein buildup on contact lenses, not sight-threatening
Pathophysiology	Mast cell/ IgE-mediated, atopy Mast cell/ eosinophil infiltration of conjunctiva and substantia propria	Mast cell/ IgE-mediated, atopy Mast cell/ eosinophil infiltration of conjunctiva and substantia propria	IgE- or non-IgE-mediated. Increased mast cells and eosinophils in conjunctival epithelium and substantia propria. Eosinophil major basic protein deposition in conjunctiva, increased collagen	IgE- or non-IgE-mediated. Increased mast cells and eosinophils in conjunctival epithelium and substantia propria Epithelial cell hypertrophy, increased collagen	Giant papillae, conjunctival thickening, increased mast cells and eosinophils in conjunctival epithelium

The presentation may be months to years after starting contact lens use. Small papillae are seen in mild cases, due to contact lenses riding high on the eyeball and irritating the conjunctival lining of the eyelid. Tear deficiency contributes to the progression of these changes, with redness of the inside of the upper eyelid being the first finding, followed by opacification and development of characteristic papillae. History of contact lens use is the main differentiating factor from vernal keratoconjunctivitis which presents similarly.

A summary of the high yield features of these ocular allergic disorders can be found in Table 1.

12 Differential Diagnosis of Ocular Allergy

Clinical allergists and immunologists should be able to differentiate allergic conditions from other immune-mediated ocular disorders as various inflammatory conditions of the eye may present with red eyes. A thorough clinical examination is the first step in this process. It begins with the external component that surrounds the eye, and then the eye itself. Clinical signs to note when examining the eyelids and eyelashes are erythema of the lid margin, scaling, thickening or swelling (blepharitis and dermatitis), periorbital discoloration (allergic shiners, heliotrope), blepharospasm, or ptosis. The conjunctiva should be examined for chemosis (swelling), hyperemia (injection), palpebral and bulbar papillae, and cicatrization (scarring). Assessment of the ocular discharge helps differentiate different forms of conjunctivitis (clear in allergic or viral thick and purulent in bacterial). When deeper tissues such as sclera, vitreous, or choroid are involved, injection and pain are the notable findings. Autoimmune disorders, such as systemic lupus erythematosus and rheumatoid arthritis, cause scleritis associated with pain. An important sign of intraocular inflammation (iritis, uveitis) is the formation of a ring of erythema at the limbal junction of the cornea (ciliary flush), whereas there is characteristic sparing of the peri-limbal area in conjunctivitis. Chronic inflammatory response in the conjunctiva and substantia propria leads to the formation of papillae seen in AKC, VKC, and GPC. A comprehensive list of differentials to consider based on their anatomic location can be seen in Fig. 5.

The eyelid skin is soft, pliable, and thin. Inflammation of the eyelid margins is called blepharitis and is often misdiagnosed as an ocular allergy. Like atopic dermatitis patients, colonization with Staphylococcus aureus is seen. Patients complain of symptoms such as burning, itching, tearing, and a feeling of dryness primarily in the mornings. The eye may be glued shut when the patient awakens in the morning due to crusted exudate. Scales and collarettes of exudative material around the eyelash bases are seen. Blepharitis can be controlled by improved hygiene with detergents such as non-stinging baby shampoos. Allergic contact blepharoconjuncti-

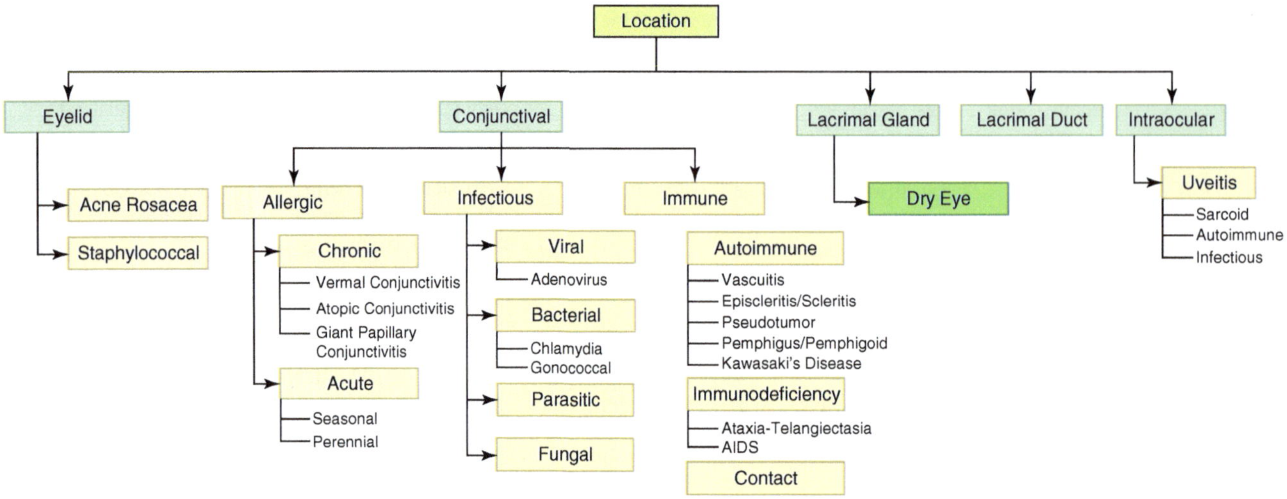

Fig. 5 Differential diagnosis of the red eye. (Reproduced with permission from Bielory L, Delgado L, Katelaris CH, Leonardi A, Rosario N, Vichyanoud P. ICON: Diagnosis and management of allergic conjunctivitis. Ann Allergy Asthma Immunol. 2020 Feb;124(2):118–134)

vitis may be caused by cosmetics applied to the hair (hair dye), face, fingernails (nail paint), or periorbital area. Ninety percent of patients with this condition are female. It is predominantly lymphocyte-mediated hypersensitivity reaction that affects the periocular skin, eyelids, or ocular surface. These reactions linked to cosmetics are often the result of sensitization to preservatives, fragrances, and additives in the products. Contact lens solutions have also been implicated in the development of allergic contact blepharoconjunctivitis.

Infectious conjunctivitis can be divided into bacterial or viral. Acute bacterial conjunctivitis is characterized by symptoms localized primarily to one eye initially, subsequently involving the other side often several days later. Typical symptoms include ocular irritation, conjunctival redness, and mucopurulent discharge. Inflammation is mediated largely by polymorphonuclear cells (neutrophils). A preauricular node may sometimes be palpable. Mucopurulent discharge of chlamydial infection lasts for more than 2 weeks. Intracytoplasmic inclusion bodies may be seen on conjunctival scrapings. Viral conjunctivitis presents with watery discharge, conjunctival injection, and chemosis. If the cornea is involved, there may be associated ocular pain, which should prompt referral to an ophthalmologist.

Tear film has a trilaminar structure composed of an outer lipid layer, an aqueous middle layer, and an inner mucin layer. True dry eye disease develops from decreased tear production, increased tear evaporation, or an abnormality of the tear film. These changes cause the tear film to become hyperosmolar, which in turn results in irritation and inflammation of the conjunctiva. Dry eye syndromes are commonly reported in computer users, long-term contact lens wearers, perimenopausal and postmenopausal women, and patients with acquired immunodeficiency syndrome. Patients may present with mild conjunctival injection and potentially with excessive tearing, leading to misidentification as seasonal or perineal allergic conjunctivitis. Patients often experience a gritty, sandy feeling in the eye along with itching and burning. These symptoms worsen throughout the day as the aqueous tear film evaporates. Winter months are associated with exacerbation of the symptoms due to decrease in relative humidity with indoor heating in the households. Of key clinical importance, long-term use of medications with anticholinergic properties can cause dry eye syndrome. These include first-generation antihistamines, tricyclic antidepressants, β-blockers, and chemotherapeutic agents.

Keratitis can present with red eye, foreign body sensation, and photophobia. Bacterial keratitis is most often seen with contact lens use. Herpes infection can cause viral keratitis. Scleritis typically presents with severe pain and is usually associated with an underlying systemic illness. The pain is often worse at night and early morning and radiates into the face. There may be accompanying redness, watery discharge, and photophobia. Episcleritis also presents with redness, clear discharge, and engorged blood vessels on a clear scleral background. It is seen in middle-aged females. Unlike scleritis, episcleritis is painless and does not cause visual impairment. It may sometimes be associated with autoimmune disease, although most cases are isolated events. Iritis typically presents with isolated limbal redness and constricted pupil. It is usually due to underlying infectious or autoimmune conditions. Acute closure glaucoma is a rare cause of red eye and is usually associated with headache, nausea, and vomiting. This is an ocular emergency characterized by a distinctive ciliary flush.

Drug-induced conjunctivitis is caused mostly by preservatives such as benzalkonium chloride, thimerasol, parabens, ethylenediaminetetraacetic acid (EDTA), and chlorobutanol, in eye drops, with prolonged use exceeding 3 months. The classic presentation is worsening of ocular symptoms after initial improvement with the medication. Discontinuation of the medication leads to the resolution of symptoms. Reactions that occur in response to prolonged use of topical preparations are often seen in the lower eyelid and inferior conjunctiva due to the pooling of the liquid therapeutics in this area. The typical signs are red-colored inflamed conjunctiva, papillae development, keratitis, and chemosis. Vasoconstricting eye drops may also lead to conjunctivitis medicamentosa characterized by increased conjunctival injection and rebound hyperemia.

Occupational conjunctivitis occurs in response to airborne substances in the workplace, inciting inflammation via allergic or nonallergic pathways. Examples of such substances include laboratory animal antigens grain, organic chemicals and irritants, wool, plants, and detergent protease. Occupational rhinitis and asthma may often coexist.

Nonallergic Perennial (Vasomotor) Conjunctivitis is a perennial, chronic condition in which the symptoms are neither immunologic nor infectious in origin and are not associated with ocular eosinophilia. While it is more common in the elderly population, it is seen also in athletes, especially those exposed to chlorinated swimming pools. Patients with vasomotor conjunctivitis complain of excessive tearing with exposure to tobacco smoke, fumes, and perfumes, leading to conjunctival injection. Among the elderly, the cause is physiologic age-related mechanical and anatomic changes.

13 Management of Ocular Allergy

The first step in managing ocular allergies is to identify the causative stimulus and avoid it if possible. Identification can be done via skin prick testing or serum IgE testing, followed by simple allergen avoidance measures such as wearing protective eyewear, keeping windows closed in both the home and while driving, and showering before going to bed. Frequent washing of clothing and bedding is also helpful. Cold compresses and refrigerated artificial tears also provide symptomatic relief. Individuals who use contact lenses may need to switch to eyeglasses during their peak seasons or use daily disposable lenses. During administration of topical medications, contact lenses should be removed to avoid interaction between contact lenses and preservatives in ophthalmic preparations. The lenses may be replaced after 10 min of topical medication administration. Daily disposable lenses are ideal during the pollen season as allergens can accumulate on the lens; allergens can also become trapped between the conjunctival surface and the contact lens.

Artificial tears can help by washing away allergens and diluting inflammatory mediators on the ocular surface. It can also serve to mitigate tear deficiencies if present. Preservative-free preparations should be used when possible, minimizing the risk of irritant effects of the preservatives. Refrigeration of ocular medications has also been shown to improve the tolerability of their application as well as the overall patient perception of symptom improvement. Ocular lubricants are available at different viscosity levels. The low viscosity preparations are effective for mild conditions, in terms of symptom relief and tolerance. For more severe cases, high-viscosity tears are used but may cause blurring of vision

Antihistamines are the first-line drugs often used by patients on their own prior to seeking medical attention. However, overuse of this class of medications, especially the first-generation antihistamines can worsen dry eye syndrome, and may even worsen allergic conjunctivitis by decreasing the barrier effect of natural tears film. Topical antihistamines are preferred due to their rapid onset of action, and lack of systemic side effects. By acting as an inverse agonist on the histamine receptor, topical antihistamines block the acute phase of allergic response. They have a high safety margin however repeat dowing is necessary as the duration of action is often short. Some second-generation antihistamines, such as emadastine, have a longer duration of action and are better tolerated. Cetirizine is available as an ophthalmic solution and has shown good efficacy and tolerability. Topical antihistamines should not be used in patients with acute angle closure glaucoma.

Agents mitigate the late phase of an allergic response by inhibiting the release of mast cell mediators. Single-action mast cell stabilizers usually require a preloading period of 3–5 days but up to 2 weeks and frequent instillation. This often leads to poor compliance. Some agents in this category are cromolyn, lodaxamide, nedocromil, and pemirolast. With the availability of dual agents and multi-action agents, the use of single-action mast cell stabilizers is limited. They are sometimes used as add-on agents. Newer topical ophthalmic agents have both antihistamine and mast cell stabilizing actions. Their duration of action is longer requiring less frequent instillation, improving compliance. Moreover, they have fewer side effects than pure antihistamine agents, improving tolerability. The dual agents target both the early and late phases of allergic response. Epinastine has been shown to increase comfortable wearing time among contact lens wearers, with decreased need for rewetting drops.

Other agents include ocular decongestants and topical nonsteroidal anti-inflammatories. The main effect of ocular decongestants is to improve the appearance of ocular erythema. They do not improve the sensation of ocular itch. This class of medication may be associated with worsening of narrow-angle glaucoma, rebound hyperemia, conjunctivitis pigmentosa, and tachyphylaxis. Therefore, long-term use of ocular decongestants is discouraged and should be avoided in patients with angle closure glaucoma. They should also be used with caution in those with cardiovascular disease, diabetes, and hyperthyroidism. The only FDA-approved ophthalmic nonsteroidal anti-inflammatory agent for treatment of allergic conjunctivitis is ketorolac. It blocks prostaglandin synthesis. Its use is limited to intermittent use or as an add-on medication due to the availability of other preparations with higher efficacy and tolerability.

Topical corticosteroids decrease the recruitment of inflammatory cells and subsequent cytokine production. However, their use is associated with cataract formation and increase in intraocular pressure (IOP). Their use is therefore limited to severe or refractory cases. Older formulations such as prednisolone, dexamethasone, and fluorometholone are known to induce cataract. Newer preparations such as loteprednol do not increase the risk of cataract but are associated with an increase in IOP at higher concentrations. Therefore, the use of topical corticosteroids should be limited to 2 weeks. Patients who need to be on these medications for longer than that should be monitored for adverse effects, under the care of an ophthalmologist. Topical corticosteroids are sometimes used for lid eczema however mild preparation such as hydrocortisone 1% are recommended for this indication. Intranasal corticosteroids (INS) have also been shown to have a beneficial effect on ocular allergy, possibly via inhibition of ocular-nasal reflex. Prolonged use of INS can potentially increase the risk of glaucoma and an increase in IOP compared with placebo has been demonstrated; however, glaucomatous optic disc changes have not been noted. Commonly prescribed INS such as fluticasone and mometasone have not been found to increase the risk of cataract or glaucoma and there is minimal to no systemic absorption. Patients using INS, especially high doses, on a prolonged base should have an annual eye examination and IOP measurement. Patients with VKC or AKC may need short bursts of systemic steroids for exacerbations.

Calcineurin inhibitors, such as topical cyclosporine or tacrolimus, can be used for severe or refractory allergic conjunctivitis, AKC and VKC. Moreover, topical tacrolimus 0.03% or 0.1% ointment and pimecrolimus 1% ointment may be used for eyelid eczema, VKC, and AKC. Their use presents the risk of local infections such as molluscum contagiosum, papilloma virus, and herpes. These agents are often associated with a burning sensation, affecting compliance. Systemic cyclosporine may be used for sight-threatening AKC. A topical cyclosporine preparation (0.05%) is FDA-approved for dry eye syndrome.

To date, allergen immunotherapy is the only therapy available that has been shown to modify the disease process. For treatment of SAC or PAC, AIT has been studied in both its subcutaneous form (SCIT) and sublingual form (SLIT). Both have been shown to offer improvement in conjunctival symptoms and medication utilization scores; however, SCIT is more efficacious than SLIT. SLIT has been found to have more frequent, milder adverse effects—especially at the start of therapy; however, it has a lower risk of anaphylactic reaction. As SLIT has been shown to have higher rates of discontinuation than SCIT, the few pharmacoeconomic analyses to date have favored SCIT in favor of SLIT. Both forms will help treat ocular allergy as well as other forms of allergy such as nasal symptomatology. As both options have their own risks and benefits, the decision remains patient-directed with adherence being the greatest hurdle for either method.

14 Emerging Therapies

With longer pollen seasons characterized by rising pollen counts and increasingly allergenic pollen, we are seeing seasonal allergies becoming more problematic to manage. With the late phase reaction thought responsible for worsening conjunctival symptoms and decreased medication efficacy throughout a pollen season, there is increased interest in the use of biologics for the treatment of allergic conjunctivitis. Biologics have also been of research interest for other ocular allergic disorders. Omalizumab, a recombinant humanized monoclonal antibody directed toward circulating IgE, has been shown to improve symptoms of allergic conjunctivitis in at least two randomized control studies evaluating the treatment of allergic rhinitis. Although no such studies exist for AKC or VKC, case reports exist suggesting the possible benefit of omalizumab after its initiation for a different indication. Despite having evidence of Th2-mediated inflammation, Dupilumab (a fully human monoclonal antibody that blocks IL-4 and IL-13) is associated with new-onset conjunctivitis in 9–28% of patients in clinical trials and 25–50% of patients in observational studies. The mechanism for this adverse effect has yet to be elucidated; however, one theory is that conjunctival inflammation is the result of IL-13 inhibition and subsequent goblet cell apoptosis and decreased mucin production. Severe atopic dermatitis, preexisting conjunctivitis, and low dupilumab levels are associated with an increased risk of conjunctivitis. The anti-IL-5 biologic agents mepolizumab and reslizumab, as well as the anti-IL-5 receptor agent benralizumab, have not been studied in the context of ocular allergy. Janus Kinase inhibitors are of increasingin-terest in the treatment of atopic disease, with tofacitinib showing efficacy in a recent 2020 murine model of allergic conjunctivitis. Although biologics are likely to hold a major role in the future treatment of ocular allergic disease, neither have been extensively studied to date nor have been approved for use for the indication of ocular allergy.

Contact lens-based drug delivery systems are also an area of interest. A contact lens-based drug delivery system for ketotifen has been shown to be comparable to direct topical drug delivery. Epinastine-releasing soft contact lenses have been shown to have prolonged drug delivery and higher efficacy compared to epinastine hydrochloride eye drops. Such innovations would be promising options for providing vision correction as well as providing treatment of ocular allergy for contact lens wearers.

Bibliography

1. Ackerman S, Smith LM, Gomes PJ. Ocular itch associated with allergic conjunctivitis: latest evidence and clinical management. Ther Adv Chronic Dis. 2016;7(1):52–67. https://doi.org/10.1177/2040622315612745.
2. Bielory L. Allergic and immunologic disorders of the eye. Part II: ocular allergy. J Allergy Clin Immunol. 2000;106(6):1019–32. https://doi.org/10.1067/mai.2000.111238.
3. Bielory L. Differential diagnoses of conjunctivitis for clinical allergist-immunologists. Ann Allergy Asthma Immunol. 2007;98(2):105–14; quiz 114-107, 152. https://doi.org/10.1016/S1081-1206(10)60681-3.
4. Bielory L, Delgado L, Katelaris CH, Leonardi A, Rosario N, Vichyanoud P. ICON: diagnosis and management of allergic conjunctivitis. Ann Allergy Asthma Immunol. 2019; https://doi.org/10.1016/j.anai.2019.11.014.
5. Bielory L, Schoenberg D. Emerging therapeutics for ocular surface disease. Curr Allergy Asthma Rep. 2019;19(3):16. https://doi.org/10.1007/s11882-019-0844-8.
6. Caillaud D, Martin S, Segala C, Besancenot JP, Clot B, Thibaudon M, French Aerobiology N. Effects of airborne birch pollen levels on clinical symptoms of seasonal allergic rhinoconjunctivitis. Int Arch Allergy Immunol. 2014;163(1):43–50. https://doi.org/10.1159/000355630.
7. Caillaud DM, Martin S, Segala C, Besancenot JP, Clot B, Thibaudon M, French Aerobiology N. Nonlinear short-term effects of airborne Poaceae levels on hay fever symptoms. J Allergy Clin Immunol. 2012;130(3):812–814 e811. https://doi.org/10.1016/j.jaci.2012.04.034.
8. Caillaud DM, Martin S, Segala C, Vidal P, Lecadet J, Pellier S, et al. Airborne pollen levels and drug consumption for seasonal allergic rhinoconjunctivitis: a 10-year study in France. Allergy. 2015;70(1):99–106. https://doi.org/10.1111/all.12522.
9. Durham SR, Nelson HS, Nolte H, Bernstein DI, Creticos PS, Li Z, Andersen JS. Magnitude of efficacy measurements in grass allergy immunotherapy trials is highly dependent on pollen exposure. Allergy. 2014;69(5):617–23. https://doi.org/10.1111/all.12373.
10. Komi DEA, Wöhrl S, Bielory L. Mast cell biology at molecular level: a comprehensive review. Clin Rev Allergy Immunol. 2020;58(3):342–65. https://doi.org/10.1007/s12016-019-08769-2.
11. Feo Brito F, Mur Gimeno P, Carnes J, et al. Grass pollen, aeroallergens, and clinical symptoms in Ciudad Real. Spain J Investig Allergol Clin Immunol. 2010;20(4):295–302.
12. Janeczko P, Norris MR, Bielory L. Assessment of receptor affinities of ophthalmic and systemic agents in dry eye disease. Curr Opin Allergy Clin Immunol. 2021;21(5):480–5. https://doi.org/10.1097/ACI.0000000000000773.
13. Kiotseridis H, Cilio CM, Bjermer L, Tunsater A, Jacobsson H, Dahl A. Grass pollen allergy in children and adolescents-symptoms, health related quality of life and the value of pollen prognosis. Clin Transl Allergy. 2013;3:19. https://doi.org/10.1186/2045-7022-3-19.
14. Mendelson S, Quintanilla J, Norris MR, Bielory L. Psychosocial impact of ocular surface allergic inflammatory disorders. Curr Opin Allergy Clin Immunol. 2021;21(5):465–71. https://doi.org/10.1097/ACI.0000000000000776.
15. Norris MR, Bielory L. Cosmetics and ocular allergy. Curr Opin Allergy Clin Immunol. 2018;18(5):404–10. https://doi.org/10.1097/ACI.0000000000000474.
16. Barney NP. Chapter 38. Allergic and immunologic diseases of the eye. In: Adkinson NF, Bochner BS, Busse WW, Holgate ST, Lemanske RF, Simons FER, editors. Middleton's allergy. 7th ed. Philadelphia, PA: W.B. Saunders; 2010. p. 605–23.
17. Prince A, Norris MR, Bielory L. Seasonal ocular allergy and pollen counts. Curr Opin Allergy Clin Immunol. 2018;18(5):387–92. https://doi.org/10.1097/ACI.0000000000000475.
18. Rodrigues J, Kuruvilla ME, Vanijcharoenkarn K, Patel N, Hom MM, Wallace DV. The spectrum of allergic ocular diseases. Ann Allergy Asthma Immunol. 2021;126(3):240–54. https://doi.org/10.1016/j.anai.2020.11.016.
19. Valentine L, Norris MR, Bielory L. Comparison of structural components and functional mechanisms within the skin vs. the conjunctival surface. Curr Opin Allergy Clin Immunol. 2021;21(5):472–9. https://doi.org/10.1097/ACI.0000000000000775.
20. Waisel Y, Mienis Z, Kosman E, Geller-Bernstein C. The partial contribution of special airborne pollen to pollen induced allergy. Aerobiologia. 2004;20(4):197–208.

Allergic Rhinitis

Meera R. Gupta and Jessica Palmieri

Case 1

A 20-year-old male presents to your clinic with complaints of congestion, clear rhinorrhea, sneezing, itchy watery eyes, and throat clearing. He states that his symptoms are present year-round but he does note worsening of symptoms in the spring and summer. He does not have a long history of seasonal allergies. He moved to your town 2 years ago for college from out of state and he started noticing these symptoms a few months after starting school. He has a remote history of eczema as a child which is now resolved. He does not have a history of asthma and he denies allergies to foods, drugs, vaccines, and Hymenoptera. He has tried taking over-the-counter loratadine 10 mg daily and cetirizine 10 mg daily for 1–2 months at a time respectively, but he did not see any improvement in his symptoms with either medication.

The patient states that he is having daily symptoms and is having difficulty falling and staying asleep 4–5 nights per week. He is subsequently having trouble focusing in class and is concerned about his grades. He would like to start a new medication regimen as he has no relief with daily oral antihistamines. In preparation for today's appointment, he has withheld all oral antihistamines for 1 week and he does not take any additional medications or supplements.

On your physical exam, you note significant bilaterally boggy, pale, and hypertrophied nasal turbinates. Cobblestoning of the oropharynx is present as are palatal petechiae. Dark circles are present under his eyes and his sclera are faintly red. Skin is normal texture and hydrated. Lungs are clear to auscultation without wheezes. The remainder of the physical exam is unremarkable.

Skin prick testing is done in the office. The results are as follows:

Allergen	Wheal/flare in mm
Histamine (+ control)	7/25
Saline (– control)	0/0
Dust mite DF	11/40
Dust mite DP	20/35
Cockroach Mix	10/30
Alternaria	5/5
Russian Thistle	10/10
Timothy Grass	25/45
Bermuda Grass	8/30
Johnson Grass	10/32
Bahia Grass	10/20

Question 1

Based on his reported symptoms, how would you best classify this patient's allergic rhinitis?

A. Moderate/severe persistent nonallergic rhinitis
B. Moderate/severe intermittent allergic rhinitis
C. Mild persistent allergic rhinitis
D. Moderate/severe persistent allergic rhinitis

Question 2

What treatment plan would be appropriate to start this patient on after your first visit? (monotherapy?)

A. Leukotriene receptor antagonist monotherapy
B. Second generation antihistamine + inhaled nasal corticosteroid
C. Inhaled nasal corticosteroid + inhaled nasal antihistamine spray
D. A 7–10 day course of inhaled nasal decongestant monotherapy followed by a second-generation antihistamine

M. R. Gupta (✉) · J. Palmieri
Department of Pediatrics, Division of Immunology, Allergy, and Retrovirology, Baylor College of Medicine, Houston, TX, USA
e-mail: mrgupta@bcm.edu

M. Mahmoudi (ed.), *Absolute Allergy and Immunology Board Review*, https://doi.org/10.1007/978-3-031-12867-7_5

Question 3
Your patient returns to your office in 6 weeks for follow-up. He has been properly adhering to the medication regimen you prescribed him however he has only noted mild improvement in his symptoms. Physical exam continues to show significant mucosal edema. What add-on therapy would be most appropriate?

A. Addition of a second-generation antihistamine to your leukotriene receptor antagonist
B. Addition of a 3-day course of an inhaled nasal decongestant to your inhaled nasal corticosteroid + inhaled nasal antihistamine spray regimen
C. Addition of a second-generation antihistamine to your inhaled nasal antihistamine
D. Stopping his current regime and starting a 3–5 day course of pseudoephedrine to help with turbinate hypertrophy

Question 4
Cytokines driving the inflammatory response in allergic rhinitis include

A. Il-1 and IL-6
B. Il-10
C. Il-4, Il-5, Il-13
D. Soluble Il-2 and Il-18

Question 5
False positive can be seen in epicutaneous skin testing (SPT) with which of the following?

A. Dermatographic patients
B. Low potency of extracts
C. Recent antihistamine use
D. Insufficient (low) pressure from operator during SPT placement

Answers: D, C, B, A, C

1 Allergic Rhinitis

Allergic Rhinitis (AR) is an IgE-mediated inflammatory disease that is characterized by one or more of the following symptoms: nasal congestion, anterior or posterior rhinorrhea, sneezing, and itching and is frequently accompanied by symptoms involving the eye, ear, or throat. AR affects up to 60 million people in the United States annually, with surveys requiring a physician-confirmed diagnosis reporting a prevalence of 14% in adults and 13% in children. Uncontrolled AR has a significant effect on quality of life including impairments in physical and social functioning, disturbed sleep, daytime somnolence, fatigue, irritability, memory defects, depression, and decreased attention and learning, as well as lost work and school days. Appropriate therapy can substantially reduce societal costs and significantly improve quality of life (QOL).

1.1 Classification

AR can be classified by the frequency and duration of symptoms, the temporal pattern and context of exposure, or by symptom severity. Appropriate classification can assist in the selection of treatment strategies for an individual patient. Symptom frequency has been divided into intermittent, occurring <4 days/week or <4 consecutive weeks/year, or persistent, occurring ≥4 days/week and ≥ 4 consecutive weeks/year. Of note, this strict definition has some limitations, as a patient with symptoms <4 days a week year-round would technically be classified as intermittent, though they might more closely resemble a patient with more persistent symptoms. The temporal pattern may be seasonal, perennial, or episodic from exposures to allergens not normally encountered in the patient's environment. Patients may have both seasonal AR (SAR) and perennial AR (PAR) as depending on the climactic region, aeroallergens may be either seasonal or perennial. Additionally, as most patients are polysensitized to both perennial and seasonal allergens, the SAR/PAR classification can be independent from frequency. However, recognition of the pattern of sensitizations and exposures may help guide the administration of medications concurrent with or in anticipation of a defined seasonal exposure. Symptom severity is defined predominantly by its impact on parameters of QOL, with mild rhinitis defined as symptoms that are not interfering with QOL, whereas moderate/severe rhinitis is defined when symptoms have a negative effect on QOL parameters. A variety of validated QOL questionnaires, either generic or rhinitis-specific, can be used to assess AR severity and response to treatment.

1.2 Testing

Diagnostic testing for allergen-specific IgE (sIgE) can be performed via epicutaneous tests or blood tests to confirm the diagnosis of AR and determine which aeroallergens may be contributing to symptoms. Skin prick testing (SPT) remains the most economical and reliable test for the identification of sIgE. SPT is generally safe, and though no fatalities have been reported, rare systemic reactions may occur with infancy, multiple positive tests, extensive eczema, and uncontrolled asthma or peak flow <70% as significant risk factors for systemic reactions. Although multiple commercially available devices have been developed to reduce trauma to the skin, the procedure for each is similar; a sharp instrument is passed through the allergen extract or control solutions and is applied to the skin on the upper back or volar aspect of the forearms to create a small break in the epidermis through which the allergen solution penetrates. After 15–20 min, the wheal and flare size is measured and recorded in millimeters as a marker of immediate mast cell responses with a positive result typically defined as a wheal ≥3 mm at its largest diameter. Concomitant evaluation of positive and negative controls is imperative to rule out false negatives and false positives. False positives typically arise due to dermatographism, whereas false negatives can arise due to operator technique, low potency of extracts, or medications and conditions that suppress the histamine response (Table 1).

Table 1 Inhibitory effect of common FDA-approved drugs on IgE-mediated skin prick tests

Drug	Number of days to withhold prior to SPT
H1 antihistamines-oral	
Chlorpheniramine	3–6
Cyproheptadine	9–11
Diphenhydramine	2–5
Hydroxyzine	5–8
Meclizine	14
Prochlorperazine	14
Promethazine	3–5
H2 antihistamines-oral	
Cetirizine	3–5
Desloratadine	7
Fexofenadine	3–5
Levocetirizine	Unknown
Loratadine	7
H2-receptor antagonists	
Famotidine	2
Antihistamines-Intranasal/ocular	
Azelastine	2
Olopatadine	0
Ketotifen (oral or topical)	>5
Corticosteroids	
Systemic, short term	0
Systemic, long term	0
Inhaled	0
Topical (>7 days)	Test on different area
Calcineurin Inhibitors-topical	
Pimecrolimus	0
Tacrolimus	7
Leukotriene receptor antagonists	0
Omalizumab	6 months after last dose
Clonidine	Limited data
Benzodiazepines	5–7
Cromolyn (oral or topical)	0
Cyclosporine	0
Methotrexate (high dose)	Serum sIgE testing preferred
Selective Norepinephrine/Serotonin Reuptake Inhibitors	0
Theophylline	0
Tricyclic Antidepressants	2–14, serum sIgE testing preferred

Overall, SPT is both highly specific (70–90%) and sensitive (80–90%) for the diagnosis of inhalant allergens. However, the utility of intradermal testing (IDT) to inhalant allergens remains controversial. Studies suggest that positive intradermal tests to aeroallergens in the setting of negative SPT have a low positive predictive value and are unlikely to identify the presence of clinically significant sensitivity but may have utility in increasing the negative predictive value of testing.

In vitro tests also identify allergen-specific IgE (sIgE) with similar specificity and accuracy as compared to epicutaneous testing methods. Second-generation clinical assays using fluorescent anti-sIgE in fluorescence enzyme immunoassays provide quantitative, reproducible results able to detect down to <0.10 kU/L of sIgE in blood. More recently, component resolved diagnostic tests using proteins to identify the patient's sIgE reactivity to recombinant allergic proteins rather than the whole allergens have been devised to allow for closer examination of the specific molecules of risk and prognostically significant sensitizations. Moreover, in vitro testing is not affected by medications or conditions that blunt histamine responses, though false positives may occur in patients with conditions associated with high baseline IgE such as atopic dermatitis. Accordingly, SPT may offer a very slight advantage in sensitivity over in vitro testing to whole allergens, particularly when considering allergen immunotherapy.

1.3 Pathophysiology of Allergic Rhinitis

The process of allergen sensitization is initiated in the nasal tissue after the first exposure to an allergen. Dendritic cells in the nasal mucosa take up and process allergen for presentation via MHC II to naive $CD4^+$ T cells in the lymph nodes. Antigen-laden DCs activate naïve $CD4^+$ T cells, which then differentiate into allergen-specific Th2 cells (type 2 helper T cells), followed by clonal expansion of allergen-specific Th2 cells and production of IL-4 and IL-13. These cytokines induce B cell activation and IgE class switching, which further leads to B cell differentiation into plasma cells that produce allergen-specific IgE. IgE in the circulation binds to the high-affinity IgE receptor (FcεRI) on the surface of mast cells, basophils, and antigen-presenting cells thus sensitizing them to specific allergens. During this process, allergen-specific memory Th2 cells and B cells against the sensitizing antigen are also formed.

Upon future exposures, the allergen binds to the allergen-specific IgE on mast cells and basophils in the nasal mucosa, resulting in IgE and FcεRI cross-linking and subsequent mast cell activation and degranulation leading to inflammation. The inflammatory response in the nasal mucosa can be divided into an early- and late-phase response, both of which contribute to the clinical presentation of allergic rhinitis. In the early phase, mast cell activation and degranulation release preformed mediators such as histamine, cysteinyl leukotrienes (LTC_4, LTD_4, LTE_4), and prostaglandin D_2 that interact with the nasal sensory nerves, glands, and vasculature resulting in acute AR symptoms such as sneezing, itching, nasal congestion, and mucus secretion. Other mediators such as tryptase, TNF-α, and bradykinin are also released in this early stage, though their role in symptom generation is not well defined. Symptoms produced immediately after exposure to allergen tend to peak within a few minutes and tend to dissipate within an hour. In the late phase, the influx of inflammatory cells and concomitant activation of allergen-specific memory Th2 cells results in the production of large amounts of IL-4, IL-5, IL-13, eotaxin, and RANTES leading to the recruitment of eosinophils, basophils, monocytes, macrophages, and lymphocytes into the inflammatory milieu causing tissue injury, increased mucus production, vascular leakage, remodeling, and nasal hyperresponsiveness. Lipid mediators such as LTC4, thromboxane A2, and platelet-activating factor along with toxic granule products such as major basic protein (MBP), eosinophil cationic protein (ECP), and eosinophil peroxidase (EPO) produced by eosinophils in particular cause significant epithelial damage. Occurring 4–6 h after allergen exposure, symptoms in the late-phase response are characterized predominantly by nasal congestion lasting anywhere from 18 to24 h.

1.4 Treatment

The selection of pharmacotherapy for AR is largely dependent on the severity and frequency of symptoms (Fig. 1a and b). Oral antihistamines (OAH) are of established benefit in reducing symptoms of itching, sneezing, and rhinorrhea associated with AR for mild or intermittent symptoms. Second-generation antihistamines should be preferentially chosen over first-generation antihistamines to reduce potential side effects including sedation, performance impairment, poor sleep quality, and anticholinergic-mediated symptoms associated with first-generation antihistamines. However, it is possible that OAH do not reach high enough concentrations in the nasal mucosa to inhibit histamine-stimulated cytokine release and other mediators of early and late-phase allergic reactions (see Case 2 discussion). Intranasal antihistamines (INAH) ensure drug delivery to the nasal mucosa and may offer added benefit over OAH in reducing nasal congestion. For relief of nasal symptoms of seasonal or intermittent allergic rhinitis, INAH are equal or superior to OAH with a more rapid onset of action, increased efficacy in reducing nasal congestion and total symptom score, and a more favorable safety profile. In double-blind placebo-controlled studies comparing the effectiveness and QOL of azelastine with either fexofenadine, loratadine, or cetirizine as compared to treatment with either loratadine or cetirizine alone, treatment with azelastine was associated with a significant improvement in total nasal symptom scores (TNSS) ranging from 19% to 29% in the azelastine treated group as well as improved QOL (27–36%). The most common adverse effects in patients receiving INAH were bitter taste, headache, epistaxis, and somnolence. The data is less clear on intranasal inhaled corticosteroids (INCS), with some showing equal efficacy and some demonstrating superiority. INAH have a rapid onset of action ranging from 15 to 30 min, as compared to 1–3 h of INCS, and reduce nasal symptoms of rhinorrhea, sneezing, itching, and nasal congestion and thus may be preferable for intermittent or mild symptoms. Leukotriene receptor antagonists (LTRA) are modestly effective in the treatment of AR, though their effectiveness is close to that of oral antihistamines and less effective than monotherapy with INCS. Thus, LTRA monotherapy is not recommended due to its reduced efficacy when compared with that of other agents, combined with the potential for serious neuropsychiatric side effects.

For moderate/severe symptoms or persistent symptoms of any severity, treatment with INCS remains the most effective monotherapy for AR due to their efficacy in controlling both nasal and allergic ocular symptoms. INCS have potent anti-inflammatory properties that reduce symptoms of sneezing, itching, rhinorrhea, and congestion, and with continuous use reduce the mediator and cytokine release upon exposure to aeroallergens. Data also suggests that INCS can reduce allergic conjunctivitis symptoms such as itching, tearing, redness, and puffiness. While continuous use is more efficacious than intermittent use, intermittent use is better than placebo in reducing TNSS.

For nasal symptoms which are moderate/severe persistent or not well controlled with monotherapy with either ICS or INAH alone, a combination of INCS and INAH either as a single medication or as two separate medications given concomitantly is recommended. Double-blind placebo-controlled

Fig. 1 (**a**) Algorithm for the treatment of mild allergic rhinitis (AR). Mild AR is defined as symptoms not significantly affecting quality of life with intermittent symptoms occurring <4 days/week or <4 consecutive weeks/year and persistent symptoms occurring ≥4 days/week or ≥4 consecutive weeks/year. *INAC* inhaled cromolyn, *INAH* inhaled antihistamine, *INCS* inhaled nasal corticosteroid, *IND* inhaled decongestant, *LTRA* leukotriene receptor antagonist, *OAH* oral antihistamine, *PSE* pseudoephedrine. (**b**) Algorithm for the treatment of moderate/severe allergic rhinitis (AR). Moderate/severe AR is defined as symptoms significantly affecting quality of life with intermittent symptoms occurring <4 days/week or <4 consecutive weeks/year and persistent symptoms occurring ≥4 days/week or ≥4 consecutive weeks/year. *INAC* inhaled cromolyn, *INAH* inhaled antihistamine, *INCS* inhaled nasal corticosteroid, *IND* inhaled decongestant, *LTRA* leukotriene receptor antagonist, *TX* treatment, *OAH* oral antihistamine, *PRN* as needed, *PSE* pseudoephedrine

a

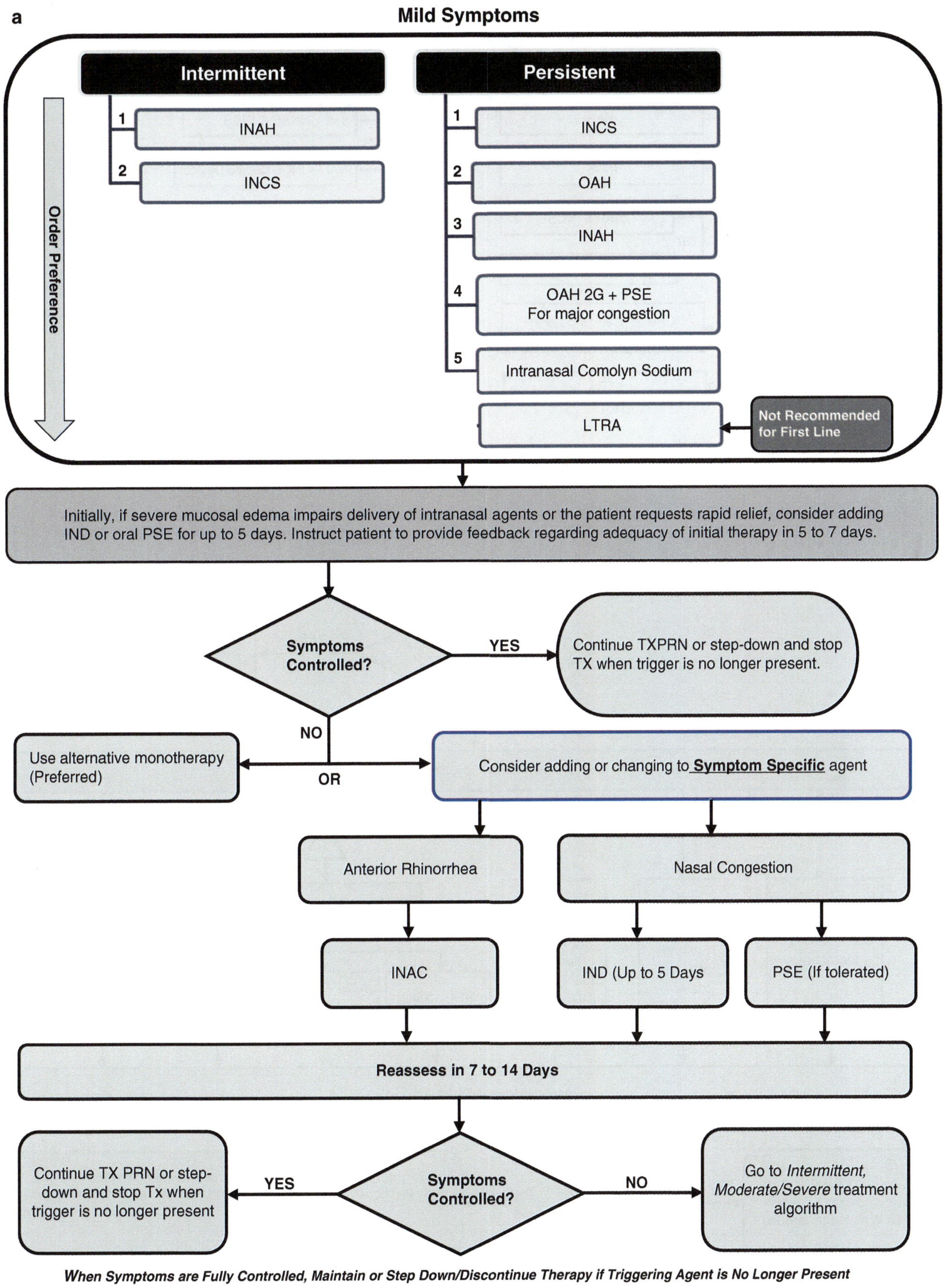
Mild Symptoms
Intermittent
1 INAH
2 INCS
Persistent
1 INCS
2 OAH
3 INAH
4 OAH 2G + PSE For major congestion
5 Intranasal Comolyn Sodium
LTRA
Not Recommended for First Line
Order Preference
Initially, if severe mucosal edema impairs delivery of intranasal agents or the patient requests rapid relief, consider adding IND or oral PSE for up to 5 days. Instruct patient to provide feedback regarding adequacy of initial therapy in 5 to 7 days.
Symptoms Controlled?
YES
Continue TXPRN or step-down and stop TX when trigger is no longer present.
NO
Use alternative monotherapy (Preferred)
OR
Consider adding or changing to Symptom Specific agent
Anterior Rhinorrhea
Nasal Congestion
INAC
IND (Up to 5 Days
PSE (If tolerated)
Reassess in 7 to 14 Days
Continue TX PRN or step-down and stop Tx when trigger is no longer present
YES
Symptoms Controlled?
NO
Go to Intermittent, Moderate/Severe treatment algorithm
When Symptoms are Fully Controlled, Maintain or Step Down/Discontinue Therapy if Triggering Agent is No Longer Present

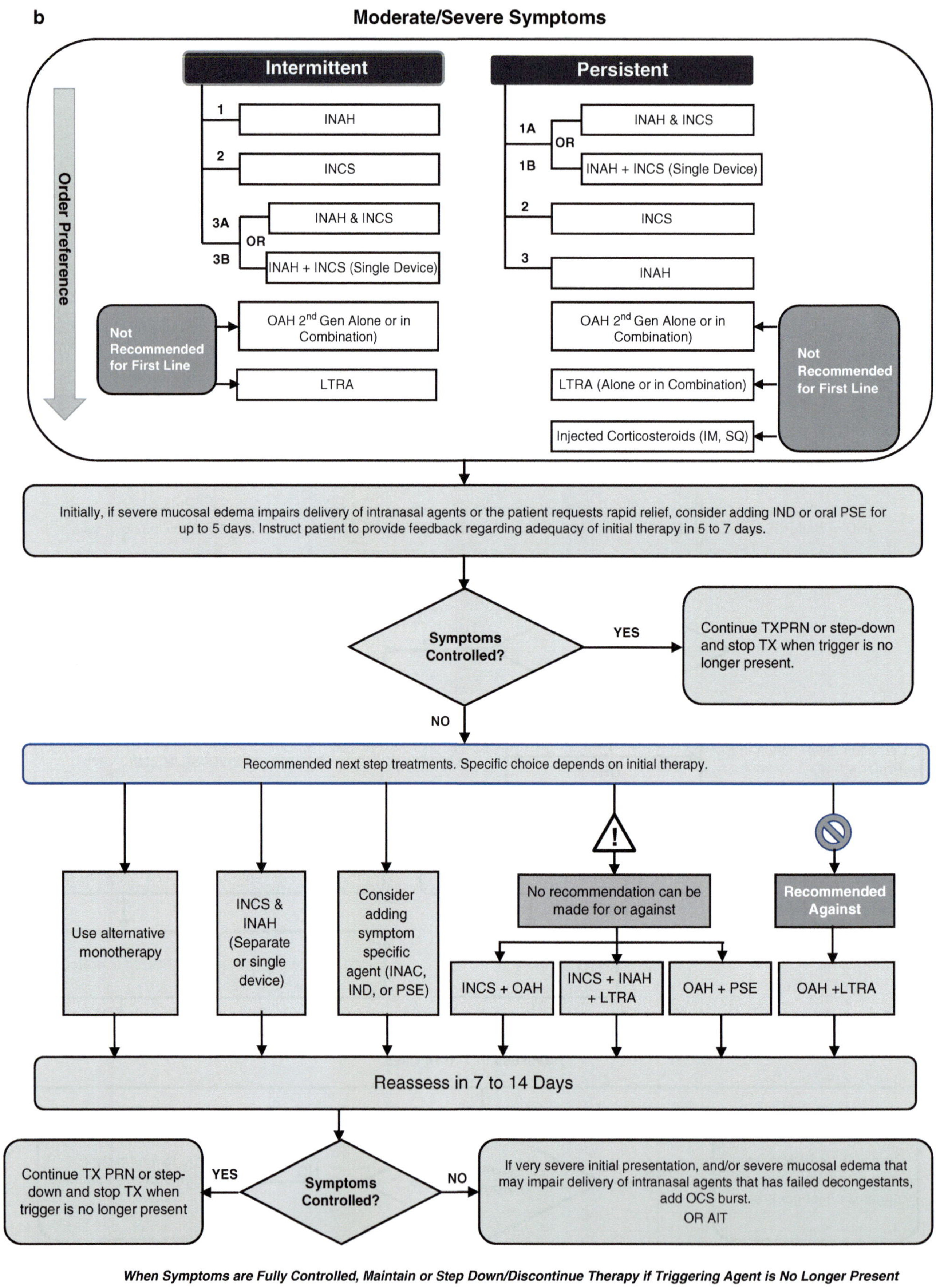

Fig. 1 (continued)

studies have demonstrated increased efficacy at reducing symptoms with a faster onset of action in combination than with individual medications alone. There is no clinical benefit of using a combination of OAH with INCS as compared to INCS alone. Additionally, there is no strong evidence to support the addition of a LTRA to an INCS, with only one study demonstrating increased efficacy in controlling nighttime symptoms, but similar efficacy in controlling total symptom score. However, in patients with asthma or in those whom INAH is unlikely to be tolerated due to its side effects, the risks and benefits of add-on treatment with a LTRA may be considered as part of shared decision-making process.

Other classes of oral or intranasal medications may be utilized as adjunctive treatment in patients with specific symptoms or disease processes. Intranasal decongestants may be considered for use in patients with severe mucosal edema that impairs the delivery of other intranasal agents. These medications have a rapid onset of action (30 s) and reduce nasal resistance for up to 10 h. However, these medications do not block allergen-provoked mediator release, sneezing, itching, or nasal secretions, and should not be used for more than 3–4 days due to the potential development of tachyphylaxis and rhinitis medicamentosa. Oral decongestants are also effective at relieving nasal congestion, though they can be associated with significant side effects and should be used with caution in patients with cerebrovascular or cardiac disease, hypertension, hyperthyroidism, closed angle glaucoma, bladder outlet obstruction, and Tourette's syndrome. Additionally, the role of pseudoephedrine as a key ingredient in making methamphetamines has led to the substitution of oral phenylephrine for pseudoephedrine in many medications, which has been shown to be less effective than pseudoephedrine at treating nasal congestion. Intranasal ipratropium bromide is effective for the treatment of rhinorrhea but does not have significant effects on congestion or sneezing. It is particularly beneficial in the treatment of anterior rhinorrhea, and an additive benefit has been demonstrated when combined with INCS or OAH. Cromolyn has been shown to stabilize mast cells, and intranasal cromolyn administered just before allergen exposure can reduce the development of AR symptoms.

1.5 Subpopulations with Special Considerations

Most pharmacologic treatments for AR are approved for children down to 5 years of age, with many down to even 2 years of age or younger. Special consideration to dose adjustment, side effects, tolerability, and long-term safety should be given. Although OAH has commonly been used as a first-line treatment, there has been a shift towards a broader approach using other agents such as INCS. However, it is important to remember that AR as a cause of chronic rhinitis in children below 2 years of age is unusual, and other differential diagnoses such as structural abnormalities, foreign body, or infections rhinosinusitis should be considered.

Although there is little evidence of prospective randomized trials supporting the use of pharmacologic agents during pregnancy, cohort studies and clinical reviews support the safety of OAH, INCS, and LTRAs in this population. As a class, OAH continues to appear to be safe for use in pregnancy. Cetirizine has not been associated with an increased rate of major malformations or teratogenic risk. Intranasal budesonide has been listed as safe in pregnancy (old classification B). Though mometasone and fluticasone have been classically listed as old classification C (adverse effect on the fetus in animal reproduction studies or no animal reproduction studies and no well-controlled human studies), new data support their safety for use during pregnancy. However, it should be noted that triamcinolone nasal preparations have been associated with higher rates of congenital respiratory defects and should be avoided if possible. Though there is little data on the safety of INAH in pregnancy, oral antihistamines have not been associated with increased rates of major malformations on teratogenic risk. LTRAs have also been listed as safe in pregnancy, with data from asthma and other large observational studies finding no increased risk of congenital malformations.

1.6 Conclusion

Allergic rhinitis can cause significant impairments in physical and social functioning resulting in increased societal costs and diminished quality of life. Classification of symptoms by severity and frequency can assist in the selection of appropriate treatment strategies, with the identification of relevant allergens helping to guide administration in anticipation of a seasonal exposure. Both epicutaneous and serum testing for allergen-specific IgE can be performed with similar sensitivity and specificity.

Case 2

The patient from Case 1 returns to your office after being seen by you regularly for 12 months. He has been properly adhering to his medication regimen and has taken steps to decrease allergen exposure including dust mite covers, using hot water and air when washing and drying his sheets every week, and showering and changing clothes after being outdoors. He has noted some improvement on a combination nasal corticosteroid/antihistamine spray plus leukotriene receptor antagonist regimen, but he is continuing to have mild symptoms 3–4 days a week with sleep interruption 3 nights a week. While pleased that he has made some progress with his symptoms he is wondering if there are any additional therapies that may be of further benefit to him. He would prefer not to continue adding medications if possible.

Your physical exam shows mild improvement from his first visit and there are no new acute findings.

You decide to discuss allergen immunotherapy (AIT) with your patient. After reviewing the time commitment, risks, and benefits with him he chooses to move forward with starting AIT.

Question 1
Which of the following best describes appropriate vial mixing for the patient discussed in Case 1 and 2 based on his skin prick test results.

A. 2 vials needed: vial 1 = grasses and dust mites, vial 2 = mold and cockroach
B. 3 vials needed: vial 1 = grasses and mold, vial 2 = cockroach, vial 3 = dust mites
C. 2 vial needed: vial 1 = cockroach and grasses, vial 3 = mold and dust mites
D. Only 1 vial is needed: all his allergens can be combined

Question 2
In which of the following situations would you likely decide against administering immunotherapy injections?

A. A 5-year-old patient with atopic dermatitis and dust mite allergy not controlled with allergen avoidance and ICS/INAH
B. A patient who discovers she is pregnant while at her maintenance immunotherapy dose
C. A patient who had a fever with URI symptoms 1 week ago but is no longer symptomatic
D. A 22-year-old patient with persistent asthma and baseline FEV1 of less than 70% of predicted value

Question 3
Immune changes from AIT include

A. An increase in both TH1 and TH2 responses
B. Decreased production of B regulatory cells
C. Increased production of T regulatory cells
D. A sustained increase in allergen-specific IgE and decreased IgG4 production

Question 4
Standardized extracts include all of the following except

A. Mountain cedar
B. Dust mite
C. Ragweed
D. Cat hair/pelt

Question 5
Cytokine changes associated with AIT include:

A. Increased production of Il-4, IL-5, and IL-13
B. Increased production of IL-10 and TGF-β
C. Increased production of soluble Il-2 and Il-18
D. Increased production of IL-1 and IL-6

Answers: A, D, C, A, B

2 Allergen Immunotherapy

Allergen Immunotherapy has been used to treat allergic rhinitis ever since Leonhard Noon and John Freeman published their novel works on allergen-specific immunotherapy (AIT) using grass pollen extracts over a century ago. AIT remains the only disease-modifying intervention for allergic rhinitis with the capacity to offer long-lasting therapeutic efficacy even after cessation of treatment. Evidence suggests that 3–5 years of SCIT results in clinical benefit and sustained allergen-specific tolerance for at least 2–3 years after treatment cessation with some patients experiencing tolerance for 7–12 years. AIT in both adults and children is effective, safe, well tolerated, and there is no absolute upper or lower age limit for the initiation of treatment. Allergen immunotherapy is efficacious for the treatment of inhalant allergy-related syndromes including allergic rhinitis, allergic conjunctivitis, and allergic asthma due to pollen, mold/fungi, animal allergen, dust mite, and cockroach hypersensitivity. Additionally, studies have shown that AIT may prevent sensitization to new allergens in monosensitized individuals, the development of asthma in children with allergic rhinitis, and improve disease control in patients with atopic dermatitis and concomitant aeroallergen sensitization.

AIT involves repeated administration of increasing doses of the sensitizing allergen(s) over time until the optimal maintenance dose in the range of 5–20 μg of major allergen is achieved, followed by continued administration of maintenance dosing typically for a period of 3–5 years with some individuals requiring longer treatment courses. Although subcutaneous injection is the most common route of delivery, it is important to note that allergen extracts can be administered through alternative routes such as sublingual immunotherapy (SLIT) in tablets or liquid drops. In the United States, only SLIT tablets are approved by the Food and Drug Administration (FDA) for the treatment of ragweed, northern pasture grasses, and dust mites; liquid drops are not FDA-approved. As such, we will predominantly be covering principles related to SCIT. Allergen immunotherapy will be referred to as AIT or subcutaneous immunotherapy (SCIT) herein.

2.1 Immune Changes Associated with AIT

Allergic responses arise due to a failure of the immune system to suppress dysregulated inflammatory responses against non-harmful antigens. The goal of AIT is to induce clinical and immunological allergen-specific tolerance by altering antigen-specific T cell and/or B cell responses, allergen-specific IgE and IgG antibody levels, and activation threshold levels of innate immune cells (mast cell, basophil, and dendritic cells). Immediately after AIT initiation, a decrease in susceptibility of mast cell and basophil activation is observed, likely related to hypo-sensitization which alters the magnitude of mediator release. The release of low quantities of inflammatory mediators may affect the subsequent threshold of activation of mast cells and basophils required in anaphylaxis. Long-term AIT is associated with a significant reduction in the late-phase response in the nasal and bronchial mucosa as well, decreasing the number of mast cells and eosinophils at the tissue sites of allergen exposure and increasing the threshold of allergen required to provoke responses and decreasing hyperreactivity to nonspecific stimuli. Nonallergic individuals have been shown to maintain a higher ratio of allergen-specific T regulatory cells (Treg) to allergen-specific Th1 or Th2 cells with allergic individuals showing a significant reduction in IL-10 as compared to nonallergic persons.

The first days to weeks after AIT initiation show increased production of Tregs that suppress both Th1 and Th2 responses. These allergen-specific Tregs produce anti-inflammatory cytokines such as IL-10 and TGF-β, which induce T cell tolerance. IL-10 is a general inhibitor of proliferation and cytokine responses in T cells, and an increase in IL-10 leads to several responses: a decrease in B cell antigen-specific IgE production, an increase in IgG4 levels, reduced pro-inflammatory cytokine release from mast cells, eosinophils, and T cells, and induction of T cell tolerance via selective inhibition of the CD28 costimulatory pathway. TGF-β also induces IgA isotype switching and allergen-specific IgA induces IL-10 release from monocytes. The number of allergen-specific B regulatory cells (Bregs) also increases. Bregs regulate the development, proliferation, and maintenance of $CD4^+$ T cells and support Treg differentiation.

Late immune changes include an allergen-specific shift from a Th2 to a Th1 cytokine profile. AIT induces a reduction in allergen-stimulated lymphocyte proliferation, leading to reduced production of IFN-y, IL-5, and IL-13. Additionally, AIT leads to a reduction in the release of immediate phase mediators like histamine from basophils and mast cells and suppresses late-phase inflammatory responses in the skin and respiratory tract. Although specific IgE (sIgE) levels transiently increase with the start of AIT, over several years these levels gradually decrease to or below a patient's baseline levels, particularly during the pollen season when IgE levels typically increase. Breg-derived IL-10 also drives the increase in allergen-specific IgG4 antibodies, a non-inflammatory isotype thought to protect from allergic reactions by capturing the allergen before reaching the cell-bound IgE thus preventing the activation of mast cells and basophils. Driven by TGF-β, levels of IgG1 and IgA also increase over time; however, changes in serum IgA and IgG isotypes have not consistently correlated with clinical benefit but instead may be interpreted as markers of change in the immune response during AIT.

2.2 Indications for Initiation and Efficacy of AIT

Aeroallergen immunotherapy is indicated for the treatment of allergic rhinitis or rhino-conjunctivitis, allergic asthma, and atopic dermatitis in patients with demonstrable and relevant allergen-specific IgE antibodies (sIgE). Patients should have both clinical symptoms upon natural exposure to the allergen and the presence of sIgE to that allergen demonstrated through either epicutaneous or serum sIgE testing. AIT should be considered in patients unable to achieve symptomatic relief with standard pharmacotherapy and allergen avoidance, as well as for those who wish to potentially avoid long-term pharmacotherapy or experience unacceptable adverse effects of medications. It is important to ensure that patients are aware of the time commitment associated with AIT and it should be emphasized that AIT will not be immediately effective for symptomatic relief.

Patient should be evaluated every 6–12 months while receiving AIT and monitored for adherence, adverse reactions, and efficacy. Clinical response generally occurs shortly after reaching maintenance. If no clinical improvement occurs after 1 year of maintenance therapy and reasons for lack of efficacy are not found, discontinuation of AIT should be considered. Reasons for AIT failure include inadequate dosing of allergens, failure to identify and include all relevant allergens in the prescription, continued high levels of environmental allergen exposure, and significant exposure to nonallergic triggers such as tobacco smoke and pollution. The typical duration for AIT is 3–5 years, though there are those who may choose to continue treatment after this time. Some patients may experience sustained clinical remission of their allergic disease after discontinuing immunotherapy and others may have a relapse in symptoms occurring within 1 or more years after discontinuation. The optimal duration of treatment for AIT has not been extensively studied, though current studies demonstrate a higher relapse rate in patients treated for less than 35 months as compared to those treated for more than 36 months. There are inadequate diagnostic tools to identify which patients will experience a sustained response; thus, the duration of treatment should be individualized to each patient.

2.3 AIT Vial Preparation

Aeroallergens determined to be clinically relevant on IgE hypersensitivity testing should be included for treatment in AIT vials. The American practice is generally to treat for all sensitivities identified as clinically relevant on SPT using mixed allergen extracts, whereas in Europe the typical practice is to treat with a single allergen. Skin testing has been the primary diagnostic tool used in clinical studies of allergen immunotherapy and should be used to identify allergen-specific IgE antibodies in most patients. However, serum testing for allergen-specific IgE antibodies can alternatively be used in patients where skin testing cannot be performed.

Allergen extracts are comprised of a solution of elutable materials derived from allergen source materials and consist of complex mixtures of proteins and glycoproteins to which antibodies can bind. They are available as either nonstandardized or standardized extracts; nonstandardized extracts are labeled as weight/volume (grams of allergen/ml of buffer) or protein nitrogen units and standardized extracts are labeled as BAU (biologically active units) determined by the mean dilution of an extract that induces erythema equaling 50 mm in diameter in 15–20 highly sensitive subjects. FDA-approved standardized allergens include extracts for cat hair, cat pelt, dust mites (*D. pteronyssinus*, *D. farinae*), short ragweed, Bermuda grass, Kentucky bluegrass, perennial ryegrass, orchard grass, timothy grass, meadow fescue, red top, sweet vernal grass, and Hymenoptera venoms (yellow jacket, honeybee, wasp, yellow hornet, and white-faced hornet). The remainder of allergen extracts are not yet standardized. Standardized extracts should be used when available and can be mixed with nonstandardized extracts.

The preparation of allergen immunotherapy vials should be performed by trained and certified personnel experienced with appropriate mixing protocols in an aseptic environment following federal USP 797 pharmaceutical compounding guidelines. When preparing mixed allergen extracts for AIT, it is important to consider the following: (1) extracts with high proteolytic activity such as cockroach and mold/fungi should not be mixed with pollen, dander, and dust mite to prevent loss of potency from protease degradation of some or all of the components in the extracts, (2) many related plant pollens contain allergens that are cross-reactive and in these cases, a physician may choose to select a single cross-reactive pollen or reduce the amount of multiple cross-reacting allergens in the vial, and (3) the maintenance concentration should be optimized to the highest effective and tolerated concentration and formulated to deliver a therapeutic dose for each allergen contained.

2.4 Dosing Administration, Safety, Adverse Effects

Administration of allergen immunotherapy is usually given following a standard schedule where maintenance dosing is slowly achieved over several months of AIT injections. In the build-up phase, AIT extracts starting at 1000-fold or 10,000-fold dilution of the maintenance dose are administered in increasingly concentrated doses given 1–3 times per week. Following standard dosing protocols, patient will typically reach maintenance dosing in 3–6 months. In the maintenance phase where the effective therapeutic dose is achieved, the interval between allergy injections is lengthened, generally between 2 and 4 weeks for the remainder of treatment. The dose mostly remains unchanged between injections unless modifications are indicated such as with the start of a new vial, severe reactions, or discomfort from persistent reactions.

Adverse effects of allergen immunotherapy range from localized reactions at the injection site to severe systemic reactions and anaphylaxis. Large local reactions (LLRs) include pain, swelling, and induration at the injection site of varying size and intensity. The rate of systemic reactions per injection with conventional schedules is approximately 0.2%, with most systemic reactions occurring within 30 min after injection. Although systemic reactions can occur >30 min post-injection, studies have shown that life-threatening anaphylactic reactions after the first 30 min are rare. As such, it is standard of care to recommend that patients be observed in a physician's office for 30 min after receiving an injection. Biphasic reactions can also occur but are typically less severe than the initial symptoms of the first phase reaction. Patients should be counseled on signs and symptoms of delayed and biphasic reactions and be instructed to carry injectable epinephrine.

Alternative dosing schedules involve accelerated AIT administration designed to achieve maintenance dosing rapidly by accelerating the build-up phase. Indications for accelerated SCIT include a desire by the patient due to convenience and/or to experience clinical benefit sooner than would be seen with conventional dosing. In cluster dosing, two or more injections are given per visit on nonconsecutive days. The injections are typically given at 30-min intervals and allow patients to reach maintenance dosing in as little as 4 weeks. In rush dosing, injections are administered to achieve maintenance dosing within a few hours to 6 days, most often within 1–2 days. Studies have shown that the cluster schedules are associated with either a similar or slightly increased frequency of systemic reactions compared with standard dosing. However, rush dosing is associated with a significantly increased risk of systemic reactions of up to 34%, as compared to conventional or cluster protocols. Despite this increased risk, most reactions to rush immunotherapy are not severe with the most common reaction being flushing. Of note, systemic reactions with rush schedules have been reported to occur up to 2 h after the final injection so patients should be monitored for longer than the 30 min recommended for conventional dosing.

2.5 Clinical Considerations When Administering AIT

On the day a patient is to receive immunotherapy, it is important to ensure that the patient is healthy with appropriate control of all concomitant atopic disease processes and without acute illness. The provider should also question patients regarding antihistamine use, exacerbation of asthma symptoms, β-blocker use, change in health status including pregnancy, and adverse reactions to the previous allergen immunotherapy injection. Additionally, there may be an increased risk of systemic reactions from AIT in patients who are sensitized when relevant pollen counts are elevated. Pretreatment with oral antihistamines at least 30 min before injection has been shown to be effective in decreasing the frequency of systemic reactions in all dosing protocols, as well as decreasing the frequency of large local reactions in accelerated schedules.

Patients with severe or uncontrolled asthma are at increased risk for systemic reactions and AIT should not be initiated until asthma is well controlled. Moreover, assessment of asthma control should be performed at each injection visit via history and/or peak flow measurement. If a patient is found to have exacerbation of asthma symptoms, then immunotherapy should not be given that day. Although concomitant treatment with β-adrenergic blocking medications does not increase the frequency of systemic reactions in AIT, their use is a risk factor for more serious and treatment-resistant anaphylaxis. β-blockade can enhance mediator release, intensify pulmonary, cardiovascular, and cutaneous end-organ effects of mediators, and has been associated with increased mortality in experimental anaphylaxis. Treatment with epinephrine may also be less effective and may paradoxically worsen anaphylaxis through unopposed α-adrenergic and vagotonic effects. Glucagon should be kept on-hand in patients receiving AIT on β-blockers to improve epinephrine refractory hypotension in β-blocker-associated anaphylaxis. Allergen immunotherapy is not usually initiated in pregnant patients due to concerns regarding the potential adverse effects of systemic reactions on the mother or fetus. If patients become pregnant during the build-up phase, holding the dose or discontinuation of AIT should be considered early in the course of treatment; however, if a patient becomes pregnant during the maintenance phase, AIT can be continued.

2.6 Conclusion

Subcutaneous immunotherapy for allergic disease involves the gradual administration of increasing amounts of allergens to sensitized patient and remains the only intervention capable of inducing immune tolerance to sensitized allergens. SCIT is effective and safe for the treatment of allergic rhinitis, allergic conjunctivitis, and allergic asthma in both children and adults. SCIT can help prevent the progression of allergic disease, the development of asthma in children with allergic rhinitis, and improve disease control in patients with atopic dermatitis. Treatment with SCIT should be considered in patients with allergic rhinitis, allergic conjunctivitis, and/or allergic asthma due to pollen, mold/fungi, animal allergen, dust mite, and cockroach hypersensitivity as it may result in symptomatic relief and sustained allergen-specific tolerance for years after dosing completion.

Bibliography

1. Ansotegui IJ, Melioli G, Canonica GW, Caraballo L, Villa E, Ebisawa M, et al. IgE allergy diagnostics and other relevant tests in allergy, a World Allergy Organization position paper. World Allergy Organ J. 2020;13(2)
2. Bernstein IL, Li JT, Bernstein DI, Hamilton R, Spector SL, Tan R, et al. Allergy diagnostic testing: an updated practice parameter. Ann Allergy Asthma Immunol. 2008;100(3 Suppl 3):S1–S148.
3. Bousquet J, Anto JM, Bachert C, Baiardini I, Bosnic-Anticevich S, Walter Canonica G, et al. Allergic rhinitis. Nat Rev Dis Primers. 2020;6:1.
4. Bousquet PJ, Combescure C, Neukirch F, Klossek JM, Méchin H, Daures JP, et al. Visual analog scales can assess the severity of rhinitis graded according to ARIA guidelines. J Allergy. 2007;62(4):367–72.
5. Cox L, Nelson H, Lockey R, Calabria C, Chacko T, Finegold I, et al. Allergen immunotherapy: a practice parameter third update. J Allergy Clin Immunol. 2011;127(1 Suppl):S1–S55.
6. Daigle BJ, Rekkerth DJ. Practical recommendations for mixing allergy immunotherapy extracts. Allergy Rhinol (Providence). 2015;6(1):1–7.
7. Dykewicz MS, Wallace DV, Amrol DJ, Baroody FM, Bernstein JA, Craig TJ, et al. Rhinitis 2020: a practice parameter update. J Allergy Clin Immunol. 2020;146(4):721–67.
8. Dykewicz MS, Wallace DV, Baroody F, Bernstein J, Craig T, Finegold I, et al. Treatment of seasonal allergic rhinitis: an evidence-based focused 2017 guideline update. Ann Allergy Asthma Immunol. 2017;119(6):489–511.e41.
9. Frew AJ. Allergen immunotherapy. J Allergy Clin Immunol. 2010;125(2 Suppl. 2):S306–S13.
10. Heinzerling L, Mari A, Bergmann KC, Bresciani M, Burbach G, Darsow U, et al. The skin prick test - European standards. Clin Transl Allergy. 2013;3(1):1–10.

11. Hossenbaccus L, Linton S, Garvey S, Ellis AK. Towards definitive management of allergic rhinitis: Best use of new and established therapies. Allergy Asthma Clin Immunol. 2020;16(1):1–17.
12. Kowalski ML, Ansotegui I, Aberer W, Al-Ahmad M, Akdis M, Ballmer-Weber BK, et al. Risk and safety requirements for diagnostic and therapeutic procedures in allergology: World Allergy Organization Statement. World Allergy Organ J. 2016;9(1):1–42.
13. Ledford DK, Lockey RF. Controversies in allergy: intradermal aeroallergen skin testing. J Allergy Clin Immunol Pract. 2018;6(6):1863–5.
14. Patel HD, Chambliss JM, Gupta MR. Utility and comparative efficacy of recombinant allergens versus allergen extract. Curr Allergy Asthma Rep. 2017;17(9):63.
15. Penagos M, Eifan AO, Durham SR, Scadding GW. Duration of allergen immunotherapy for long-term efficacy in allergic rhinoconjunctivitis. Curr Treat Options Allergy. 2018;5(3):275–90.
16. Sin B, Togias A. Pathophysiology of allergic and nonallergic rhinitis. Proc Am Thorac Soc. 2011;8(1):106–14.
17. Wallace DV, Dykewicz MS, Bernstein DI, Blessing-Moore J, Cox L, Khan DA, et al. The diagnosis and management of rhinitis: an updated practice parameter. J Allergy Clin Immunol. 2008;122(2 Suppl):1–84.

Otitis Media

Elizabeth L. Wisner and Kenneth Paris

Case 1
A 10-month-old female presents to the pediatric clinic for a sick visit. Her parents report that she has been increasingly fussy over the past 24 h. They also report that she has felt warm to the touch and has not been eating very well. One week prior to this acute presentation, she was seen in clinic for rhinorrhea and the family was counseled on supportive care. Other past medical history includes one episode of acute otitis media diagnosed and treated 3 weeks prior to this presentation. Immunizations are up-to-date. She attends daycare 3 days/week.

On physical exam, you note that she is fussy but consolable. Her temperature is 39.5 °C. She has moist mucous membranes, a clear oropharynx, and a normal cardiopulmonary exam. You proceed with an otoscopic exam.

Question 1
On physical exam, the most specific finding leading to the diagnosis of acute otitis media is:

A. Moderate-to-severe bulging of the tympanic membrane
B. Erythema of the tympanic membrane
C. Retraction of the tympanic membrane
D. Impaired mobility of the tympanic membrane via pneumatic otoscopy

Answer: The correct answer is A

The pneumatic otoscope is the standard tool used to diagnose otitis media. Pneumatic otoscopy allows one to assess the contour (normal, retracted, bulging, full), color, translucency, and mobility of the tympanic membrane. The normal tympanic membrane should be translucent, pearly gray, and with a ground-glass appearance.

E. L. Wisner (✉) · K. Paris
Division of Allergy and Immunology, Department of Pediatrics, Louisiana State University Health Sciences Center, Children's Hospital New Orleans, New Orleans, LA, USA
e-mail: elind2@lsuhsc.edu; kparis@lsuhsc.edu

Moderate-to-severe bulging of the tympanic membrane is the most specific finding used to diagnose acute otitis media (choice a) and differentiates AOM from OME. AOM can also be diagnosed in cases of tympanic membrane perforation with acute purulent otorrhea after excluding a diagnosis of otitis externa. In the absence of a bulging tympanic membrane or purulent otorrhea, the diagnosis of acute otitis media may also be made in patients with signs of acute inflammation (marked erythema of the tympanic membrane, fever, otalgia) and middle ear effusion.

Although patients with bulging tympanic membranes have decreased or absent motility (choice d), pneumatic otoscopy can be painful in children with AOM and thus is not necessary for diagnosis. Choice d is also incorrect as this would not differentiate AOM from OME.

Question 2
Which of the following is associated with an increased risk for acute otitis media in this patient?

A. Recent viral upper respiratory tract infection
B. Female gender
C. Daycare attendance
D. A and C
E. All of the above.

Answer: The correct answer is D

As in this patient, AOM occurs most commonly as a consequence of a viral upper respiratory tract infection. Such infections lead to eustachian tube dysfunction or inflammation, negative middle ear pressure, and movement of secretions from the upper respiratory tract in the nasopharynx into the middle ear cleft. The consequence of these changes includes an increased density of colonization of the nasopharynx with the most common bacterial pathogens implicated in otitis media including *Streptococcus pneumoniae*, non-typeable *Haemophilus influenzae*, and *Moraxella catarrhalis*.

M. Mahmoudi (ed.), *Absolute Allergy and Immunology Board Review*, https://doi.org/10.1007/978-3-031-12867-7_6

Male gender, daycare attendance, additional siblings in the home, craniofacial abnormalities such as cleft lip or palate, immunocompromised state, and eustachian tube dysfunction are all additional risk factors for the development of acute otitis media.

Question 3

The patient's previous episode of otitis media 3 weeks ago resolved completely after taking high-dose amoxicillin. The most appropriate treatment for this patient is:

A. Ceftriaxone
B. Amoxicillin-clavulanate
C. Azithromycin
D. Cefazolin
E. Observation

Answer: The correct answer is B

The decision of whether or not to treat with antibiotics is based on the age and clinical presentation of the patient (see Table 1). The importance of obtaining an accurate diagnosis to avoid unnecessary use of antibiotics cannot be overstated.

Appropriate use of antibiotic therapy has been shown to decrease the duration of pain, analgesic use, days of school missed, and missed days of work for parents/guardians. First-line therapy for the treatment of acute otitis media in patients at low risk for amoxicillin resistance is amoxicillin (90 mg/kg/day divided into two doses; maximum of 3 g/day). Duration of treatment for children <2 years is 10 days and for those ≥2 years, 5–7 days.

An antibiotic with additional ß-lactamase coverage (amoxicillin-clavulanate) should be prescribed in patients at an increased risk of beta-lactam resistance. Such patients include those who have received amoxicillin in the preceding 30 days, have a history of recurrent otitis media unresponsive to amoxicillin, and/or those with concomitant purulent conjunctivitis (suggesting infection with non-typeable *Haemophilus influenzae*). The dose is 90 mg/kg/day of amoxicillin and 6.4 mg/kg/day of clavulanate in two divided doses with a maximum daily amoxicillin component of 3 g/day.

Reassessment of the patient should occur if the child's symptoms have worsened or failed to respond to initial antibiotic treatment within 48–72 h.

Table 1 Treatment guidelines for acute otitis media based on age and presentation

Age	Physical exam and symptoms	Treatment
<6 months	Unilateral or bilateral AOM	Antibiotics
≥6 months	Unilateral or bilateral AOM with severe signs or symptoms[a]	Antibiotics
<24 months	Bilateral AOM without severe signs or symptoms[a]	Antibiotics
6–23 months	Unilateral AOM without severe signs or symptoms[a]	[b]Observation or antibiotics
≥24 months	Unilateral or bilateral AOM without severe signs or symptoms[a]	[b]Observation or antibiotics

[a] Severe signs or symptoms include moderate or severe otalgia, otalgia >48 h and/or temperature ≥ 39 °C (102.2 °F)

[b] Observation may be considered for patients as long as an appropriate follow up can be ensured based on joint decision-making with the parent(s)/caregiver. If the child worsens or fails to improve within 48–72 h of the onset of symptoms, antibiotic therapy should be initiated.

Question 4

In addition to starting antimicrobial therapy, you are most likely to counsel the family that:

A. Fever may persist >24 h, even after starting antibiotics
B. Antibiotics will help with otalgia within the first 12 h of initiation
C. Topical benzocaine may be considered for analgesia
D. Analgesia with acetaminophen or ibuprofen should be started only after 24 h of antibiotics

Answer: The correct answer is A

Most symptoms of acute otitis media resolve within 72 h, although children younger than 2 years of age may take longer to show clinical improvement. Regardless of the use of antibiotics, the management of acute otitis media should include an assessment of pain. Antibiotic therapy does not provide symptomatic relief of pain in the first 24 h and pain and/or fever may persist in young children for 3–7 days. Analgesia can be achieved with acetaminophen or ibuprofen using dosing recommendations appropriate for the age and weight of the patient.

Topical agents such as lidocaine and naturopathic agents may provide some relief, although effects tend to be brief. Topical agents should not be used in children with tympanic membrane perforation. In addition, topical benzocaine should be avoided in children <2 years of age due to the risk of methemoglobinemia.

Question 5

The family calls 7 days after starting antibiotics to report the development of a diffuse, non-urticarial, erythematous, and pruritic rash. There is no history of angioedema or bronchospasm. Pending evaluation with an allergist, if she were to develop another episode of otitis media requiring antibiotics, which of the following would be appropriate?

A. Cefdinir
B. Cefpodoxime
C. Ceftriaxone
D. All of the above

Answer: The correct answer is D

Antibiotic treatment for penicillin-allergic patients depends on the type of previous reaction. As in our patient, those with mild delayed reactions to penicillin may be treated with cephalosporins such as cefdinir, cefpodoxime, cefuroxime, and ceftriaxone. It is important to note that oral cephalosporins such as cefdinir, cefpodoxime, and cefuroxime do not achieve sufficient concentration in the middle ear to eradicate penicillin-resistant *S. pneumoniae*.

In patients with immediate, IgE-mediated reactions or serious delayed reactions (i.e., Stevens-Johnson syndrome, toxic epidermal necrolysis, etc.), macrolide or lincosamide antibiotics may be used. Macrolide and lincosamide resistance is common among isolates of *S. pneumoniae*. Macrolides and lincosamides are also not effective in eradication of *H. influenzae*.

Case 2

A 4-year-old male presents to the clinic secondary to parental concern for attention deficit hyperactivity disorder. His parents state that his teachers have reported that he seems inattentive and has difficulty maintaining attention when assigned tasks. His past medical history is significant for recurrent otitis media, with the most recent episode occurring 3 months prior to presentation. He also has a history of cleft palate which has been repaired. He has no history of allergic rhinitis or adenoid hypertrophy.

On physical exam, he is found to be in no acute distress. His oropharynx is clear except for postsurgical changes in the soft palate. He has no pain upon manipulation of either ear. An otoscopic exam reveals a normal right tympanic membrane. Findings on examination of the left tympanic membrane are below (Fig. 1). The left tympanic membrane has reduced mobility when gentle pressure is applied via pneumatic otoscopy. A hearing screen is subsequently ordered.

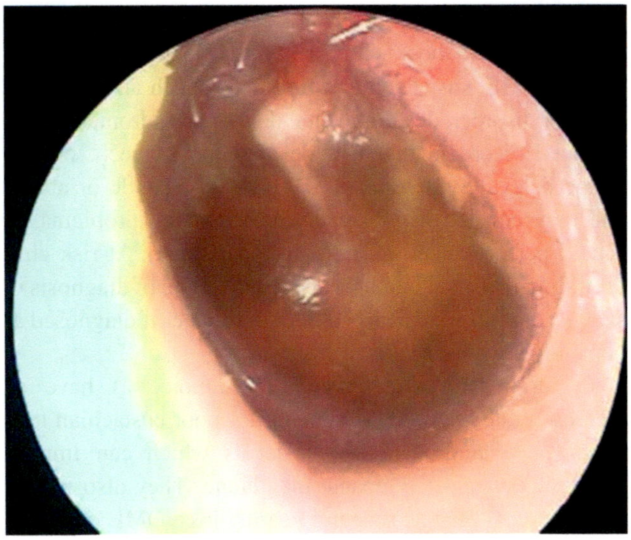

Fig. 1 Otoscopic findings of a left tympanic membrane. (Image courtesy of Belinda Mantle, MD. LSU Health Sciences Center and Children's Hospital New Orleans, LA)

Question 1

The history and physical exam are most consistent with a diagnosis of:

A. Acute otitis media
B. Cholesteatoma
C. Tympanic membrane perforation
D. Otitis media with effusion

Answer: The correct choice is D

Otitis media with effusion (OME) is characterized by the accumulation of fluid in the middle ear in the absence of acute inflammation. It most commonly occurs in children between the ages of 6 months and 4 years. OME may occur during an upper respiratory infection, as an inflammatory response following an episode of acute otitis media, or as a consequence of eustachian tube dysfunction. Otoscopic findings may include an air-fluid level, amber-colored or clear middle ear fluid, and normal or retracted tympanic membrane position. As opposed to AOM, bulging of the tympanic membrane is not a feature of OME.

Clinicians should document the presence of middle ear effusion with pneumatic otoscopy when diagnosing OME. Pneumatic otoscopy should also be performed in children with otalgia, hearing loss, or both. A normal eardrum moves briskly with applied pressure. Diagnosis of OME should occur when the movement of the tympanic membrane is sluggish, dampened, or restricted; complete absence of mobility is not required. If the diagnosis is uncertain after pneumatic otoscopy is performed or attempted, professional tympanometry should be performed. In OME, vibration during tympanometry is impaired resulting in a flat, or nearly flat, tracing.

Question 2

Which of the following conditions places a child with OME at increased risk for speech, language, or learning problems?

A. Down Syndrome
B. Blindness or uncorrectable visual impairment
C. Autism spectrum disorder
D. Cleft palate
E. All of the above

Answer: The correct answer is E

The above conditions likely place affected children with middle ear effusions at an increased risk for speech, language, or learning problems because of baseline sensory, physical, cognitive, or behavioral factors. Additional at-risk children include those with permanent hearing loss independent of OME, suspected or confirmed speech and language delay or disorder, developmental delay, or syndromes or craniofacial disorders that include cognitive, speech, or language delays. Symptoms of OME may be subtle or absent and may consist of poor balance, behavioral problems, or school performance issues (as in this patient). At-risk children should be evaluated for OME at the time of diagnosis of such a condition and at 12–18 months of age (if diagnosed as being at-risk prior to this age).

Children with Down syndrome (choice a) have an increased rate of AOM, chronic OME, poor eustachian tube dysfunction, and stenotic ear canals which can impede examination of the tympanic membrane. They also have a risk of mixed or sensorineural hearing loss. OME occurs in nearly all infants and children with cleft palate (choice d) because of abnormal insertions of the tensor veli palatini, which restricts opening of the eustachian tube.

Question 3
Which of the following treatments should be offered to this patient?

A. Intranasal steroids
B. Systemic antibiotics
C. Systemic steroids
D. Diphenhydramine
E. None of the above

Answer: The correct answer is E

Current evidence does not support the routine use of steroids (intranasal or systemic), antimicrobials, antihistamines, or decongestants as therapy for OME. There may be a short-term benefit of topical intranasal steroids in children with adenoidal hypertrophy. Intranasal steroids may also be considered in patients with concurrent allergic rhinitis, which may be a contributing factor to OME. Antihistamines and decongestants have well-recognized adverse effects and have not been shown to aid in the resolution of OME.

Question 4
In addition to our patient, tympanostomy tube insertion should be recommended in which of the following patients:

A. A patient with bilateral OME ≥ 3 months without documented hearing difficulty
B. A patient with recurrent acute otitis media without a middle ear effusion at the time of assessment
C. A patient with Down Syndrome and chronic OME
D. A patient with a single episode of OME < 3 months duration

Answer: The correct answer is C

The treatment for OME first consists of reexamination 3 months after onset with the addition of an age-appropriate hearing test in those with persistent OME. Hearing tests should also occur for any at-risk child (as previously defined) with OME of any duration. Chronic OME is defined as OME lasting ≥3 months. Serial evaluation should occur every 3–6 months until the effusion resolves, hearing loss is documented, or structural changes of the tympanic membrane or middle ear are noted. Otherwise healthy children with persistent OME who do not have any at-risk criteria can be safely observed for 6–12 months with a low risk of developing sequelae or reduced quality of life.

Current recommendations for tympanostomy tube placement include:

- Children with bilateral OME ≥ 3 months AND documented hearing difficulties.
- Children with unilateral or bilateral OME ≥ 3 months AND symptoms attributable to OME (e.g., vestibular problems, poor school performance, behavioral problems, ear discomfort, or reduced quality of life).
- Children with recurrent AOM who have unilateral or bilateral middle ear effusions at the time of assessment for tube candidacy.
- At-risk children with unilateral or bilateral OME which are unlikely to resolve quickly as reflected by a type B (flat) tympanogram or persistence of effusion ≥3 months.

Question 5
Adenoidectomy should be considered in which of the following patients:

A. A 2-year old with chronic OME with need for repeat surgery
B. A 6-year old with chronic OME
C. A 3-year old with chronic OME and adenoid hypertrophy
D. B and C
E. All of the above

Answer: The correct answer is D

Original recommendations suggested a role for adenoidectomy when repeat surgery was needed for OME after prior tympanostomy tubes in children as young as 2 years, but this recommendation has since changed with a new threshold of 4 years (making choice an incorrect). Adenoidectomy can be considered in patients <4 years of age when another indication exists such as chronic adenoiditis or nasal obstruction secondary to adenoid hypertrophy (choice c).

In children ≥4 years, adenoidectomy, tympanostomy tube insertion, adenoidectomy plus myringotomy (without tubes), or adenoidectomy plus tympanostomy placement can be recommended. Shared decision-making should occur between the clinician and caregiver regarding each option. The primary benefits of adenoidectomy are to reduce failure rates, reduce time with middle ear effusions, and decrease the need for repeat surgery. These benefits may be related to improved microflora in the nasopharynx and are independent of adenoid volume.

Bibliography

1. Coker TR, Chan LS, Newberry SJ, Limbos MA, Suttorp MJ, Shekelle PG, Takata GS. Diagnosis, microbial epidemiology, and antibiotic treatment of acute otitis media in children: a systematic review. JAMA. 2010;304(19):2161–9.
2. Griffin G, Flynn CA, Bailey RE, Schultz JK. Antihistamines and/or decongestants for otitis media with effusion (OME) in children. Cochrane Database Syst Rev. 2006;4
3. Lieberthal AS, Carroll AE, Chonmaitree T, Ganiats TG, Hoberman A, Jackson MA, Joffe MD, Miller DT, Rosenfeld RM, Sevilla XD, Schwartz RH. The diagnosis and management of acute otitis media. Pediatrics. 2013;131(3):e964–99.
4. Rosenfeld RM, Shin JJ, Schwartz SR, Coggins R, Gagnon L, Hackell JM, Hoelting D, Hunter LL, Kummer AW, Payne SC, Poe DS. Clinical practice guideline: otitis media with effusion (update). Otolaryngol Head Neck Surgery. 2016;154(1_suppl):S1–41.
5. Rosenfeld RM, Schwartz SR, Pynnonen MA, Tunkel DE, Hussey HM, Fichera JS, Grimes AM, Hackell JM, Harrison MF, Haskell H, Haynes DS. Clinical practice guideline: tympanostomy tubes in children. Otolaryngol Head Neck Surg. 2013;149(1_suppl):S1–35.
6. Shaikh N, Hoberman A, Kaleida PH, Rockette HE, Kurs-Lasky M, Hoover H, Pichichero ME, Roddey OF, Harrison C, Hadley JA, Schwartz RH. Otoscopic signs of otitis media. Pediatr Infect Dis J. 2011;30(10):822–6.
7. Shaikh N, Hoberman A, Rockette HE, Kurs-Lasky M. Development of an algorithm for the diagnosis of otitis media. Acad Pediatr. 2012;12(3):214–8.
8. Takata GS, Chan LS, Morphew T, Mangione-Smith R, Morton SC, Shekelle P. Evidence assessment of the accuracy of methods of diagnosing middle ear effusion in children with otitis media with effusion. Pediatrics. 2003;112(6):1379–87.
9. Wald ER, DeMuri GP. Antibiotic recommendations for acute otitis media and acute bacterial sinusitis: conundrum no more. Pediatr Infect Dis J. 2018;37(12):1255.

Asthma

Sylvette Nazario

Abbreviations

% pred	Percent predicted
AERD	Asthma exacerbated respiratory disease
API	Asthma predictive index
ED	Emergency Department
FcERI	High-affinity IgE receptor
FDA	Federal Drug Administration
FeNO	Fractional exhaled nitric oxide
FEV1	Forced expiratory volume in 1 min
ICS	Inhaled corticosteroid
ICS/LABA	Inhaled corticosteroid/long-acting B agonist
IgE	Immunoglobulin E
LAMA	Long-acting muscarinic agent
LAMA	Long-term muscarinic agents
LTRA	Leukotriene receptor antagonists
NAEPP	National Asthma Education and Prevention Program
SABA	Short-acting Beta agonists
SMART	Single maintenance and reliever therapy
T2	Type 2

1 Introduction

Asthma affects millions worldwide and causes significant morbidity. Asthma is one of the leading causes of hospitalizations and one of the main causes of emergency department admission among children.

The National Asthma Education and Prevention Program (NAEPP) published asthma diagnosis and treatment guidelines. The Global Initiative for Asthma (GINA) and NAEPP updated these recommendations. The guidelines emphasize the importance of addressing inflammation, comorbidities, and patient education. The most important recommendations include the use of an inhaled corticosteroid (ICS)/formoterol combination for persistent asthma in subjects 4 years and older as rescue and maintenance therapy, the use of a short course of inhaled corticosteroids for viral-associated exacerbation in infants younger than 4 years of age, fractional exhaled nitric oxide (FeNO) as an adjuvant to asthma diagnosis, and the use of long-acting muscarinic agents (LAMAs) as an add-on therapy in subjects with uncontrolled asthma despite ICSs/long-acting B agonists (LABAs).

Case 1
A 19-year-old African American woman with childhood-onset asthma arrives at the clinic for asthma management. She visited the emergency department (ED) due to asthma at least twice and was hospitalized once last year. She uses a short-acting B agonist (SABA) once a day and wakes with symptoms 4 nights per week. Her forced expiratory volume in 1 min (FEV1) was 80% of the predicted value.

Question 1
How severe is her asthma?

A. Intermittent
B. Mild persistent
C. Moderate persistent
D. Severe persistent
E. Labile asthma

Correct answer: C

Expert Panel Report-3 divides asthma into 4 severity criteria: intermittent, mild, moderate, and severe persistent based on the frequency of diurnal and nocturnal symptoms, interference with daily activities, and lung function. The criterion with the highest impairment determines the severity category. These categories are affected by age. The basic framework is similar. Enclosed is a simplified version of the

S. Nazario (✉)
Allergy Immunology Section, Department of Internal Medicine, University of Puerto Rico Medical Sciences Campus, San Juan, Puerto Rico
e-mail: sylvette.nazario@upr.edu

M. Mahmoudi (ed.), *Absolute Allergy and Immunology Board Review*, https://doi.org/10.1007/978-3-031-12867-7_7

Table 1 Asthma severity classification

Asthma severity: > = 12 years old				
Parameter	Intermittent	Mild	Moderate	Severe
	Impairment			
Symptom/SABA	1–2/week	3–6/week	Daily	>1/day
Night awakening	0–2/month	1/week	2–6/week	Nightly
Interference	None	Mild	Moderate	Severe
PFT[a]	Normal	Normal	FEV1 60–80% pred FEV1/FVC decrease <5%	FEV1 < 60% pred FEV1/FVC decrease >5%
	Risk			
Exacerbations requiring systemic steroids	0–1/year	> = 2/year	> = 2/year	> = 2/year

[a]FEV1/FVC 8–19 year 85%; 20–39 year 80%; 40–59 year 75%; 60–80 year 70%

Asthma 5–11-years old				
Parameter	Intermittent	Mild	Moderate	Severe
	Impairment			
PFT	Normal	Normal	% pre. FEV1 60–80% FEV1/FVC 75–80%	FEV1 < 60% FEV1/FVC <75%

Asthma severity: 0–4 years old				
Parameter	Intermittent	Mild	Moderate	Severe
	Impairment			
Night awakening	0	1–2/month	1/week	> = 2 week
PFT	Unreliable			
	Risk			
Exacerbations requiring systemic steroids	0–1/year	> = 2 in 6 months OR > =4 wheezing days+ risk factors	> = 2 in 6 months OR > =4 wheezing days+ risk factors	> = 2 in 6 months OR > =4 wheezing days+ risk factors

Expert Report, using 12 years of age or older as a reference. The differences at younger ages are described. Assume a similar classification as in a 12 y/o if not listed (Table 1).

In this case, the patient required SABA use every day, which indicates moderate; nighttime awakenings 4/week are also moderately persistent. Her FEV1 is 80% predicted or normal for age, which fits into intermittent or mild persistent. She visited the emergency department and was hospitalized at least twice in the last year, so she is not intermittent but persistent. Based on the overall impairment and risk categories, her asthma is **moderate persistent**.

Question 2

How would you treat her?

A. Formoterol 2 puffs bid
B. Fluticasone propionate 44 μg 2 puffs bid
C. Budesonide 180 μg 1 puff bid
D. Fluticasone/Salmeterol 500/50 μg 1 puff bid
E. Budesonide/Formoterol 80/4.5 μg 1 puff bid

Correct answer: E

Patients with moderate persistent asthma should receive low-dose ICS/LABA (Budesonide/Formoterol 80/4.5 1 puff bid, Mometasone/Formoterol 100/5 μg 1 puff bid, Fluticasone/Salmeterol 100/50 1 puff bid) or medium-dose ICS (Beclomethasone dipropionate 80 μg 3–5 puffs/day, Fluticasone propionate 110 μg 1–2 puffs bid, Mometasone furoate 200 μg 2 puffs daily, Budesonide DPI 200 μg 2–4 puffs/day) as the preferred choice. The revised guidelines introduced the term single maintenance and reliever therapy (SMART) as the preferred choice for step 3 (moderate) and 4 (severe) subjects older than 4 years of age. SMART consists of a low-dose ICS/Formoterol combination, used as maintenance and rescue therapy 1–2 puffs qd-bid and as needed, up to 12 puffs per day for 12 y/o and older and up to 8 puffs/day for those 5–11 years of age.

Choice A, LABA alone, should not be used in the treatment of asthma due to an increased risk of death, particularly among subjects homozygous for arginine at codon 16 of the β_2 agonist receptor. Choice B, fluticasone 44 μg 2 puffs bid, or 176 μg/day, is a low dose recommended for mild persistent patients. Similarly, 360 μg/day budesonide, choice C, is also a low-dose steroid, requiring 2 puffs bid to achieve medium potency. Choice D would correspond to high-dose ICS/LABA, a medium dose would be 250/50 1 puff bid and 100/50 would be low-dose ICS/LABA. According to new guideline revisions, Salmeterol-based ICS/LABA would not be recommended for use as rescue and maintenance due to the long onset of action of Salmeterol.

Table 2 shows the medication dosage recommended according to severity criteria incorporating the revised guideline recommendations. Adequate inhaler technique education is an important part of asthma care.

Table 2 Asthma treatment recommendations according to severity and age

Asthma treatment: > = 12 years old					
Intermittent	Mild	Moderate	Severe	Severe+	++
Preferred					
SABA prn	Low-dose ICS qd and SABA prn or **ICS + SABA prn**	**Low-dose ICS/ formoterol qd and prn***	**Med. dose ICS/ formoterol qd and prn***	Med-high ICS/ LABA + **LAMA** and SABA prn	High-dose ICS/ LABA + OCS + SABA prn
Consider	Immunotherapy	Immunotherapy	NA	**Omalizumab, anti-IL-5, IL-5R, Il4 and Il-13**	**Omalizumab, anti-IL-5, IL-5R, Il4 and Il-13**
Beclomethasone HFA 80 μg	*2–3 puffs bid*	*4–6 puffs bid*	*+7puffs bid*		
Fluticasone 110 μg	*1 puff bid*	*2 puffs/bid*	*3+puffs bid*		
Mometasone DPI 200 μg	*1 puff bid*	*2 puffs/bid*	*3+puffs bid*		
Budesonide DPI 180 μg	*1 puff bid*	*2 puffs/bid*	*3+puffs bid*		
Budesonide/ formoterol 160/4.5	*N/A*	*1 puff bid*	*2 puffs bid*	*+2 puffs bid*	
Mometasone/ formoterol 200/5	*N/A*	*1 puff bid*	*2 puffs bid*	*+2 puffs bid*	
Alternative					
N/A	LTRA, chromones, zileuton, theophylline, and SABA prn	Medium ICS+ SABA prn or Low ICS/LABA, low ICS/LAMA, low ICS + LTRA, low ICS + Theo, low ICS + zileuton qd, and SABA prn	Medium ICS/LABA or med ICS + LAMA and SABA prn or Med ICS + LTRA, med ICS + Theo, med ICS + zileuton qd, and SABA prn	Med-high ICS/LABA or high ICS + LTRA and SABA prn	

Up to 12 puffs/day

Asthma treatment: 5–11 years old					
Intermittent	Mild	Moderate	Severe	Severe+	++
Preferred					
SABA prn	Low-dose ICS qd and SABA prn	Low-dose ICS/formoterol qd and prn*	Med. Dose ICS/ formoterol qd and prn*	High-dose ICS/ LABA and SABA prn	High-dose ICS/ LABA + OCS + SABA prn
Consider	Immunotherapy	Immunotherapy		Omalizumab	Omalizumab
Beclomethasone HFA	40 μg bid		80 μg bid	>2 puff bid	
Fluticasone HFA	44 μg bid	80/4.5 bid	110 μg 1–2 puffs bid	220 μg 2 puff bid	
Mometasone DPI	110 μg qd	50/5 bid	220 μg 1–2 puffs qd	>2 puffs bid	
Budesonide DPI	90 μg bid		180 μg 1–2 puffs bid	>2 puffs bid	
Budesonide/ formoterol				2 puff bid	
Mometasone/ formoterol				100/5 bid	
Alternative					
N/A	LTRA qd, chromones, theophylline, and SABA prn	Medium ICS + SABA prn or Low ICS/LABA, low ICS + LTRA, low ICS+ Theo qd, and SABA prn or	Medium ICS/LABA and SABA prn or Med ICS + LTRA, or Theo qd and SABA prn	High ICS + LTRA or High ICS + Theo and SABA prn	High ICS + LTRA + OCS qd or High ICS + Theo + OCS and SABA prn

Up to 8 puffs/day

Asthma treatment: 0–4 years old					
Intermittent	Mild	Moderate	Severe	Severe+	++
Preferred					
SABA prn	Low-dose ICS qd and SABA prn	Low-dose ICS/LABA and SABA prn or medium-dose ICS and SABA prn or Low-dose ICS/formoterol qd and prn*	Medium-dose ICS/ LABA and SABA prn or Medium-dose ICS/ formoterol qd and prn*	High-dose ICS/ LABA and SABA prn	High-dose ICS/ LABA + OCS + SABA prn
Fluticasone Budesonide Respules Budesonide/ formoterol* Mometasone/ formoterol*	44 μg 2 puffs bid 0.25–0.5 mg/day	80/4.5 bid 50/5 bid	110 μg 1 puff bid 0.5–1 mg/day	110 μg 2 puff bid >1 mg/day 2 puff bid 100/5 bid	
Alternative					
N/A	LTRA qd, chromones, and SABA prn		Med ICS + LTRA and SABA prn	High-dose ICS + LTRA and SABA prn	High-dose ICS + LTRA + OCS qd and SABA prn

Only for 4 years old
*5y/o and older

Question 3

She persisted uncontrolled with an asthma control test score of 12, reporting dyspnea on exertion and nocturnal asthma symptoms. After reviewing the inhaler technique, which comorbidities should be addressed.

A. Obesity
B. Atopy
C. Chronic rhinitis
D. Depression
E. All the above

Correct answer: E

Asthma comorbidities are directly linked to uncontrolled asthma. Atopy is particularly prevalent in childhood-onset asthma and among non-Hispanic Blacks and Puerto Ricans. It is crucial to address the atopic component, particularly allergic rhinitis, and to recognize the "atopic march" or infants with atopic dermatitis and/or food allergies that later develop asthma and/or allergic rhinitis. Physiologically, the permeable fragile epithelia in the skin and airway have increased susceptibility to sensitization, and, thereafter, allergy development. Integrated allergy-specific environmental control measures and allergy desensitization, if indicated, are part of treatment.

Obesity is another important asthma comorbidity. Increased body weight is associated with asthma exacerbations, increased asthma morbidity, and decreased response to corticosteroids and β-agonists.

Chronic rhinitis is also associated with poorly controlled asthma. The nasal and pulmonary epithelia share common cellular components and mediators. Rhinitis, allergic or chronic, and nasal polyposis are associated with asthma. Intranasal steroids and/or antihistamines are the drugs of choice. Asthma exacerbated respiratory disease (AERD) is a subgroup of patients with high asthma severity. Multidisciplinary evaluations including intranasal steroids, evaluation for surgical polypectomy, aspirin desensitization, and/or treatment with IL-4 and Il-13 antagonists or IgE antagonists are recommended.

Depression and stress are also associated with increased asthma morbidity and exacerbations through the increased release of inflammatory cytokines such as IL-1, IL-6, and IL-8; stimulation of the adreno-pituitary axis leading to increased cortisol; and altered β-agonists and steroid receptor response.

Education about how to identify triggers and exacerbation, having an asthma action plan, and knowing how and when to use rescue and maintenance medications is important. Influenza, pneumococcal, zoster, and coronavirus-19 immunization are also important parts of asthma treatment.

Question 4

On follow-up, she still reported nocturnal awakenings every night and required a short-term bronchodilator 2–3 times/ day. Assuming adequate medication adherence and technique and attention to comorbidities, which is the best next step of therapy?

A. Add leukotriene inhibitor
B. Add systemic steroids
C. Add long-term antimuscarinic agent
D. Add theophylline
E. Double ICS/LABA dose

Correct answer: C

The patient is still uncontrolled despite her treatment with an ICS/LABA. NAEPP guidelines provide several alternative therapies, including systemic steroids, theophylline, and leukotriene inhibitors. The most recent Expert Panel recommends the addition of long-term antimuscarinic agents (LAMAs) as they are synergistic with ICS/LABA in asthma treatment instead of doubling the ICS/LABA dose. Treatment improves asthma control but does not improve quality of life or decrease exacerbations. Therapy should not be recommended in subjects with glaucoma or urinary retention. Education on the inhaler technique is essential and is intended to be used as maintenance, not as rescue therapy.

LAMA can also be added to subjects uncontrolled on ICS, but LABA is still a preferred alternative, particularly among African Americans.

Information provided was based on the use of tiotropium, as sufficient information on umeclidinium was not available at the time of the publication. No recommendation was made for subjects younger than 12 years of age.

Question 5

Four weeks after the initial evaluation, the patient is still using albuterol 2–3 times per day and reported nocturnal awakenings almost every night. She was sensitized to mites and roaches, her IgE was 323, and her eosinophil count was 125 μg/ml. She weighs 80 kg. Assuming the patient is compliant with treatment, uses adequate technique, and that comorbidities are controlled, which would be the next step in therapy?

A. Double dose of ICS
B. Start Omalizumab
C. Start Benralizumab
D. Start Dupilumab
E. Start Mepolizumab

Correct answer: B

Omalizumab is a humanized monoclonal antibody directed against the third constant region domain of immunoglobulin E (IgE). It prevents binding to the high-affinity IgE receptor (FcERI), downregulating the receptor on mast cells and basophils. It is dosed according to weight and IgE levels and is the preferred biologic for atopic subjects who have a weight and IgE level within the therapeutic range for medication. It improves the quality of life and reduces exacerbations, ED visits, and hospitalizations. Some subjects have a mild improvement in lung function. Omalizumab side effects include headaches, injection site reactions, and, in rare cases, anaphylaxis. It is recommended that subjects carry epinephrine autoinjectors and that they wait 2 h for the first 3 occasions and 30 min thereafter. The risk of anaphylaxis decreases with time. It is approved for asthma in children 6 years of age and older.

Doubling the dose of ICS does not improve symptom control in subjects already on ICS/LABA.

Mepolizumab is a monoclonal humanized antibody directed against IL-5. It requires a minimum of 150 eosinophils per milliliter to start therapy. Among the approved monoclonals for eosinophilic asthma, mepolizumab requires the least number of eosinophils and does not require subjects to be allergic. It decreases exacerbation frequency, hospitalizations, and ED visits by half. Side effects include headaches, injection site reactions, and zoster exacerbations. It can be self-administered at home and is dosed at a fixed dose of 100 mg SC q 4 weeks. Recently, mepolizumab was approved by the Federal Drug Administration (FDA) for hypereosinophilic syndrome and eosinophilic granulomatous polyangiitis at 300 mg SC q 4 weeks. It is also approved for children 6 years of age and older.

Benralizumab is a humanized monoclonal antibody directed against the IL-5 receptor. It induces the death of eosinophils by inducing antibody-dependent cytotoxicity of eosinophils by natural killer cells. It requires a minimum count of 300 eosinophils/dl. Unlike mepolizumab, it leads to a prompt depletion of eosinophil counts, allowing dosing every 2 months after the first 3 monthly injections. Side effects include local injection site reactions.

Dupilumab is a human monoclonal antibody against the common alpha chain of the IL-4 and IL-13 receptors. It is approved for steroid-dependent asthma, nasal polyposis, severe persistent asthma, and atopic dermatitis. It is dosed every 2 weeks at 200 or 300 mg SC depending on the indication and the subject's age. It is approved for patients aged 12 and up with asthma. Side effects include conjunctivitis, mostly described in the setting of atopic dermatitis and eosinophilia.

Biologic therapies require prior evaluation to exclude parasitic diseases, particularly in places where parasitic infections are prevalent.

Question 6

Which is a long-term pulmonary change seen in asthma?

A. Epithelial fibrosis
B. Decreased airway vasculature
C. Decreased smooth muscle mass
D. Mucus hypersecretion
E. Decreased goblet cells

Correct answer: D

Asthma induces structural changes in the airway called airway remodeling. These changes include:

- Airway epithelial thickening affecting small and large airways. The epithelia suffer damage and become hyperplastic with sloughing and the formation of mucus plugs.

- Thickening of the reticular basement membrane due to subepithelial fibrosis.
- Increased airway smooth muscle due to hypertrophy or hyperplasia.
- Mucus hypersecretion due to mucous gland hyperplasia and increased goblet cells.
- Increased neovascularization.

These anatomic changes lead to functional changes reflected by loss of airway elasticity and increased stiffness, leading to breaking of alveolar septa and further bronchoconstriction.

Childhood-onset asthma may lead to impaired lung function in adults, particularly among those with low baseline FEV1; severe, persistent, early-onset symptoms; those with lower bronchodilator response; and increased airway hyperreactivity. Adult-onset asthma poor lung function also correlated with long duration, severity, frequent exacerbations, bronchial hyperreactivity, smoking, continued exposure to allergens, and elevated IgE and eosinophil count.

Case 2

John, a 3 y/o boy, has frequent wheezing episodes and attends daycare.

Question 7

Which are risk factors for persistent wheezing in children?

A. Female sex
B. Parents with asthma
C. Lack of atopy
D. Parasitic infection
E. Brother with asthma

Correct answer: B

Several studies have evaluated predictive risk factors for asthma among children who wheeze. The best known, the Tucson Longitudinal study, evaluated newborns initially through 3 years of age and later expanded to 6- and 18-years old. Almost half of the children never wheezed, 20% wheezed as an infant and resolved by 6 years (transient), 14% had persistent wheezing since infancy (persistent), and 15% started wheezing between 3 and 6 years of age (late wheezers). Transient wheezers had no association with atopy but rather with maternal smoking, daycare attendance, or older siblings at school. Atopic persistent wheezers were more likely to continue wheezing through adolescence and had higher total IgE, atopic dermatitis, parents with asthma, and food allergies. Their lung function, which was originally normal, worsened over time.

Several longitudinal or cross-sectional studies confirmed the Tucson study and identified risk factors for asthma. These include:

- Atopy: multiallergen, Alternaria.
- Boys have a greater risk than girls.
- Decreased lung function was linked to prolonged or early wheezing.
- Upper respiratory tract infections by virus: *Rhinovirus* (particularly *Rhinovirus C*), *Respiratory Syncytial Virus, Influenza, Parainfluenza, Metapneumovirus*; or bacteria: *Streptococcus pneumonia, Hemophilus, Moraxella.*

The **Asthma Predictive Index (API),** derived from the Tucson Longitudinal Asthma Study, intends to predict the risk of asthma in the future among 3 y/o children who had a history of recurrent wheezing episodes. Parental asthma or physician-diagnosed atopic dermatitis are required for a positive score, or 2 minor criteria, including rhinitis, wheezing besides cold, or eosinophil count of 4%. The **modified API** increased the number of wheezing episodes to 4 and added aeroallergen sensitization as a major criterion. The minor criteria of allergic rhinitis were substituted for food sensitization. The API score has a large negative predictive value; thus, negative results predict that wheezing will resolve over time.

Question 8

Which of the following is an inciting factor for asthma?

A. Early-onset cat exposure
B. Coronavirus infection
C. Increased interferon-gamma production at birth
D. Exposure to *Rhinovirus* infection
E. High birth weight

Correct answer: D

Immune dysregulation at birth with Type 2 (T2) predominance over Type 1 response increases the risk of lower respiratory tract infections, wheezing, allergen sensitization, and asthma onset. It is unclear whether this immune imbalance is primary or secondary to other epigenetic or confounding factors, including pollution, stress, obesity, low birth weight, allergen exposure, or dietary factors, which could impinge on the immune response.

Exposure to traffic pollution, particularly nitric oxide and dioxide, and particulate matter increase the risk of asthma inception and could be an important explanation for the increased asthma prevalence in urban dwellers.

Allergic sensitization, particularly to cockroaches and rodents, is associated with asthma morbidity, particularly among inner-city children. A dose-response of dust mite exposure to sensitization has been shown, although the association with asthma morbidity is ambivalent. The associations between cat and dog exposure, sensitization, and allergies are complex. Studies suggest that early exposure to pets among children protects against asthma development.

Viral infections in infancy can also affect the risk of asthma inception, particularly with *Rhinovirus*, by inducing a T2 immune shift (Answer D).

Stress and depression in caregivers have been associated with asthma inception and morbidity. Obesity is also associated with increased asthma prevalence in children and adults but not in newborns. Preterm birth is associated with asthma.

Question 9
John developed increased nasal rhinorrhea and a low-grade fever. The next day he started wheezing. Which would be the best treatment recommendation?

F. Amoxicillin
G. Albuterol inhalations q 4 h
H. Prelone for 4 days
I. Double ICS dose
J. Inhaled budesonide 1 mg bid for 7 days

Correct answer: E

The expert panel revision addresses the treatment of infants and toddlers 4 years of age or younger who have had three or more wheezing episodes in their life or two episodes of wheezing in the last year, associated with respiratory tract infections, who were asymptomatic the rest of the time. They recommend the use of budesonide inhalation suspension bid for 7 days at the onset of respiratory infection symptoms, along with as-needed SABAs as rescue therapy. This treatment is started at home by the caregivers, following the recommendations in a written action plan described by the provider. Inhaled steroids at the onset of respiratory tract infection decrease the need for systemic steroids and have better outcomes than SABAs alone. In children 4–11 years of age, increasing the inhaled corticosteroid dose, even five times, did not reduce exacerbations or improve quality of life scores. Most exacerbations in infants are viral-induced, and antibiotics are not indicated. Oral corticosteroids can have a negative effect on growth, and inhaled corticosteroids are preferably used.

Question 10
What is the role of fractional exhaled nitric oxide (FeNO) in the evaluation of asthma?

A. Can be used in isolation to assess asthma control
B. Predicts future exacerbations
C. Assesses asthma exacerbation severity
D. It is an adjunct to history, physical exam, and spirometry in asthma evaluation
E. Levels less than 25 parts per billion are consistent with T2 inflammation and asthma diagnosis

Correct answer: D

FeNO is a noninvasive biomarker that measures T2 inflammation. It is a safe procedure recommended as an adjunct to asthma monitoring in subjects 5 years of age and older. The value of FeNO measures in younger children has not yet been established.

The FeNO value increases in allergic conditions such as allergic rhinitis and should be interpreted with caution in this setting. Levels less than 25 ppb if older than 12 y/o or < 20 ppb in children 5–12 years exclude active T2 inflammation but could be seen in non-eosinophilic asthma, smoking, COPD, bronchiectasis, vocal cord dysfunction, and rhinosinusitis. **Levels > 50 ppb or > 35 ppb in children 5–12 years of age are consistent with active T2 inflammation and response to steroids** and support a diagnosis of asthma. Levels in the intermediate range of 25–50 or 20–35 in children are nondiagnostic.

FeNO measurement is an **adjunct** to asthma diagnosis and by itself does not establish the diagnosis. It **predicts response to corticosteroids** and can be used to monitor asthma in patients on ICS, ICS/LABA, montelukast, or omalizumab. FeNO monitoring **decreases asthma exacerbations** but does neither predict future exacerbations unless it is used every 2–3 months, nor does it predict asthma severity. It is not intended to measure adherence.

2 Conclusion

The updated asthma management builds on the NAEPP guidelines, incorporating recommendations to use single maintenance and reliever therapy on mild and moderate asthmatics older than 4 years of age. It updates the use of long-acting muscarinic agents and other biologics that have been approved for T2 inflammation. This publication makes recommendations on the role of FeNO as an adjunct for the diagnosis of asthma and supports the role of subcutaneous immunotherapy in atopic controlled asthmatics and comprehensive environmental control measures geared to specific sensitization of subjects with asthma. Incorporating the asthma guideline recommendations is intended to improve asthma care and quality of life and reduce asthma morbidity and mortality.

Bibliography

1. Patel SJ, Teach SJ. Asthma. Pediatr Rev. 2019;40(11):549–65.
2. Nhlbi. National Heart, Lung, and Blood Institute National Asthma Education and Prevention Program Expert Panel Report 3: Guidelines for the Diagnosis and Management of Asthma Full Report 2007. 2007.

3. The Global Initiative for Asthma is supported by unrestricted educational grants from: Visit the GINA website at www.ginaasthma.org. Global strategy for asthma management and prevention Visit the GINA website at www.ginasthma.org © 2018 Global Initiative f [Internet] 2014. www.ginasthma.org
4. Cloutier MM, Baptist AP, Blake KV, Brooks EG, Bryant-Stephens T, DiMango E, et al. 2020 Focused Updates to the Asthma Management Guidelines: A Report from the National Asthma Education and Prevention Program Coordinating Committee Expert Panel Working Group. J Allergy Clin Immunol. 2020;146(6):1217–70.
5. McGarry ME, Castellanos E, Thakur N, Oh SS, Eng C, Davis A, et al. Obesity and bronchodilator response in black and hispanic children and adolescents with asthma. Chest. 2015;147(6):1591–8.
6. Barnthouse M, Jones BL. The impact of environmental chronic and toxic stress on asthma. Clin Rev Allergy Immunol. 2019;57:427–38.
7. Middleton's Allergy: Principles and Practice- ClinicalKey [Internet]. https://datubazes.lanet.lv:5250/#!/browse/book/3-s2.0-C20161002419
8. O'Byrne P, Fabbri LM, Pavord ID, Papi A, Petruzzelli S, Lange P. Asthma progression and mortality: the role of inhaled corticosteroids. Eur Respir J. 2019;54:1900491.
9. Khreis H, Kelly C, Tate J, Parslow R, Lucas K, Nieuwenhuijsen M. Exposure to traffic-related air pollution and risk of development of childhood asthma: a systematic review and meta-analysis. Environ Int. 2017;100:1–31.
10. Poowuttikul P, Saini S, Seth D. Inner-city asthma in children. Clin Rev Allergy Immunol. 2019;56:248–68.

Cough and Allergic Diseases

Satoshi Yoshida

Abbreviations

AC	Atopic cough
BHR	Bronchial hyperresponsiveness
CF	Cystic fibrosis
CVA	Cough variant asthma
GERD	Gastroesophageal reflux disease
ICS	Inhaled corticosteroids
LABA	Long-acting inhaled β2 agonists
NAEB	Non-asthmatic eosinophilic bronchitis
PC20	Load concentration (mg/mL) that reduces the amount of forced exhalation for 1 s before and after the test by 20%
PNDS	Postnasal drip syndrome
SABA	Short-acting inhaled β2 agonists
UACS	Upper airway cough syndrome

1 Introduction

Cough is a very common complaint for which patients seek medical attention. A multicenter study by general practitioners has also reported that cough was the most frequent reason (11.7% of all) that made patients visit clinics. Clinical findings frequently indicate a specific cause. Cough is both an important physiological reflex protecting the airways, and a frequent complaint associated with virtually all pulmonary and several extrapulmonary diseases. Cough is also a contributing factor in the spreading of infectious disease.

Strictly speaking, even if you cough in one word, there are various forms of infection transmission depending on the pathogen. Cough commonly forms "droplets" that contain a variety of pathogens that cause inhaled infections. And they form smaller "microaerosols" that cause infections similar to airborne infections. For this reason, cough can contribute to the spread of various infectious diseases. Diseases that cause droplet infections include, for example, mycoplasma and pertussis, which are frequently transmitted to young people and children. COVID-19 is a disease that is spread by not only droplet but also microaerosols. In addition, *tubercle bacilli* are among those that are transmitted by the more infectious "airborne infection."

Acute cough due to the common cold is one of the most frequent causes of primary care consultations. Cough is a reflex that helps clear the airways of secretions, protects the airway from foreign body aspiration, and can be the manifesting symptom of a disease. Reflexes are characterized by complexity and plasticity and are caused by physical and chemical stimuli. Stimulatory and C-fiber receptors are activated in the respiratory tract, pleura, pericardium, and esophagus. The impulse is then transmitted to the brainstem cough generator circuit via the vagus nerves. There is also a connection to the cortex, allowing voluntary control of both eliciting and, to a limited degree, inhibiting cough. Efferent innervations reach the effector muscles (diaphragm, abdominal, intercostals, back, and muscles of the larynx and upper airway). Mucociliary clearance is the primary means of clearing the bronchial system. To a certain degree, cough can compensate for impaired mucociliary clearance (e.g., that caused by the effects of smoking). If mucociliary clearance is overwhelmed by aspiration, an intact cough reflex protects the lungs effectively. However, an impaired cough reflex, e.g., after stroke, results in life-threatening aspiration pneumonia. The clearing competence of the cough reflex depends on several conditions: obstruction of the airways, bronchial collapsibility, lung volumes, respiratory muscle and laryngeal function, and the amount and viscosity of the mucus. Cough is productive (wet) if the amount of the daily expectoration is ≥30 mL (two tablespoons' worth). The phlegm can be mucous, serous, purulent, or bloody. Bronchial casts can also be produced. The cough reflex arc consists of five parts: (1) cough receptors; (2) afferent nerves; (3) brainstem

S. Yoshida (✉)
Department of Continuing Medical Education, Harvard Medical School, Boston, MA, USA

M. Mahmoudi (ed.), *Absolute Allergy and Immunology Board Review*, https://doi.org/10.1007/978-3-031-12867-7_8

cough generator circuit; (4) efferent nerves; and (5) effector organs including muscles. Hypersensitivity of the cough nociceptors elicits pathological cough. Numerous respiratory and other diseases cause cough nerve hypersensitivity and thus produce cough. In a considerable number of clinical cases however only cough nerve sensitivity is affected without another "specific" cause. Such patients are suffering from idiopathic cough.

Chronic cough is defined as symptoms lasting longer than 8 weeks, whereas acute cough lasts less than 3 weeks and subacute cough from 3 to 8 weeks. When persistent and excessive, cough can seriously impair quality of life and lead to vomiting, muscle pain, rib fractures, urinary incontinence, tiredness, syncope, and depression. It also has psychosocial effects, such as embarrassment and negative impact on social interactions. Etiologies of Chronic Cough in Adults and Children are shown in (Table 1).

Table 1 Etiologies of chronic cough in adults and children

Adults
Most common
Protracted bacterial bronchitis (i.e., postinfection cough)
Environmental exposures
Smoking status
Airway hyperresponsiveness
Upper airway cough syndrome
Postnasal drip syndrome
Synobronchial syndrome
Upper airway intoxication
Chronic pharyngitis
Chronic laryngitis
Bronchial asthma
Cough variant asthma
Atopic cough
Non-asthmatic eosinophilic bronchitis
Chronic obstructive pulmonary disease
Pneumothorax
Gastroesophageal disease
Laryngopharyngeal reflux disease
Angiotensin-converting enzyme inhibitor use
Less common
Chronic aspiration
Bronchiectasis
Obstructive sleep apnea
Pertussis
Pulmonary embolism
Postinfectious bronchospasm
Sleep apnea
Least common
Reactive airways dysfunction syndrome
Irritant-induced asthma
Arteriovenous malformation
Bronchiolitis
Bronchogenic carcinoma
Chronic interstitial lung disease
Irritation of external auditory canal
Persistent pneumonia
Tuberculosis
Sarcoidosis
Vocal cord dysfunction
Somatic cough syndrome
Children
Most common
Postinfectious cough
Aspiration
Airway hyperresponsiveness
Upper airway cough syndrome (in children older than 6 years)
Postnasal drip syndrome
Synobronchial syndrome
Bronchial asthma (cough variant asthma)
Cystic fibrosis
Less common
Protracted bacterial bronchitis (i.e., postinfection cough)
Environmental triggers
Foreign body (in younger children)
Gastroesophageal reflux disease
Pertussis
Postinfectious bronchospasm
Non-cystic fibrosis bronchiectasis
Tracheobronchomalacia
Least common
Chronic aspiration
Congenital abnormality
Immunodeficiency
Primary ciliary dyskinesia
Somatic cough syndrome
Tourette syndrome/tic
Isolated orphan airways disease

Case Presentation 1

36-year-old male. Four weeks ago, he visited a primary care outpatient clinic because he had symptoms such as fever of 100.0F (37.8 °C), headache, sore throat, runny nose, cough, and sputum. He was prescribed azithromycin 500 mg/day for 3 days after the examination. The fever went down and symptoms such as runny nose, headache, and sore throat disappeared within a week. However, he went to an outpatient clinic because his coughing did not stop and he could not sleep at night, especially because he coughed at bedtime. At the time of consultation, he had a body temperature of 96.8 °F (36.1 °C). He breathes 14 breaths/minute. Blood pressure 134/76, Sp02 98%, COVID-19 and influenza antigen rapid test were both negative. On auscultation, he had no abnormal breath sounds and did not hear any secondary noise. There was no abnormality in the findings on the chest X-ray. In the spirometry test, FVC: 94%, FEV1.0: 86%. He usually does not have paroxysmal coughing, wheezing, or dyspnea. There is no heartburn or coughing after eating. He has no dyspnea or coughing after exercise, no history of sinusitis, and no history of smoking.

Question 1

What is the most probable diagnosis for this patient?

A. Exacerbation of bronchial asthma after infection
B. Prolonged cough after infection
C. Sinus bronchial syndrome
D. Chronic bronchitis
E. Bronchiolitis

Answer: B

Prolonged cough after infection.

Cough as a symptom is attributed to distinct diseases and is categorized as either acute (lasting ≤8 weeks) or chronic (lasting >8 weeks). Of course, these limits are arbitrary. Acute cough (due to common cold) usually lasts only 2–3 weeks. Common cold is the most common cause of cough and usually subsides spontaneously, in otherwise healthy persons, after 2–3 weeks. Enhancing the cough nerve sensitivity following diseases can cause acute cough. Various acute infections, (e.g., *Mycoplasma pneumoniae* and adenoviruses) however can elicit cough lasting more than 8 weeks; for example, *Bordetella pertussis* can cause cough lasting around 3 months. In otherwise healthy individuals, acute infections of the upper and/or lower airways, the most common cause of cough, are self-limiting. Medical history and physical examination are usually sufficient in the diagnosis, and over-the-counter (OTC, i.e., nonprescription) remedies for the treatment. However, a few special circumstances require an immediate, full diagnosis of acute cough.

In contrast, immediate diagnostic workup is essential in all patients presenting with chronic (>8 weeks duration) cough; a chest radiograph and lung function test should be performed immediately. This is consistent with the recommendations in all published guidelines on cough. If the chest radiograph proves inconclusive, the lung function test is unremarkable and cough is the only presenting symptom, it will always be difficult to establish the diagnosis.

A dry cough is annoying and destructive, but it has some common root causes, mainly allergies such as allergic rhinitis, colds, and the flu. Many guidelines define chronic cough as a cough that lasts for more than 8 weeks. As a result, chronic cough can lead to poor quality of life. The underlying cough reflex is the same whether the patient suffers from a cough associated with an allergy or infection. A cough begins when a special nerve that ends in the airways detects some disorder and informs the brain that something is happening in the respiratory system. When the brain processes enough signals to determine that something may be wrong, the brain sends back chemical signals to start coughing. Its lung-brain communication remains the same no matter what the cause of the cough. Allergy-related cough can be caused by inflammation of the airways caused by an inappropriate or overly sensitive immune response to relatively harmless particles (such as pollen), whereas cold or flu-related inflammation is caused by a viral infection. If the cough is due to an allergy rather than a cold or flu, there are some differences in the timing of symptoms. If the allergen is present in certain weather conditions and not in other weather conditions, an allergy-related cough may begin in response to seasonal changes. Patients may also notice more coughing in some settings, but not in others. For example, a patient may feel good in the office but begins to cough at night due to household allergens such as animal dander and smoke. Allergy-related coughs can affect you for weeks or months at a time, and symptoms can vary in intensity from day to day. In the case of a cold or flu cough, the symptoms worsen in a few days and gradually relieve as they improve. Patients may notice some variation in timing symptoms such as a more severe night cough than during the day but may not have the same variability that is seen with allergic coughing. Patients may be able to decipher the cough by looking at the other symptoms they experience with the cough. Colds, flu, and allergies share common symptoms, such as sneezing, stuffy nose, and runny nose, but they usually differ in other ways. Allergies can cause itching of the eyes, palate, itching of the throat, and dark circles under the eyes. These bears, also known as allergy shiners, are associated with chronic untreated allergies. If the cough is due to a cold or flu, the patient may experience fatigue, pain, fever, and an upset stomach. Cough can change from dry to moist mucus-filled and vice versa as the patient develops from a cold or flu and recovers.

Question 2

What is the most important pathophysiology of this patient?

A. Eosinophilic inflammation
B. Persistent purulent inflammation due to infection
C. Sensitization to antigen
D. Exposure to irritant chemicals
E. Bronchial hyperresponsiveness

Answer: E

Eosinophilic inflammation is important as the pathophysiology of allergic inflammatory diseases of the airways, such as bronchial asthma, and it is important to distinguish it from purulent inflammation caused by neutrophils that control biodefense immunity. Also, lymphocytes are involved in two pathophysiologies, allergic inflammation and purulent inflammation. In this case, sensitization to the antigen is not related to the condition, as this patient does not have an allergic cough. When coughing persists after airway infection, it is characterized by persistent coughing even after the inflammation of the respiratory tract has subsided. "Bronchial sensitivity" is a term that refers to the hypersensitivity of the airways caused by specific antigen stimulation. The responsiveness of the airways to nonspecific stimuli such as exhaust fumes, cigarette smoke, or air pollution is called "bronchial hyperresponsiveness." Accurate understanding of the difference in meaning between these two terms, "airway responsiveness" and "airway hyperresponsiveness," can lead to airway infections such as acute bronchitis, or different airway inflammations such as bronchial asthma and chronic obstructive pulmonary disease (COPD). It is very important to understand the difference in pathophysiology of diseases and to make a differential diagnosis of various respiratory tract diseases.

Question 3

Which test is appropriate for assessing BHR?

A. Allergen-specific IgE test
B. Methacholine inhalation test
C. Kveim test
D. Drug-induced lymphocyte stimulation test
E. Pre- and post-ß2-agonist inhalation spirometry test

Answer: B

BHR can be assessed with a bronchial challenge test. Experimentally, BHR is expressed by the leftward shift of the concentration-response curves following aerosol administration of histamine or methacholine. These chemicals trigger bronchospasm in normal individuals as well, but people with BHR have a lower threshold. In asthmatic patients, BHR results in a significant decrease in the provocative concentration of histamine or methacholine causing a 20% decrease in forced expiratory volume in 1 s (PC20). The cutoff value of PC20, which indicates increased BHR, is 8 mg/mL for both histamine and methacholine. BHR referred to as airway hyperresponsiveness is the occurrence of excessive bronchoconstriction in response to a variety of inhaled stimuli, both chemical and physical. BHR is the expression of an exaggerated bronchopulmonary response associated with airway inflammation, involving vascular alterations, increase in bronchial secretions, recruitment, and activation of inflammatory cells. Widely used as an objective measure of fluctuating airflow, BHR is considered as a hallmark, a defining feature and most characteristic clinical feature of asthma. However, BHR is also found in a spectrum of other lung diseases from COPD to cystic fibrosis. It is often detected in atopic individuals, in patients with rhinitis but without pulmonary symptoms, in smokers and ex-smokers, smoking history of 20 pack-years, or in smokers older than 45 years) that suggest a serious underlying cause of cough. Also, after respiratory infections and following acute inhalation exposure to irritant chemicals. It is also found in asymptomatic nonsmoking members of the general population. From a clinical perspective, BHR testing has played an important role in the diagnosis of airway diseases. It can serve as a predictor of future lung disease and disease progression, and it has been shown to be an independent risk factor for irreversible loss of lung function in patients with and without pulmonary symptoms. In addition, the effect of treatments on BHR provides a practicable measure for titrating medication doses in asthmatics and others, and from the perspective of pathophysiology, study of BHR provides insights into the capacity of the normal lung to control airway function and the pathological mechanisms that lead to airway dysregulation and disease. BHR is an important feature of asthma and is characterized by a nonspecific exaggerated response to bronchoconstrictor agents, such as histamine and acetylcholine. As mentioned above, BHR is expressed by the leftward shift of the concentration-response curves following aerosol administration of histamine or methacholine. These chemicals trigger bronchospasm in normal individuals as well, but people with BHR have a lower threshold. In asthmatic patients, BHR results in a significant decrease in the provocative concentration of histamine or methacholine causing a 20% decrease in PC20. BHR is a hallmark of asthma but also occurs frequently in people suffering from COPD. In the Lung Health Study, BHR was present in approximately two-thirds of patients with nonsevere COPD, and this predicted lung function decline independently of other factors.

Kveimtest is a specific intradermal reaction for diagnosing sarcoidosis. Pre- and post-ß2-agonist inhalation spirometry is a test to evaluate airway reversibility.

Question 4

What is this patient's first-line drug?

A. Inhaled corticosteroids
B. Inhaled anticholinergic agent
C. Short-duration antitussives
D. Antileukotrienes
E. Inhaled long-acting $\beta2$ agonist

Answer: C

It is probable that the respiratory tract infection did not prolong because the respiratory tract infection developed 4 weeks ago, and azithromycin was effective. However, it is estimated that airway hyperresponsiveness is high in about 10% of the population. That is, the airways are highly responsive to nonspecific stimuli, and the respiratory tract infection can damage the airway mucosal epithelium due to severe cough, resulting in prolonged cough and sometimes sputum over several months. Patients often go to a medical institution because their cough does not go away even after taking over-the-counter antitussives and they cannot sleep due to severe coughing. In general, abuse of antitussives interferes with sputum discharge and may cause atelectasis, so its instant use during coughing is not recommended, especially in children. However, long-term coughing after an airway infection can damage the walls of the bronchial mucosa, which can lead to more severe and prolonged coughing. Persistent cough causes multiple damages to the airway mucosa, which can lead to further coughing, leading to prolonged airway mucosal damage, leading to a vicious cycle. Furthermore, there is concern about the possibility of recurrent airway inflammation due to secondary infection of the damaged airway mucosa. Especially, patients with BHR are expected to treat persistent cough by promoting healing of airway mucosal damage. Antitussives may be indicated for prolonged coughing that may interfere with daytime activity, or insomnia for more than 4 weeks. It aims to suppress the severe cough reflex by keeping the bronchial wall at rest and repairing damaged airway mucosa. The appropriate treatment is to prescribe antitussives and expectorants that stop coughing, promote bronchial mucosal repair, and promote sputum output for the minimum required period. Repeated and prolonged damage to the airway mucosa due to prolonged cough causes the damaged bronchial wall to become fibrotic through substrate formation, resulting in remodeling of the bronchi and bronchioles. The development of airway remodeling due to mechanical stress to the airways exerted by long-standing coughing may result in a vicious cycle of cough and remodeling, and imply the importance of control of coughing.

Question 5
What is this patient's second-line drug?

A. Inhaled corticosteroids
B. Inhaled anticholinergic agent
C. Short-duration antitussives
D. Antileukotrienes
E. Inhaled long-acting β2 agonist

Answer: A

Inhaled corticosteroids may relieve inflammation, but their anti-inflammatory effect on persistent severe airway inflammation and damage to the airway mucosal epithelium during the acute phase is inadequate. If necessary, concomitant use of expectorants and short-term cough suppressants may also be effective in relieving acute symptoms. In addition, if the respiratory tract infection persists, such as yellow sputum being excreted at the first diagnosis, it may be necessary to consider antibiotics such as azithromycin. Even if these treatments cure bronchial inflammation, increased airway hypersensitivity can prolong coughing. That is the clinical course of this case. If cough persists after the airway infection has completely healed, short-term administration of cough medications to rest the bronchi to promote repair of the damaged airway mucosa, and inhaled corticosteroids for the treatment of potentially related persistent inflammation may be effective for treating prolonged and complicated cough.

Case Presentation 2
28-year-old female. She had a history of childhood asthma until she was under the age of five, but she has never had an asthma attack since she was an adult. She has been allergic to cedar pollen for 15 years and has symptoms such as runny nose, nasal congestion, sneezing, eye congestion, tearing, and itching every year from late January to mid-April. She went to my family physician's office every year. From late January, her symptoms, such as runny nose, nasal congestion, and sneezing, as well as coughing, caused the patient to purchase and take a common cold medicine at a drugstore. As all her symptoms disappeared quickly, she self-diagnosed that she had acute upper respiratory tract inflammation and decided to continue taking her common cold medicine for some time. While taking the cold medicine, the symptoms disappeared, but when the medicine was stopped after 4 weeks or more after the onset, symptoms such as runny nose, nasal congestion, sneezing, coughing, and redness in the eyes recurred. It was also accompanied by hyperemia and itching. Her family doctor referred her to a respiratory department, especially because she had a severe dry cough without sputum. According to her findings at the time of her visit, she had a body temperature of 96.8F (36.0 °C). She breathes 16 breaths/minute. She had a blood pressure of 122/70, Sp02 of 95%, and a negative COVID-19 rapid antigen test. On auscultation, she had no abnormal breath sounds and she did not hear the second respiratory sounds. Her chest X-ray findings showed no abnormalities. Her spirometry test showed FVC: 102% and FEV1.0: 82%. The FEV1.0 improvement test with short-acting bronchodilators (SABA) was negative (<10%). Bronchial responsiveness to methacholine was measured. The provocative concentration of methacholine required to cause a 20% fall from baseline FEV 1 (PC 20) was 20 mg/mL. The capsaicin cough threshold, measured was 0.98 μ mol/L. This suggested cough reflex hypersensitivity. Bronchodilator therapy, inhaled salbutamol sulfate 200 μg on demand, was entirely ineffective for this patient's daily coughing. No history of food allergies. She has no dyspnea or coughing after exercise. She has no history of sinusitis, and no history of smoking.

Question 6
What is the most probable diagnosis for this patient?

A. Post respiratory infectious cough
B. Postnasal drip syndrome
C. Cough Variant Asthma (CVA)
D. Non-asthmatic eosinophilic bronchitis (NAEB)
E. Atopic Cough (AC)

Answer: E

AC is defined as an isolated chronic bronchodilator-resistant unproductive cough with atopic constitution, eosinophilic tracheobronchitis, and airway cough receptor hypersensitivity without BHR. However, there is an absence of BHR and variable airflow obstruction. AC is characterized by a lack of eosinophilia in bronchoalveolar lavage fluid during bronchial fiber examination.

Postinfectious cough persists >3 weeks after an acute, often viral airway infection and resolves after <8 weeks. As mentioned above, epithelial damage after infection, such as *B. pertussis* or *M. pneumoniae* infection, or a transient increase in BHR, later subsiding spontaneously, are responsible for postinfectious cough. The most important thing in confirming the diagnosis is to listen carefully to the clinical course.

PNDS was described as a common condition in Britain in the nineteenth century and was so prevalent in the United States that it was called the "American catarrh." American thoracic doctors use PNDS as the most common cause of chronic cough. The relationship between PNDS and chronic cough was unacceptable to British thoracic physicians who preferred to use the term "nasal sinusitis" instead of PNDS. In the United States, the diagnosis of PNDS was associated with a response to treatment with sedative antihistamines and decongestants, but British doctors doubt whether this is a particular treatment and did not accept treatment as a diagnosis of PNDS. In 2006, the American Thoracic Society replaced the term PNDS with upper airway cough syndrome, and some British otolaryngologists suggested replacing PNDS with nasal sinusitis. PNDS is now being replaced by a more general description of upper respiratory tract disease, and its causal link to chronic cough is being contested. PNDS can be caused by a hypersecretory phenotype that occurs after the airways are chronically exposed to particulate matter, allergens, irritants, and pathogens. Current research on the treatment of excess airway mucus in the lower respiratory tract may be applicable to PNDS. Physiological and pathological studies have revealed that this condition, formerly known as PNDS is part or synonymous with Upper Airway Cough Syndrome (UACS). PNDS is now called UACS and is caused by chronic rhinitis (allergic and nonallergic) or chronic sinusitis. Patients usually show runny nose, stuffy nose, and itchy throat. Purulent discharge and facial pain may indicate complications of sinusitis. A careful physical examination of the posterior oropharyngeal cavity may reveal the appearance of the cobblestones. Allergic rhinitis usually responds to antihistamine treatment and nasal steroid courses for at least 2 weeks. However, the management of persistent allergic rhinitis, which is common in Singapore, may require longer term treatment. If the cough does not subside and chronic sinusitis is suggested, the patient should be referred to an otolaryngologist for further investigation and management.

CVA is characterized by eosinophilic inflammation of the whole lower airways. Further, NAEB and AC also present with chronic cough and involve eosinophilic inflammation. However, patients with atopic cough rarely progress to asthma or suffer from chronic airflow obstruction. Treatment of AC could therefore be symptomatic and short-term. In contrast, long-term anti-inflammatory treatment with inhaled corticosteroids is considered essential and recommended by guidelines for patients with CVA, and this might also be the case with NAEB, because chronic airflow obstruction, ascribed to airway remodeling, and classic asthma may develop in both conditions.

NAEB is eosinophilic inflammation of the respiratory tract, without any bronchospasm. Eosinophilia is present in all induced or spontaneous sputum samples of NAEB patients. NAEB patients and asthmatic patients have similar airway inflammation. Remarkably, NAEB mainly occurs in the lower airways. Unlike asthma, mast cells in NAEB are active in the bronchial epithelium. Diagnosis is based on the clinical, radiological, and spirometric measurements of other causes of chronic cough and the assessment of inflammation in the lower respiratory tract. Airway inflammation can be assessed by sputum induction. The main treatment is anti-inflammatory therapy with inhaled corticosteroids and taking protective measures if inflammation is due to occupational exposure or allergen inhalation. If NAEB is untreated, it may be transient, episodic, or persistent in patients, long-term oral steroid treatment may be required.

Question 7

Which is the specific finding involved in AC?

A. Intermittent fever
B. Serum IgE elevation
C. BHR
D. Response to H1 antagonist
E. Sputum eosinophilia

Answer: D

Eosinophilic tracheobronchitis and cough hypersensitivity are pathological and physiological characteristics of AC. The clinical features of AC are as follows: (1) chronic bronchodilator-resistant nonproductive cough with "tickle" in the throat and/or "a sensation of irritation in the throat" lasting for more than 8 weeks; (2) absence of wheezing, dyspnea, hemoptysis, or pleurisy, and no adventitious lung sounds on physical examination; (3) presence of one or more global atopic findings, including past history and/or complication of allergic diseases except BA, family history of allergic diseases, peripheral blood eosinophilia, elevated total IgE level in the serum, positive specific IgE antibody to common aeroallergens, and positive allergen skin test; (4) presence of eosinophils in hypertonic saline-induced sputum and/or submucosa of biopsied trachea and/or bronchi; (5) normal limits of forced expiratory volume in 1.0 s (FEV1.0), forced vital capacity (FVC), and FEV1.0/FVC ratio; (6) no bronchial reversibility defined as less than a 5% increase in FEV1 after inhalation of 300 μg salbutamol following 250 mg aminophylline injection; (7) bronchial responsiveness within normal limits; (8) increased airway cough reflex sensitivity; and (9) complete relief of the cough upon treatment with histamine H1 antagonists. In addition, the main site of complaints in most patients with AC, "a sensation of irritation in the throat," is around the trachea in front of the neck region.

Question 8

If this patient tests positive in the methacholine inhalation test, which disease should be differentially diagnosed?

A. Chronic bronchitis
B. Hypersensitivity pneumonitis
C. Non-asthmatic eosinophilic bronchitis
D. Cough Variant Asthma (CVA)
E. Diffuse panbronchiolitis (DPB)

Answer: D

The methacholine inhalation test is the most widely used BHR test and is a very useful and safe test for the differential diagnosis of airway diseases.

Chronic Bronchitis is defined as a chronic cough and sputum production for at least 3 months a year for 2 consecutive years. Chronic bronchitis is classified as one of the types of chronic obstructive pulmonary disease (COPD). The COPD spectrum ranges from Emphysema to Chronic Bronchitis. Many patients have characteristics of both, putting them somewhere along the spectrum. Chronic bronchitis is thought to be caused by overproduction and hypersecretion of mucus by goblet cells. Epithelial cells arranged in the airways, which respond to injurious, infectious stimuli by releasing inflammatory mediators and proinflammatory cytokines. During an acute exacerbation of chronic bronchitis, the bronchial mucous membrane becomes hyperemic and edematous with diminished bronchial mucociliary function. This, in turn, leads to airflow impediment because of luminal obstruction to small airways. The airways become clogged by debris and this further increases the irritation. The characteristic cough of bronchitis is caused by the copious secretion of mucus in chronic bronchitis.

Hypersensitivity pneumonitis is one of the most common interstitial lung diseases, that presents unique challenges for a confident diagnosis and limited therapeutic options. The disease is triggered by exposure to a wide variety of inciting antigens in susceptible individuals which results in T cell hyperactivation and bronchioloalveolar inflammation. Antigen removal and treating the inflammatory process is crucial in the progression of the disease since chronic persistent inflammation seems to be one of the mechanisms leading to lung fibrotic remodeling. Fibrotic hypersensitivity pneumonitis has a few therapeutic options but evidence of efficacy is still scanty.

In CVA, cough is the sole presenting symptom. CVA is characterized by BHR. CVA remains the most common cause of chronic cough followed by sinus bronchial syndrome, gastroesophageal reflux disease, and AC. In the majority of patients with CVA, cough can be controlled with ICS. Especially, ICS not only control cough in CVA but also prevent the development of wheezing, airway remodeling, and chronic airflow obstruction. CVA is characterized by BHR and responsiveness to bronchodilators, but the presence of BHR is only consistent with, but not diagnostic of, CVA. Improvement of cough with bronchodilators such as beta-agonists is the essential diagnostic feature of CVA. Based on these features, several international cough management guidelines consider responsiveness to bronchodilators as the key diagnostic feature of CVA. Sputum eosinophilia suggests a diagnosis of CVA, as well as AC or NAEB. Exhaled NO levels may also be elevated and useful in the diagnosis of CVA. In addition, although CVA (in which the cough responds to bronchodilators) is recognized as a precursor of typical asthma, AC is not a precursor of asthma.

Diffuse panbronchiolitis (DPB) is characterized by chronic sinobronchial infection and diffuse bilateral micronodular pulmonary lesions consisting of inflammatory cells. Studies on disease etiology point to a genetic predisposition unique to Asians. Early therapy for DPB was largely symptomatic. Various pathophysiology has been proposed for this disease, but the most prevailing theory today suggests that DPB is a subtype of chronic bronchitis that causes multiple granulomas formation in the bronchi and bronchioles. The advent of macrolide antibiotics, including erythromycin, roxithromycin, and clarithromycin, has strikingly changed disease prognosis. Low-dose, long-term macrolide therapy for DPB originated from detailed observations of response to therapy in a single patient. The bactericidal activity of macrolides, particularly erythromycin, is not a significant factor for their clinical efficacy in DPB. Firstly, irrespective of bacterial clearance, clinical improvement is observed in patients treated with erythromycin. Secondly, even in cases with bacterial superinfection with Pseudomonas aeruginosa resistant to macrolides, treatment has proved effective. Thirdly, the recommended dosage of macrolides produces peak levels in tissue that are below the minimum inhibitory concentrations for major pathogenic bacteria that colonize the airway. In the last two decades, the possible mechanism underlying the effectiveness of macrolide therapy has been extensively studied. The proposed mechanism of action includes inhibition of excessive mucus and water secretion from the airway epithelium, inhibition of neutrophil accumulation in the large airway, inhibition of lymphocyte and macrophage accumulation around the small airway, and modulation of bacterial virulence.

Question 9

What is this patient's first-line drug?

A. Inhaled long-acting β2 agonist
B. Inhaled anticholinergic agent
C. Antihistamines
D. Antileukotrienes
E. Antitussives

Answer: C

Histamine H1 antagonists are effective against AC but not against other diseases. Eosinophilic airway inflammation and increased cough sensitivity without BHR are the pathologic and physiologic features of bronchodilator-resistant nonproductive cough-associated global atopic tendency, abbreviated herein as atopic cough. Histamine H1 receptor antagonists are effective in relieving the cough in nearly 60% of patients with atopic cough. Some patients presenting with chronic bronchodilator-resistant nonproductive cough have a global atopic tendency and cough hypersensitivity without BHR, abbreviated as AC. As mentioned above, AC does not show BHR, so not only is the methacholine inhalation test negative, but the histamine inhalation test can induce cough, but FEV1 is not reduced by more than 20%.

Question 10

What is this patient's second-line drug?

A. Antitussives
B. Corticosteroids
C. Antileukotrienes
D. Inhaled anticholinergic agents
E. Inhaled long-acting β2 agonists

Answer: B

Many cases of paroxysmal cough in AC can be successfully treated by oral or inhalation of histamine H1 antagonists. Moreover, in severe cases, inhalation or oral corticosteroids are effective when histamine H1 antagonists are ineffective or inadequate. However, it should be noted that the action of glucocorticoids is not fast-acting and is often not sufficient to stop acute paroxysmal cough. For this reason, it is also common to administer a combination of an H1 antagonist and a corticosteroid, aiming for a synergistic effect between the two. It is also noteworthy that inhalation of bronchodilators such as β-agonists does not suppress cough and does not improve FEV1.

AC is a major clinical problem. The causes of chronic cough can be categorized into eosinophilic and noneosinophilic disorders, and approximately 30–50% of people with chronic cough have eosinophilic airway inflammation. CVA is characterized by eosinophilic inflammation of the whole lower airways. Further, NAEB and atopic cough also present with chronic cough and involve eosinophilic inflammation. Patients with atopic cough rarely progress to asthma or suffer from chronic airflow obstruction. Treatment of atopic cough could therefore be symptomatic and short-term. In contrast, long-term anti-inflammatory treatment with ICS is considered essential for patients with CVA, and this might also be the case with both classic asthma and NAEB because chronic airflow obstruction ascribed to airway remodeling. Airway diseases that require differential diagnosis from AC are shown in (Table 2).

Case Presentation 3

A 9-year-old woman, her mother is Japanese, her father is German, and she lives in Shanghai for her father's business. She has been prone to catching colds since childhood, and she is seen by a nearby family doctor several times a year. She has a history of meconium ileus, and she is still prone to constipation daily. In her family history, her paternal grandfather died of cholestasis-type cirrhosis, and her sister died of acute respiratory distress due to pneumonia at the age of four. She returned to her mother's home in Tokyo during the summer vacation, but the week after her trip, she suffered from acute upper respiratory tract symptoms such as coughing, sore throat, and slight fever in the 37 °C range. She went

Table 2 Differential diagnosis of eosinophilic airway disorders

	Symptoms	Atopy[a]	BHR	Response to SABA/LABA	Response to H1 antagonist	Sputum eosinophils(>3%)
Asthma	Wheeze, SOB cough	60–80%	+	+	±	+
CVA	Cough only	40–80%	+	+	±	±
AC	Cough only	40–50%	–	–	+	±
NAEB	Cough and upper airway symptoms	20–70%	–	–	±	++

CVA cough variant asthma, *AC* atopic cough, *NAEB* non-asthmatic eosinophilic bronchitis, *AHR* airway hyperresponsiveness, *SABA* short-acting beta agonist, *LABA* long-acting beta agonist, *SOB* shortness of breath

[a]Defined by the presence of at least one positive serum-specific IgE or skin test response to common aeroallergens

to the emergency walk-in clinic. Her COVID-19 PCR test was negative.

Question 11
What are the most likely illnesses in this patient?

A. Charg-Strauss syndrome
B. Cystic fibrosis
C. Celiac diseases
D. Hypersensitivity pneumonitis
E. Diffuse panbronchiolitis

Answer: B
Cystic fibrosis (CF) is an inherited disease of the exocrine glands affecting primarily the gastrointestinal and respiratory systems. It leads to chronic lung disease, exocrine pancreatic insufficiency, hepatobiliary disease, and abnormally high sweat electrolytes. CF is the most common life-threatening genetic disease in the white population. In the US, it occurs in about 1/3300 white births, 1/15,300 black births, and 1/32,000 Asian American births. It is rarely seen in Asians. In the case of mixed-race children, such as this patient, the frequency of occurrence is different, and in the case of mixed-race with a race with a low incidence, such as Japanese, the problem is that the diagnosis rate is low. Approximately 50% of patients with cystic fibrosis are adults in the United States because advances in treatment have increased the life expectancy of people with cystic fibrosis. The incidence of cystic fibrosis is similar in boys and girls. Treatment is supportive through aggressive multidisciplinary care along with small-molecule correctors and potentiators targeting the cystic fibrosis transmembrane conductance regulator protein defect. Because of improved treatment and life expectancy, about 54% of patients in the US with CF are adults. CF is carried as an autosomal recessive trait by about 3% of the white population. Although the lungs are generally histologically normal at birth, most patients develop pulmonary disease beginning in infancy or early childhood. Mucus plugging and chronic bacterial infection, accompanied by a pronounced inflammatory response, damage the airways, ultimately leading to bronchiectasis and respiratory insufficiency. The course is characterized by episodic exacerbations with infection and progressive decline in pulmonary function.

Churg-Strauss syndrome is a disorder characterized by inflammation of blood vessels. This inflammation limits blood flow to organs and tissues and can sometimes cause permanent damage. This condition is also known as eosinophilic granulomatosis (EGPA) with polyangiitis. The most common symptom of Churg-Strauss syndrome is bronchial asthma.

Celiac disease is a condition in which the immune system attacks your tissues when you eat gluten. This damages your intestines (small intestine), so you cannot get the nutrients. Celiac disease can cause a variety of symptoms, including diarrhea, abdominal pain, and bloating. It is not associated with symptoms such as respiratory tract infections.

Hypersensitivity pneumonitis is an immune system disorder in which the lungs become inflamed as an allergic reaction to inhaled microorganisms, animal and plant proteins or chemicals. It may develop with changes in the environment, but there are no recurrent respiratory tract infections or genetic predisposition.

Diffuse panbronchiolitis (DPB) is an idiopathic inflammatory disease that is well recognized in Japan and primarily affects respiratory bronchioles, causing progressive purulent and severe obstructive respiratory disease because it is one of the diseases that require differential diagnosis from cystic fibrosis and is considered to be a type of chronic bronchitis with multiple granulomatosis formation. If left untreated, DPB progresses to bronchiectasis, respiratory failure, and death.

Question 12
Which is the most frequent symptom of this disease?

A. Cleft lip palate
B. Autism spectrum
C. Hiatal hernia
D. Nasal polyps
E. Funnel chest

Answer: D
Fifty percent of patients not diagnosed through newborn screening present with pulmonary manifestations, often beginning in infancy. Recurrent or chronic infections manifested by cough, sputum production, and wheezing are common. Cough is the most troublesome complaint, often accompanied by sputum, gagging, vomiting, and disturbed sleep. Intercostal retractions, use of accessory muscles for respiration, a barrel-chest deformity, digital clubbing, cyanosis, and a declining tolerance for exercise occur with disease progression. Upper respiratory tract involvement includes nasal polyposis and chronic or recurrent rhinosinusitis.

Question 13
Which is correct for this patient?

A. Decreased airway mucus secretion
B. Increased sweating
C. Present with chronic diarrhea
D. Exocrine pancreatic function is normal
E. Impaired glucose tolerance

Answer: A

In typical cases, meconium ileus often occurs shortly after birth. After that, poor digestion and malabsorption due to exocrine pancreatic insufficiency occurred, and respiratory tract infections were repeated, resulting in respiratory failure. The reabsorption of chloride ions in the sweat glands is impaired, resulting in high salinity in sweat. Organs are damaged and their severity vary, but in some patients only a single organ is damaged. Meconium ileus is found in 40–50% of CF patients. The highly viscous mucus impedes meconium excretion and causes impaired passage at the terminal ileum. Respiratory symptoms are found in almost all CF patients. After birth, highly viscous mucus accumulates in the bronchioles, and when pathogenic bacteria settle, bronchiolitis and bronchitis are repeated, resulting in respiratory failure. It causes purulent sputum production, coughing, and dyspnea. It is characterized by persistent infection with Pseudomonas aeruginosa. Exocrine pancreatic insufficiency is found in 80–85% of CF patients. The small pancreatic duct is occluded by an acidic secretion with a high protein concentration, and the pancreatic parenchyma gradually falls off. Changes begin in the womb, typically around the age of two with exocrine pancreatic insufficiency, resulting in steatorrhea, malnutrition, and underweight. Imaging findings often present with pancreatic atrophy or fat replacement. Cholestasis-type cirrhosis is found in 20–25% of CF patients.

Question 14
Which is the respiratory complication of this disease?

A. Bronchial asthma
B. Pulmonary fibrosis
C. Recurrent pneumothorax
D. Bronchiectasis
E. Lymphangioleiomyomatosis (LAM)

Answer: D

Pulmonary complications include pneumothorax, nontuberculous mycobacterial infection, hemoptysis, allergic bronchopulmonary aspergillosis (ABPA), and right heart failure secondary to pulmonary hypertension.

Abortive forms can manifest in adulthood for the first time through cough, bronchial infections, and bronchiectasis. LAM is a rare pulmonary disease characterized by low-grade cancer cell growth in the smooth muscle tissue of the lungs and abdomen. LAM affects almost only women around 35 years.

Bronchiectasis is defined by permanent and abnormal widening of the bronchi. As bronchiectasis is an acquired disorder, its pathophysiology is commonly described as distinct phases of infection and chronic inflammation. Bronchiectasis is also characterized by mild to moderate airflow obstruction. The interaction between these phases establishes a vicious circle in which the end result is the destruction of the bronchi and the accompanying clinical symptoms. Pulmonary damage is probably initiated by a diffuse obstruction in the small airways by abnormally thick mucus secretions. Bronchiolitis and mucopurulent plugging of the airways occur secondary to obstruction and infection. Chronic inflammation secondary to the release of proteases and proinflammatory cytokines by cells in the airways also contributes to lung injury. Airway changes are more common than parenchymal changes, and emphysema is not prominent. About 50% of patients have bronchial hyperreactivity that may respond to bronchodilators. In patients with advanced pulmonary disease, chronic hypoxemia results in muscular hypertrophy of the pulmonary arteries, pulmonary hypertension, and right ventricular hypertrophy. The lungs of most patients are colonized by pathogenic bacteria. Early in the course, Staphylococcus aureus is the most common pathogen, but as the disease progresses, Pseudomonas aeruginosa is frequently isolated. A mucoid variant of P. aeruginosa is uniquely associated with CF and results in a worse prognosis than nonmucoid P. aeruginosa. In the US, the prevalence of methicillin-resistant S. aureus (MRSA) in the respiratory tract is now about 27%. Patients who are chronically infected with MRSA have a more rapid decline in pulmonary function and lower survival rates than those who are not. Colonization with *Burkholderia cepacia* complex occurs in about 2.6% of patients and may be associated with more rapid pulmonary deterioration. *Nontuberculous mycobacteria*, including *Mycobacterium avium* complex and *M. abscessus*, are potential respiratory pathogens. Prevalence varies with age and geographic location and probably exceeds 10%. Differentiating infection from colonization can be challenging. Other common respiratory pathogens include *Stenotrophomonas maltophilia*, *Achromobacter xylosoxidans*, and *Aspergillus* species.

Question 15
Which assay is valuable in diagnosing this disease?

A. HER2 gene test
B. BRCA gene test
C. DaTscan
D. MIBG-myocardial scintigraphy
E. CFTR gene test

Answer: E

Diagnosis of CF is by sweat test or identification of cystic fibrosis-causing mutations in patients with a positive newborn screening test result or characteristic clinical features. CF is caused by a mutation in the CFTR (Cystic fibrosis transmembrane conductance regulator gene). The CFTR protein is the major anion channel of the luminal organs throughout the body. In CF, CFTR dysfunction impairs the transport of chloride (chloride ions) and water through the epithelium/mucosa of the airways, intestines, pancreatic ducts, bile ducts, sweat ducts, and vas deferens. As a result, the mucus/secretion in the lumen becomes excessively viscous, the lumen is occluded and infection is likely to occur, resulting in damage to multiple organs. There are more than 1900 genetic mutations reported so far, which vary by race and country. Even patients with the same gene mutation have different organs and severity of damage, and there are many unclear points in the mechanism of pathogenesis. The responsible gene has been localized on the long arm of chromosome 7. It encodes a membrane-associated protein called the cystic fibrosis transmembrane conductance regulator (CFTR). The most common gene mutation, F508del, occurs in about 85% of CF alleles; >2000 less common CFTR mutations have been identified. CFTR is a cyclic adenosine monophosphate (cAMP)-regulated chloride channel, regulating chloride, sodium, and bicarbonate transport across epithelial membranes. A number of additional functions are considered likely. Disease manifests only in homozygotes. Heterozygotes may show subtle abnormalities of epithelial electrolyte transport but are clinically unaffected. The CFTR mutations have been divided into six classes based on how the mutation affects the function or processing of the CFTR protein. Patients with class I, II, or III mutations are considered to have a more severe genotype that results in little or no CFTR function, whereas patients with 1 or 2 class IV, V, or VI mutations are considered to have a milder genotype that results in residual CFTR function. However, there is no strict relationship between specific mutations and disease manifestation, so clinical testing (i.e., of organ function) rather than genotyping is a better guide to prognosis. CFTR mutations can be frameshift (deletion or insertion in a DNA sequence that shifts the way a sequence is read) or nonsense (stop) mutations.

Human epidermal growth factor receptor 2 (HER2) is a member of the epidermal growth factor receptor family having tyrosine kinase activity. Dimerization of the receptor results in the autophosphorylation of tyrosine residues within the cytoplasmic domain of the receptors and initiates a variety of signaling pathways leading to cell proliferation and tumorigenesis. Amplification or overexpression of HER2 occurs in approximately 15–30% of breast cancers and 10–30% of gastric/gastroesophageal cancers and serves as a prognostic and predictive biomarker. HER2 overexpression has also been seen in other cancers like ovary, endometrium, bladder, lung, colon, and head and neck. The introduction of HER2-directed therapies has dramatically influenced the outcome of patients with HER2-positive breast and gastric/gastroesophageal cancers; however, the results have been proved disappointing in other HER2 overexpressing cancers.

One of the most important risk factors for breast cancer is family history of the disease, indicating that genetic factors are important determinants of breast cancer risk. About 5–10% of breast cancers are thought to be hereditary, caused by abnormal genes passed from parent to child. A number of breast cancer susceptibility genes have been identified, the most important being *BRCA1* and *BRCA2*. However, it is estimated that all the currently known breast cancer susceptibility genes account for less than 25% of the familial aggregation of breast cancer. In this chapter, we review the evidence for other breast cancer susceptibility genes arising from twin studies, pedigree analysis, and studies of phenotypes associated with breast cancer, and the progress towards finding other breast cancer susceptibility genes through linkage and association studies. Taken together, the available evidence indicates that susceptibility to breast cancer is mediated through variants in many genes, each conferring a moderate risk of the disease. Such a model of susceptibility has implications for both risk prediction and for future gene identification studies.

In 2011, the US Food and Drug Administration (FDA) approved a brain imaging test called DaTscan (DAT-SPECT: Dopamine transporter-single photon emission computed tomography) to help diagnose Parkinson's disease (PD). For some people, DaTscan can be a useful addition to the doctor's examination in diagnosing Parkinson's. A DaTscan is an imaging technology that uses small amounts of a radioactive drug to help determine how much dopamine is available in a person's brain. A SPECT scanner is used to measure the amount and location of the drug in the brain.

Iodine 123 (^{123}I) metaiodobenzylguanidine (MIBG) is the first-line functional imaging agent used in neuroblastoma imaging. MIBG uptake is seen in 90% of neuroblastomas, identifying both the primary tumor and sites of metastatic disease. The addition of single photon emission computed tomography (SPECT) and SPECT/computed tomography to ^{123}I-MIBG planar images can improve identification and characterization of sites of uptake. During scan interpretation, use of MIBG semiquantitative scoring systems improves description of disease extent and distribution and may be helpful in defining prognosis. Therapeutic use of MIBG labeled with iodine 131 (^{131}I) is being investigated as part of research trials, both as a single agent and in conjunction with other therapies. ^{131}I-MIBG therapy has been studied in patients with newly diagnosed neuroblastoma and those with relapsed disease.

2 Conclusion

Treatment of cough should always seek a causal treatment. However, if this approach is not possible on the first visit, such as an acute viral respiratory infection, or if it tends to be delayed, such as tuberculosis, cough treatment tends to always be empirical. Symptomatic treatment is unavoidable, especially if the causative approach, such as an acute respiratory infection, is not possible at the first visit, or if the test tends to be delayed due to cough due to a chronic condition. By coating the cough receptors in the throat, mucus removers are thought to have antitussive effects. Cough syrup, lozenges and drops, and honey share sugar as a common ingredient. Efficacy, if any, is time-limited to contact between the sugar and the receptor. This is usually 20–30 min. Systemic α-adrenaline agonists for nasal congestion are popular in the United States but are virtually unused internationally. Fixed combinations with older anticholinergic agents and centrally active antihistamines (i.e., chlorpheniramine or dexbrompheniramine) are not readily available in Europe. In addition, there is a lack of evidence of their effectiveness from randomized controlled trials. Antibiotics are only effective against cough caused by a bacterial infection characterized by purulent sputum (purulent bronchiectasis, bronchiectasis, exacerbation of COPD, purulent rhinitis, sinusitis). Antibiotics are not indicated for acute bronchitis. Inhaled and nasal corticosteroids and oral leukotriene antagonists reduce asthma, eosinophil bronchitis, postinfection cough with BHR, and rhinitis cough. Local anesthetics negate the electrophysiological activity of receptors and afferents (such as during bronchoscopy). They are increasingly used off-label in idiopathic cough and palliative medicine. Drugs that affect the central mechanism of cough (antitussives) include systemic morphine or codeine, as well as natural and synthetic derivatives (dextromethorphan, dihydrocodeine, noscapine, pentoxyverine). Some nonaddictive herbal remedies (Ribwort Plantain, Drosera rotundifolia) claim a central antitussive effect, which has not been proven in clinical studies. Opiates are recommended for effective symptomatic treatment of dry debilitating cough. They have a limited effect on the treatment of cough caused by the common cold. Expectorants are the most common drugs used for respiratory disease, such as ambroxol and N-acetylcysteine, which reduce the irritation of cough receptors by accumulating mucus by "coughing"). Efficacy is difficult to assess because there is no suitable method. Inconsistent evidence of the relative efficacy of various expectorants is present throughout the published literature. Symptomatologic use of expectorants is recommended to relieve cough in the case of mucous secretions (COPD and bronchiectasis). Many patients also report positive subjective efficacy using self-medication for acute bronchitis. Combined with phytotherapy, it can reduce the duration of acute coughing or colds. In bronchiectasis of CF, inhaled Dornase α relieves cough. Inhaled anticholinergic agents (i.e., ipratropium and tiotropium) are thought to reduce mucus production. However, their antitussive effects are inconsistent. Theophylline and β2-adrenergic agents increase mucous fimbria clearance but have no effect on coughing. Despite being a clinical routine in both hospital and outpatient care, and rehabilitation, there is insufficient evidence of the effectiveness of physiotherapy for cough. Pulmonary rehabilitation may be indicated depending on the pathophysiology of the disease. The purpose of physical therapy is (1) For patients with a productive but ineffective cough, use effective coughing techniques to increase sputum. (2) To spontaneously suppress unproductive cough. (3) Acapella® (DHD Healthcare, Wampsville, NY, USA), Flutter® (Desitin/ ScandipharmVarioRaw SA, Birmingham, AL, USA), RC Cornet® (BoniCur, East Court, UK). Quite a lot of involvement in allergy and immunology has emerged in the pathology of cough. In some cases, it has become clear that the mechanism of allergy and immunology is involved in the pathophysiology of diseases that were previously thought to be unrelated to allergy and immunology. Future progress in research should be ardently expected.

Bibliography

1. Irwin RS. Introduction to the diagnosis and management of cough: ACCP evidence-based clinical practice guidelines. Chest. 2006;129(1 Suppl):25S–7S.
2. Irwin RS, et al. Overview of the management of cough. Chest guideline and expert panel report. Chest. 2014;146(4):885–9.
3. Cockcroft DW, Davis BE. Mechanisms of airway hyperresponsiveness. J Allergy Clin Immunol. 2006;118:551–9.
4. Tarlo SM, et al. Diagnosis and management of work-related asthma: American College of Chest Physicians Consensus Statement. Chest. 2008;134:1S.
5. Irwin RS, et al. Chronic persistent cough in the adult: the spectrum and frequency of causes and successful outcome of specific therapy. Am Rev Respir Dis. 1981;123(4 pt 1):413–7.
6. Pratter MR. Chronic upper airway cough syndrome secondary to rhinosinus diseases (previously referred to as postnasal drip syndrome): ACCP evidence-based clinical practice guidelines. Chest. 2006;129(1 Suppl):63S–71S.
7. Reddel HK, et al. American Thoracic Society/European Respiratory Society Task Force on Asthma Control and Exacerbations. An official American Thoracic Society/European Respiratory Society statement: asthma control and exacerbations: standardizing endpoints for clinical asthma trials and clinical practice. Am J Respir Crit Care Med. 2009;180:59–99.
8. Mukae H, et al. The Japanese respiratory society guidelines for the management of cough and sputum (digest edition). Respir Investig. 2021;59(3):270–90.
9. Morice AH, et al. The diagnosis and management of chronic cough. Eur Respir J. 2004;24:481–92.
10. Morice AH, et al. ERS guidelines on the assessment of cough. Eur Respir J. 2007;29:1256–76.
11. Wilkes J. ACCP provides updated recommendations on the management of somatic cough syndrome and tic Cough. Am Fam Physician. 2016;93:416.
12. Irwin RS, Rosen MJ, Braman SS. Cough. A comprehensive review. Arch Intern Med. 1977;137:1186–91.
13. Irwin RS, Madison JM. The diagnosis and treatment of cough. N Engl J Med. 2000;343(23):1715–21.
14. Kardos P. Management of cough in adults. Breathe. 2010;7:122–1332.
15. Bolser DC. Cough suppressant and pharmacologic protussive therapy: ACCP evidence-based clinical practice guidelines. Chest. 2006;129:238S–49S.

Hypersensitivity Pneumonitis

Ria Gripaldo and Avanthika Thanushi Wynn

Abbreviations

AEP	Acute eosinophilic pneumonia
ANA	Antinuclear antibody
ANCA	Antineutrophilic cytoplasmic antibody
ARDS	Acute respiratory distress syndrome
BAL	Bronchoalveolar lavage
CBC	Complete blood count
CEP	Chronic eosinophilic pneumonia
CXR	Chest radiograph
EGPA	Eosinophilic granulomatosis and polyangiitis
HP	Hypersensitivity pneumonitis
HR	Heart rate
HRCT	High resolution computed tomography
IgE	Immunoglobulin E
IgG	Immunoglobulin G
IL	Interleukin
ILD	Interstitial lung disease
IPF	Idiopathic pulmonary fibrosis
MPO	Myeloperoxidase antibody
MUC5B	Mucin 5b
NTM	Nontuberculous mycobacterium
OSA	Obstructive sleep apnea
Th1	T helper cell type 1
Th17	T helper cell type 17
Th2	T helper cell type 2
UIP	Usual interstitial pneumonia

R. Gripaldo (✉) · A. T. Wynn
Division of Pulmonary, Allergy, Critical Care and Sleep Medicine, Department of Internal Medicine, Emory University School of Medicine, Atlanta, GA, USA
e-mail: ria.gripaldo@emory.edu; a.t.wynn@emory.edu

1 Introduction

Hypersensitivity pneumonitis, previously termed extrinsic allergic alveolitis, is a complex immune-mediated syndrome that occurs in susceptible individuals who have had an occult or overt exposure to antigens small enough to reach the alveoli (<5 μm). These antigens include microorganisms (bacteria, fungi, mycobacteria), animal proteins, and chemicals to which the individual is sensitized and then develops an exaggerated immune response that leads to pronounced pulmonary inflammation. It should be noted that HP can also occur after antigen exposure that may not be inhalational (gastrointestinal and skin) although this is less common. Both humoral and cellular mechanisms appear to contribute to the pathogenesis of HP. The susceptibility of the individual is an important factor as these antigens are universal and can be found in a variety of home, occupational and recreational environments yet only a proportion of exposed individuals develop the disease. A two-hit hypothesis has been proposed where a preexisting genetic or environmental susceptibility increases the risk for HP after exposure. Although numerous antigens have been identified to cause HP, the clinical presentation is similar regardless of the antigen but the clinical course for each patient is variable. Key clinical features of the disease include dyspnea, cough, and mid-inspiratory squeaks which may or may not be accompanied by systemic symptoms such as fever, chills, and weight loss. Clubbing may be identified in some patients with fibrotic disease. The onset may be acute or insidious and exacerbations may be recurrent. The clinical course may range from improvement and resolution of symptoms in some patients to progressive decline and the development of respiratory failure in others. HP has traditionally been classified as having acute, subacute, and chronic forms with the subacute form having considerable overlap with the acute and chronic forms. However, it is the presence of radiographic or histopathologic fibrosis that primarily determines prognosis hence HP has recently been classified as either fibrotic or nonfibrotic (2020 guidelines from the American Thoracic Society, Japanese Respiratory Society and Asociación Latinoamericana de

M. Mahmoudi (ed.), *Absolute Allergy and Immunology Board Review*, https://doi.org/10.1007/978-3-031-12867-7_9

Tórax) recognizing that some patients may have overlapping features. Although there is no clear consensus among experts regarding the definition, diagnostic criteria, and management of patients with HP, there is a consensus regarding the importance of a multidisciplinary approach to the evaluation and management of these patients. A thorough clinical history, compatible high-resolution chest imaging findings, and consistent histopathology when available are all important components of evaluation. Management includes antigen avoidance and immunomodulation. However, in a proportion of patients (exceeding 50% in some series), the causative antigen cannot be identified. Patients with fibrotic disease tend to have an unidentified antigen. Recently, antifibrotic therapy has been approved for chronic fibrosing interstitial lung diseases which includes progressive fibrotic HP. Lung transplantation may also be an option for select patients.

Case 1

A 71-year-old man with a history of OSA presented with cough, fatigue, fever, headache, shortness of breath, and a 15-pound weight loss occurring in 1 month. He noted that his respiratory symptoms would improve outdoors. He was initially evaluated by his primary care physician who presumptively diagnosed COVID-19 infection and recommended supportive care. His symptoms became progressive prompting an ER visit where he had an abnormal chest CT scan showing ground glass opacities.

2 Past Medical History

OSA on CPAP.

3 Social History

Former smoker.

20-pack-year history of tobacco abuse and quit 3 months prior.

Car salesman.

Denied any significant environmental exposures.

No pets.

4 Physical Exam

Blood Pressure: 145/75 mmHg.

HR: 106 bpm.

Respiratory Rate: 24.

Oxygen Saturation: 85%.

Temperature: 101.2 F.

Chest exam: Bronchial breath sounds bilaterally with mid-inspiratory squeaks.

5 Investigations

COVID-19 PCR—Negative.

Extended viral panel—Negative.

CBC—Mildly elevated white blood cell count with neutrophilia.

Procalcitonin—Normal.

C-reactive protein—Elevated.

PFT—FEV1 57%, FVC 50%, FEV1/FVC 84, TLC 53%, DLCO 33%.

CT chest—Bilateral airway-based ground glass nodules with peripheral ground glass opacities.

The appearance of his CT chest was highly suspicious for nonfibrotic acute HP. He was started on methylprednisolone at 1 mg/kg with a plan to taper slowly over several weeks. He had a rapid response to steroid therapy and by his 2-month outpatient follow up, his chest CT scan and PFTs had normalized (Figs. 1 and 2).

It should be noted that a careful inspection of his home led to the discovery of water damage beneath his carpet and visible mold on his flooring which was remediated. He has not had any further symptoms.

Question

Which of the following clinical features distinguishes between nonfibrotic and fibrotic HP on presentation?

A. Inspiratory squawks
B. Presence of hypoxemia
C. Presence of a known positive exposure
D. Systemic symptoms

Answer and Explanation

Answer: D

Nonfibrotic HP can present with prominent systemic symptoms including fever, malaise, myalgias, and weight loss. Fibrotic HP tends to have primarily pulmonary symptoms without systemic manifestations.

5.1 Clinical Presentation of Nonfibrotic HP

The systemic inflammatory reaction in HP can be quite pronounced with presenting features of fever, malaise, anorexia, weight loss, and tachycardia. Severe hypoxia can occur within hours to days of antigen exposure. Patients can also present with acute respiratory failure similar to ARDS that may require endotracheal intubation and mechanical ventilation. Patients are often unaware of their positive exposure. A thorough history focusing on their home environment, work exposures, hobbies, and pets can provide the key in identifying the potential antigen. Questionnaires listing known risk factors, antigens, and occupational associations can be of

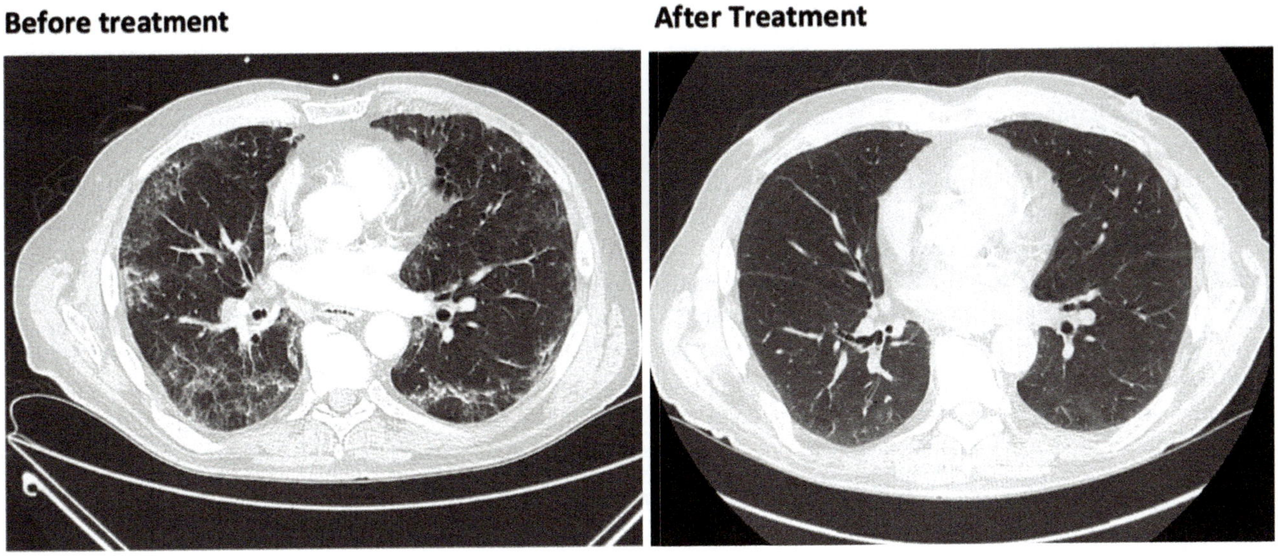

Fig. 1 HRCT scan of the chest showing bilateral airway-based ground glass nodules with peripheral ground glass opacities representing alveolitis that completely resolved after treatment with steroids

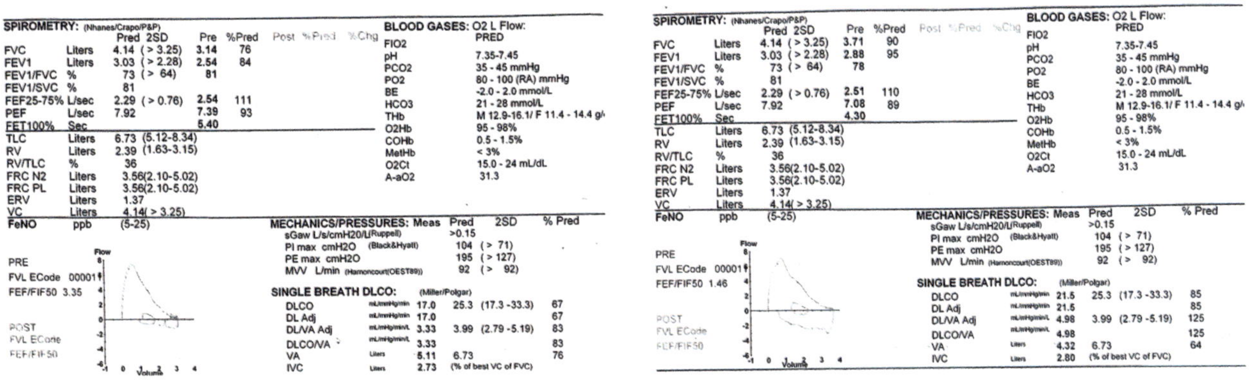

Fig. 2 Pulmonary function tests showing moderately severe restriction with severely reduced diffusion capcity that normalized after treatment with steroids

value. Table 1 shows a few examples of known HP antigens and associated HP syndromes.

The physical examination in nonfibrotic HP may include tachypnea, diffuse fine crackles, wheezing that is usually described as a high-pitched end-inspiratory wheeze known as a "squawk," which may be due to rapid oscillation during the opening of the small airways. If there is extensive parenchymal involvement, bronchial breath sounds can be heard as well.

Table 1 Examples of Antigens and associated Hypersensitivity Pneumonitis syndromes

Antigen	HP syndrome
Saccharopolyspora species	Farmer's lung
Alternaria species, wood dust	Woodworker's lung
Thermoactinomyces species	Humidifier lung Bagassosis (sugarcane) Air conditioner lung Mushroom worker's lung
Isocyanates (with host proteins) Haptens	Polyurethane foams and plastic workers
Penicillium species	Cheese washer's lung
Mycobacterium species	Machine operator's lung (contaminated metal) Wind instrument lung Hot tub lung

Question 2

Which of the following pulmonary diseases mimics nonfibrotic HP?

A. Community Acquired Pneumonia
B. ARDS
C. Viral pneumonitis

D. AEP
E. All of the above

Answer and Explanation
Answer: E
Onfibrotic hypersensitivity pneumonitis shares clinical and radiological features with multiple clinical entities including infectious pneumonia (whether viral, bacterial, or fungal). Unless there is a direct correlation between the exposure and symptom onset, nonfibrotic HP is almost always initially treated as pneumonia. Failure to respond to appropriate antimicrobials should raise the possibility of an alternative diagnosis. The presence of hypoxia, and physical exam findings of bronchial breathing and wheezing are common to many acute respiratory conditions. The imaging findings of patchy ground glass opacities and consolidations which may be present in acute HP are also nonspecific and may occur in autoimmune lung disease, sarcoid, drug-induced lung disease, acute eosinophilic pneumonia, and ARDS. A trained ILD radiologist or pulmonologist can help identify radiological patterns and help with narrowing the differentials.

Question 3
What of the following is the least helpful diagnostic modality in early acute hypersensitivity pneumonitis?

A. High-resolution CT scan
B. Serum precipitins
C. Bronchoscopy and bronchoalveolar lavage
D. Surgical lung biopsy
E. Pulmonary function testing

Answer and Explanation
Answer: D
Surgical lung biopsy is usually not performed in the hyperacute setting. During this period, a surgical lung biopsy may show only nonspecific inflammation without granulomas and may then cloud the diagnostic picture. In addition, patients may decompensate while getting a surgical lung biopsy during this period.

5.2 Diagnosis of Nonfibrotic HP

In terms of chest imaging, diffuse ground glass opacities may be the only feature evident in the acute alveolitis phase. Airway-centric ground glass nodules which are indicative of airway inflammation may eventually appear. The nodules can become denser over time and later exhibit features of parenchymal fibrosis.

The diagnostic value of serum precipitating antibodies is debatable as they can be negative in acute disease or seasonal disease (such as in farmer's lung). The panel is also not comprehensive and can underrepresent some exposures. A positive HP panel may also merely indicate exposure and sensitization rather than causation and should be interpreted with caution. Relying on a single positive test to identify the causative antigen can lead to misdiagnosis.

Pulmonary function testing typically shows restrictive physiology. The diffusing capacity may be severely reduced. A rapid improvement with steroids and continued antigen avoidance is not unusual since there is no fibrin deposition and architectural distortion at this stage. Imaging studies and pulmonary function tests can return to normal which is the case in the patient described in the vignette.

The bronchoalveolar lavage (BAL) in nonfibrotic HP is usually lymphocytic but may also show eosinophils although usually not to the degree that meets the diagnostic criteria for eosinophilic pneumonia.

Question 4
All of the following are associated with nonfibrotic HP except:

A. Smoking
B. CPAP machine use
C. Hot tub use
D. Humidifier
E. House with basement

Answer and Explanation
Answer: A
Nicotine has a paradoxical effect on HP through unproven mechanisms but possibly through the effects on nicotine receptors on lymphocytes causing a decrease in Th1 and Th17 responses and an increase in Th2 response. When HP does occur in smokers, it tends to result in more severe disease.

5.3 Exposure History

Eliciting a thorough exposure history is important. Although the exposure in this case was initially negative, a careful examination uncovered mold exposure which highlights the value of constant reevaluation of the home and workplace and even potential exposures from the patient's hobbies.

CPAP machine circuits when not changed and cleaned as directed can grow mold such as Aspergillus and Thermoactinomyces that are known as HP antigens. Humidifiers can similarly grow aspergillus and candida that can aerosolize. Mold can also be found under carpets, basements, behind wallpaper, and HVAC systems. Hot tub lung is a known HP entity that occurs from exposure to inhalation of water aerosol containing NTM and has been increasingly identified in residences and hotels with spa facilities.

Question 5
Which of the following have been implicated in the pathogenesis of HP?

A. HLA-DR and DQ polymorphisms
B. Over expression of HLA-DR2(15)
C. MUC5B promoter polymorphism
D. Short telomeres
E. All of the above

Answer and Explanation
Answer: E

5.4 Pathophysiology of HP

HP develops in susceptible individuals after exposure to one or more inciting agents. Less than 10% of patients develop HP after exposure to the same antigen which highlights the existence of predisposing factors inherently present in the patient. MHC class II polymorphisms have been identified to be present in patients susceptible to HP, specifically HLA-DR and DQ that have been associated with increased risk. In addition, the overexpression of HLA-DR2(15) was also seen in a small study of patients who developed the disease. Proteasomes, transporter proteins, and tissue inhibitors of matrix metalloproteinases have been studied with variable associations. In susceptible individuals who have been exposed and sensitized, the immune reaction is mediated by both humoral and cellular responses. Antigen-specific IgG antibodies and Th1 responses lead to a lymphocytic and granulomatous inflammation predominant in the non-fibrotic form of HP. Neutrophilic inflammation may also play a role in early disease but predominate in fibrotic disease. It is possible that the switch from Th1 to Th2 along with impaired T regulatory cell response are responsible for the development of fibrosis. The MUC5B promoter polymorphism and the presence of short telomeres may be associated with rapid progression from nonfibrotic to fibrotic HP.

5.5 Treatment of Nonfibrotic HP

Antigen avoidance and corticosteroids are the hallmarks of treatment. Alveolitis occurring within hours of exposure may respond rapidly to intravenous or oral steroids. Patients presenting within 1 week of onset who receive corticosteroid therapy may have complete resolution without any further need for treatment. In severe cases, mechanical ventilation may be necessary while awaiting steroid response. In general, patients who present acutely with rapid symptom onset and flu-like symptoms tend to respond to lower doses of steroids (less than 1 mg/kg doses) and will not require pulse doses. Patients who present sub-acutely may need higher doses with a longer duration of steroid taper.

Our patient presented after 4 weeks and was started on 1 mg/kg dose of methylprednisolone. He was discharged on 60 mg of prednisone which was tapered over 8 weeks. He still experienced complete resolution of his symptoms. His pulmonary function abnormalities, and pulmonary impairment and radiologic abnormalities also resolved with no further episodes after remediating the mold in his home.

Case 2
A 58-year-old woman with a history of mild intermittent asthma developed persistent dry cough and progressive shortness of breath over 3 years. She later presented to the hospital after developing a facial droop which resolved in several hours. During the admission, she was noted to have a mild peripheral eosinophilia, an elevated IgE level at 258 kU/L, and an abnormal chest computed tomography (CT) scan. Bronchoscopy was performed with bronchoalveolar lavage (BAL) showing 15% eosinophils. She was presumptively diagnosed with eosinophilic granulomatosis with polyangiitis (EGPA) and subsequently started on steroids and mepolizumab. There were no other signs and symptoms of vasculitis. Her lung disease continued to progress to the point where she required oxygen supplementation. Azathioprine was then added to her therapeutic regimen.

6 Past Medical History

Mild intermittent asthma.

7 Social History

Never smoker.

Has a home cleaning business and works with chemicals but uses appropriate precautions.

Has a 13-year history of parakeet ownership but recently gave them away.

8 Physical Exam

Blood Pressure: 125/65 mmHg.

HR: 92 bpm.

Respiratory Rate: 20.

Oxygen Saturation: 88%.

Chest exam: There were mild inspiratory squawks anteriorly and basal crackles posteriorly.

Extremity exam: Finger clubbing was present bilaterally.

9 Investigations

CBC—No peripheral eosinophilia.

BMP—Normal renal function.

Fig. 3 Pulmonary function test showing severe restriction

	Pre-Bronch				Post Bronch		
	Actual	Pred	%Pred	LLN	Actual	%Chng	%Pred
— SPIROMETRY—							
FVC (L)	0.80	2.90	27	2.21			
FEVI (L)	0.80	2.31	34	1.75			
FEV1/FVC (%)	100	80	125	68			
FEF 25% (L/sec)	2.84	4.71	60	2.57			
FEF 75% (L/sec)	1.98	0.63	312	0.25			
FEF 25-75% (L/sec)	2.76	2.21	125	1.15			
FEF Max (L/sec)	3.22	5.71	56	3.88			
FIVC (L)	0.74						
FIF Max (L/sec)	2.06						
—LUNG VOLUMES—							
SVC (L)	1.09	2.90	37	2.21	1.23	+12	42
IC (L)	0.56	1.93	28		0.82	+48	42
ERV (L)	0.54	0.96	55		0.22	−59	22
TGV (L)	1.43	2.58	55	1.53			
RV (Pleth) (L)	0.89	1.80	49	1.04			
TLC (Pleth) (L)	1.98	4.60	43	3.53			
RV/TLC(Pleth) (%)	45	38	116	28			
Trapped Gas (L)							
—AIRWAYS RESISTANCE -							
Raw (cmH2O/L/s)	0.73	1.86	39	1.15			
Gaw (L/s/cmH20)	1.42	1.03	137				
sRaw (cmH20*s)	1.06	<4.76					
sGaw (l/cmH2O*s)	1.00	0.20	486	0.14			

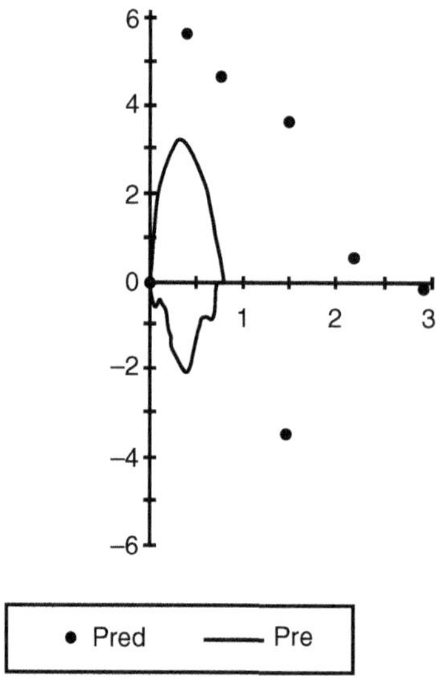

ANCA—Negative.

HP panel—Negative.

Avian panel—Positive for parakeet sera.

PFT—FEV1 34%, FVC 27%, FEV1/FVC 100, TLC 43% (Fig. 3).

CXR—Bilateral interstitial changes (Fig. 4).

HRCT scan of the chest—Bilateral ground glass opacities with reticulation along with mosaic attenuation and traction bronchiectasis (Fig. 5).

Question 1

What is the most likely diagnosis?

A. EGPA
B. Fibrotic HP
C. Idiopathic pulmonary fibrosis (IPF)
D. Chronic eosinophilic pneumonia (CEP)

Answer and Explanation

Answer: B

This patient's clinical history, exposure history, and her imaging studies are consistent with chronic/fibrotic HP.

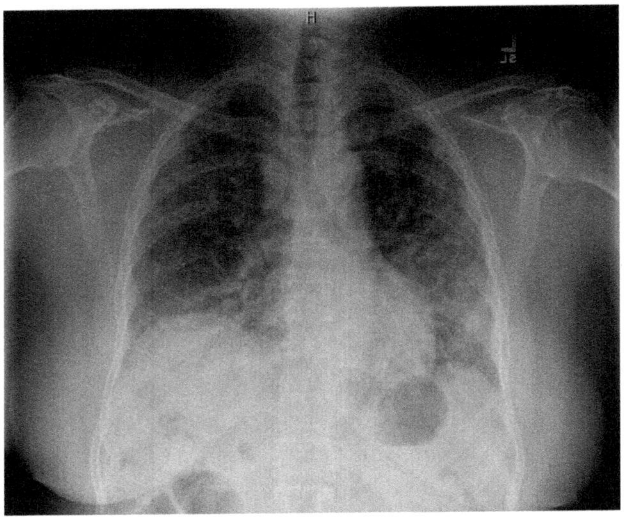

Fig. 4 Chest radiograph showing bilateral interstitial opacities

9.1 Clinical Presentation of Fibrotic HP

This patient has a history of progressive cough and shortness of breath in the setting of parakeet ownership. She had precipitating antibodies to parakeets which indicate sensitization. The latency period between exposure and sensitization followed by the development of hypersensitivity pneumonitis is variable and the 13-year history of bird ownership prior to disease development is still in keeping with HP. Her chest HRCT scan showed features consistent with fibrotic HP namely upper to mid lung zone involvement, airway-based disease process with pronounced mosaic attenuation, ground glass opacities and areas of normal lung (three density appearance/head cheese sign) There was also reticulation, and traction bronchiectasis indicating a chronic process.

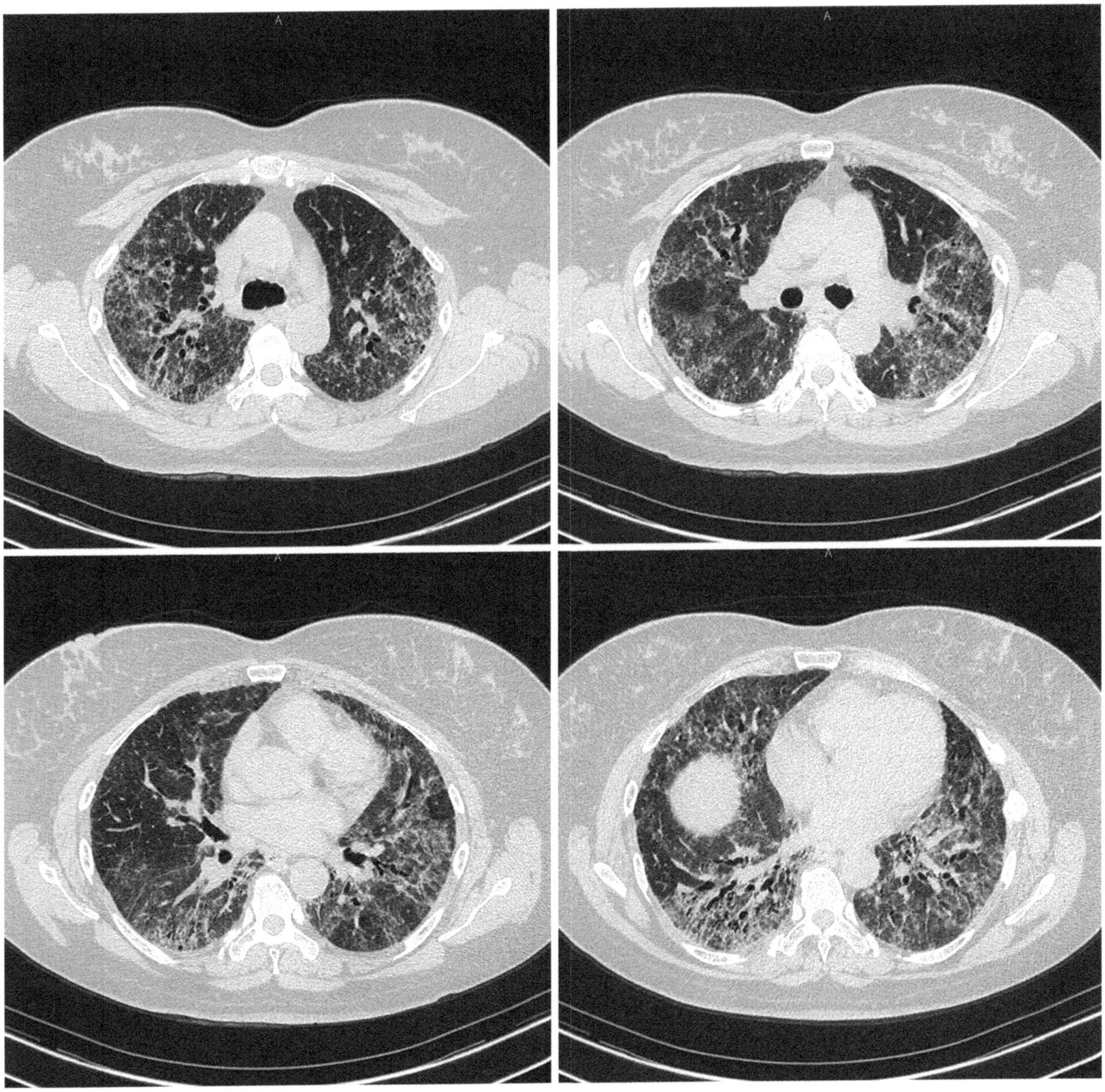

Fig. 5 High resolution computed tomography (HRCT) scan of the chest showing bilateral airway-based ground glass opacities with mosaic attenuation, reticulation, and traction bronchiectasis involving all lung zones

She was initially diagnosed with EGPA. The typical presentation of EGPA in patients who are in the vasculitic phase usually includes poorly controlled asthma, sinus disease, signs and symptoms of systemic vasculitis, marked peripheral and bronchoalveolar eosinophilia, highly elevated IgE (frequently upwards of 1000 kU/L), and histopathologic evidence of necrotizing eosinophilic granulomatous vasculitis. A proportion of patients will also be P-ANCA/MPO positive. The diagnosis of EGPA is a red herring in this case. This was diagnosed based on transient neurologic symptoms along with her history of asthma which was mild at that time. In addition, she had only mild peripheral and bronchoalveolar eosinophilia, and a modest elevation in her IgE. Although she had neurologic symptoms, there were no other clinical signs and symptoms of vasculitis. It is also notable that despite appropriate therapy for EGPA (steroids and mepolizumab) leading to eosinophil depletion, she still had progression of her lung disease.

Question 2
The following disease entities are possible differential diagnoses for HP except:

A. Pulmonary sarcoid
B. IPF
C. CEP
D. Bronchial asthma

Answer and Explanation
Answer: D
The exception is bronchial asthma. Asthma is an airway-based disease process and typically does not involve the pulmonary parenchyma and would not present with interstitial radiographic abnormalities.

9.2 Differential Diagnoses

Sarcoidosis is a multisystem granulomatous disorder of unknown etiology that commonly affects the lung. Pulmonary sarcoid can have both an interstitial and airway component. The radiologic appearance of pulmonary sarcoid can mimic that of HP in the upper lobe predominance, nodular changes, and bronchial involvement. Pathologically both sarcoid and HP demonstrate noncaseating granulomas except that the sarcoid granuloma is usually compact while the HP granuloma is usually poorly formed.

IPF is the most common type of idiopathic interstitial pneumonia usually occurring in older patients with a male predominance. The diagnosis of IPF requires the exclusion of other known causes of ILD (e.g., connective tissue disease, drug toxicity, domestic and occupational environmental exposures,). The distribution of fibrotic changes is basal predominant. Fibrotic HP can appear radiologically similar to IPF particularly in the later stages of disease when fibrosis can also involve the bases and can demonstrate hallmark features of IPF such as honeycomb change, reticulation, and traction bronchiectasis. IPF has the histological pattern of UIP. Fibrotic HP can also have a UIP appearance but the appearance of loose granulomas amidst the typical UIP pattern along with bronchocentric involvement helps distinguish between the two. Chronic eosinophilic pneumonia is an idiopathic disorder characterized by the marked abnormal accumulation of eosinophils in the interstitial and alveolar spaces of the lung. The radiologic appearance of CEP includes bilateral peripheral pleural-based consolidative and ground glass abnormalities that can proceed to fibrosis if untreated. The typical peripheral radiologic abnormalities for CEP occur in only 20% of patients. It can also present with upper lobe predominant bronchovascular changes similar to HP.

Question 3
The pathogenesis of HP involves immune dysregulation that results in the following features except:

A. Lymphocytic inflammation
B. Neutrophilic inflammation
C. Eosinophilic inflammation
D. Granulomatous inflammation

Answer and Explanation
Answer: C
Marked eosinophilic inflammation is not a prominent feature of HP.

9.3 Pathophysiology of HP

The pathophysiology of HP involves immune dysregulation in sensitized individuals who are exposed to an offending antigen. Humoral and cell-mediated responses both play a role in the immune reaction which consists of antigen-specific IgG antibody responses and Th1 responses (with contributions from TKR 2 and 9 pathways) that lead to a lymphocyte-mediated delayed hypersensitivity reaction with a resulting granulomatous inflammation. Neutrophilic inflammation also plays a role in the disease particularly in the fibrotic form where the persistence of neutrophils in the alveolar spaces and alveolar septa may contribute to ongoing lung injury by releasing potentially injurious proteolytic enzymes and oxygen free radicals. IL-17- and IL-22-secreting Th17 cells are also involved along with the impaired immunomodulatory function of regulatory T cells. A genetic predisposition has been suggested based on reports of familial clustering and associations between HP and the presence of specific MHC class II alleles. MUC5B rs35705950 minor alleles may also increase the risk for the development of fibrotic HP. In addition, telomere-related genes associated with short telomere lengths have been implicated in shortened transplant-free survival among fibrotic HP patients.

Question 4

The following features are specific for a diagnosis of HP except:

A. Sensitization to a known offending agent
B. Chest HRCT scan showing upper lobe predominant, airway-centric disease with nodularity and air-trapping
C. Pathologic findings of granulomatous inflammation
D. None of the above

Answer and Explanation

Answer: D

While all these features may be present in a patient with HP, none of them in isolation can make the diagnosis.

9.4 Diagnosis of HP

There are also no standardized diagnostic criteria for the diagnosis of HP therefore a comprehensive multidisciplinary approach is essential. However, the key features that consistently increase the likelihood of an HP diagnosis include (1) exposure to a known offending agent (based on clinical history with or without a questionnaire, serum IgG testing against potential antigens associated with HP, and/or specific inhalational challenge), (2) typical HRCT findings, and (3) consistent BAL/histopathology findings. Other features that have been associated with an HP diagnosis include female sex, the absence of smoking history, mid-inspiratory squeaks, and a restrictive or mixed spirometry pattern. However, these features are nonspecific and therefore have reduced diagnostic utility.

Although both fibrotic and nonfibrotic HP follow the same diagnostic algorithm, these patients have differing characteristics. Patients with nonfibrotic HP tend to have a more acute presentation and frequently also have systemic symptoms. They frequently have an identifiable offending antigen and their CT scans usually show centrilobular nodularity suggestive of small airway disease along with ground glass opacities and mosaic attenuation. They also tend to have lymphocytosis on BAL. On the other hand, patients with fibrotic HP tend to have a chronic clinical course with minimal to absent systemic symptoms. CT scans show fibrotic features such as traction bronchiectasis and honeycombing on a background of ground glass opacities and air-trapping. Their BAL analysis may show a nonspecific differential cell pattern.

Table 2 Distinguishing features between nonfibrotic and fibrotic HP

Nonfibrotic/acute HP	Fibrotic/chronic HP
B cell-mediated	T cell mediated
Acute onset of symptoms hours after exposure	Insidious onset with chronic symptoms
SIRS increased inflammatory markers	Inflammatory markers usually normal
Hypersensitivity panel can be negative	Higher possibility of HP panel being positive
Signature syndrome—farmer's lung	Signature syndrome—Bird fancier's lung
PFTs—Restriction and reduced DLCO	PFTs—Restriction and reduced DLCO
CT with diffuse infiltrates, airway-centric ground glass nodules	CT with airway-centric nodular opacities with reticulations and traction bronchiectasis
Rapid response to steroid and antigen avoidance	Insidious response to steroids and can continue years after the antigen is fully avoided

Histopathology also has some differences between nonfibrotic and fibrotic HP. The common features between the two entities include the presence of a cellular interstitial pneumonia, a cellular chronic bronchiolitis, a granulomatous inflammation that is comprised of poorly formed granulomas. However, in patients with fibrotic disease, there is also evidence of fibrosis that may overlap with a UIP pattern (presence of fibroblastic foci and subpleural honeycombing) (Table 2).

Question 5

Which of the following is associated with improved survival in HP?

A. Removal and avoidance of antigen
B. Absence of a smoking history
C. Judicious use of immunosuppression
D. Absence of fibrosis on chest HRCT scan

Answer and Explanation

Answer: A

The hallmark of management is removal and avoidance of the offending antigen which is associated with improved survival.

9.5 Management and Prognosis of HP

It has also been shown that the amount of antigen exposure influences outcomes. Patients with avian-related HP are more likely to deteriorate over time when exposed to higher amounts of antigen. Meanwhile, patients who are exposed to lower amounts of antigen tend to achieve disease stability. However, it should be noted that progression of lung disease can occur in some patients even with successful antigen

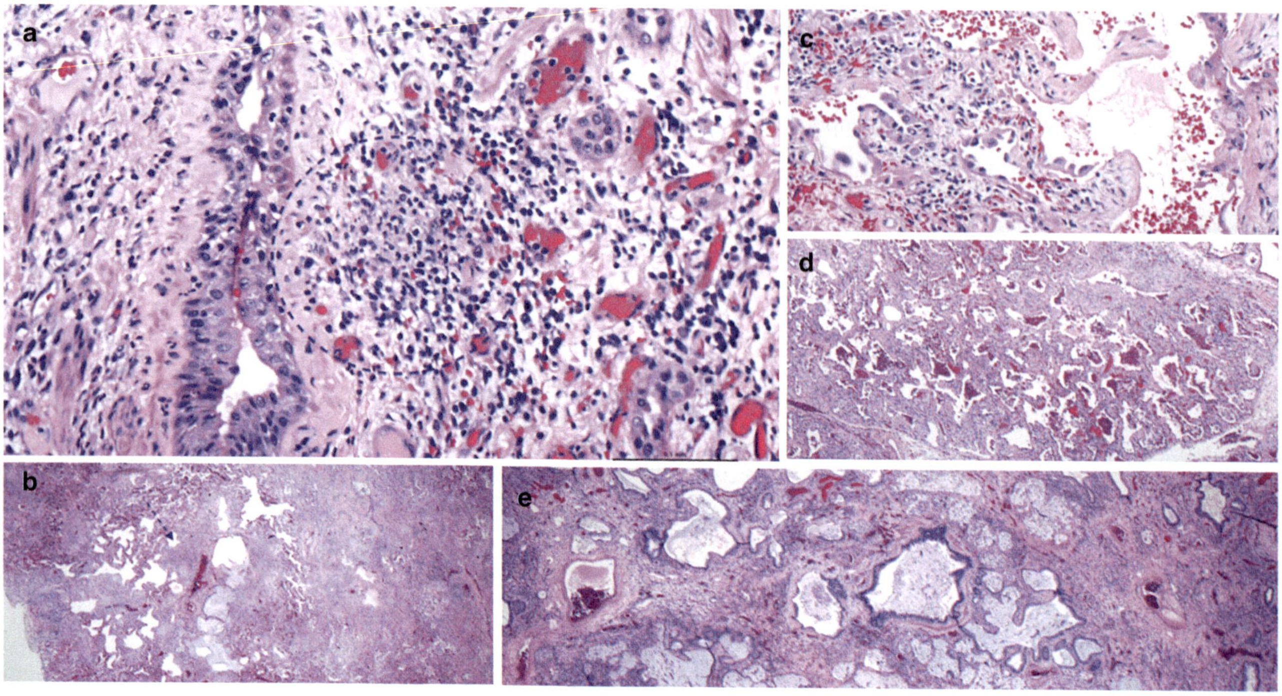

Fig. 6 (**a**) Poorly formed granuloma; (**b**) Lobule with airway-centered injury (arrow pointing at bronchovascular bundle in the center of the lobule); (**c**) Subacyte lung injury (with reactive pneumocyte hyperplasia); (**d**) NSIP pattern; (**e**) UIP pattern

avoidance. Antigen identification also confers a better prognosis, but this occurs in only 50% of patients.

HP is a disease that results from an exaggerated immune response hence glucocorticoid therapy is often employed to modulate this response particularly in patients who have inflammatory features (BAL lymphocytosis, ground glass opacities on HRCT and histopathologic features of cellular interstitial pneumonia, organizing pneumonia, or granulomatous inflammation). The initial dose is usually 0.5 mg/kg to taper over 8–12 weeks based on observational studies and expert opinion. In patients who have clinical deterioration upon weaning of glucocorticoids, a steroid-sparing agent may be added to aid in weaning. Common steroid-sparing agents include mycophenolate mofetil and azathioprine. Tacrolimus and cyclosporine have also been used. Rituximab, a humanized monoclonal antibody that binds to CD20 and depletes B cells, has also been reported to have some benefits in refractory chronic HP which highlights the humoral component of the disease.

In patients lacking inflammatory features, the benefit of immunosuppressive therapy is unclear. Patients may be considered for clinical trials, antifibrotic therapy (recently FDA approved Nintedanib), and lung transplantation although it is notable that HP may recur in the allograft of some patients.

The patient described in the vignette had already been started on glucocorticoids and azathioprine (for her presumed EGPA diagnosis). Although her chest HRCT scan showed ground glass changes, there is clear evidence of fibrosis which is unlikely to be reversible. In addition, her PFTs showed severe restriction and she was already oxygen-dependent. Although the offending agents had been removed and immunosuppression had been initiated, the decision was made to refer her for lung transplantation. She subsequently underwent a successful double lung transplant. The native lungs showed pathologic findings of chronic fibrosing lung disease with features of UIP (honeycomb change), some areas of non-specific interstitial pneumonia and also airway-centered injury. Rare non-necrotizing granulomas were also seen. Overall, these findings are consistent with chronic hypersensitivity pneumonitis (Fig. 6).

Bibliography

1. Ojanguren I, Ferraro V, Morisset J, Muñoz X, Fink J, Cruz MJ. Assessment and management of occupational hypersensitivity pneumonitis. J Allergy Clin Immunol Pract. 2020;8(10):3295–309.
2. Raghu G, Remy-Jardin M, Ryerson CJ, Myers JL, Kreuter M, Vasakova M, et al. Diagnosis of hypersensitivity pneumonitis in adults. An official ATS/JRS/ALAT clinical practice guideline. Am J Respir Crit Care Med. 2020;202(3):e36–69.
3. Selman M, Pardo A, King TE Jr. Hypersensitivity pneumonitis: insights in diagnosis and pathobiology. Am J Respir Crit Care Med. 2012;186(4):314–24.
4. Hariri LP, Mino-Kenudson M, Shea B, Digumarthy S, Onozato M, Yagi Y, et al. Distinct histopathology of acute onset or abrupt exacerbation of hypersensitivity pneumonitis. Hum Pathol. 2012;43(5):660–8.
5. Wahlström J, Berlin M, Lundgren R, Olerup O, Wigzell H, Eklund A, et al. Lung and blood T-cell receptor repertoire in extrinsic allergic alveolitis. Eur Respir J. 1997;10(4):772–9.

6. Spagnolo P, Rossi G, Cavazza A, Bonifazi M, Paladini I, Bonella F, Sverzellati N, Costabel U. Hypersensitivity pneumonitis: a comprehensive review. J Investig Allergol Clin Immunol. 2015;25(4):237–50.
7. Lacasse Y, Girard M, Cormier Y. Recent advances in hypersensitivity pneumonitis. Chest. 2012;142:208–17.
8. De Sadeleer LJ, Hermans F, De Dycker E, Yserbyt J, Verschakelen JA, Verbeken EK, et al. Effects of corticosteroid treatment and antigen avoidance in a large hypersensitivity pneumonitis cohort: a single Centre cohort study. J Clin Med. 2018;8:14.
9. Flaherty KR, et al. Nintedanib in progressive fibrosing interstitial lung diseases. N Engl J Med. 2019;381:1718–27.
10. Kern RM, Singer JP, Koth L, Mooney J, Golden J, Hays S, Greenland J, Wolters P, Ghio E, Jones KD, Leard L, Kukreja J, Blanc PD. Lung transplantation for hypersensitivity pneumonitis. Chest. 2015;147(6):1558–65.

Allergic Bronchopulmonary Aspergillosis

Mariam Majzoub and Rauno Joks

1 Introduction

Pulmonary Aspergillosis is a constellation of diseases caused by the fungus Aspergillus fumigatus. The host's immunity and presence of underlying lung diseases usually define the illness.

Invasive Pulmonary Aspergillosis is usually seen in immunocompromised hosts. Those with chronic lung diseases may suffer from Chronic Pulmonary Aspergillosis such as Aspergillus Nodule or Aspergilloma. Allergic bronchopulmonary aspergillosis (ABPA) should be always considered in the differential diagnosis of uncontrolled asthma or worsening cystic fibrosis. In this chapter, we present two cases of ABPA.

Using a case-based discussion, we will review the characteristics of Aspergillus fumigatus species and definition of ABPA. Then we will review the risk factors for ABPA, its pathogenesis, and how it presents clinically and radiographically. After that we will go over the differential diagnosis. Given the multiple diagnostic criteria, we will present the most recent and used criteria. Two tables will summarize the staging of ABPA. Finally, the management section provides valuable information on treatment and follow up.

Case 1

Mrs. Smith is a 51-year-old schoolteacher with hypertension, diabetes mellitus, and severe persistent asthma presenting to your clinic for evaluation of her allergies.

Over the past year, she was hospitalized twice on medical service for asthma exacerbations. She had multiple Emergency Department (ED) visits for asthma exacerbation and required more than five outpatient prolonged oral steroid courses. She states that her asthma was well controlled prior to this year. Mrs. Smith is adherent to her medications and demonstrates proper inhaler technique. She was diagnosed with asthma at age of 12, has lived in the same house for the past 30 years, and has no pets. Today she is doing fine and has no complaints.

Medications: Montelukast, Mometasone furoate/ Formoterol fumarate Inhaler once daily, albuterol HFA PRN and Ipratropium during exacerbations.

Question 1

Allergic bronchopulmonary aspergillosis (ABPA) is on your differential. Which of the following renders Mrs. Smith at increased risk of ABPA?

A. Female gender
B. Asthma
C. Increased exposure to aspergillus species (spp.)
D. Recurrent use of oral steroids

Answer and Explanation
The correct answer is B

2 Definition

ABPA is a progressive pulmonary disease caused by the colonization and infection of airways by the fungus Aspergillus fumigatus accompanied by a vigorous combined Type I and Type IV hypersensitivity responses.

ABPA occurs in almost 12.9% of asthmatics and 8.9% of cystic fibrosis (CF) patients. It usually manifests between third and fourth decades of life with no gender predilection.

No clear relationship has been established between exposure to Aspergillus species and occurrence of ABPA. High levels of outdoor spore counts may be associated with ABPA exacerbation.

Aspergillus fumigatus is an asexual, fast-growing, thermophilic species that habitats in soil, decaying vegetation, and stored grains. It grows through the production of hyphae from which sprout conidiophores. It secretes extracellular

M. Majzoub (✉) · R. Joks
Division of Allergy and Immunology, Department of Medicine, Center for Allergy and Asthma Research, SUNY Downstate Health Sciences University, Brooklyn, NY, USA
e-mail: rauno.joks@downstate.edu

M. Mahmoudi (ed.), *Absolute Allergy and Immunology Board Review*, https://doi.org/10.1007/978-3-031-12867-7_10

proteolytic enzymes such as proteases and elastases that assist in colonization of lung tissue.

Conidia (2.5–3.5 mm in diameter) are ubiquitously inhaled into the lower airways and alveoli where alveolar macrophages clear them in healthy subjects. In susceptible individuals, spores adhere to airway epithelium and initiate an inflammatory response.

Question 2
Which of the following genetic polymorphisms has NOT been associated with ABPA?

A. HLA-DR2 expression
B. HLA-DQ2 expression
C. Polymorphisms in the pulmonary surfactant protein A2
D. Mutations in the CF transmembrane conductance regulator gene

Answer and Explanation
The correct answer is B

3 Risk Factor for ABPA

1. Impaired local immunity: preexisting airway damage; presence of biofilm; impaired mucociliary clearance as seen in Cystic Fibrosis (CF), bronchiectasis, and asthma all contribute to increased adherence of spores to airway epithelium and impaired spore removal.
2. Inherited immune response properties/ genetics:
   - HLA-DR2 and/or HLA-DR5 expression.
   - HLA-DQ2 lack of expression.
   - IL-10 promoter polymorphisms.
   - Polymorphisms (ala91pro, arg94arg) in the collagen region of pulmonary surfactant protein A2.
   - Mutations in the CF transmembrane conductance regulator gene in subjects with asthma who did not have a diagnosis of CF.
   - IL4 polymorphisms.
   - TLR 9 polymorphisms.

Question 3
Which of these is NOT characteristic in the pathogenesis of ABPA?

A. Dendritic cells recognize Dectin 1 in the aspergillus cell wall
B. Th1 skewed inflammation is characteristic
C. Aspergillus fumigatus is a virulent fungus that secretes proteases involved in the airway damage
D. If left untreated, ABPA leads to airway remodeling with pulmonary fibrosis

Answer and Explanation
The correct answer is B

4 Pathogenesis (Fig. 1)

Question 4
All of the following clinical features are common in ABPA EXCEPT:

A. Severe hemoptysis
B. Worsening productive cough
C. Production of thick brownish sputum
D. Fatigue and weight loss
E. Fever

Answer and Explanation
The correct answer is A

5 Clinical Features

Symptoms
- Cough productive of thick brown sputum or mucus plugs.
- Systemic symptoms: fever, malaise, and weight loss are common.
- Wheezing.
- Dyspnea.
- Hemoptysis may occur but is rarely severe.

Asthma is reported in >90% of patients with ABPA.

Question 5
Which of the following radiographic features is pathognomonic for ABPA:

A. Transitory infiltrates on chest Xray or Ct scan
B. Bronchiectasis
C. High attenuation mucoid impaction on CT
D. Finger in glove on chest X-ray

Answer and Explanation
The correct answer is C

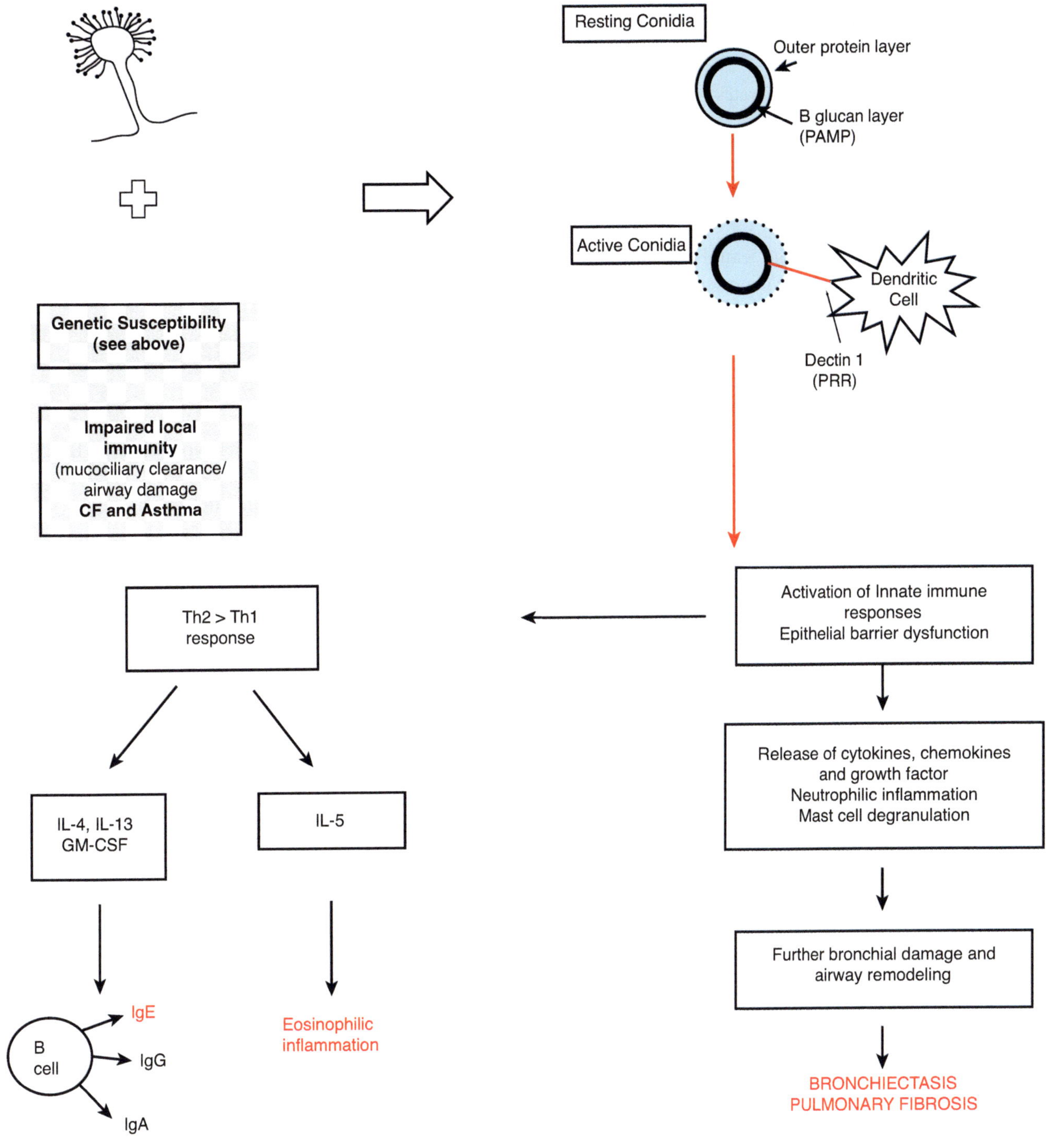

Fig. 1 Pathogenesis of ABPA—In genetically susceptible individuals and those with impaired local lung immunity, Aspergillus fumigatus spores are not cleared by alveolar macrophages. Growing (active) conidia disrupt the epithelial barrier and activate the innate immunity through antigen recognition by dendritic cells. A Th2 skewed inflammation occurs with release of cytokines that mediate eosinophilic inflammation and IgE production by B cells. Continued inflammation ultimately leads to neutrophilic inflammation and further bronchial damage characterized by pulmonary fibrosis and bronchiectasis. PAMP (pathogen-associated molecular pattern), PRR (pattern recognition receptor, Th1 and 2 (T helper cells type 1 and 2, respectively), IL (interleukin)

6 Radiographic Abnormalities Seen in ABPA

- Transitory pulmonary infiltrates—fleeting shadows (early).
- Lobar collapse.
- Mucus impaction.
- Bronchiectasis on HRCT (high-resolution CT scan).

Bronchiectasis involving large central airways with a predilection for the upper lobes are diagnostic of ABPA.

HAM: high attenuation mucoid impaction is pathognomonic for ABPA, usually seen on HRCT.

Case 2

A 60-year-old man with mild persistent asthma and allergic rhinitis on maintenance allergen immunotherapy with ragweed and dust mite presenting to your clinic for his monthly subcutaneous injection.

Patient is not feeling well today. He is coughing with the production of yellowish-brown tinged sputum. He also reports feeling more tired than usual. Symptoms have been going on for two weeks now.

He is using albuterol inhaler 2–3 times daily and reports medication adherence.

Patient just completed a course of oral steroids 4 weeks ago. Over the past 6 months, he was treated twice in the ED for asthma exacerbations and completed four courses of oral steroids. Fluticasone propionate/Salmeterol was added to his regimen with minimal improvement. He is on Montelukast too.

Prior to that patient had 1–2 exacerbations per year usually around ragweed season treated with oral steroids.

On your exam patient has scattered wheezing and oxygen saturation is 96% on room air.

Given that his asthma is uncontrolled, you decided to order some tests to aid you with diagnosis.

Initially, you performed a skin prick test and was reactive to Aspergillus fumigatus.

Question 1

Given Aspergillus fumigatus reactivity on skin prick, your patient is less likely to have something else other than ABPA.

A. True
B. B- False

Answer and Explanation
Correct answer is B

Aspergillus fumigatus reactivity on skin prick aids but does not confirm the diagnosis of ABPA.

7 Differential Diagnosis

1. SAFS: Severe Asthma with Fungal Sensitization.
   Occurs exclusively in severe asthma.
2. Asthma or cystic fibrosis exacerbation secondary to bacterial or viral pneumonia.
3. Acute or chronic eosinophilic pneumonia.
4. Churg-Straus Syndrome.

Question 2

What is the single test that will help you rule out ABPA?

A. Nonreactive Aspergillus skin prick testing
B. Normal chest radiography
C. Low total serum IgE level
D. Normal eosinophil count on CBC

Answer and Explanation
The correct answer is C

8 Diagnosis of ABPA

Absence of sensitization to *Aspergillus* excludes ABPA.

Presence of *Aspergillus*-specific IgE alone is insufficient for the diagnosis.

Question 3

Your patient's total IgE is 1280 IU/ml ruling out SAFS. He has asthma and skin reactivity to Aspergillus fumigatus.

Which of these is required for the definitive diagnosis of ABPA?

A. Absolute eosinophil count >500 on CBC
B. Aspergillus fumigatus-specific IgE > 0.35 KUA/ L
C. Aspergillus fumigatus-specific IgG > 27 mgA/L
D. Bronchiectasis on CT scan

Answer and Explanation
The correct answer is B

9 Diagnostic Criteria of ABPA

Modified ISHAM Criteria for ABPA Diagnosis in Asthmatics (2020)

All of the following:
- Asthma
- Aspergillus Fumigatus specific IgE > 0.35 KUA/L
- 3- Serum Total IgE > 500 IU/ml

2 or more of the following:
- Aspergillus specific IgG >27 mg A/L
- Eosinophil count >500 cell/ml
- Bronchiectasis on chest CT
- Mucus Impaction on chest CT

ABPA diagnosis in cystic fibrosis
Clinical deterioration (increased cough and sputum production, wheezing, decreased pulmonary function)
Total serum IgE >1000 KU/L (unless receiving systemic steroids)
Elevated IgG antibodies to aspergillus (or precipitating antibodies)
Sensitivity to aspergillus fumigatus (positive skin prick test or elevated serum aspergillus-specific IgE)
Radiographic abnormalities: Pulmonary infiltrates, mucus plugs

10 Staging

10.1 Clinical Staging

Stage	Radiographic findings	Total serum IgE
I—Acute	Homogeneous infiltrates, tree in bud findings, mucus plugs, consolidations, lobar collapse, bronchiectasis	Elevated
II—Remission	No infiltrates	Normal or elevated (less than stage I)
III—Exacerbation	As in stage I	Elevated (double that of stage II)
IV—Steroid-dependent asthma	No infiltrates Atelectasis, hyperinflation	Elevated or normal
V—End stage-fibrotic	Scarring, hyperinflation, fibro-cavitary lesions	Normal or elevated

10.2 Radiological Staging

Stage	Clinical features	Airflow obstruction
ABPA-S	Serologic Pt meets all criteria except for central bronchiectasis	Mild
ABPA-CB	Serologic + central bronchiectasis	Moderate
ABPA-CB-ORF	Central bronchiectasis + pulmonary fibrosis/ground glass opacities, fibro-cavitary lesions	Severe

Question 4

You diagnosed your patient with ABPA, what is the treatment of choice?

A. Inhaled corticosteroids
B. Antifungals
C. Systemic steroids
D. Exposure prevention

Answer and Explanation
The correct answer C

11 Management

11.1 Treatment

1. Oral Corticosteroids:
   - Mainstay of treatment.
   - Common strategy: 2 weeks of daily prednisone 0.5 mg/kg, followed by 6–8 weeks of alternate day therapy and then tapering by 5–10 mg every 2 weeks.
   - More aggressive approaches with higher doses of oral steroids or pulsed IV methylprednisolone have been used too.
   - High doses of [inhaled corticosteroids] alone have no role in the management of ABPA-S and should not be used as first-line therapy.
2. Antifungals:
   - Adjunctive treatment.
   - Recommended for steroid-dependent patients and relapses.
   - Benefits:
     - Decreases the burden of fungal colonization.
     - Significantly reduces total serum IgE, sputum eosinophils.
     - Improves symptoms.
     - **Reduces the requirement for prolonged high-dose systemic corticosteroids**

- Usual regimen: Itraconazole 200 mg twice daily for 4–6 months then tapered over the next 4–6 months.
- Alternatives: voriconazole, posaconazole.

3. Biologics:
   - Possible benefits of omalizumab and dupilumab based on randomized control trials and studies.

Question 5:
You initiated oral steroids on your patient. You contact him after 2 weeks and he reports improvement in his symptoms. You schedule a follow up visit in 4 weeks. Which of these is the most sensitive indicator of remission?

A. Decline in serum total IgE of 35% after 6 weeks
B. Resolution of pulmonary infiltrates on HRCT after 4–8 weeks of initiation of steroid treatment
C. Normalization of eosinophil count on CBC
D. Nonreactive skin testing for A fumigatus

Answer and Explanation
The correct answer is A

Monitoring and Follow Up

- The most sensitive indicator of disease progression is serial measurements of total IgE.
- A decline in serum total IgE of 35% is considered diagnostic of achieving remission of ABPA.
- Doubling of serum total IgE is considered diagnostic of relapse of ABPA, especially in CF patients.
- Patients are considered in remission when they remain without pulmonary infiltrates and/or eosinophilia for 6 months after oral steroid withdrawal.

How to Follow Up?

- Obtain Total serum IgE every 6–8 weeks after the initiation of oral steroids and for 1 year thereafter.
- Obtain chest imaging, either by CXR or HRCT after 4–8 weeks of initiation of oral steroid therapy to assess resolution of infiltration.
- Spirometry is a useful tool to objectively assess response to therapy.

Key Points

- Always think of ABPA when you have a patient with uncontrolled asthma.
- Diagnosis: asthma or cystic fibrosis, elevated total and Aspergillus-specific serum IgE.
- Absence of Aspergillus sensitization excludes ABPA.
  Bronchiectasis of large airways with upper lobe involvement is characteristic of ABPA.
  Oral Steroids are the mainstay of treatment. Antifungals are adjunctive.
- Total serums IgE:
  35% decrease indicates remission.
  Doubling indicates relapse.

Bibliography

1. Hew M, Douglass JA, Tay TR, O'Hehir RE. Middleton's allergy e-book, 9th Edition - Chapter 58: Allergic bronchopulmonary aspergillosis, hypersensitivity pneumonitis, and epidemic thunderstorm asthma. Elsevier - OHCE; 2019.
2. Greenberger PA, Bush RK, Demain JG, Luong A, Slavin RG, Knutsen AP. Allergic bronchopulmonary aspergillosis. J Allergy Clin Immunol. 2014;2(6):703–8.
3. Latgé JP, Chamilos G. Aspergillus fumigatus and aspergillosis in 2019. Clin Microbiol Rev. 2019;33(1):e00140–18. https://doi.org/10.1128/CMR.00140-18. PMID: 31722890; PMCID: PMC6860006
4. Bains SN, Judson MA. Allergic bronchopulmonary aspergillosis. Clin Chest Med. 2012;33(2):265–28.
5. Saxena P, Choudhary H, Muthu V, Sehgal IS, Dhooria S, Prasad KT, et al. Which are the optimal criteria for the diagnosis of allergic bronchopulmonary aspergillosis? A latent class analysis. J Allergy Clin Immunol Pract. 2021;9:328–35.

Food Allergy

Catherine A. Popadiuk and Doerthe A. Andreae

Case 1

Four-month-old monozygotic twin girls are referred to an allergist after their pediatrician became concerned about a peanut allergy in both twins. No testing was completed by the pediatrician. Both twins were born at term, to a healthy 37-year-old mother without complications. There is no family history of atopic or allergic disease. Moderate eczema has been noted in both children since the age of 2 months. Their parents have been controlling flares with topical steroids that are used about twice weekly in addition to daily lukewarm baths and application of a thick moisturizing cream after bathing. About 6 weeks prior to the visit with the allergist, one twin developed hives around her mouth and neck after being kissed by her grandmother who was eating peanuts at the time. She developed no other symptoms and the hives resolved within a few minutes of receiving diphenhydramine 1 mg/kg. She had not ingested peanuts or tree nuts prior to that incident. Neither of the twins has eaten peanut or tried any tree nuts since this event because of parental concern about an allergy to peanut and tree nuts.

Question 1

Which of the following symptoms are commonly associated with the presentation of a food allergy?

A. Urticaria and angioedema
B. Vomiting and diarrhea
C. Rhinorrhea, conjunctivitis, and sneezing
D. Wheezing
E. All of the above

The correct answer is E

Question 2

Which of the following scenarios warrant further testing for an IgE mediated food allergy?

A. All infants with moderate to severe atopic dermatitis
B. Patients with symptoms suggestive of anaphylaxis within an hour of ingesting a highly allergenic food
C. The sibling of a child with atopic dermatitis who has not had any symptoms concerning a food allergy
D. A patient with lactose intolerance
E. A and B
F. All of the above

The correct answer is E

Awareness of food allergies has increased within the last few decades, especially in the Western hemisphere. Incidence varies across age groups and in populations worldwide. Recent studies have documented the prevalence of food allergies as high as 10% in most populations worldwide with food allergies far more prevalent in children. By survey sampling, the prevalence of food allergy in US adults has been estimated at 10.8% with the most common allergens reported in order of prevalence as shellfish, peanut, cow's milk, tree nuts, and finned fish. By contrast, food allergen prevalence in US children is estimated at 7.6% with 40% of these children having more than one food allergy. The most common food allergens reported in US children include peanut and cow's milk followed by shellfish, tree nuts, egg, and finned fish.

For comparison, in recent European survey studies, prevalence of food allergy in adults was as high as 5.6% in Switzerland and lowest at 0.3% in Greece with just as wide a variation in the most common food allergen. Also, unlike the

C. A. Popadiuk
Division of Pulmonary, Allergy and Critical Care, Department of Medicine, Penn State College of Medicine, Milton S. Hershey Medical Center, Hershey, PA, USA
e-mail: cpopadiuk@pennstatehealth.psu.edu

D. A. Andreae (✉)
Division of Allergy and Immunology, Department of Dermatology, University of Utah, Salt Lake City, UT, USA
e-mail: adriana.andreae@hsc.utah.edu

M. Mahmoudi (ed.), *Absolute Allergy and Immunology Board Review*, https://doi.org/10.1007/978-3-031-12867-7_11

US findings, fruits and vegetables predominated as the major causative food allergens across Europe.

Numerous studies have sought to identify risk factors for the development of food allergies and understanding of this has improved in recent decades. The most studied and best-described risk factors include skin barrier compromise as in conditions like atopic dermatitis as well as age at first ingestion of a highly allergenic food. Other risk factors to consider which may or may not be modifiable include the intestinal microbiome, exposure to environmental allergens, and maternal dietary habits during pregnancy and lactation.

In the simplest of explanations, food allergy is the result of the loss of immunologic tolerance to an ingested food. The clinical manifestations we typically associate with a food allergy are mediated by IgE, a Type I hypersensitivity reaction. While the loss of immunologic tolerance may result in IgE mediated, non-IgE-mediated, or mixed IgE and non-IgE-mediated reactions, the focus of this chapter will be on IgE mediated food allergies. Sensitization to a food allergen is the process of developing food antigen-specific IgE. Repeated exposure to the allergen then induces an IgE-mediated hypersensitivity reaction. Symptoms manifest quickly following ingestion of the culprit food. We will first briefly discuss the development of tolerance and sensitization followed by a clinical presentation and diagnosis of food allergies.

Tolerance can be induced via the gastrointestinal tract or often through the skin, respiratory system, and oral cavity. For the purposes of this chapter, we will focus on the gastrointestinal tract and skin. In the gastrointestinal tract, food proteins are broken down by gastric acid and digestive enzymes leaving behind a few food proteins usually in the form of peptides which are transported across the gut lumen to dendritic cells. Dendritic cells will internalize and present these peptides as food antigens to naïve T cells in the lymphatic tissue of the gut. Under regular circumstances, this process leads to tolerance of the food proteins. If this process causes T cell release of cytokines leading to B cell proliferation, isotype switching, and the formation of IgE to food antigens it will lead to sensitization to the food proteins, now considered food allergens.

Loss of immunologic tolerance also occurs due to breakdown in skin barriers. The twins presented in the above case both suffer from atopic dermatitis (eczema). Skin breakdown especially in patients with atopic dermatitis allows for sensitization to occur via the skin as opposed to the GI tract due to the breakdown of filaggrin which allows for bacterial colonization or infection, epidermal water loss, and allergen penetration through the skin leading to sensitization.

Sensitization allows for the development of IgE antibodies to a food allergen, usually a protein. IgE will bind to high-affinity receptors on mast cells and basophils. When this IgE encounters a food allergen, it mediates cross linking of mast cells or basophils which then release mediators leading to a hypersensitivity reaction. These mediators include histamine, tryptase, leukotrienes, prostaglandins, and platelet-activating factors. Together, they are responsible for the signs and symptoms of allergic reactions including urticaria, angioedema, and smooth muscle contraction which may cause wheezing and/or abdominal pain.

Diagnosis of food allergy can be complex. In the US, guidelines for diagnosis and management of food allergies are published by the National Institute of Allergy and Infectious Diseases (NIAID) to promote best practices nationwide. As discussed above, food allergy is usually suspected when typical signs and symptoms of an immediate IgE mediated hypersensitivity reaction occur rapidly following ingestion of a likely food allergen. IgE-mediated reactions are usually quick in onset (minutes to almost 2 h after ingestion) and can manifest with a variety of symptoms depending on the organ system involved. As previously discussed, symptoms are triggered by an IgE-mediated reaction and typically affect the skin, gastrointestinal system, respiratory, and/or cardiovascular system. When the onset of symptoms occurs within minutes to 2 h following ingestion of an allergenic food it is highly suspicious of an IgE-mediated food allergy and warrants further investigation by an allergist.

A thorough history is imperative to correctly diagnose a food allergy. Clinical history and evidence of IgE-mediated sensitization to the food or foods in question are required for establishing a diagnosis of food allergy. For example, a child who is otherwise well presenting with a history of urticaria and vomiting within 30 min of ingestion of cashew for the first time would be concerning for a new onset tree nut allergy. However, a child who previously tolerated tree nuts now presenting with symptoms of viral illness for several days including vomiting and urticaria would be less concerning for a new onset food allergy and his/her presentation is more consistent with hives triggered by viral illness. Because of a high probability of false positive results, food allergy screening is not recommended and testing for any food allergy should only be performed in the appropriate clinical scenario based on history.

In our case, Twin A presents with new onset urticaria in the setting of what the parents believe is her first exposure to peanut without other symptoms; leading to concern for a food allergy. It is therefore imperative to note, that all testing for confirmation of a food allergy needs to be properly based on a thorough history which is suggestive of a food allergy.

Due to the twins' concerning history in the setting of moderate atopic dermatitis, skin prick testing for peanuts is performed in the clinic: Twin A, because of her history of peanut reaction and Twin B, because of her moderate atopic dermatitis. Results are reflected in the table below along with positive (histamine) and negative (diluent) controls:

	Peanut SPT (mm)	Diluent (mm)	Histamine (mm)	Serum IgE to Peanut (kU/L)
Twin A	5/15	0/0	10/50	0.8
Twin B	15/45	2/5	20/50	34

As discussed in the above section, all patients with symptoms suggestive of an IgE-mediated reaction following ingestion of a culprit food warrant further testing. In addition to the presence of moderate to severe atopic dermatitis in infants, current guidelines advise further testing even prior to the introduction of egg and peanut in certain cases. We will explore these indications as well as avenues for testing in the next paragraphs.

When the history is concerning for an IgE-mediated food allergy, further testing to confirm the presence of said allergy as well as to assess the extent of the allergy should be pursued. This is typically performed by an allergist. When the diagnosis is in question, a food allergy can be confirmed or excluded with an oral food challenge. An oral food challenge consists of administering the food in question to a patient suspected of having an allergy to that food in incremental amounts while being monitored for signs and symptoms of an IgE-mediated reaction, which if present, are diagnostic for a food allergy. Oral food challenges remain the gold standard for diagnosis of food allergy; however, most food allergies can be diagnosed on the basis of history and skin prick and/or serum IgE testing to the suspected allergen without an oral food challenge.

Diagnostic testing for food allergies includes skin prick and/or serum IgE testing for the suspected allergen. Skin prick testing, as the name implies, involves the cutaneous administration of the allergen extract followed by a bloodless skin prick in order to allow antigen exposure to cutaneous mast cells. This is administered along with positive (histamine) and negative (saline) controls. A resulting wheal and flare at the area of allergen extract application is a measure of food sensitization relative to the size of the negative control. A wheal of 3 mm or more than the negative control is considered a positive test result indicative of sensitization to that particular food allergen. Skin prick testing has been the preferred method for aid in food allergy diagnosis since the 1950s. It should be noted that the specificity and sensitivity of skin prick testing vary by age, study, prevalence of selected food allergy in a given population and by technique. It is also subject to operator error in terms of administration and result interpretation. Current guidelines provide cut-off limits for the size of the wheal/flare that should be taken into consideration when determining if a patient is a candidate for an oral food challenge based on testing results and history.

In a similar manner, serum IgE testing for a food allergen is also widely utilized. Current guidelines do not recommend serum IgE testing alone as a means for diagnosing food allergies. Serum IgE testing has been used since the 1990s and is easily reproduceable. The serum IgE immunoassay, more commonly known as ImmunoCAP testing, is the most commonly used platform. In this form of allergy testing, serum antibodies bind to the allergen and are then quantified with enzyme-labeled anti IgE. Results are reported in allergen-specific kilounits per liter (kU/L) and indicate sensitization to the food allergen tested. Atopic comorbidities such as sensitization to aeroallergens or atopic dermatitis may decrease the pretest probability of serum IgE testing and this should be taken into account when interpreting the results. Sensitivity and specificity vary depending on the food tested. An advantage of serum IgE testing is the possibility of food allergen component testing. Foods contain many allergenic proteins and serum IgE testing allows for the assessment of many of these allergenic proteins in a specific food individually, which skin prick testing does not. Some of these proteins found in certain allergenic foods are elevated due to cross-sensitization with pollens in a patient with concomitant seasonal allergic rhinitis. We will discuss pollen food allergy syndrome also known as oral allergy syndrome later on in this chapter. Component testing will allow the clinician to risk stratify and separate those patients at risk for a severe allergic reaction to food versus those who are cross-sensitized with pollens but not at high risk for a severe food allergy reaction. While both skin prick testing and serum IgE testing have limitations, when used together they provide useful information for determining the severity of food allergies in an individual patient.

Going back to our case, the skin prick testing to peanut is positive for both twins. In order to better assess the extent of peanut allergy in each twin, serum IgE testing for peanut is obtained. Twin A's serum IgE level to peanut is mildly elevated at 0.8 kU/L and Twin B's is elevated at 34 kU/L. As discussed above, the serum IgE level and skin prick testing must be considered together with clinical history in determining if an oral food challenge to evaluate for the presence or absence of a food allergy is appropriate. Twin A's skin prick and serum IgE testing are within acceptable limits to offer an oral challenge to peanut; however, her sister's results are not. The family chooses to pursue an oral food challenge to peanut for Twin A and she passes. Peanut is introduced into this twin's diet regularly and she tolerates it without any adverse reactions.

Question 3

Commonly known allergenic foods include which of the following?

A. Peanuts
B. Egg
C. Fruits

D. Tree Nuts
E. A, B, D
F. All of the above

The correct answer is E

Question 4
What is the most appropriate management of a confirmed food allergy?

A. Strict avoidance of the food the patient is allergic to
B. Continued consumption of the food in small amounts
C. Serial testing with consumption of the food in small amounts
D. Immunotherapy

The correct answer is A

Peanut, egg, and tree nuts are among the most widely prevalent food allergens in the Western hemisphere. According to recent guidelines, the most common food allergens include cow's milk, egg, peanut, tree nuts, shellfish, finned fish, wheat, and soy listed in order of prevalence. It should be noted this list will differ among regions and worldwide as the prevalence of food allergy is studied most often in Western nations and is dependent on the foods most commonly eaten by a specific population. These eight foods listed above are responsible for more than half of all food allergies in the United States. In recent years, peanut has emerged as more prevalent, especially in children. Studies cite the prevalence of childhood food allergy as around 7% with peanut as the most common food allergy followed by cow's milk.

Peanut allergy has been evaluated closely over the last few decades, especially with respect to comorbid atopic dermatitis which is the case for our twin infants presented in this chapter. Atopic dermatitis is characterized by an erythematous, papulovesicular rash that usually develops in early infancy. Over time, we have learned about its association with other atopic diseases. Of note, roughly 85% of patients with atopic dermatitis have elevated IgE levels and more often than not these IgE antibodies are to food allergens. Studies have demonstrated that food ingestion can lead to worsening of atopic dermatitis and skin clearing when foods the patients are sensitized to (determined either by skin prick testing or serum IgE testing) were removed from the diet. While initial management guidelines recommended preemptive food avoidance, studies have since demonstrated that avoidance does not prevent the development of food allergy even in patients with severe atopic dermatitis. Rather, in these patients, sensitization to the food might often occur first through the breakdown of the skin which is the result of severe uncontrolled atopic dermatitis. Further studies promote early introduction (before sensitization occurs) of allergenic foods into the diets of infants even those with atopic dermatitis as a means of preventing food allergy. However, this too has come with its own challenges.

Perhaps the most well-known study regarding peanut allergy and atopic dermatitis is the Learning Early About Peanut Allergy or LEAP. In this study, infants with severe atopic dermatitis were randomized to two groups—either early peanut consumption or peanut avoidance. The researchers followed these infants from age 11 months through 4-years old, at which time they were tested for sensitization to peanut. The study ultimately concluded that introducing peanut early into the infant's diet correlated with a decreased prevalence of peanut allergy later. In a similar manner, the Beating Egg Allergy Trial (BEAT) evaluated the early introduction of egg in infants with severe atopic dermatitis. Based on these studies, infants with moderate to severe eczema are now tested for sensitization to egg and peanut in order to preserve a window of opportunity for early introduction of these foods to prevent the development of a food allergy. However, avoidance of the foods in patients who are allergic presents a narrow window for introduction and testing must be completed often for the best chance of preventing the development of food allergy.

Getting back to our case, the family of the twins is subsequently lost to allergy follow-up due to several moves; however, return to an allergist's care when the twins are 5-years old. The parents are eager to see if they can introduce peanut into Twin B's diet. She undergoes skin prick and serum IgE testing which is depicted in the table below along with positive (histamine) and negative (diluent) controls:

	Peanut SPT (mm)	Diluent (mm)	Histamine (mm)	Serum IgE to Peanut (kU/L)
Twin B	10/30	0/0	15/35	18

Twin B's skin prick testing and serum IgE testing remain too elevated to safely perform an oral food challenge to peanut. She repeats this testing at a 1-year follow up, but her skin prick test and serum IgE levels remain elevated. She does not outgrow her peanut allergy. Her sister continues to tolerate peanut.

Once a food allergy is confirmed by diagnosis, the food must be strictly avoided to prevent a life-threatening reaction. Children often outgrow food allergies and serial testing for sensitization whether by skin prick testing or serum IgE testing can be performed to see if the child can safely tolerate an oral food challenge. An oral food challenge is the gold standard of food allergy diagnosis and will determine if the child has or has not outgrown the food allergy. Adults typically do not outgrow food allergies. All patients with a food allergy should be instructed on strict avoidance measures and appropriate use of an epinephrine autoinjector to treat anaphylaxis.

Immunotherapy is the incremental administration of increasing doses of the allergen with the ultimate goal of raising the threshold dose which will cause an allergic response. In a patient with food allergies, oral immunotherapy (OIT) can prevent life-threatening reactions.

Oral immunotherapy has recently been introduced as a treatment for peanut allergy and is meant to reduce the chance of a life-threatening allergic reaction to peanut by achieving sustained unresponsiveness. It does not make peanut safe for peanut-allergic patients to eat at will. The largest, most well-known study to date, ultimately lent itself to the FDA approval of peanut OIT in 2020. Study participants were patients between ages 4 and 17 years with known peanut allergy and randomly assigned to receive the peanut OIT or placebo—67% of the participants receiving OIT were able to consume a minimum of 600 mg peanut protein without symptoms at the end of the study and a follow-on study demonstrated continued safety and desensitization up to 2000 mg peanut protein. This study and the subsequent FDA approval of peanut OIT, generic name AR 101, are a tremendous breakthrough in the management of food allergy. OIT has been investigated as a treatment option for other food allergens, but more studies are needed to assess efficacy and safety.

Question 5

Twin B, now in kindergarten, has decided to trade lunches with a friend who has a peanut butter and jelly sandwich. After two bites, she begins gagging and develops hives on her face and arms. What is the likely diagnosis and best treatment option?

A. She is probably choking, umbilical thrusts
B. She is having an allergic reaction, cetirizine 5 mg by mouth
C. She is having an allergic reaction, her epinephrine autoinjector should be administered by an adult immediately
D. She probably does not like the taste, watchful waiting

The correct answer is C

This patient is known to be allergic to peanut and has been witnessed ingesting part of a peanut butter and jelly sandwich. She has developed generalized hives and appears to be choking; the involvement of two organ systems meets the criteria for anaphylaxis and epinephrine, usually in the form of an autoinjector, should be administered immediately to stop the reaction. The patient should then be monitored for any worsening symptoms or a biphasic reaction.

Food allergy is best managed by avoidance of the allergenic food and immediate treatment of any allergic reactions caused by accidental ingestion. Of course, avoidance is not always easily done. Consumers and parents of pediatric patients with food allergens as well as caregivers (babysitters, coaches, teachers, etc.) will need to familiarize themselves with food labels to avoid cross-contamination or accidental ingestion. Food labeling laws in the United States mandate that all major food allergens are identified. This includes milk, egg, peanut, tree nuts, fish, crustaceans, wheat, and soy. It should be noted that many allergenic foods are labeled with ambiguous phrases such as "may contain" or "processed in a facility with" which are known to be ignored by about 40% of patients with food allergies. It should be noted that while traveling, differences in labeling laws elsewhere in the world including the European Union, may create even higher risk for the patient.

Despite strict adherence to a food avoidance diet, cross-contamination can occur. This is especially a concern at schools, daycare centers, and camps where children often eat together. This can also occur at places where food is prepared by a person unfamiliar with the patient's food allergy such as a communal meal among neighbors or at a restaurant. In addition, contact with food other than ingestion can lead to less severe reactions such as skin rashes in food handlers who are allergic to what they are preparing. Other potential sources of contact include poor cleaning of food preparation surfaces, sharing utensils without cleaning, and kissing which can trigger a reaction caused by accidental ingestion when an allergen is transferred in saliva. Uncertainties in food preparation and handling of all kinds necessitate ready identification of an allergic reaction and preparation of the patient and family members or at times, facility staff to intervene appropriately. As alluded to above, even patients adhering to an elimination diet can encounter many factors putting them at risk of accidental ingestion. Therefore, the patient, as well as family and caregivers, must be knowledgeable about recognizing the signs and symptoms of an allergic reaction and understand how to intervene in order to prevent or treat a life-threatening reaction including anaphylaxis.

Recognizing an allergic reaction is critical to early intervention which can be life-saving for a patient with a food allergy. Symptoms can appear discretely but progress rapidly and range from localized urticaria, a tingling sensation in the lips or tongue, swelling of the tongue, difficulty breathing, wheezing, skin rashes, eczema flares, vomiting, diarrhea, and/or lightheadedness. Anaphylaxis must be recognized promptly, and intramuscular epinephrine should be administered immediately. The typical dose is 0.01 mg/kg (to a maximum dose of usually 0.5 mg) delivered intramuscularly, though commercial epinephrine autoinjectors apply standardized doses for pediatric and adult patients.

Parents and other caregivers should be educated on the appropriate use and administration of intramuscular epinephrine for pediatric patients. All adults with food allergies should be taught to recognize their symptoms and to use an epinephrine autoinjector.

Risk factors for predisposition to anaphylaxis in a food allergic patient have been well documented in the literature. Between 1994 and 2006, 63 fatalities from food allergy were studied by the Food Allergy and Anaphylaxis Network/American Academy of Allergy, Asthma and Immunology. The majority of these fatalities included foods introduced to the patient at a restaurant. Most of these patients had had a previous reaction to the same food implicated in the fatal reaction. The two most common risk factors in this population were concomitant asthma and age less than 30 years old. Teenagers and young adults are likely to exhibit more risky behaviors and/or be away from home without parental supervision. Oftentimes, they do not have access to their epinephrine autoinjector or unfortunately, their peers do not know how to administer it or otherwise assist. Lastly, peanut allergy was implicated in 87% (55 out of 63) of these fatalities. Avoidance of peanuts is not always an easy task given peanuts are used in a variety of cuisines worldwide. This study demonstrates the need for proper education of the patient and family by the allergist and primary care clinician on food allergy avoidance measures as well as early recognition and treatment of allergic reactions.

While anaphylaxis must be recognized and treated early, it is possible for patients with food allergy to have more minor reactions following accidental ingestion of the allergen. These can manifest as more subtle symptoms often skin rashes or a few hives in a patient who has had skin contact with the food allergen. Antihistamines can be given to patients with minor symptoms such as localized urticaria. These patients should continue to be monitored for progression of their symptoms and the administration of epinephrine should not be delayed if there is any concern for anaphylaxis.

Case 2

A 16-year-old boy is evaluated for possible food allergies. Over the last 3 years, he has noticed numbness and tingling of the lips and tongue when eating fresh fruits, particularly raw apples, pears, and peaches. He used to enjoy these fruits regularly, but his parents have removed them from his diet out of concern for a food allergy. The symptoms do not occur when he eats the fruits in cooked form such as peach cobbler and apple pie. While eating these fruits he has never developed hives, lip-, or tongue-swelling, wheezing, difficulty breathing, vomiting, or diarrhea. His symptoms resolved almost immediately after using an oral antihistamine. The patient has chronic rhinitis and ocular pruritus that are present year-round but worsen in the springtime. This becomes especially bothersome as he is a baseball catcher for his high school team and has difficulty playing when his symptoms flare. His symptoms are fairly well controlled when he remembers to use his over-the-counter oral antihistamine and steroid nasal spray. He is otherwise healthy and does not take any other medications. There is no family history of other atopic diseases, asthma, or medication allergies.

Question 6

What is the next appropriate step in this patient's work up/what is his likely diagnosis?

A. Skin prick testing for the implicated foods/food allergy
B. No further workup/food intolerance
C. Skin prick testing to perennial and seasonal allergens/pollen food syndrome and chronic allergic rhinitis and conjunctivitis
D. Skin prick testing for perennial and seasonal allergens/nonallergic rhinitis

The correct answer is C

Question 7

Which of the following pollens are implicated in pollen food allergy syndrome?

A. Bermuda grass
B. Mugwort
C. Birch
D. Ragweed
E. All of the above

The correct answer is E

Question 8

The patient undergoes skin prick testing which determines he is sensitized to various trees, grasses, and weeds including birch pollen. He is started on a daily intranasal steroid and as-needed oral antihistamine to treat his allergic rhinoconjunctivitis. What is the best treatment option for his food sensitization?

A. Complete avoidance of the foods causing his pollen food allergy syndrome
B. Consuming these foods in the cooked form only
C. Prescribing an epinephrine autoinjector
D. No absolute avoidance is necessary but may be indicated based on symptoms

The correct answer is D

The patient is suffering from allergy symptoms worsening in the spring. This is concerning for seasonal allergic rhinitis. His lip pruritus while ingesting certain raw fruits and vegetables is indicative of pollen food allergy syndrome (also known as oral allergy syndrome) which is caused by initial sensitization to plant proteins from inhaled pollens. Pollen food allergy syndrome is an IgE-mediated process; however, symptoms are most often limited only to the lips and mouth (areas which come into direct contact with the food) and very rarely progress to more systemic symptoms including anaphylaxis. Patients present with seasonal allergic rhinitis as seen in our case, initially tolerating all fruits and vegetables without adverse reactions. A few years after the onset of the environmental allergies, ingestion of the cross-reactive raw plant foods starts to lead to symptoms such as tingling or itching limited to the mouth, lips, throat, and/or tongue. Patients can usually tolerate the same plant-derived food cooked without these symptoms since the protein has been denatured.

It is important to note a distinction between foods sensitized through the gut versus the respiratory tract, as food allergens whose initial sensitization is through the gastrointestinal tract are stable to heat and digestive enzymes therefore producing immediate type symptoms including anaphylaxis. However, food allergens whose initial sensitization is through inhalation as in pollen food allergy syndrome are cross-reactive with pollens and generally susceptible to degradation by heat and digestive enzymes making systemic symptoms rare.

One of the most commonly implicated aeroallergens leading to pollen food allergy syndrome is birch pollen. Birch pollen is cross-reactive with a variety of fruits and vegetables including apple, pear, sweet cherry, peach, plum, apricot, celery, carrot, kiwi, hazelnut, almond, mango, and chili pepper. Sensitization to birch pollen through inhalation therefore can manifest later in life as pollen food allergy syndrome to any number of these fruits, vegetables, or tree nuts. Cross-reactivities of common foods implicated in pollen food allergy syndrome are summarized in the Table 1.

Because of the underlying mechanism of prior sensitization to pollen, pollen food allergy syndrome is not often seen in young children, but rather predominates in adolescents and adults. Studies suggest a prevalence of about 10% in US adults though can vary based on geographic location both in the US and worldwide depending on the presence or absence of certain aeroallergens. For example, in areas with a high prevalence of birch trees, prevalence of pollen food allergy syndrome in the population has been estimated as high as 90%.

Table 1 Examples of cross-reactivity patterns between specific pollen and selected natural foods

Birch	Apple
	Pear
	Cherry
	Plum
	Apricot
	Celery
	Carrot
	Kiwi
	Mango
	Chili pepper
	Hazelnut
	Peanut
Grass	Watermelon
	Tomato
	Potato
	Kiwi
	Orange
Ragweed	Watermelon
	Cantaloupe
	Zucchini
	Cucumber
	Banana
Mugwort	Celery
	Carrot
	Mango

Patients with pollen food allergy syndrome are initially sensitized through inhalation of pollen and then develop symptoms to foods due to the cross-reactivity with pollen. While the risk of anaphylaxis in patients with pollen food allergy syndrome is low, about 3% of these patients may develop a systemic reaction and 1.7% develop anaphylactic shock. For this reason, treatment practices can vary, though roughly 70% of allergists practicing in the US prescribe an epinephrine autoinjector for patients with pollen food allergy syndrome due to a history of symptoms, mostly those involving the lower respiratory tract or pharynx, concomitant asthma, facial angioedema and/or hives. Current practice parameters do recommend prescribing an epinephrine autoinjector to patients with severe symptoms such as laryngeal swelling or respiratory distress in addition to avoidance of the raw food. Our patient did not have severe symptoms and therefore was educated on the risks of systemic reactions including anaphylaxis and advised to avoid the food in the raw form if symptoms were bothersome. He may continue to ingest the foods in cooked form.

Social Aspects of Food Allergy

Food allergies affect children, adolescents, and adults. They can have a significant impact on the quality of life and the socialization of children and adolescents among peers. Management and treatment are a team approach and must include not only the patient and family, but their allergist and numerous other players including teachers, coaches, food handlers, and other personal caregivers. We will end this chapter briefly by looking at the social aspects of food allergy with special regard to food allergy management.

Question 9

Recalling our first case, one twin did not outgrow her peanut allergy. When she reaches school age, which factors would contribute to a potentially fatal reaction for the patient at school?

A. Use of school lunches instead of homemade lunches
B. Lack of school staff to recognize anaphylaxis
C. Lack of a food allergy action plan
D. All of the above

The correct answer is D

Question 10

Your peanut allergic patient has now started middle school. She and her parents report bullying because of her food allergy, and you proceed to counsel the family. Which of the following statement regarding the social aspects of food allergy is incorrect?

A. A third of children with food allergy have experienced bullying
B. Open communication and education are effective strategies to address bullying
C. Bullying is rarely targeted at food allergic patients because the disease is not visible
D. Allergic reactions due to bullying with the food allergen have been reported
E. Addressing food allergy-related bullying requires a team approach

The correct answer is C

As discussed earlier, food allergy has become more prevalent in recent decades, especially in Western societies. The increase in prevalence suggests a need for increased awareness of prevention and management of adverse reactions due to food allergies. Studies on near-fatal and fatal reactions caused by anaphylaxis from food allergies have concluded that a lack of adequately recognizing anaphylaxis leads to a delay in epinephrine administration. This, in combination with inadequate or sometimes incomplete prevention strategies have contributed to near-fatal and fatal outcomes. Children and teenagers are at an elevated risk of potentially life-threatening reactions due to multiple factors. Young age and lack of appropriate education or carelessness around the severe consequences of a food allergy as well as a lack of understanding or misreading of food labels confounded by the desire to fit in with a peer group can lead to unsafe eating habits. Peer pressure may increase risky behaviors leading to severe and potentially fatal consequences. When and if these patients, by choice or accident, eat the food to which they are allergic, they may become severely ill and unable to self-administer epinephrine. Friends may be unfamiliar with the presentation of a severe allergic reaction and/or the use of an epinephrine autoinjector making them unable to assist. Parents and teenagers need to be educated on food allergies and the risks associated with ingestion of the food as well as the need for access to an epinephrine autoinjector in case of an accidental ingestion. For younger children, the risk of a reaction from accidental ingestion is higher in settings like schools and daycares due to food sharing which is increased by the child's lack of understanding. Significant adverse reactions in school-aged children are most often attributed to an adult's inability to recognize and appropriately manage a reaction, stressing the need for formal teaching. The allergist plays a critical role in educating patients and their caregivers so they may educate others in the child's life.Higher rates of anxiety have been reported in patients with food allergies, especially children with food allergies and their caregivers This includes anxiety, depression, social isolation, and bullying. In the United States, bullying is defined as harassment with the intention to harm another person, usually taking place when there is an imbalance of power between two people. It can involve physical or and/or verbal attacks intended to harm.Lieberman et al. conducted a study in 2010 specifically surveying caregivers of children with food allergy about the bullying associated with food allergy in children and adults. The investigators found that roughly one quarter (24%) of 353 respondents indicated they or their child have been a victim of bullying due to their food allergy and the most common food allergy reported in cases of bullying was peanut. Excluding children under 5-years old, the group least likely to be bullied, the number of respondents who reported bullying because of food allergy increased to 35.2% overall. The most common setting for bullying to occur was in school (82.4%). It should be noted that the majority of food allergic persons included in this study were school-aged children between the ages of 4 and 11 years old (55%).Lieberman's findings are not unique as more recent studies have demonstrated similar findings. The FORWARD study compared bullying in black children with food allergy to bullying in white children with food allergy. Surveys were completed by 252 families and the study found that 18.7% of all families (black and white) of children ages 4–15 years old reported their child had been bullied because of their food allergy. However, the majority of food bullying was targeted in older children (>11 years) at 33.3% among both black and white families combined. No significant racial differences related to bullying were identified in the study. Among children with food allergies, survey studies have identified reports of social exclusion, being forced to eat, or make skin contact with the allergic food or taunting with the allergic food.Bullying because of food allergy has been correlated with a decreased quality of life in children with food allergies and their caregivers. Parents, teachers, and other adults who care for children with food allergies need to be aware of the risk of bullying so they can appropriately intervene as necessary. This is yet another aspect of food allergy that the primary care physician and allergist will need to assist their patients in managing.We have presented in this chapter a brief introduction to the epidemiology, pathogenesis, diagnosis, treatment, and social implications of food allergy. Further research in the field will expand our knowledge and understanding of food allergy laying the groundwork for more innovative treatments. A basic understanding of the concepts presented in this chapter is vital for any physician or other clinician type involved in the management of patients with food allergy.

Bibliography

1. Calvani M, Anania C, Caffarelli C, Martelli A, Miraglia Del Giudice M, Cravidi C, et al. Food allergy: an updated review on pathogenesis, diagnosis, prevention and management. Acta Biomed. 2020;91(11-S):e2020012.
2. Sicherer SH, Sampson HA. Food allergy: a review and update on epidemiology, pathogenesis, diagnosis, prevention, and management. J Allergy Clin Immunol. 2018;141(1):41–58.
3. Lopes JP, Sicherer S. Food allergy: epidemiology, pathogenesis, diagnosis, prevention, and treatment. Curr Opin Immunol. 2020;66:57–64.
4. Gupta RS, Warren CM, Smith BM, Jiang J, Blumenstock JA, Davis MM, et al. Prevalence and severity of food allergies among US adults. JAMA Netw Open. 2019;2(1):e185630.
5. Gupta RS, Warren CM, Smith BM, Blumenstock JA, Jiang J, Davis MM, et al. The public health impact of parent-reported childhood food allergies in the United States. Pediatrics. 2018;142(6)
6. Sicherer SH, Warren CM, Dant C, Gupta RS, Nadeau KC. Food allergy from infancy through adulthood. J Allergy Clin Immunol Pract. 2020;8(6):1854–64.
7. Lyons SA, Burney PGJ, Ballmer-Weber BK, Fernandez-Rivas M, Barreales L, Clausen M, et al. Food allergy in adults: substantial variation in prevalence and causative foods across Europe. J Allergy Clin Immunol Pract. 2019;7(6):1920–8.e11.
8. Du Toit G, Roberts G, Sayre PH, Bahnson HT, Radulovic S, Santos AF, et al. Randomized trial of peanut consumption in infants at risk for peanut allergy. N Engl J Med. 2015;372(9):803–13.
9. Sampson HA, O'Mahony L, Burks AW, Plaut M, Lack G, Akdis CA. Mechanisms of food allergy. J Allergy Clin Immunol. 2018;141(1):11–9.
10. Oriel RC, Wang J. Diagnosis and management of food allergy. Pediatr Clin N Am. 2019;66(5):941–54.
11. Nowak-Wegrzyn A, Burks AW, Sampson HA. Reactions to foods. Middleton's allergy, principles and practice. 9th ed. Philadelphia, PA: Elsevier; 2020.
12. LaHood NA, Patil SU. Food allergy testing. Clin Lab Med. 2019;39(4):625–42.
13. Boyce JA, Assa'ad A, Burks AW, Jones SM, Sampson HA, Wood RA, et al. Guidelines for the diagnosis and management of food allergy in the United States: summary of the NIAID-sponsored expert panel report. Nutr Res. 2011;31(1):61–75.
14. Sampson HA, Aceves S, Bock SA, James J, Jones S, Lang D, et al. Food allergy: a practice parameter update-2014. J Allergy Clin Immunol. 2014;134(5):1016–25.e43.
15. Bettcher CM, Rockwell PG, Ravikumar R. Managing food allergy in children: an evidence-based update. J Fam Pract. 2020;69(7):336;8;40;43
16. Sampson HA. The evaluation and management of food allergy in atopic dermatitis. Clin Dermatol. 2003;21(3):183–92.
17. Fisher HR, Du Toit G, Bahnson HT, Lack G. The challenges of preventing food allergy: lessons learned from LEAP and EAT. Ann Allergy Asthma Immunol. 2018;121(3):313–9.
18. Wei-Liang Tan J, Valerio C, Barnes EH, Turner PJ, Van Asperen PA, Kakakios AM, et al. A randomized trial of egg introduction from 4 months of age in infants at risk for egg allergy. J Allergy Clin Immunol. 2017;139(5):1621–8.e8.
19. Sicherer SH, Lack G, Jones SM. Food allergy management. Middleton's allergy, principles and practice. 9th ed. Philadelphia, PA: Elsevier; 2020.
20. Vickery BP, Vereda A, Casale TB, Beyer K, du Toit G, Hourihane JO, et al. AR101 Oral immunotherapy for peanut allergy. N Engl J Med. 2018;379(21):1991–2001.
21. Vickery BP, Vereda A, Nilsson C, du Toit G, Shreffler WG, Burks AW, et al. Continuous and daily oral immunotherapy for peanut allergy: results from a 2-year open-label follow-on study. J Allergy Clin Immunol Pract. 2021;9(5):1879–89.
22. Wood RA. Oral immunotherapy for food allergy. J Investig Allergol Clin Immunol. 2017;27(3):151–9.
23. Muñoz-Furlong A, Weiss CC. Characteristics of food-allergic patients placing them at risk for a fatal anaphylactic episode. Curr Allergy Asthma Rep. 2009;9(1):57–63.
24. Kim JH. Insights into pediatric pollen food allergy syndrome. Clin Exp Pediatr. 2020;63(12):483–4.
25. Sicherer SH. Clinical implications of cross-reactive food allergens. J Allergy Clin Immunol. 2001;108(6):881–90.
26. Jeon YH. Pollen-food allergy syndrome in children. Clin Exp Pediatr. 2020;63(12):463–8.
27. Carlson G, Coop C. Pollen food allergy syndrome (PFAS): a review of current available literature. Ann Allergy Asthma Immunol. 2019;123(4):359–65.
28. Young MC, Muñoz-Furlong A, Sicherer SH. Management of food allergies in schools: a perspective for allergists. J Allergy Clin Immunol. 2009;124(2):175–82, 82.e1–4; quiz 83-4
29. Ravid NL, Annunziato RA, Ambrose MA, Chuang K, Mullarkey C, Sicherer SH, et al. Mental health and quality-of-life concerns related to the burden of food allergy. Psychiatr Clin North Am. 2015;38(1):77–89.
30. Services USDoHaH. What is Bullying https://www.stopbullying.gov/bullying/what-is-bullying.
31. Egan M, Sicherer S. Doctor, my child is bullied: food allergy management in schools. Curr Opin Allergy Clin Immunol. 2016;16(3):291–6.
32. Lieberman JA, Weiss C, Furlong TJ, Sicherer M, Sicherer SH. Bullying among pediatric patients with food allergy. Ann Allergy Asthma Immunol. 2010;105(4):282–6.
33. Brown D, Negris O, Gupta R, Herbert L, Lombard L, Bozen A, et al. Food allergy-related bullying and associated peer dynamics among black and white children in the FORWARD study. Ann Allergy Asthma Immunol. 2021;126(3):255–63.e1.
34. Cooke F, Ramos A, Herbert L. Food allergy-related bullying among children and adolescents. J Pediatr Psychol. 2022;47(3):318–26.
35. Herbert L, Shemesh E, Bender B. Clinical management of psychosocial concerns related to food allergy. J Allergy Clin Immunol Pract. 2016;4(2):205–13.

Eosinophilic Disorders

Taha Al-Shaikhly, Matthew R. Norris, Heather Stern, and Gisoo Ghaffari

Abbreviations

DOCK8	Dedicator of Cytokinesis 8
EGID	Eosinophilic gastrointestinal disorder
EGPA	Eosinophilic granulomatosis with polyangiitis
EoE	Eosinophilic esophagitis
HE	Hypereosinophilia
HE_{FA}	Familial hypereosinophilia
HE_{R}	Reactive hypereosinophilia
HES	Hypereosinophilic syndrome
HE_{US}	Hypereosinophilia of undetermined significance
L-HES	Lymphocytic variant of hypereosinophilic syndrome
M-HES	Myeloid variant of hypereosinophilic syndrome
STAT3	Signal transducer and activator of transcription-3
Th2	Type-2 helper

1 Introduction

Eosinophils are white blood cells of myeloid origin, contain granules, which produce many molecules and chemicals through which they participate in host defense against certain infections and are involved in various allergic and immunological responses. Understanding the nature, function, and regulatory mechanisms of eosinophils is essential to allergists-immunologists. Eosinophilic disorders are most commonly secondary to an underlying disease process. Eosinophils could also be the primary source of the disorder, most commonly due to aberrant production of eosinophilopoietic cells. Recognizing the differences between these and understanding the molecular basis of these conditions are important for proper diagnosis and management of these conditions, which are frequently encountered by practicing allergists-immunologists.

In this chapter, we will discuss eosinophil biology, definition of eosinophilia and its severity, common causes of secondary eosinophilic disorders, and eosinophilic gastrointestinal disorders with a particular focus on eosinophilic esophagitis and hypereosinophilic syndromes. These will be discussed through case-based questions. Two cases are presented, with five questions related to each and explanations of the answers both briefly and in more detail.

Case 1

A 26-year-old white male presents to the allergy clinic for evaluation of a few years' history of intermittent difficulty swallowing particularly with solid foods. His symptoms, which were initially sporadic, are getting worse in the past 6 months. He was recently at a family dinner when for the first time, food got stuck in his throat while eating turkey. Induction of vomiting did not resolve the symptom. He was taken to an emergency department where a food bolus was removed via endoscopy, and he was told that his esophagus looked inflamed. He had experienced milder episodes while at college after eating peas, pork, beef, and chicken. These symptoms do not occur every time while eating one of these foods and are not associated with any ocular, nasal, respiratory, or dermatologic symptoms. His medical history included childhood asthma not requiring daily inhaler therapy and seasonal rhinitis for which he takes antihistaminic pills on as-needed basis. He has a history of eczema, which cleared by age 5 and had an IgE-mediated milk allergy until the age of 7. His mother and sister have asthma and allergic rhinitis. He is a medical student, does not smoke, drinks alcohol in moderation, and does not use any illicit drugs. He lives in an apartment, which has carpeting, and last year during the pandemic, he obtained a pet dog.

T. Al-Shaikhly · M. R. Norris · H. Stern · G. Ghaffari (✉)
Division of Pulmonary, Allergy and Critical Care Medicine, Department of Medicine, Pennsylvania State College of Medicine, Penn State Health-Milton S Hershey Medical Center, Hershey, PA, USA
e-mail: talshaikhly@pennstatehealth.psu.edu; mnorris3@pennstatehealth.psu.edu; hstern@pennstatehealth.psu.edu; gghaffari@pennstatehealth.psu.edu

M. Mahmoudi (ed.), *Absolute Allergy and Immunology Board Review*, https://doi.org/10.1007/978-3-031-12867-7_12

He does have occasional heartburn and regurgitation of food. He admits to being the last person to finish his meals when eating with others and tends to drink plenty of water with meals to help with swallowing foods.

After the visit to the emergency department, he was referred to a gastroenterologist for an esophagogastroduodenoscopy. Gross endoscopy revealed linear furrows and esophageal biopsy showed intraepithelial eosinophilic infiltration (20 eosinophils per high power field in proximal esophagus, 25 in mid esophagus, and 30 in distal esophagus). No eosinophilic inflammation was identified in either gastric or duodenal biopsies taken during the procedure. H-pylori was not identified. He was told that he most likely has an allergic condition known as eosinophilic esophagitis and was referred to an allergist to further help with the management plan.

Question 1

The patient remembers from basic science years in medical school, that eosinophils are seen in parasitic infestations and is wondering if that is the reason for this pathological finding. The allergist explains to him the nature of eosinophils and how he might have gotten them in his esophagus. Which of the following is incorrect?

A. Eosinophils are primarily tissue leukocytes and can be normally seen in some tissues, including the esophagus
B. Eosinophilic granules produce chemicals which are involved in protection against parasitic infestations
C. Increased number of eosinophils are commonly seen in allergic conditions
D. Eosinophils can be seen in an increased amount in tissues without being at higher levels in the circulation
E. Interleukin-5 is an important cytokine in the process of eosinophilopoiesis

Answer: A

Eosinophils are normally found in many tissues including many parts of the gastrointestinal tract; however, normally the esophagus should be devoid of any eosinophils.

2 Eosinophil Biology

Eosinophils are derived from multipotent CD34$^+$ hematopoietic stem cells in the bone marrow. Eosinophil differentiation from the myeloid lineage is dependent upon the transcription factors GATA-1, PU.1, and the C/EBP family, and is instructed by the eosinophil-promoting cytokines including interleukin-3 (IL-3), interleukin-5 (IL-5), and granulocyte-macrophage colony-stimulating factor (GM-CSF). The eosinophil life cycle includes bone marrow, blood, and tissue stages. However, eosinophils are primarily tissue-dwelling, and in humans, they have a tissue-to-blood ratio of approximately 100:1. They have a circulating half-life of approximately 8–18 h while their life span in tissue ranges from 2 to 5 days, and this may be prolonged by cytokines. Eotaxins, which are chemokines that bind to the eosinophil chemokine receptor CCR3, play a key role in the recruitment of eosinophils into tissues.

Eosinophils contain several types of granules. The most important are specific or secondary granules, which contain major basic protein (MBP), eosinophil cationic protein (ECP), eosinophil-derived neurotoxin (EDN), and eosinophil peroxidase which are capable of inducing tissue damage. Charcot-Leyden crystals, which can be found in tissues where there is eosinophilic inflammation, are located in the primary granules of eosinophils. Eosinophils also produce lipid mediators including cysteinyl leukotrienes, prostaglandins, and platelet-activating factor (PAF) that play a role in allergic disease, specifically related to their effects on vascular permeability, vascular and bronchial smooth muscle tone, and chemotaxis. Even when eosinophils are no longer present in a specific tissue, the presence of their granules and mediators is evidence of their existence. To detect evidence of prior eosinophil degranulation in addition to intact eosinophils, immunofluorescent identification can be used, which is more sensitive than other staining techniques.

3 Eosinophilia

Eosinophilia is an increase in the number of eosinophils in blood and/or tissue. Greater than 450–550 cells/μL in the periphery is typically considered elevated. Although knowing the absolute eosinophil count is more important, greater than 5% of eosinophils on a differential is generally considered abnormal. The severity of eosinophilia does not necessarily predict end-organ damage; however, tissue damage is more likely to occur when the peripheral eosinophil count is greater than 1500 cells/μL.

4 Secondary Versus Primary Eosinophilic Disorders

When evaluating a patient with eosinophilia, a practical approach involves considering if the eosinophilia is secondary to another pathologic process or if it is related to a primary process in which the eosinophils are the main source of tissue and/or organ damage. Common secondary causes of eosinophilia include atopic conditions, chronic rhinosinusitis, drug-induced eosinophilia, and infection-related eosinophilia. Significant eosinophilia in a patient with asthma should raise concern for possible allergic bronchopulmonary aspergillosis or eosinophilic granulomatosis with polyangiitis.

Drug-induced eosinophilia can occur with a prescription or nonprescription medications including herbal supplements. In areas of the world where helminth infections are uncommon, persistently elevated eosinophil levels are most commonly attributed to drug reactions. Drug-induced eosinophilia may occur alone or in conjunction with other manifestations, such as rash and/or fever. Various types of drug reactions can be associated with eosinophilia including but not limited to drug-induced hepatitis, which can occur with tetracyclines or semisynthetic penicillins, and drug-induced interstitial nephritis, which has been associated with cephalosporins or semisynthetic penicillins. Drug reaction with eosinophilia and systemic symptoms (DRESS) is a severe adverse drug reaction that can have a heterogeneous clinical presentation associated with an extensive skin rash. It often presents with fever and lymphadenopathy, and organ failure can result. Commonly implicated medications include antibiotics, anticonvulsants, and allopurinol.

Helminth infections are a leading cause of eosinophilia worldwide. Strongyloides is a common culprit. Other helminths that can be contracted in the United States without foreign travel include Trichinella, Ascaris, hookworm, and visceral larva migrans. Ectoparasites (particularly scabies), certain fungi (such as coccidioidomycosis and cryptococcosis), and rarely viruses (such as human T cell lymphocytic virus-1) can result in eosinophilia while bacterial infections are not typically associated with eosinophilia.

Eosinophilia may occur in the setting of certain malignancies such as Hodgkin's lymphoma, adenocarcinomas, and thyroid cancer whereas some malignancies are eosinophil-derived, such as acute and chronic eosinophilic leukemias. Systemic mastocytosis is another condition that can be associated with peripheral eosinophilia, which occurs in up to 28% of patients with the condition. Eosinophils can also be present on bone marrow biopsy even if there is not significant peripheral eosinophilia. There are various immunologic disorders associated with peripheral and/or tissue eosinophilia including but not limited to Omenn Syndrome, DOCK8 deficiency, and Wiskott-Aldrich Syndrome. Lastly, connective tissue and autoimmune disorders can also be associated with eosinophilia. Eosinophils can be found in specific organs without being present in the circulation. Organ-specific eosinophilia can be seen in many tissues including the respiratory and gastrointestinal tracts, skin, and mucous membranes.

5 Eosinophilic Gastrointestinal Disorders

Eosinophils can be found normally in all parts of the gastrointestinal tract, except for the esophagus. The upper limit of normal in the various sections of the gastrointestinal tract is not well-defined. However, their presence in excess is almost always abnormal and raises a broad differential diagnosis. Eosinophilic gastrointestinal disorders (EGIDs) are diseases characterized by eosinophil-rich inflammation of the gastrointestinal tract. Symptoms of these diseases are often heterogeneous and nonspecific. EGIDs can have a significant impact on quality of life and are associated with the need for extensive dietary restrictions. Current literature supports the role of immunological, environmental, and genetic factors in the pathogenesis of EGIDs.

6 Eosinophilic Gastritis, Eosinophilic Gastroenteritis, and Eosinophilic Colitis

The remaining EGIDs are poorly characterized and infrequently encountered in clinical practice. They are pathologically characterized by eosinophilic infiltration of their respective tissue of the gastrointestinal tract, with symptoms often relating to the tissue level affected. There are the mucosal, the muscularis, and the serosal forms of EGID. Using eosinophilic gastritis and gastroenteritis as an example, the mucosal form will be mostly characterized by nausea, vomiting, abdominal pain, and malabsorption syndromes, such as anemia and protein-losing enteropathies. The muscularis form often results in thickening of the bowel wall that results in decreased motility and obstructive symptoms. The serosal form, a relatively rare variant compared to the mucosal and muscularis forms, is characterized by higher peripheral eosinophil counts and ascites. The differential diagnoses for these conditions can be extensive, and there is no consensus on an ideal treatment approach. With eosinophilic colitis, in particular, the accumulation of eosinophils in the colon has a very wide differential, and their presence should prompt investigation of secondary causes of eosinophilic inflammation, such as inflammatory bowel diseases.

Question 2
Which of the following may have conferred a predisposition to developing eosinophilic esophagitis as the cause of dysphagia?

A. Patient's gender
B. Patient's race
C. Patient's age
D. Patient's history of atopy
E. All of the above

Answer: E

All of the epidemiologic factors provided, and the patient's history of atopy puts him at higher risk of developing eosinophilic esophagitis. Over 75% of patients with EoE will have at least one atopic disease. History of asthma, allergic rhinitis, food allergy, and atopic dermatitis are common among EoE patients.

There have been reports of 3–5% risk of developing EoE in patients receiving oral immunotherapy for the treatment of food allergy. In these patients, remission of the EoE will occur upon cessation of the immunotherapeutic antigen

7 Epidemiology and Risk Factors for Eosinophilic Esophagitis

While EoE can be diagnosed at any age, most patients are diagnosed in their third decade of life. There is a strong male predominance in EoE with an estimated males-to-females ratio of 3:1. A recent meta-analysis showed a pooled prevalence of 53.8 per 100,000 inhabitants for male patients compared to 20.1 per 100,000 for female patients. Patients are also more likely to be Caucasian. A study of over 7000 patients with EoE found that 89% were Caucasian, 6.1% African American, and 5.6% Asian. EoE in Hispanic patients is thought to be rare, with the prevalence of EoE in Caucasian patients being sevenfold higher than in Hispanic patients. Socioeconomic status is not associated with the diagnosis of EoE in either adults or children. Furthermore, over 75% of patients with EoE will have at least one atopic disease, with asthma, allergic rhinitis, and eczema being more common among EoE patients

There appears to be a genetic component to EoE, as 10% of parents of EoE patients have a history of esophageal strictures. Dizygotic twins have a 22% disease concordance, and 2.4% of siblings of patients with EoE have EoE, which is higher than the risk of EoE in the general population (0.05%). Studies of monozygotic and dizygotic twins suggest genetic risk variants account for 15% of the phenotypic variation of disease risk. Although many candidate genes are under investigation, three genes responsible for producing thymic stromal lymphopoietin (TSLP), calpain-14, and eotaxin-3 have been identified as being altered in association with EoE. *CCL26* is the most highly expressed gene in the EoE transcriptome and is induced by IL-13.

As with many atopic diseases, the impact of altered immune system stimulation and microbiome disruption at an early age appears to increase the risk of developing EoE. Children with EoE had higher odds (odds ratio, 3.2) of being born via cesarian section when compared to controls. Antibiotic use in the first year of life was also found to be associated with 3.5 times higher odds of developing EoE. Although breastfeeding has not consistently been found to correlate with protection against EoE development, a case-control study identified a possible protective role of breastmilk in a subset of patients. Formula feeding, neonatal intensive care unit admission, prematurity, maternal fever, C-section, antibiotic use in infancy, and acid suppressant use in infancy have all been found to be early risk factors for the development of EoE, with antibiotic and proton pump inhibitor use in infancy showing the most consistent correlation.

Question 3

The student then asks his physician if there are any other steps to confirm the diagnosis. Which of the following statements is correct regarding EoE?

A. Regardless of his symptoms, presence of more than 15 eosinophils/hpf in any part of esophagus is sufficient for diagnosis
B. History of food impaction along with furrowing in the esophagus is sufficient for diagnosis
C. History of dysphagia, with more than 15 eosinophils/hpf, which is confined to esophagus, and exclusion of H-pylori are consistent with diagnosis of EoE
D. Treatment with high-dose PPI is required to exclude GERD before making the diagnosis of EoE
E. GERD and EoE are separate entities and they do not coexist together

Answer: C

EoE is a clinicopathological diagnosis. Clinical symptoms of esophageal dysfunction along with pathological findings of at least 15 eos/hpf in various parts of the esophagus are required for diagnosis. Except for GERD which may coexist, other causes of esophageal eosinophilia should be excluded either by history or through endoscopic findings.

8 Eosinophilic Esophagitis

Eosinophilic esophagitis (EoE) is a clinicopathologic diagnosis. It is defined by the presence of symptoms attributed to esophageal dysfunction (including dysphagia and food impaction in adults or feeding intolerance and gastroesophageal reflux disease-like symptoms in children) and eosinophil-predominant inflammation of ≥15 eosinophils per high power field (HPF). EoE has emerged over the past 30 years as one of the leading causes of food impaction and dysphagia in adults and children. It was first described in 1978 by Landres and colleagues as an isolated case of vigorous achalasia with marked smooth muscle hypertrophy and eosinophilic infiltration of the esophagus. Case reports continued to accumulate over the next decade, culminating in the 1989 case series entitled "Esophageal Asthma—an episodic dysphagia with eosinophilic infiltrates" by Attwood and colleagues." They went on to publish their description of these adult patients with dysphagia, normal pH monitoring, and esophageal eosinophilia in 1993, titled "Esophageal

Eosinophilia with dysphagia. A distinct clinicopathological syndrome." There were rapid discoveries of this novel disease through the 1990s, with the first guidelines for EoE being written in 2007. These guidelines were first updated in 2011 and most recently in 2020. These will be discussed in further detail in this chapter.

9 Clinical Presentation of Eosinophilic Esophagitis

The clinical presentation of EoE is nonspecific and symptoms vary by age of onset (Fig. 1). Infants and toddlers tend to present with food refusal, feeding difficulties, vomiting, and/or failure to thrive. Older children will present with nausea, vomiting, regurgitation, heartburn, and abdominal pain. The most common presenting symptoms in adults are food impaction, chest pain, and dysphagia, with dysphagia being the most common presentation of EoE in patients ≥18 years of age. This progression of symptoms is well documented, with some questions as to whether adults have a late-onset disease (i.e., a different clinical phenotype of EoE) or have asymptomatic chronic inflammation from childhood that goes unrecognized. There is growing evidence to support the second phenotype of EoE affecting patients 65 years of age or older. These patients have a higher likelihood of having gastric acid reflux and are less likely to be male.

10 Diagnosis of Eosinophilic Esophagitis

EoE is a clinicopathological diagnosis and both clinical symptoms and pathological criteria should be present at the same time. The diagnosis of EoE should be considered if the patient has a history of dysphagia and personal or family history of atopy, with or without peripheral eosinophilia. Positive allergy skin or serologic testing is not needed to diagnose EoE, as a large fraction of patients with EoE lack allergen sensitization. The gold standard for diagnosing EoE remains tissue biopsy of the esophagus showing increased intraepithelial eosinophilic inflammation. Although upper endoscopy is diagnostic in >80% of patients with mucosal involvement, upto 10% of endoscopies in patients with EoE can appear normal, particularly in children Traditionally, at least 5 biopsies are obtained at multiple esophageal levels to maximize the sensitivity of identifying intraepithelial eosinophilia. One or more of the biopsies must show eosinophil-predominant inflammation of at least ≥15 eosinophils/HPF. Proton pump inhibitor-responsive esophageal eosinophilia (PPI-REE), an entity described in the 2011 consensus guidelines, was later eliminated from the most recent consensus guidelines. PPI-REE was recategorized as being part of the EoE continuum. To this end, transcriptional profiling of PPI-REE patients demonstrated a near-identical gene expression profile when compared to PPI-unresponsive EoE counterparts. Therefore, a trial of high-dose PPI therapy is no longer required before making an EoE diagnosis.

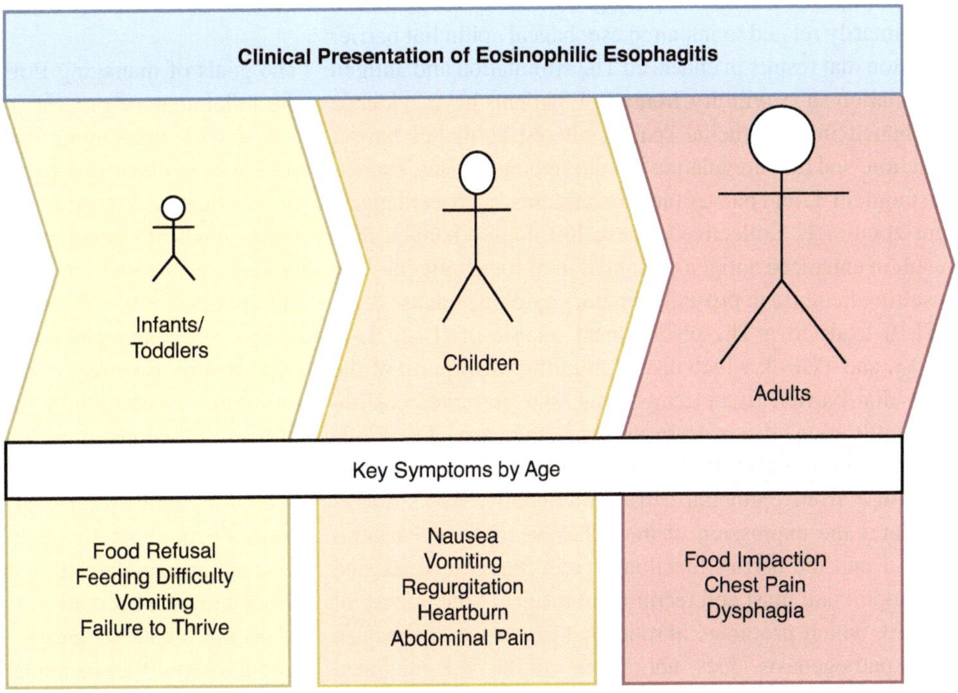

Fig. 1 Clinical presentation of eosinophilic esophagitis varies by age

Question 4

The student is curious to know how the patient developed the condition, anything he could have done to prevent that from happening and if his history of other allergic conditions has anything to do with this diagnosis.

Which statement is incorrect about the pathogenesis of EoE?

A. Genetic, epigenetic, and environmental factors play roles in the pathogenesis
B. Th-2 inflammatory cytokines are responsible for many of the findings including eosinophilic inflammation
C. EoE is an IgE-mediated allergic condition with food antigens being the main culprit
D. TGF-ß could be responsible for fibrostenotic changes in the esophagus
E. IL-13 causes downregulation of fillagrin, a protein involved in the epithelial barrier function

Answer: C

Although food antigens can trigger an inflammatory response, EoE is not an IgE-mediated food allergy.

11 Pathogenesis of Eosinophilic Esophagitis

There has been a tremendous amount of research on the pathogenesis of EoE, particularly on the role of genetic and environmental factors. The pathogenesis of EoE appears to be primarily related to impaired esophageal epithelial barrier function that results in enhanced Th2 stimulation and antigen stimulation susceptibility. Many EoE patients have evidence of dilated inter-epithelial spaces, altered epithelial barrier function, and downregulation of adhesion molecules, such as desmoglein-1, and barrier function proteins, such as filaggrin and zonulin-1. Collectively, these histologic changes may result in enhanced antigen presentation of food antigens. The resulting increase in pro-inflammatory cytokine release (e.g., TSLP) leads to a Th2-predominant release of IL-4, IL-5, IL-13, and TGF-ß, which results in further disruption of the epithelial barrier, tissue remodeling, and promotes eosinophil infiltration. IL-4 release results in increased T cell differentiation into Th2. IL-13 release induces the secretion of eotaxin-3 from epithelial cells. Additionally, IL-13 downregulates the expression of the adhesion molecule desmoglein-1 and the barrier function protein filaggrin. Increased eosinophil and mast cell recruitment results in the release of TGF-ß, which promotes fibrotic changes in the esophagus. EoE pathogenesis does not appear to be IgE-mediated because (1) skin testing does not consistently identify food triggers, (2) biologic therapy with omalizumab, a monoclonal antibody that binds to the free human immunoglobulin E, has not improved EoE symptoms and (3) EoE has been recreated in IgE-null mice. The non-IgE-mediated role for food antigens in EoE pathogenesis is supported by symptom responsiveness to dietary elimination challenges and recurrence of symptoms upon food reintroduction in both adult and pediatric populations.

Question 5

The physician then explains how to treat this condition to prevent future episodes of food impaction and the outcome of this diagnosis. Which of the following statements is correct regarding the treatment of EoE?

A. Treatment with omalizumab, which targets the IgE-mediated allergic response, is effective
B. PPI is the first line of treatment for many individuals regardless of presence or absence of symptoms of acid reflux
C. Dietary modifications are only effective in children, but not in adults
D. Since he just had an endoscopy with dilatation, no further treatment is needed
E. Barrett's esophagus can be seen in both GERD and EoE

Answer: B

Based on the recent guidelines for treatment, PPI is the first line of treatment for many individuals.

12 Treatment of Eosinophilic Esophagitis

The goals of managing EoE patients include (1) alleviating the inflammatory process, (2) providing symptomatic control, and (3) preventing fibrotic sequela. Ideally, the patient should be connected with an interdisciplinary team inclusive of an allergist, a gastroenterologist, and a nutritionist. The three medical treatment modalities currently available include dietary therapy (elemental diet or selective dietary elimination), PPI, and swallowed steroids.

In 1995, a landmark study by Kelly and colleagues sought to determine the effect of diet on eosinophilic esophagitis. In this study, 10 children with EoE who were unresponsive to PPIs or fundoplication, attained clinical and histological remission while consuming an amino acid-based formula. Symptoms and esophageal eosinophilia recurred with the reintroduction of allergenic foods, establishing a role for food antigens in the pathogenesis of the disease. Subsequent single-armed, observational studies demonstrated that ~94% of pediatric patients and ~ 80% of adult patients on an elemental diet will experience histologic remission. Harms of elemental diets in EoE include social isolation, cost of elemental formula, the burden of repeat endoscopies on food reintroduction, and the development of oral motor skill deficits. Elemental diets are therefore designed for short-term use only. For these reasons, the 2020 practice parameters for EoE suggest that patients who wish to avoid the challenges of adherence to an elemental diet and prolonged process of dietary reintroduction may reasonably decline this treatment option.

To overcome the challenges associated with an elemental diet, less-restrictive variations of elimination diets have been proposed. A six-food elimination diet of the most commonly implicated food triggers results in symptom resolution in approximately 68% of patients based on 10 single-arm, observational studies. These foods are those made with cow's milk (~75% of patients require continued elimination), wheat (~26% of patients), egg (~17% of patients), soy (~10% of patients), peanut (~6% of patients)/tree nuts and seafood (fish and shellfish). After complete cessation of all six foods for 8–12 weeks, a repeat endoscopy is needed to assess for histopathological remission. If EoE is resolved, foods are then reintroduced individually, followed by repeat endoscopy. The 2020 practice parameters for EoE carry the same notation as for the elemental diet, specifically that patients may decline the six-food elimination diet if they wish to avoid the challenges of adherence to it and the requirement for multiple endoscopic procedures. More liberal variations of the six-food elimination diet include a four-food elimination diet (milk, wheat, eggs, soy) and a milk-only elimination diet. There have been attempts to use allergy testing to guide dietary elimination; however, there is only very low-quality evidence to support such an approach. Coupled with the limited accuracy of current food allergy testing, the 2020 guidelines state that patients may reasonably prefer an alternative therapeutic approach.

Although PPI therapy no longer plays a role in the diagnostic evaluation of patients with suspected EoE, it can induce histologic remission in approximately one-third of patients. Although the mechanism behind such histologic remission has yet to be elucidated, in vitro studies showed that PPIs, independently of their acid suppressive effect, decrease IL-13-induced eotaxin-3 production, which may also account for their anti-inflammatory benefit. In patients with symptomatic esophageal eosinophilia, the 2020 guideline conditionally recommends using PPIs over no treatment. That said, direct comparison of PPIs to other therapeutic modalities is limited because patients who were responsive to PPIs were excluded from earlier clinical trials based on the previous categorization of PPI-responsive esophageal eosinophilia, the term which is no longer used.

Topical glucocorticoid therapy, having an established role in asthma, is being increasingly recognized for its ability to target the inflammatory basis of EoE. Based on several double-blinded, placebo-controlled studies assessing the benefit of topical steroids (budesonide or fluticasone), approximately 67% of patients can experience histologic remission. For this reason, practice parameters recommend using topical steroids over no treatment. In studies ≤3 months, there was no increased risk of adverse events in patients treated with topical steroids compared to placebo, but long-term studies are still lacking. Although no FDA-approved treatments exist in the United States, the European Medicines Agency (EMA) approved a budesonide tablet for EoE in 2018. The use of systemic glucocorticoids to treat EoE has further been shown to have similar therapeutic results at the expense of a greater side effect profile.

In patients who develop EoE-associated dysphagia secondary to stricture formation, the 2020 practice parameters recommend endoscopic dilation since 87% of these patients will experience symptomatic improvement with such intervention. Other therapeutic modalities under investigation include the use of biologic therapies, montelukast, cromolyn sodium, immunomodulators, and anti-TNF therapies. Although some literature exists supporting their use, the 2020 guidelines recommend their use only in the setting of a clinical trial.

13 Prognosis of Eosinophilic Esophagitis

Although 81% of patients diagnosed with EoE in childhood experience regression of their symptoms as young adults, both prospective and retrospective studies indicated that spontaneous remission of EoE is uncommon. Treatment is necessary as untreated EoE is associated with fibrostenotic complications; untreated EoE has not been associated with Barret's esophagus or premalignant state.

As the guidance for management continues to evolve, the 2020 American Gastroenterological Association and the Joint Task Force on practice parameters for the management of EoE serve as the most up-to-date consensus on best practices. It is important to recognize that there is no FDA-approved treatment for eosinophilic esophagitis. It is also important to note that all therapeutics evaluated are applicable to short-term treatments only as current evidence is based on trials extending from 2 to 16 weeks. Long-term therapy with medicaments, such as topical steroids and PPIs has not been studied—either for efficacy or long-term side effects. Although side effects such as nutritional deficiencies can be managed in a patient on dietary restriction, long-term use has a significant impact on quality of life in other conditions for which it is used.

Case 2

A 54-year-old man presented for evaluation of a 6-month history of worsening pruritic skin rash. The rash is maculopapular and was unresponsive to topical corticosteroids. He was evaluated by a dermatologist and skin biopsy was performed revealing eosinophilic infiltration. Complete blood count revealed an absolute eosinophil count (AEC) of 1600 cells/μL. Six weeks later, his AEC increased to 4600 cells/μL. Patient-reported no respiratory or gastrointestinal symptoms. He is an active smoker with a 10-pack year smoking history. He reported no history of travel outside the United States. Physical examination revealed a diffuse maculopapular rash involving the trunk. No hepatosplenomegaly was noted. Chest X-ray showed no evidence of parenchymal lung abnormality. His cardiac Troponin I was within normal limit. Echocardiographic evaluation showed a normal ejection fraction with no cardiac motion abnormalities. His liver enzymes were within normal limits. Serum B12 level was 400 pg/μL (normal, 200–1100 pg/μL). Quantitative immunoglobulin levels including IgG, IgM, and IgA levels were within normal limits, but his serum IgE level was elevated at 1100 IU/L. Lymphocyte subsets analysis by flow cytometry showed a $CD3^+$ T cell count of 1000 cells/μL with $CD4^+$ count of 1000 and $CD8^+$ of 200/μL. The patient was started on prednisone 1 mg/kg/day and repeat AEC 3 days later was 300 cells/μL. His skin rash improved.

Question 1
Which of the following is the most accurate statement?

A. This patient has hypereosinophilia of undetermined significance
B. This patient has hypereosinophilic syndrome
C. This patient has idiopathic hypereosinophilia
D. This patient has clonal hypereosinophilia

Answer: B
This patient has hypereosinophilic syndrome as evidenced by his peripheral blood hypereosinophilia and the presence of end-organ damage (cutaneous involvement). This patient most likely has a form of reactive hypereosinophilia. Hypereosinophilia describes the presence of hypereosinophilia in the absence of end-organ damage.

14 Hypereosinophilia and Hypereosinophilic Syndromes

Hypereosinophilia is defined as an absolute eosinophil count of 1500 cells/microliter on two occasions at least 4 weeks apart and/or tissue hypereosinophilia. Based on the underlying basis of eosinophilia, hypereosinophilia is classified into four major categories: (1) reactive or secondary (HE_R) driven by eosinopoetic cytokines, (2) primary or clonal HE (HE_N), (3) familial (HE_{FA}), and, lastly, (4) idiopathic in which no underlying basis for eosinophilia exists (HE_{US}) (Table 1). Patients with hypereosinophilia and end-organ damage or dysfunction attributable to tissue HE are said to have hypereosinophilic syndrome (HES). Therefore, other causes for organ damage should be excluded before making the diagnosis of HES. HES is often diagnosed in the fifth decade of life (median age at diagnosis, 45 years) and the disorder has an overall slight male predominance (M:F ratio is 1.3). Eosinophils can cause tissue damage by promoting fibrosis and the clinical manifestations of HES vary depending on the organ(s) involved. The most frequently affected organs in HES are the skin, gut, and lungs with gastrointestinal involvement being more common in pediatric patients whereas pulmonary involvement is more common in adult patients. Serious manifestations such as those secondary to neurological and cardiovascular involvements are relatively less frequent. Further, eosinophils can promote thrombosis, yet another sign of end-organ damage, which is thought to be related to the excess expression of tissue factor in the eosinophils of patients with HES. These various causes of HE and the wide clinical manifestations of HES call for a thorough evaluation to identify evidence of end-organ damage and the basis of HE, as timing and type of therapy differ by the extent of tissue damage and the underlying variant of HES.

Table 1 Mechanism of the different classes of hypereosinophilia (HE) and hypereosinophilic syndromes (HES)

Mechanism of HE	HE classification	HES variant
Excess IL-5 production	HE_R	Associated HES Overlap HES L-HES
Clonal expansion of eosinophils	HE_N	M-HES
Unknown	HE_{US}	Idiopathic HES
Unknown	HE_{FA}	Familial HES

Abbreviations: *HE_R* reactive or secondary, *L-HES* lymphocytic variant of HES, *HE_N* neoplastic or clonal HE, *M-HES* myeloid variant of HES, *HE_{FA}* familial HE, *HE_US* HE of undetermined significance

Question 2
Which of the following is the most likely diagnosis in the patient described above?

A. L-HES
B. M-HES
C. Associated HES
D. EGPA

Answer: A
The patient is likely to have L-HES. His age at onset, the indolent course (6-month duration), predominant cutaneous manifestation, elevated IgE level, and the presence of a likely abnormal T cell clone as suggested by $CD3^+$ count being less than the sum of $CD4^+$ and CD8+ are all feature suggestive of L-HES.

15 Hypereosinophilic Syndrome Variants

Associated variant of HES: In this IL-5-dependent variant of HES, the cause of hypereosinophilia is secondary to a well-defined allergic, immunologic, infectious, or neoplastic disease and that resolution is expected with the treatment of such underlying cause (Table 2). While it is arguable that these patients represent a separate entity, their clinical presentation can be indistinguishable from patients with idiopathic HES. Among the common causes of the associated variant of HES are drug hypersensitivity, helminthic infection, sarcoidosis, and neoplasms, such as lymphoma. In the pediatric age group, HE can be secondary to atopic diseases, such as atopic dermatitis or allergic asthma, or can occur due to primary immunodeficiency disorders, such as hyper IgE syndrome, due to dominant-negative mutation in STAT3, CARD11, or recessive mutation in DOCK8. Other rare etiologies of associated HES variant are IgG4-related disease that may have overlapping clinical features with L-HES (see below) and sarcoidosis. Elevated IgE level is common to many of the secondary causes of HE and is a useful clinical clue, but resolution of hypereosinophilia with the treatment of the underlying cause is necessary to confirm the diagnosis of this variant of HES (Table 3).

Myeloid Variant of HES (M-HES): In this variant of HES, excess production of eosinophils by the bone marrow is independent of IL-5 and results from clonal expansion of mature eosinophils. The World Health Organization (WHO) categorizes M-HES as chronic eosinophilic leukemia. Interstitial deletion in chromosome 4q12 that results in FIP1L1-PDGFRA rearrangement, a fusion protein with a constitutive tyrosine kinase activity or less common cytogenetic abnormalities, such as rearrangements involving the PDGFRB, FGFR1, or JAK2 genes, account for the clonal expansion of eosinophils in M-HES. Clonal hypereosinophilia can also be observed in other myeloproliferative disorders, such as BCR-ABL1-positive chronic myeloid leukemia, polycythemia rubra vera, and systemic mastocytosis. M-HES has a male predominance, an aggressive clinical course with higher odds of cardiac and neurological involvement, and may be unresponsive to glucocorticoids. Akin to other myeloproliferative disorders, patients often have hepatomegaly, splenomegaly, and abnormalities in other cell lines, such as anemia, thrombocytopenia, and elevated serum vitamin B12 levels (Table 3). Higher serum tryptase level is noted in these individuals; however, levels are not as high as in systemic mastocytosis patients. Bone marrow biopsy with cytogenetic study is key to establish the diagnosis and to guide therapeutics.

Lymphocytic or Lymphoid Variant (L-HES): In this IL-5-dependent variant of HES, there is an abnormal clone of T cells producing excess amount of IL-5 cytokine which drives the overproduction of eosinophils (i.e., clonal Th2 cells). The immunophenotype of such an aberrant IL5-producing T cell clone is variable, but CD3-CD4+ is the most common. Other reported immunophenotypes include CD3+CD4+CD7− and CD3+CD4−CD8− TCRαβ+. Unlike M-HES, L-HES equally affects men and women and has an indolent course with the prediction for skin and soft tissue involvements. Cutaneous manifestations are variable and

Table 2 Causes of associated variant of HES

Most common
• Helminthic infections (e.g., strongyloidiasis, toxocariasis, scabies)
• Drug hypersensitivity
• Neoplasms (e.g., lymphoma, solid tumors)
Less common
• Atopic diseases especially in children
• Allergic bronchopulmonary aspergillosis
• Primary immunodeficiency disorders (e.g., omen syndrome, job syndrome, DOCK8 deficiency)
• Human immunodeficiency virus (HIV) infection
• Graft-versus-host disease
• Sarcoidosis
• IgG4-related disease
• Inflammatory bowel diseases

Table 3 Key features of the different HES variants

Variant of HES	Key features	Response to steroid	Response to anti-IL5
Associated variant	Elevated IgE levels, resolves with treatment of the underlying cause	N/A[a]	N/A[b]
Overlap variant	Single organ involvement (often gastrointestinal and pulmonary), positive ANCA (in EGPA)	+++	+++
Lymphoid variant	Predilection for cutaneous involvement, abnormal T cell clone, elevated TARC, and IgE levels, possibly elevated IgM levels, episodic angioedema (in Gleich syndrome), risk of T cell lymphoma transformation	++	++
Myeloid variant	Male predominance, aggressive course, corticosteroid-resistant, cardiac/neurologic involvement, cytopenia, promyelocytes on peripheral blood smear, hepatosplenomegaly, elevated B12 levels, elevated tryptase levels	−/+	−
Familial variant	Positive family history, mostly asymptomatic (diagnosis of exclusion)	No data available	No data available
Idiopathic variant	No specific features, negative family history (diagnosis of exclusion)	+++	++

[a] Systemic glucocorticoid therapy may not be necessary
[b] Anti-IL5 therapy use and response will be dependent on the underlying cause

range from diffuse pruritus, urticarial plaques, and macular to maculopapular rash. Lymphadenopathy, gastrointestinal, and pulmonary involvement are also common in these patients. In keeping with the cytokine profile of this variant of HES, elevated IgE and TARC levels are observed in these patients who may additionally exhibit atopic diathesis (Table 3). A unique subset of the lymphocytic variant HES, characterized by cyclic angioedema and eosinophilia, is known as Gleich syndrome. Clonal expansion of CD32 + CD41+ T cells and elevated serum IgM levels are observed in this unique subset of L-HES patients. L-HES should be distinguished from IgG4-related disease, a rare cause of associated HES variant, that exhibits overlapping clinical phenotype, but the presence of elevated IgG4 levels and the absence of aberrant T cell phenotype support the former. Diagnosis of L-HES remains challenging. Lymphocyte subset analysis by flow cytometry may offer a clue to the presence of an abnormal T cell clone, such as CD3-CD4+, but T cell receptor rearrangement study remains necessary to detect the presence of T cell clonality. However, the sensitivity of such testing remains limited. While the clinical course of lymphoid variant HES is indolent in the majority of patients, it is complicated by those patients being at risk of developing T cell lymphoma

Overlap Variant of HES: This is yet another IL-5-dependent HES variant in which hypereosinophilia exists in conjunction with a single organ eosinophilic disorder, such as eosinophilic gastrointestinal disorder, eosinophilic pneumonia, or eosinophilic skin diseases. Eosinophilic granulomatosis with polyangiitis (EGPA) is a unique subset of overlap HES, in which blood vessel involvement results in multisystem manifestations (Table 3). As these single organ eosinophilic disorders only occasionally manifest hypereosinophilia, the presence of such high levels of peripheral blood eosinophils makes them earn the "overlap" title. Because it is often difficult to rule out clonal eosinophilia in these patients, bone marrow biopsy is advisable, and biopsy of the involved organ is often necessary to establish a definitive diagnosis

Familial Hypereosinophilia (HEFA) and Familial HES: This rare variant of HES is inherited in an autosomal dominant fashion (MIM 131400). Genome-wide linkage analysis mapped familial hypereosinophilia to chromosomal region 5q31q33 where a cluster of cytokine genes including genes for IL-5, IL-3, and GM-CSF are located. The levels of these cytokines including serum level of IL-5 are normal in these patients. Eosinophils from patients with familial hypereosinophilia exhibit normal morphology; however, there are increased serum levels of eosinophil granular proteins and increased surface expression of CD69, CD25, and HLA-DR, which are markers of eosinophil activation. The observed increased markers of eosinophil activation and serum eosinophil granular protein levels are less than what is observed in other variants of HES, which may explain why patients with familiar hypereosinophilia are often asymptomatic (Table 3). End-organ damage including endomyocardial fibrosis and neurological involvement has been reported in few affected family members. Factors associated with disease progression in this small subset of patients with familial hypereosinophilia remain unknown. Familial HES is a diagnosis of exclusion and should be considered when other family members report hypereosinophilia

Idiopathic Variant of HES: A thorough evaluation to identify a cause for HES is non-revealing in over 50% of patients, and hence these cases are known as idiopathic HES. Idiopathic HES should be distinguished from HEUS, in which persistent hypereosinophilia exists without end-organ damage and without features to suggest familial or an alternative basis for hypereosinophilia such as HEN or HER. Watchful waiting in HEUS to identify early signs and symptoms of end-organ damage is necessary. Symptomatic patients with idiopathic HES require treatment to prevent progressive end-organ damage

Question 3

Which of the following is a feature of L-HES?

A. Elevated serum tryptase level
B. Elevated serum B12 level
C. Elevated serum TARC level
D. Low IgM level

Answers: C

Elevated TARC level. In keeping with the type-2 cytokine profile of L-HES, elevated IgE and TARC levels are observed in these patients.

Question 4

Which of the following is the best test to confirm the diagnosis of L-HES?

A. Bone marrow biopsy and cytogenetic study for *FIP1L1-PDGFRA* rearrangement
B. T cell receptor rearrangement study
C. Serum IgG4 level
D. Gain of function mutation in a gene that codes for CD113

Answer: B

T cell receptor rearrangement study will assist in detecting the presence of T cell clonality, a unique feature of L-HES.

Bone marrow biopsy and cytogenetic study for FIP1L1-PDGFRA rearrangement will assist in diagnosing M-HES. L-HES should be distinguished from IgG4-related disease, a rare cause of associated HES variant, that exhibits overlapping clinical phenotype, but the presence of elevated

IgG4 levels and the absence of aberrant T cell phenotype support the former. Gain of function mutation in a gene that codes for CD113 or c-kit mutation is associated with systemic mastocytosis.

Question 5

In discussing treatment options for patients with HES, which of the following HES variants is least responsive to corticosteroids?

A. Lymphocytic variant HES
B. Overleap variant HES
C. Myeloid variant HES
D. Idiopathic variant HES

Answer: C

Myeloid variant HES can be resistant to steroids. All other variants are responsive to corticosteroids.

16 Approach to Patients with Hypereosinophilia and HES

Patients presenting with hypereosinophilia should be evaluated for the presence of end-organ dysfunction (Fig. 2). This evaluation includes performing detailed history, physical examination, and diagnostic testing. Diagnostic tests to evaluate for underlying end-organ damage may include electrocardiogram, echocardiograph, cardiac enzymes, chest x-ray and/or computerized tomography (CT) of the chest, pulmonary function testing, and liver and renal function tests. In patients with hypereosinophilia with no evidence of end-organ damage, identifying and addressing an underlying treatable cause such as helminthic infection or drug hypersensitivity (i.e., HE_R) should be sought to prevent disease progression. This evaluation includes a careful history and directed laboratory testing. Particular attention should be paid to allergic disorders, travel, and drug history. Atopic diseases, such as allergic asthma and atopic dermatitis can cause hypereosinophilia, especially in children (Table 2). Remote or recent travel history, especially to tropical regions is relevant. Both prescription and nonprescription medications should be reviewed in relationship to disease onset and unnecessary medications should be discontinued. For those where no obvious secondary cause for hypereosinophilia or if treatment of the presumed underlying cause failed to resolve patient's eosinophilia, evaluation of underlying myeloid or lymphocytic variant HE should be sought especially when there are clinical features to support these variants (Table 3). Peripheral blood smears, serum levels of vitamin B12, and tryptase levels should be checked. The presence of immature granulocytes, elevated B12, or tryptase levels may support the presence of a myeloproliferative process. Peripheral blood analysis for *FIP1L1-PDGFR*A by fluorescence in situ hybridization (FISH) or RT-PCR should be considered. Bone marrow biopsy and cytogenetic studies may also be needed when the suspicion remains high. Lymphocyte subset analysis by flow cytometry and T cell receptor rearrangement may identify abnormal T cell clones, typical of L-HES. In patients with an otherwise HE_{US}, watchful waiting with interval follow-up for end-organ damage can be advised.

In patients with evidence of end-organ damage, urgent treatment is necessary for those with life-threatening manifestations, such as cardiac, pulmonary, or neurologic involvement, and corticosteroids are often the first line of therapy (prednisone 1 mg/kg/day). Because systemic corticosteroids are associated with increased mortality in patients with

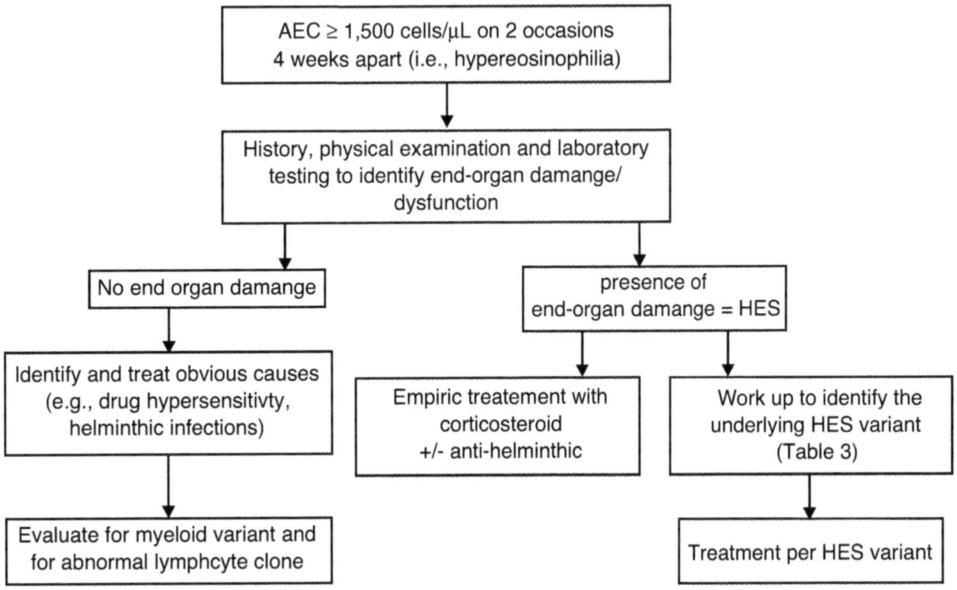

Fig. 2 General approach to patients with hypereosinophilia. Abbreviations: *AEC* absolute eosinophil count, *HES* hypereosinophilic syndrome

strongyloidiasis, those with a travel history outside the United States (even if it was remote), should also receive empiric anti-helminthic therapy with a 2-day course of ivermectin. When the clinical suspicion of M-HES is high (Table 3), imatinib (± corticosteroids for those with cardiac involvement) should be given as a first-line agent. For HES patients without life-threatening features, or those initiated on empiric therapy, further treatment can be tailored toward the underlying cause of hypereosinophilia or the variant of HES.

Treatment of associated HES variant is directed toward the underlying cause. In patients with suspected drug hypersensitivity, the offending drug should be discontinued, and in patients with helminthic infections, an anthelminthic agent, such as ivermectin, should be attempted. Neoplasms, such as lymphoma, should be treated according to established guidelines. If no response is observed with proper treatment of the presumed underlying cause of HE, alternative variants of HES should be considered.

As discussed above, in the M-HES, there is clonal expansion of eosinophils that is independent of the IL-5 cytokine and is driven by acquired molecular abnormalities that drive eosinophil proliferation and expansion. The IL5-independent nature of M-HES explains why these patients are also resistant to anti-IL-5 therapy with mepolizumab or benralizumab. This survival advantage created by the underlying molecular abnormalities explains why these patients tend to be resistant to corticosteroid therapy and that M-HES should be suspected when high-dose corticosteroid fails to decrease the eosinophil count within 48 h. Alternatively, the approach to M-HES is dictated by the underlying cytogenetic abnormality and whether the HE is part of a WHO-defined myeloid neoplasm. In those with myeloid neoplasm with eosinophilia and abnormality of *PDGFRA, PDGFRB, FGFR1, or JAK2*, the resultant cytogenetic abnormality with intrinsic tyrosine kinase activity offers a therapeutic opportunity. For example, in the most commonly identified cytogenetic abnormality of *FIP1L1-PDGFRA* (F/P-positive M-HES), imatinib results in universal hematological and molecular remission in these patients. In some of the F/P-positive patients, a cure was attained (i.e., molecular-free remission even after cessation of imatinib). In those without cytogenetic abnormality, imatinib in high doses may be attempted, but standard treatment for acute myeloid leukemia should be offered. Patients with other WHO-defined myeloid neoplasms are treated according to the standard guidelines for these neoplasms, but this is beyond the scope of this chapter.

Patients with lymphocytic and idiopathic variants of HES are both responsive to steroids, but relapse is common. Patients who fail to respond to corticosteroids, develop debilitating side effects, or require a prolonged course, would benefit from the use of a steroid-sparing agent, such as hydroxyurea, cyclosporine, azathioprine, INF-α, or more recently, mepolizumab. As a last resort, alemtuzumab, an anti-CD52 monoclonal antibody, may be considered. The relative safety profile of mepolizumab makes it an attractive steroid-sparing agent in these patients. In a sub-analysis of the international, randomized, double-blind, placebo-controlled clinical trial to investigate the efficacy of mepolizumab as a corticosteroid-sparing agent in F/P-negative HES, mepolizumab at a dose of 750 mg intravenously was found to be effective as a corticosteroid-sparing agent in these patients. Mepolizumab at a dose of 300 mg subcutaneously every 4 weeks is an FDA-approved treatment for patients with F/P-negative HES (>6 months) including patients with lymphocytic variant HES. A phase II study of benralizumab, anti-IL5 receptor alpha monoclonal antibody, resulted in clinical and hematological response in four patients with L-HES and six other patients with idiopathic variants; however, three patients with L-HES had relapse.

Topical therapies for single organ eosinophilic disorders, such as eosinophilic gastrointestinal disorders with hypereosinophilia, may be insufficient, and systemic corticosteroids may be necessary. Patients with single organ eosinophilic disorders with HE are responsive to medium doses of corticosteroids (<20 mg/day). Anti-IL5 therapy with mepolizumab offers a steroid-sparing effect. In a study evaluating the long-term outcome of high-dose mepolizumab (750 mg intravenously every month), all the six subjects with overlap variant HES showed complete response (defined as resolution of symptoms, eosinophilia, and requirement of <10 mg/day of prednisone). Patients with multisystem involvement, such as EGPA, often require higher doses of corticosteroids. In patients with relapsing or refractory EGPA, mepolizumab at a dose of 300 mg subcutaneously every 4 weeks resulted in remission in approximately 50% of patients with a significantly lower relapse rate relative to placebo. Mepolizumab at a dose of 300 mg subcutaneously every 4 weeks is an FDA-approved treatment for patients with F/P-negative HES (>6 months) including patients with overlap variant HES. Recently, a phase II study of benralizumab, an anti-IL5 receptor alpha monoclonal antibody, resulted in clinical and hematological responses in six patients with overlap variant HES.

Given the rarity of the familial variant of HES, a consensus treatment approach is lacking. Fortunately, the majority of patients are asymptomatic, and watchful waiting is all that is necessary. In the small subset of patients who develops end-organ dysfunction, a treatment approach similar to those with an idiopathic variant of HES has been recommended.

Bibliography

1. Kita H, Bochner B. Biology of eosinophils. In: Burks A, et al., editors. Middleton's allergy principles and practice. Philadelphia, PA: Elsevier Inc.; 2019. p. 255–66.
2. Abbas AK, et al. Cellular and molecular immunology. Philadelphia, PA: Elsevier; 2018. p. 565.
3. Gonzalez-Cervera J, et al. Association between atopic manifestations and eosinophilic esophagitis: a systematic review and meta-analysis. Ann Allergy Asthma Immunol. 2017;118(5):582–590 e2.
4. Arias A, et al. Systematic review with meta-analysis: the incidence and prevalence of eosinophilic oesophagitis in children and adults in population-based studies. Aliment Pharmacol Ther. 2016;43(1):3–15.
5. Hruz P. Epidemiology of eosinophilic esophagitis. Dig Dis. 2014;32(1–2):40–7.
6. Hirano I, et al. AGA institute and the joint task force on allergy-immunology practice parameters clinical guidelines for the management of eosinophilic esophagitis. Gastroenterology. 2020;158(6):1776–86.
7. Gomez Torrijos E, et al. Eosinophilic esophagitis: review and update. Front Med (Lausanne). 2018;5:247.
8. Spergel J, Aceves SS. Allergic components of eosinophilic esophagitis. J Allergy Clin Immunol. 2018;142(1):1–8.
9. Rothenberg ME. Molecular, genetic, and cellular bases for treating eosinophilic esophagitis. Gastroenterology. 2015;148(6):1143–57.
10. Furuta GT, Katzka DA. Eosinophilic esophagitis. N Engl J Med. 2015;373(17):1640–8.
11. Davis BP. Pathophysiology of eosinophilic esophagitis. Clin Rev Allergy Immunol. 2018;55(1):19–42.
12. Dellon ES, et al. Updated international consensus diagnostic criteria for eosinophilic esophagitis: proceedings of the AGREE conference. Gastroenterology. 2018;155(4):1022–1033 e10.
13. Rank MA, et al. Technical review on the management of eosinophilic esophagitis: a report from the AGA institute and the joint task force on allergy-immunology practice parameters. Ann Allergy Asthma Immunol. 2020;124(5):424–440 e17.
14. Dellon ES, Hirano I. Epidemiology and natural history of eosinophilic esophagitis. Gastroenterology. 2018;154(2):319–332 e3.
15. Klion A. Hypereosinophilic syndrome: approach to treatment in the era of precision medicine. Hematology Am Soc Hematol Educ Program. 2018;2018(1):326–31.
16. Butt NM, et al. Guideline for the investigation and management of eosinophilia. Br J Haematol. 2017;176(4):553–72.
17. Khoury P, et al. Hypereosinophilic syndrome subtype predicts responsiveness to glucocorticoids. J Allergy Clin Immunol Pract. 2018;6(1):190–5.
18. Kuang FL, et al. Long-term clinical outcomes of high-dose mepolizumab treatment for hypereosinophilic syndrome. J Allergy Clin Immunol Pract. 2018;6(5):1518–1527.e5.
19. Roufosse F, et al. Efficacy and safety of mepolizumab in hypereosinophilic syndrome: a phase III, randomized, placebo-controlled trial. J Allergy Clin Immunol. 2020;146(6):1397–405.
20. Kuang FL, et al. Benralizumab for PDGFRA-negative hypereosinophilic syndrome. N Engl J Med. 2019;380(14):1336–46.

A Case-Based Board Review of Angioedema and Urticaria

Marisa Riley, Sally Ng, and Timothy Craig

1 Introduction

Angioedema is a clinical manifestation that can be caused by various etiologies. It can be acquired, hereditary, or medication-induced. Histamine can induce urticaria, angioedema, or both; however, urticaria does not occur with bradykinin-induced disease. To determine the type of angioedema that a patient has on presentation, a thorough history, physical exam, and, in certain situations, diagnostic tests are required to determine the proper treatment. The pathophysiology involves either bradykinin or histamine. This chapter explains the differences among the types of angioedema and the appropriate workup and management of each.

Case 1: A Family Matter
Caitlin is an 18-year-old female who was in good health until starting birth control pills for severe acne, after which she developed severe abdominal pain that progressed to nausea and vomiting. She presented to the emergency room and was found to have the CT scan results shown in Fig. 1a. Despite changes inconsistent with an appendiceal pathology, an appendectomy was still performed, which was normal on pathology. A month later, she again presented to the emergency room with hand swelling (Fig. 1b). You are on duty and responsible for the case.

M. Riley
Penn State College of Medicine, Hershey, PA, USA

S. Ng
Department of Pediatrics, Robert Wood Johnson Medical School, New Brunswick, NJ, USA

T. Craig (✉)
Medicine, Pediatrics and Biomedical Sciences, Penn State University, Hershey, PA, USA

Question 1
Based on the patient's presentation, what tests would you order to make the diagnosis?

A. C3, C4, and CH50
B. C1q, C1-inhibitor function, and protein
C. C4, C1-inhibitor function, and protein
D. C4, C2, and CH50

Based on the guidelines, the appropriate tests for a patient this age would be answer (C). If the patient first presented in middle age, acquired -angioedema with low C1-inhibitor (AA), previously called acquired-angioedema, would be a viable diagnosis, and C1q would help differentiate it from Hereditary Angioedema (HAE). C4 is the best screening test for HAE type 1 and 2 since it is readily available, easy to perform, and a sensitivity approaches 100% during an attack. CH50 is used to assess for complement deficiency and adds nothing to the diagnosis of HAE. C3 can be used to assess the severity of SLE, but C3 is not useful for HAE (Table 1).

Caitlin's tests from the last visit return with the following results:

- C4 of 5 mg/dl (normal above 12–45 mg/dl).
- C1-inhibitor functional of 45% (normal above 67%).
- C1-inhibitor protein of 10 mg/dl (normal 16–33 mg/dl).

Question 2
How long would you expect the symptoms of HAE to persist without treatment?

A. 12 h
B. 24 h
C. Symptoms may persist until treated
D. 72 h

The answer is D

M. Mahmoudi (ed.), *Absolute Allergy and Immunology Board Review*, https://doi.org/10.1007/978-3-031-12867-7_13

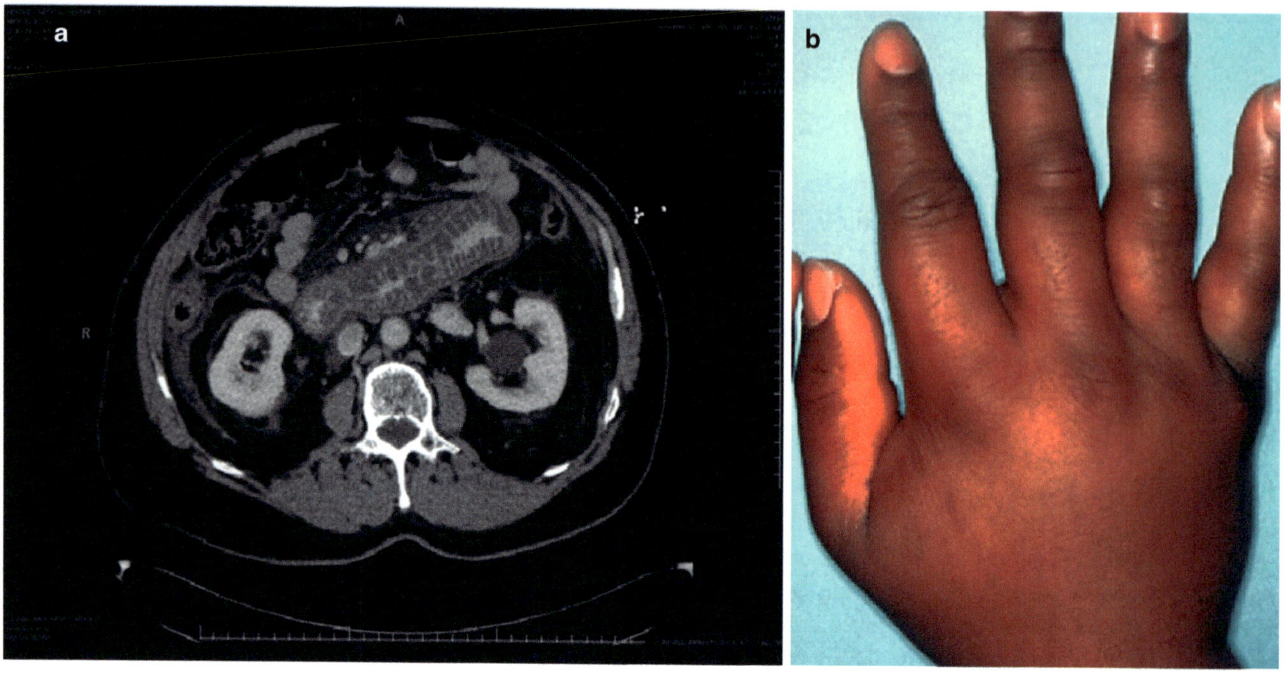

Fig. 1 (**a**) Abdominal CT during the period of pain on presentation to the Emergency Department (ED). (**b**) Hand swelling on presentation to the ED

Table 1 Tests for angioedema. HAE is Hereditary Angioedema. HAE type 3 is best referred to as HAE with normal C1-inhibitor. ACE-I is angiotensin-converting enzyme inhibitor

Four types of bradykinin induced angioedema excluding ACE-I

	HAE			Non-HAE
Parameter:	I	II	III	Acquired with low C1-inhibitor
Percentage of HAE	85	15	rare	Not HAE Rare
C4	Low	Low	Normal	Low
C1INH protein	Low	Normal	Normal	Low
C1INH activity	Low	Low	Normal	Low

Symptoms of HAE persist for 2–5 days and resolve without treatment. The guidelines stress that all attacks, even those that are not life-threatening, should be treated as early as possible. In contrast, histamine-induced angioedema is usually rapid in both onset and resolution, usually resolving in 24 h. In addition, abdominal pain is unlikely to occur with histamine angioedema but is a more frequent occurrence in hereditary angioedema and as many as 50% or more of the angioedema attacks may have abdominal pain as a symptom.

Question 3

Knowing Caitlin's laboratory tests, which of the medications would you give her in the emergency department?

A. Icatibant
B. Subcutaneous C1-inhibitor
C. Lanadelumab
D. Danazol

The answer is A

Subcutaneous C1-inhibitor, androgens, berotralstat (a new oral kallikrein inhibitor), and lanadelumab are all for long-term prophylaxis. C1-inhibitor intravenous therapy can be used for long-term prophylaxis as well as pre-procedural, often called short-term prophylaxis, and for treatment of attacks. Other "on-demand" therapies for immediate treatment of attacks include the medications ecallantide, icatibant, and recombinant C1-inhibitor. The preferred therapy for short-term prophylaxis is intravenous C1-inhibitor, but androgens and fresh frozen plasma (FFP) can also be used if C1-inhibitor is not available. FFP and intravenous C1-inhibitor are given just prior to a procedure, while androgens must be started about 5 days before the procedure since they work by inducing protein production. Notably, antihistamines, corticosteroids, and epinephrine are neither effective for treatment nor prevention of HAE.

As noted above, treatment includes on-demand therapy to treat attacks, short-term prophylaxis (STP) to prepare patients for surgery and procedures, and long-term prophylaxis (LTP) to prevent attacks. On average, most on-demand therapies have a failure rate of 10%, and for this reason patients are always advised to carry 2 doses of on-demand therapy and use it as early as possible during an attack. STP should be administered before any procedure that approximates the upper airway and would include dental extractions, medical procedures such as endoscopy and intubation. Also, any traumatic procedure could stimulate angioedema and thus for most surgeries, unless minimal, STP is given. Lastly, LTP adds significantly to the quality of life and can reduce attacks from multiple episodes a month to none in some cases. The decision to start LTP is often difficult, but in most cases if on-demand therapy cannot return a person's life to near normal, then LTP should be considered.

It can be difficult to differentiate whether a patient's angioedema is caused by histamine or by bradykinin, especially without testing and in patients with Hereditary Angioedema with normal C1-inhibitor, previously referred as type 3 HAE. This can make it difficult to determine which treatment regimens to use since treatments differ between histamine and bradykinin-induced disease. Caitlin's case was obvious when the laboratory tests returned, but certain exam findings helped hint at the diagnosis as well.

Question 4

Which feature of angioedema suggests the etiology to be histamine or bradykinin?

A. Histamine-induced angioedema may present with erythema marginatum
B. Bradykinin angioedema is associated with urticaria
C. With hereditary angioedema up to 75% have a family history
D. Histamine-induced angioedema is frequently inherited

The answer is C

Certain exam findings and historic information can often favor bradykinin or histamine-induced disease (Table 2). Histamine-induced angioedema may occur with or without urticaria. If urticaria is present, the angioedema is not secondary to bradykinin. Hereditary angioedema is autosomal dominant, and family history is apparent in 75% of cases while the additional 25% are new mutations. Two strong triggers for HAE are estrogen and ACE-inhibitors, and both are relatively contraindicated in HAE. Trauma is also a trigger for HAE. In some individuals, a rash prodrome called erythema marginatum occurs with HAE. Erythema marginatum may precede the angioedema or occur with the angioedema, and on occasion may occur without angioedema (Table 2).

Table 2 Comparing symptoms of histamine to bradykinin-induced angioedema

Comparing histamine to bradykinin induced angioedema

	Histamine Induced	Bradykinin Induced
Associated with urticaria?	often	no
Duration of swelling?	hours to a day	2-5 days
Frequency of bowel symptoms?	infrequent	50% of attacks
Family history?	no	yes
Trigged by trauma?	no	yes
Responsive to corticosteroids and epinephrine?	yes	no
Triggered by estrogen?	possible	Often

Case 2: The Chronic Conundrum

Ethan is a 45-year-old male with a recurrent rash for many months. The rash is itchy, transient, and does not cause pigment changes. The rash persists for less than 24 h and is replaced by a similar rash on different parts of the body. He denies physical causes of urticaria, except dermatographia. His medical history is otherwise benign, and his exam is normal except for his dermatographia. You diagnose him with spontaneous urticaria. He wants to know what is causing the rash.

Question 5

What laboratory tests are indicated for chronic spontaneous urticaria?

A. IgE for foods, ANA, CBC, ESR, TSH, chemistry panel, and UA
B. No tests are required unless indicated by history or exam
C. Tryptase, CBC, LFT, UA and CRP
D. C4, CBC, ESR, TSH, chemistry panel, and UA

The answer is B

The guidelines are clear that testing is not required unless indicated by history and exam. The physician may order a CBC, chemistry panel, IgE and TPO as a general health screen, but this is not an essential part of the workup. Biopsy is indicated if the rash has atypical findings, such as persistence for days, pain, or purpura. Skin testing for aeroallergens and foods is not indicated unless there is a strong suspicion of food causing recurrent hives.

Ethan did not require any testing. He requested skin testing for aeroallergens, but you explained it was not indicated and the etiology, if not suggested by the history or exam, is usually autoimmune and referred to as "spontaneous" since the underlying cause of chronic urticaria is rarely found. You start him on cetirizine 10 mg a day and see him back in a month. At that time, he continues to have urticaria.

Question 6

Since one 10 mg cetirizine tablet did not control his urticaria, what would your next therapeutic suggestion be?

A. Add montelukast 10 mg a day
B. Add an H2-blocker such as famotidine
C. Add a first-generation antihistamine
D. Increase the dose of cetirizine 10 mg to two tablets twice a day

The answer is D

The guidelines suggest initial therapy is a second-generation antihistamine, such as loratadine 10 mg, desloratadine 5 mg, cetirizine 10 mg, or fexofenadine 180 mg, orally once a day. If single tablet therapy fails, the second-generation antihistamine, referred to as low or non-sedating, should be quadrupled to the dose of two tablets bid. The guidelines place little emphasis on adding montelukast or H2-blockers secondary to a lack of objective data demonstrating efficacy. First-generation antihistamines should be avoided because of adverse effects such as sedation.

Ethan fails to respond to cetirizine 20 mg twice a day. You discuss the benefits and risks of omalizumab and cyclosporine. During this discussion, you review the side effects and benefits of both.

Question 7

What are the correct benefits and side effects of omalizumab and cyclosporine?

A. Because the dose of cyclosporine is low, serum levels of cyclosporine are not indicated
B. Because the dose of cyclosporine is low, renal function tests and blood pressure are not essential to follow
C. Because the risk of anaphylaxis to omalizumab is 3–4%, observation for allergic reaction should be 2 h after every dose
D. The risk of cancer associated with omalizumab is significant, and therefore informed consent should be obtained

The answer is A

The guidelines suggest that the next step if a patient fails high-dose second-generation antihistamines is to use omalizumab 300 mg subcutaneous once a month. Omalizumab response rate with complete control of urticaria is about 55%. However, urticaria often rebounds following treatment termination. Unlike the phase three study for omalizumab for asthma, which demonstrated a risk of neoplasms, the long-term observational studies failed to demonstrate an association between omalizumab and malignancy. The anaphylaxis rate of omalizumab is considered by most to be approximately 0.1%.

Cyclosporine is used as the last option in the management of chronic urticaria and is given in low doses of 2–5 mg/kg. Adverse effects at this dose are of minimal severity and mainly consisted of GI complaints and peripheral neuropathy. Guidance suggests that blood pressure and renal function should be followed; however, cyclosporine serum levels are not needed. The improvement with cyclosporine is 63% and benefit often persists after therapy is withdrawn (Fig. 2).

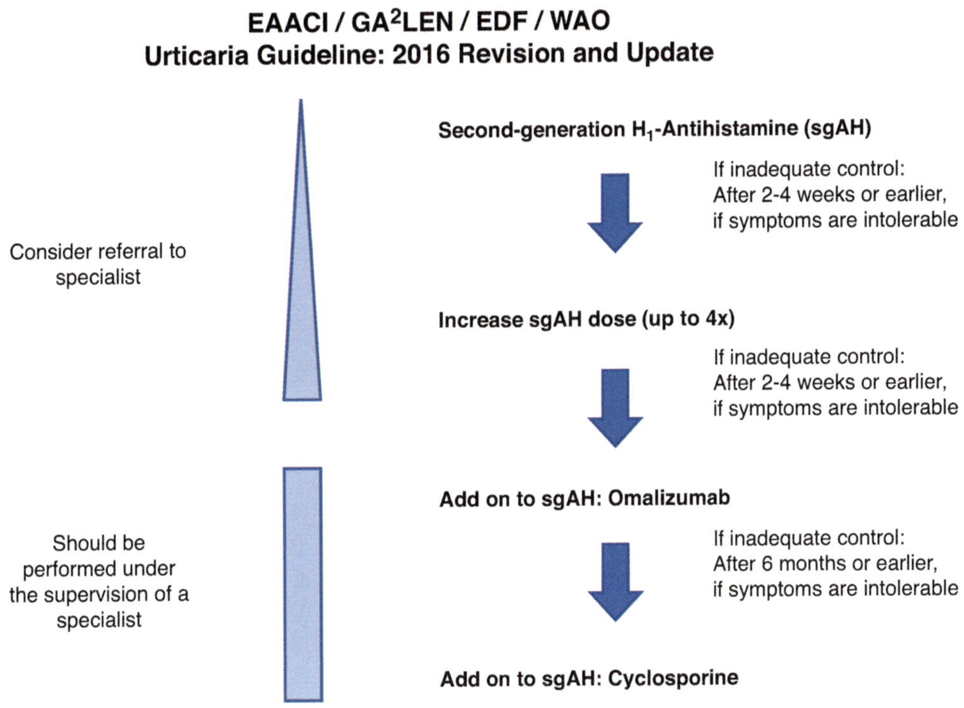

Fig. 2 Algorithm for treatment of urticaria developed from The WAO Urticaria Guideline

Case 3: The Medication Mishap

Michael is a 48-year-old male. He has hypertension, back pain, diabetes mellitus (DM), hyperlipemia, and a new history of angioedema. Last week he presented with angioedema of the tongue to the ED. After receiving antihistamines, corticosteroids, and an overnight observation, he was discharged to follow up with you. He had no other history of urticaria or angioedema, and despite his comorbidities, he states he feels well. He denies any allergies. He is on hydrochlorothiazide 12.5 mg a day, rosuvastatin 40 mg a day, ibuprofen 400 mg twice a day, and lisinopril 40 mg a day, all for over 5 years. Family history is positive for hypertension, diabetes mellitus, and hyperlipemia.

Question 8

What is your next step in managing/assessing this patient?

A. Test him for hereditary angioedema
B. Start him on cetirizine 10 mg once a day
C. Discontinue his ibuprofen
D. Stop his lisinopril

The answer is D

Approximately 40–50% of angioedema that presents to the emergency department is secondary to angiotensin-converting enzyme inhibitors (ACE-I). ACE-I-induced angioedema occurs in 0.1–0.7% of those using this therapy. The incidence is five times greater in those of African descent. ACE-I angioedema frequently involves the face and the mouth, especially the tongue. Swelling can occur shortly after starting ACE-I or after many years of use and swelling can reoccur. There are reports of abdominal swelling, but this is very infrequent.

Question 9

You stopped his lisinopril and his angioedema has not returned; however, his blood pressure is now poorly controlled and you want to protect his kidneys because of his DM. What would you do at this time?

A. Restart his lisinopril at a low dose of 2.5 mg a day
B. Use losartan for blood pressure control
C. Use an alternative ACE-I
D. Prescribe icatibant to use for angioedema and restart the lisinopril

Metabolism of Bradykinin (Fig. 3):

The answer is B

The treatment is total avoidance of all ACE-I since the angioedema is not dose-dependent, and can occur with all ACE-I. Angiotensin Receptor Blockers (ARBs) may be used as an alternative blood pressure therapy since they do not affect the catabolism of bradykinin.

Symptoms can resolve without therapy. Although the swelling was commonly treated with antihistamines and corticosteroids, medications traditionally used for HAE are now

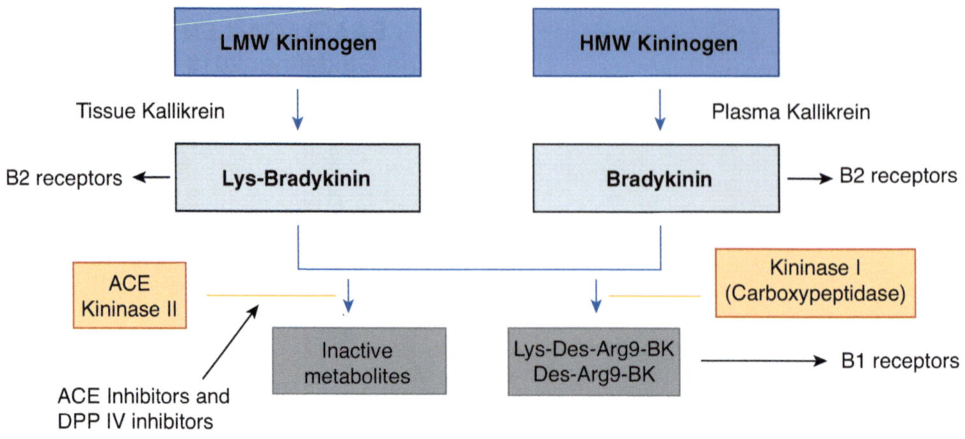

Fig. 3 This figure demonstrates the catabolism of bradykinin. LMW is low molecular weight, and HMW is high molecular weight. ACE is angiotensin-converting enzyme. DPP is dipeptidyl peptidase. The figure was modified from reference

often utilized. This change appeared obvious since ACE-I leads to decreased catabolism of bradykinin, and thus an accumulation of bradykinin. A series of clinical trials demonstrated conflicting results using icatibant, a bradykinin B2 receptor antagonist. One study failed to show a difference in efficacy between traditional therapy and icatibant. At the present time, icatibant is not FDA approved for the treatment of ACE-I-induced edema, but may be indicated off label to protect the airway.

Main take-away points:

- ACE-I-induced angioedema is a common cause of new angioedema and should be suspected in a patient using an ACE-I.
- Management includes discontinuing the medication, treating swelling with antihistamines and steroids, and possibly treating with icatibant (used off-label).

Case 4: A Late-Onset Nuisance

Jill is a 56-year-old female. She was apparently in good health until she started having episodes of recurrent angioedema without hives. She is otherwise free of constitutional symptoms and has not had any weight loss. Her angioedema failed to be controlled with cetirizine 20 mg twice a day. She is not on any medications. Your exam failed to find any abnormalities. You have no recent laboratory tests available.

Question 10

What tests would you order for Jill to determine what type of angioedema she has?

A. CH50, C4, C1-inhibitor, and C1-inhibitor function
B. C1q, C4, C1-inhibitor, and C1-inhibitor function
C. C3, C4, C1-inhibitor, and C1-inhibitor function
D. No tests are necessary

The answer is B

Most cases of HAE occur within the first two decades of life. The guidelines specify that patients over the age of 35–40 years with new onset angioedema should be tested with a similar protocol as with HAE plus a unique test, which is C1q. C1q should be normal in HAE (Fig. 4). Seventy-five percent of those with acquired angioedema with low C1-inhibitor (AAE-low-C1-I) have a low C1q secondary to an autoantibody. Similar to ACE-I-induced angioedema, the angioedema associated with AAE-low-C1-I, for unknown reasons, more often involves the face and upper airway.

The medications used for HAE are also effective for AAE-low-C1-I, but none are FDA approved. Additionally, recent clinical trials have demonstrated that rituximab is effective for treatment of AAE-low-C1-I by reducing the autoantibodies, allowing increased levels of C1-inhibitor protein with homeostasis to the contact pathway. Even when treated with rituximab, on-demand therapy for attacks and STP before procedures are needed.

Of note, patients with AAE-low-C1-I should be assessed for lymphoma and monoclonal gammopathy. On occasion, the angioedema may precede either monoclonal gammopathy or lymphoma, so serial assessment for both is important. Usually, the lymphoma is low grade and at times, treatment may be needed for the angioedema even if therapy is not indicated for the neoplasm.

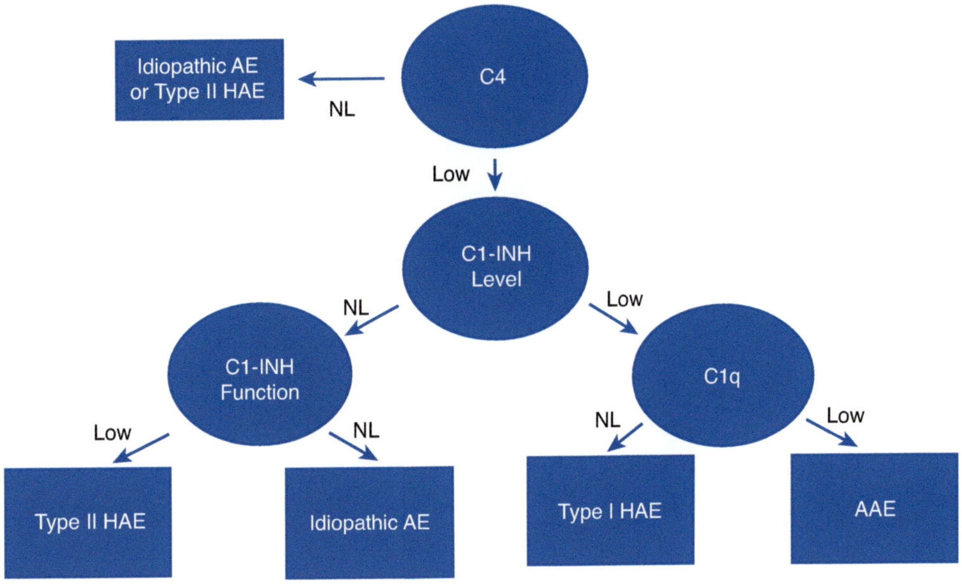

Fig. 4 Diagnostic algorithm for hereditary angioedema developed from WAO HAE Guidelines

Bibliography

1. Maurer M, et al. The international WAO/EAACI guideline for the management of hereditary angioedema. Allergy. 2022;77:1961–90. PIMD: 35006617.
2. Zuberbier T, et al. The EAACI/GA2LEN/EDF/WAO guideline for the definition, classification, diagnosis, and management of urticaria. Allergy. 2022;77:734–66.
3. Ghobrial I, Thalanayar P, Lubin F, et al. Drug induced visceral angioedema. J Communit Hosp Intern Med Prospect. 2014;4(4):PMCID 4185149.

Atopic Dermatitis

Iona Malinow

1 Introduction

Atopic dermatitis (AD) is the most common chronic and relapsing, **pruritic**, inflammatory skin disease that affects both children and adults with a prevalence of up to 18% and 7%, respectively. The atopic dermatitis pathogenesis is multifactorial: **epidermal barrier dysfunction**, immunological abnormalities as **antimicrobial peptides deficiency**, host genetics as null mutations in the gene encoding the epidermal structural protein filaggrin (FLG) and missense variants in genes encoding the T_H2 signature cytokine IL-13 and IL-6R, altered skin microbiota, and environmental factors. AD patients have an increased risk of bacterial, viral, and fungal infections, such as ***S. aureus*, *Herpes simplex*, *Molluscum*, and *Malassezia***. Some endotypes are associated with a higher risk for the **atopic march**: early onset AD, those with polysensitization, with an atopic parent, with a persistent or severe endotype, and those with a **filaggrin** mutation.

Cases 1

A 3-year-old child presents to your office with a severe, itchy, rash, involving his chest, back, upper and lower extremities, including the flexural areas, not relieved with sporadic topical (corticosteroid) CS medications given by his primary care provider (PCP). On physical exam, some lesions are frankly red and with crusts, and his nares are crusted.

Question 1

What criteria is compatible with an atopic dermatitis diagnosis?

A. Posterior subcapsular cataracts
B. Unilateral keratoconjunctivitis
C. Lesional skin with spongiosis in the epidermis
D. Parakeratosis with neutrophilic collections, spongiform pustules with neutrophils

The answer is C

The Hanfin and Rajka criteria for diagnosis AD include (Table 1).

The minor criteria for diagnosing include: (Table 2).

Table 1 Criteria for diagnosis of atopic dermatitis

Pruritus
Facial and extensor involvement in infants and children
Flexural lichenification in older children and adults
Scratching-related: Lichenification, prurigo nodularis
Chronic or relapsing dermatitis
Personal or family history of atopic disease (asthma, rhinitis, conjunctivitis, food allergy)

Table 2 Minor criteria for diagnosing AD

Xerosis
Keratosis pilaris
Hyperlinear palms
Ichthyosis
Pityriasis alba
Allergic shiners
Dennie-Morgan folds
Cheilitis
Keratoconus (conical deformity of the cornea from persistent rubbing of the eyes)
Immediate skin test reactivity
Elevated serum IgE
Food intolerance
Tendency towards cutaneous infections—*S. aureus*, *HSV*, *Molluscum*, dermatophytosis
Tendency towards nonspecific hand or foot dermatitis (especially in adults)
Nipple eczema
Itch when sweating
White dermatographism and delayed blanch response
Anterior subcapsular cataracts (may develop during adolescence or early adult life)
Intolerance to wool and lipid solvents
Course influenced by environmental/emotional factors

I. Malinow (✉)
Allergy and Immunology, Internal Medicine Department, University of Puerto Rico Medical Sciences Campus, San Juan, Puerto Rico

M. Mahmoudi (ed.), *Absolute Allergy and Immunology Board Review*, https://doi.org/10.1007/978-3-031-12867-7_14

Other related conditions are: quality of life impairment (sleep disturbance, anxiety, depression, suicidality).

Atopic keratoconjunctivitis is always bilateral and symptoms include itching, burning, tearing, and copious mucoid discharge. It is frequently associated with eyelid dermatitis and chronic blepharitis and may result in visual impairment from corneal scarring. The Hertoghe's sign is a rarefaction of the lateral eyebrow.

Acute AD skin lesions are characterized by spongiosis, or marked intercellular edema, of the epidermis. Dendritic antigen-presenting cells in the epidermis, such as Langerhans cells, with surface-bound immunoglobulin (Ig) E molecule, present allergen to Th2 cells. Perivenular T cells with occasional monocyte-macrophages are also present. Lymphocytes are CD3, CD4, CD45RO memory T cells and also CD25, HLA-DR, and CLA+ T cells (attracted by CCL27). Mast cells are found in normal numbers but in different stages of degranulation. Innate lymphocyte cells 2 (ILC2) and basophils are an important source of T2 cytokines.

The C-C chemokines RANTES (regulated on activation, normal T cell expressed and secreted), MCP-4 (monocyte chemotactic protein-4), and eotaxin are increased in AD skin lesions, resulting in chemotaxis of eosinophils, macrophages, and Th2 lymphocytes expressing their receptor (CCR3). COL6A5^{+} fibroblasts are found in the dermal-epidermal junction.

Parakeratosis with neutrophilic collections and spongiform pustules with neutrophils refer to psoriasis.

Question 2

He develops this rash, as shown in figure below

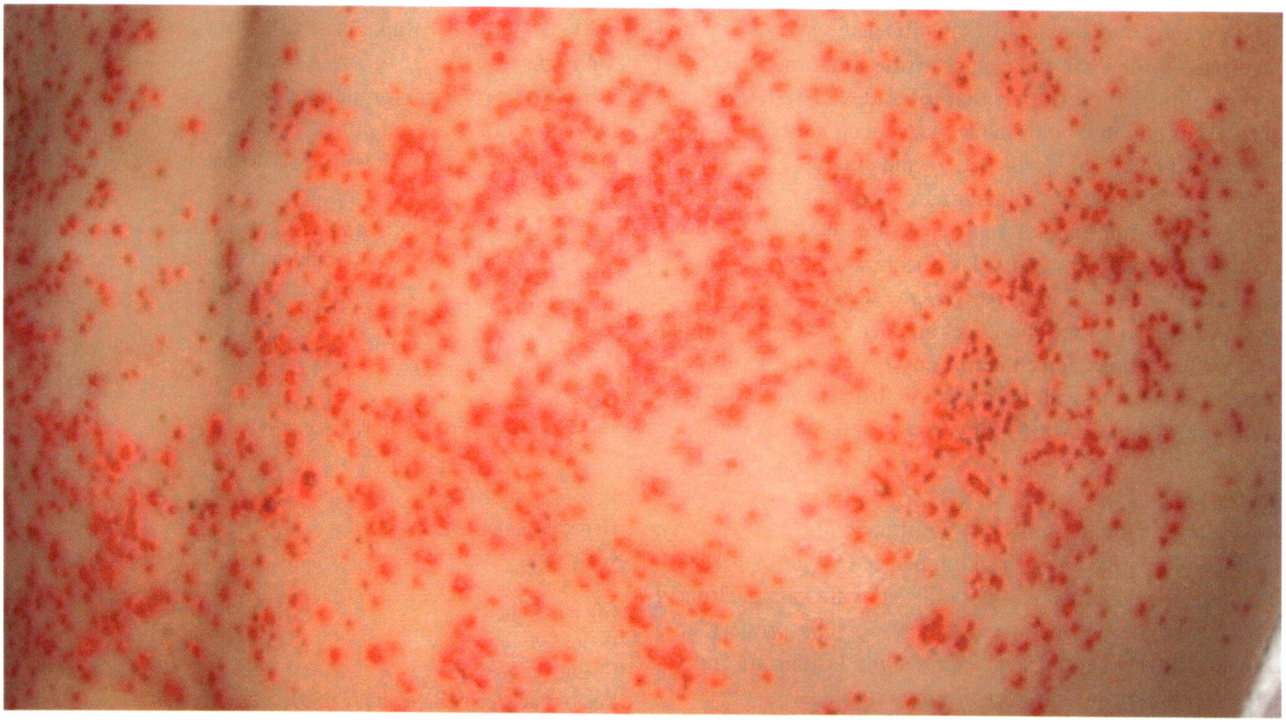

What predisposed him to this rash?

A. Filaggrin (FLG) loss-of-function mutation
B. Presence of MRSA on his nares
C. Smallpox vaccine given to his active military dad
D. STAT3 deficiency

The answer is A

The odds ratio of eczema herpeticum (EH) is more than 10 in patients with FLG mutations.

Loss-of-function (LoF) mutations in the gene encoding filaggrin, the epidermal barrier protein, are also associated strongly with early and persistent AD, asthma, and peanut allergy. FLG heterozygotes outgrow their disease, slower than those without the mutation. Th2 cytokines such as interleukins 4 and 13 (IL-4, IL-13), which are upregulated in AD, can downregulate *FLG* expression.

FLG LoF leads to decreased expression of filaggrin protein and its metabolites, urocanic acid, and pyrrolidone carboxylic acid, increasing the ph of skin, and the expression of two staphylococcal surface proteins, clumping factor B and fibronectin-binding protein, allowing *Staphylococcus aureus* to proliferate. Patients with AD with FLG LoF have a higher risk for four or more episodes of skin infections requiring antibiotics within a year than patients with AD without FLG LoF. African Americans have a 1.7 risk of having AD, and the odds of having an FLG loss-of-function variant is 2.44-fold greater in white children compared with African American children.

EH is caused by infection with herpes simplex virus (HSV-1). Patients may complain of pruritus or pain. The lesions can be vesicles, punched-out erosions, or hemorrhagic crusts.

AD patients with EH have more severe disease based on scoring systems, body surface area affected, and biomarkers (e.g., circulating eosinophil counts, serum IgE, TARC, CTACK) They also have more cutaneous infections with *Staphylococcus aureus* or molluscum contagiosum virus, and they are more likely to have a history of asthma and food and inhalant allergies.

AD patients with EH have reduced interferon γ (IFN-γ) production, and IFN-γ and receptor (IFN-γR1) single-nucleotide polymorphisms (SNPs) are significantly associated with AD and EH.

A leaky barrier in both lesional and nonlesional skin allows the entrance of irritants and allergens and increased transepidermal water loss. (TEWL).

S. aureus is the main organism isolated in patients with AD, causing impetigo, cellulitis, and skin abscesses. If warmth, edema, erythema, and tenderness in an extensive skin area are present, the patient may have erysipelas or cellulitis. Pustules or impetigo may also be due to *Streptococcus pyogenes*, which may cause infections by itself or in combination with *S. aureus*, and sometimes it may cause some lesions resembling eczema herpeticum.

STAT3 deficiency, autosomal dominant Hyper IgE syndrome (HIES), is characterized by eczema, recurrent staphylococcal skin abscesses, recurrent lung infections (resulting in the formation of cavitary lesions in the lungs or pneumatoceles), eosinophilia, and elevated serum IgE. Other frequent findings of STAT3 deficiency include a newborn rash, mucocutaneous candidiasis, connective tissue and skeletal abnormalities such as a typical facial appearance, hyperextensibility of their joints, retained primary teeth, and recurrent bone fractures secondary to even minimal trauma.

Autosomal recessive Hyper IgE syndrome (AR-HIES) with DOCK8 deficiency, presents with eczema, skin abscesses, recurrent respiratory infections, candidiasis, and other fungal infections, and severe, recurrent viral infections from *Herpes simplex, Herpes zoster, and Molluscum contagiosum.* They are also susceptible to allergic and autoimmune manifestations, including food allergy, hemolytic anemia, and vasculitis. They also may develop encephalitis and vascular brain lesions. Patients with AR-HIES do not have connective tissue or skeletal abnormalities.

A history of atopic dermatitis in the household contacts is a contraindication for smallpox vaccine in the military. The exanthem of smallpox appears 2–4 days after the onset of fever in the proximal extremities and spreads distally; lesions are vesicles with central depression, or homogeneous pustules raised, embedded in the skin.

Question 3

What correlates with disease severity?

A. Peanut and egg allergy
B. IgE against superantigens
C. Filaggrin deficiency
D. Decrease in commensal organisms *S. hominis* and *Roseomonas mucosa*

The answer is B

Ninety percent of patients with AD are colonized with *S. aureus*, and most produce enterotoxins A, B, and toxic shock syndrome toxin-1. Staphylococcal toxins can activate T cells as superantigen, a biopsy shows expansion of the T cell receptor (TCR) variable-domain β chain (Vb) in skin lesions and in CLA+ T cells, consistent with superantigen stimulation. The toxins may also induce specific IgE. Basophils from patients with antitoxin IgE release histamine on exposure to the relevant toxin.

Staph superantigens can induce glucocorticoid resistance, and can exacerbate itch by inducing IL-31 mRNA. Even though CD4+ CD25+ (Tregs) are increased in patients with AD, their function is decreased after activation of T cells by superantigens.

Disease severity is dependent on the colonization of *Staph aureus* and production of superantigens, the amount of house dust mite (HDM) IgE, a decrease in microbiome diversity, and the degree of sensitization to aeroallergens.

Beyond 3 years of age, inhalation and intranasal challenge of HDM can induce skin lesions, and T cells specific to Dpter are isolated from lesional skin.

Respiratory route is an important factor in exacerbations-inhalant allergens such as mites, animal danders, and pollens.

A decrease in microbiome diversity correlates with increased colonization of *S. aureus*; topical application of commensal organisms (*S. hominis* or *Roseomonas mucosa*) reduces AD severity, S. hominis produces antimicrobial peptides that reduce *S. aureus* colonization. Staph epidermidis found in some atopic patients damage the skin barrier by expression of a cysteine protease.

Question 4
Staph aureus is found on his nares and on his skin culture. What predisposes patients with AD to *Staph aureus?*

A. Low CD86B cells
B. Antimicrobial peptides deficiency
C. Increase IFN-g
D. Low CCL18 expressed in DC, LC, IDECS

The answer is B
AD patients are predisposed to *S. aureus*, and disseminated infections with *herpes simplex*, *molluscum human papillomavirus (HPV)*, and/or *smallpox*.

Antimicrobial peptides—Cathelicidin LL-37 and B-defensins—are part of the innate immune system responsible for the rapid response against bacteria, fungi, and viruses. Toll-like receptors-2(TLR-2)-sensing of *S. aureus* by Langerhans cells are impaired. *S. aureus* can induce release of IL-33 from human keratinocytes independent of the Toll-like receptor.

The costimulatory molecule CD86 is expressed on Langerhans cells in both the epidermis and the dermis in AD patients, and in B cells; and CD86+ B cells correlate with the amount of IgE.

CCL18 is expressed by DCs in the dermis and LCs and IDECs in the epidermis; it binds to CLA+ T cells in peripheral blood. Expression of CCL18 is induced by exposure to allergens and *S. aureus* enterotoxin B (SEB).

Low IFN-g is associated with EH. The production of IFN-g is inhibited by IL-4 and by PGE2. Keratinocytes from AD patients exhibit increased IFN-g-induced apoptosis.

Deficient NK cells and absence of Tregs may also contribute to the immune deficiency in patients with atopic dermatitis. Both myeloid and plasmacytoid dendritic cells in patients with AD produced less interferon-A.

Question 5
He develops hypopigmented areas in all the sites where her mother was applying topical corticosteroids, and you switch to non-steroidal crisaborole

What is the mechanism of action of crisaborole?

A. Phosphodiesterase (PDE4) inhibitors reduce cAMP
B. Phosphodiesterase (PDE4) inhibitors prevent the degradation of cAMP
C. Calcineurin inhibitor, blocking the dephosphorylation of NFAT
D. Inhibits intracellular signaling of IL-4, IL-13, IL-5, IL-31 IFN-g, and TSLP

The answer is B
Phosphodiesterase inhibitors increase the reduced levels of cyclic adenosine monophosphate (cAMP)of AD, suppressing the activity of NF-kB and NFAT thus inhibiting inflammatory cytokines, IFN-g, TNF-a, IL-2, and IL-5. Crisaborole is approved from 3 months of age and older. Epidermal thickness is significantly decreased compared with baseline only in crisaborole-treated lesions, improved TEWL and pruritus. Significant reductions in numbers of CD3 T cells and CD11c DCs are also seen.

Pimecrolimus and tacrolimus are topical calcineurin inhibitors that block NFAT. JAK inhibitors prevent STAT recruitment and thus inhibit the intracellular signaling of multiple cytokines—IL-4, IL-13, IL-5, IL-31, IFN-g, and TSLP.

Case 2 AD
A 23-year-old female with a history of allergic rhinitis and atopic dermatitis in the past, presents with generalized, severe dermatitis on her chest, eyelids, and lichenified flexural areas in her extremities. Her primary physician prescribed her a midpotency topical corticosteroid (CS) to the chest and extremities, which did not improve her rash.

Question 1
What will you do next?

A. Skin biopsy to r/o Sezary syndrome
B. Patch test to r/o contact allergen
C. Skin test to r/o food and/or inhalant allergy
D. Prescribe an oral CS

The answer is B
A contactant should be considered in patients whose AD does not respond to appropriate therapy, especially those with eyelid, hand, or foot dermatitis or with a history of worsening eczema after application of topical corticosteroids. Sensitizing chemicals, such as parabens and lanolin, can be irritants for patients with AD and are commonly found as vehicles in therapeutic topical agents. This patient may be reacting to the propylene glycol present in the CS preparation; to allergens from the nails that were carried to the eyelids; to sylamide, formaldehyde, methacrylate; or to fragrance/Balsam of Peru.

Eczematous dermatitis has also been reported with HIV infection and scabies. Other conditions that can be confused with AD include psoriasis, ichthyoses, and seborrheic dermatitis.

Only 33–40% of infants and young children (not adults) with moderate-severe atopic dermatitis may have an immediate IgE-mediated reaction to milk, egg, peanut, soy, wheat, fish, and nuts. The immunological reactions include immediate reactions, within 2 h; late-onset IgE reactions, within 6–10 h; and delayed reactions, 6–48 h (T cell-mediated)

Pruritic, morbilliform, or macular eruptions in the predilection sites for AD, or GI/respiratory reactions after ingestion of the food allergen is seen; and there are food-specific T cells from peripheral blood and lesional skin.

An adult who has eczematous dermatitis with no history of childhood eczema (Most cases of AD, 95% approximately, are diagnosed prior to 5 years of age) and without other atopic features may have contact dermatitis, but more importantly, cutaneous T cell lymphoma needs to be ruled out. A red, pruritic rash with adenopathy and increased atypical lymphocytes on the blood smear may be Sezary syndrome, if the rash in her chest does not improve, she needs a biopsy from three separate sites to increase the yield to identify Sezary cells, also need to r/o HIV dermatitis.

Question 2

You start her on pimecrolimus, a calcineurin inhibitor, inhibiting the activation of NFAT (nuclear factor of activated T cells), inhibits activation of helper lymphocytes and production of IL-2, to treat her eyelids with good response. Her dermatitis in the chest and extremities however do not respond to a short course of oral CS, and you start her on dupilumab

Dupilumab:

A. Improves EASI-90 score by 50%
B. It inhibits the IL-4R and the IL-13R through JAK2/STAT3 signaling
C. Most common adverse event is conjunctivitis and blepharitis in 26% of patients
D. It improves itch by 30% in patients older than 6 years of age

The answer is C

A humanized monoclonal antibody, dupilumab inhibits interleukin-4 (IL-4) and IL-13 signaling by specifically binding to the IL-4R-alpha subunit shared by the IL-4 and IL-13 receptor complexes. Dupilumab reduces the expression of genes involved in type 2 inflammation, epidermal hyperplasia, T cells, DCs.

The IL-4R and IL-13 R signal through Jak1, Jak3, Tyk2, STAT5/6.

Lower levels of IL4RA and IL13 and high IL36A expression were related to a stronger clinical response to dupilumab.

Seventy-eight percent improved EASI-50 score after 16 weeks of treatment; 33% EASI-90 score after 16 weeks of treatment.

Dupilumab improves itch by 48–57% in patients 6 years of age and older, taking 300 mg dose, every 4 weeks, by day 2–5. Both IL-4 and IL-13 interact with the itch sensory neurons. IL-4 promotes neural hypersensitivity to many pruritogens: histamine, IL-31, TSLP, and thus an inhibitor to IL-4 and IL-13 ameliorates itch.

Some patients taking dupilumab experience facial erythema. Most common adverse event seen in 26% of 908 patients with AD (14 studies) was conjunctivitis and blepharitis, seen bilaterally, and most patients did not discontinue the medication.

Biomarkers to follow AD severity and disease improvement with treatment are CTACK (CCL27) and the chemokine C-C motif ligand 17/thymus and activation-regulated chemokine(TARC), a chemoattractant of T_H2 cells.

Question 3

Her dermatitis improved but then she presents with head and neck acneiform-looking lesions and/or hypopigmented lesions. What does the picture show? How will you treat her now?

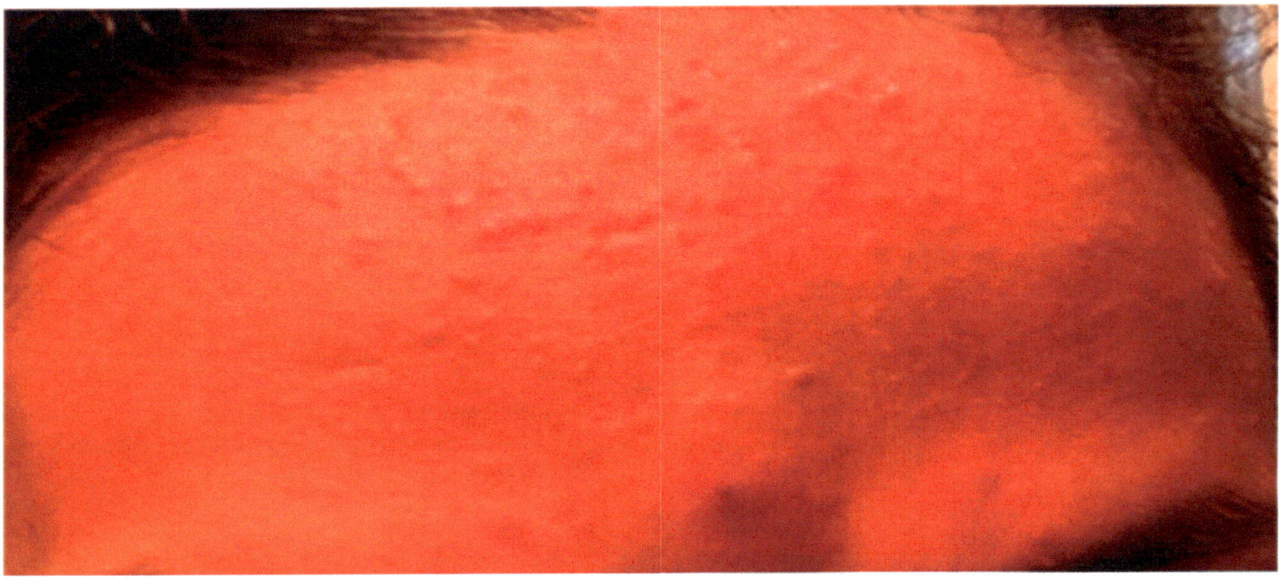

A. *Malassezia sympodialis*, treat with itraconazole 100 mg qd for 1 week, then 200 mg qd
B. Comedonal acne, treat her with benzoyl peroxide/ clindamycin gel.
C. Pityriasis alba, low potency topical CS or pimecrolimus with moisturizer
D. Contact dermatitis, pimecrolimus

The answer is A

Dermatophytosis may cause AD to flare. The opportunistic yeast *Malassezia sympodialis* (formerly *Pityrosporum ovale*) has been associated with a predominantly head and neck distribution of AD, one can order IgE to *Malassezia*. The autoantigen that may cross-react with *Malassezia*, seen in chronic AD patients is Human manganese superoxide dismutase. The superficial dermatophyte *Trichophyton rubrum* has also been associated with elevated allergen-specific IgE levels.

Question 4

What will the skin biopsy of a chronic, lichenified lesion show?

A. Degranulated mast cells
B. CD4+ Th2 cells only
C. Activated eosinophils
D. CD4+ Th1 cells only

The answer is C, activated eosinophils, releasing BMP (basic major protein)

Barrier dysfunction leads to the secretion of IL-33 and TSLP from keratinocytes, promoting type 2 inflammation activating basophils, group2 innate lymphoid cells, and Th2 cells. These cells, in turn, produce the effector cytokines IL-4, IL-5, IL-13, and IL-31. The acute inflammation in the skin is mostly due to IL-4 and chronic inflammation is mostly due to IL-5 and eosinophils. Chronic atopic dermatitis involves Th2, Th22, Th1, and Th17 cells. The biopsy will also show epidermal hyperkeratosis, increased epidermal Langerhans, monocytes, macrophages in dermis, and mast cells (intact, not degranulated) with minimal spongiosis.

The old paradigm that acute atopic dermatitis activated Th2 and chronic AD activated Th1 is changing to a range of AD endotypes with mixed T_H1, T_H2, T_H17, and T_H22 characteristics.

There are many cells responsible for the inflammation in AD: Th1, Th2, ILC2, IDECs, eosinophils, and CD8+ T cells. According to ethnicity, the immune signal differs. The African American patients are predominately Th2, Th22; the pediatric patients are Th2, Th22, Th17; the Asian Th2, Th 22, Th17, Th1, with parakeratosis(more typical for psoriasis); and the European American Th2, Th22, Th1.

Question 5

Which one of the following is true about IL-31:

A. Is the only pruritogen expressed in lesional skin
B. Induced by enterotoxin B from *S. aureus*
C. Produced mostly by keratinocytes
D. IL-31 receptor is expressed constitutively on mast cells

The answer is B

IL-31 is produced mainly by activated CD4Th2 cells, and skin homing CD45RO+CLA+T cells, basophils, ILC2, but also by keratinocytes, mast cells, eosinophils, fibroblasts, DC's, macrophages. IL-31 expression occurs in lesional and nonlesional skin of patients with AD; and its serum levels correlate with the severity of AD. The IL-31 receptors are expressed constitutively on keratinocytes, eosinophils, and neurons. The signaling involves four pathways: JAK-STAT, NF-KB, MAPK, and AKT-PI3K. *Staph aureus* colonization increases IL-31 expression.

The pruritogens (IL-4, IL-13, IL-31, TSLP, histamine, proteases, neuropeptides) are released by inflammation and by scratching, and they are released by keratinocytes, mast cells, and immune cells (T cells and eosinophils). The pruritogens bind to receptors on the sensory C-nerve fibers and A delta-nerve fibers in the epidermis and dermis, which trigger pruritus and pain. Only a small group (<5%) of skin C-nerve fibers are histamine sensitive.

Bibliography

1. Robert B, Pavel A, Diaz A, et al. Crisaborole and atopic dermatitis skin biomarkers: an intrapatient randomized trial. J Allergy Clin Immunol. 2019;144(1274):89.
2. Boguniewicz M, Leung DYM. Atopic dermatitis. Middleton's allergy: principles and practice. Immunol Rev. 2020;33:530–52.
3. Erickson S, Ver Heul A, Kim BS. New and emerging treatments for inflammatory itch. Ann Allergy Asthma Immunol. 2021;126:13–20.
4. Kumar R, Seibold MA, Burchard E. Atopic dermatitis, race and genetics. J Allergy Clin Immunol. 2020;145:108–10.
5. Oetjen LK, Mack MR, Feng J, et al. Sensory neurons co-opt classical immune signaling pathways to chronic itch. Cell. 2017;171(1):217–228e13.
6. Raap U, Wichmann K, Bruder M, et al. Correlation of IL-31 serum levels with severity of atopic dermatitis. J Allergy Clin Immunol. 2017;122(2):421–3.
7. Ständer S. Atopic dermatitis. N Engl J Med. 2021;384(12):1136–43.
8. Wang V, et al. The infectious complications of atopic dermatitis. Ann Allergy Asthma Immunol. 2021;126:3–12.

Allergic Contact Dermatitis

Ryan Steele

Abbreviations

ACD	Allergic contact dermatitis
ACDS	American contact dermatitis society
CAMP	Contact allergy management program
DTH	Delayed type hypersensitivity
ICD	Irritant contact dermatitis
IDRG	International Contact Dermatitis Research Group
NACDG	North American Contact Dermatitis Group

1 Introduction

Contact dermatitis is an inflammatory skin condition caused by exogenous substances coming into contact with the skin. Two types of contact dermatitis exist: irritant contact dermatitis and allergic contact dermatitis (ACD). The former is characterized by an external agent acting as a physical or chemical irritant to the skin causing a nonallergic inflammatory response. ACD is characterized by an external agent acting as an antigen or allergen causing an immunologic response, typically a delayed T-cell mediated (Type IV) hypersensitivity response. It can occur as a local phenomenon at the site of exposure or may even become generalized and cause a systemic ACD. Patients with contact dermatitis have reported that they have social and functional difficulties in life, including work-related difficulties resulting in sick leave, occupational change, or even permanent disability. Early identification of offending allergens has been shown to lead to rapid resolution of symptoms and improved quality of life. Patch testing is the gold standard for the identification of potential contact allergens and is a biological provocation test to investigate type IV hypersensitivity reactions. Allergens at nonirritating concentrations in an optimal vehicle are placed in chambers under occlusion on a patient's skin and secured with an adhesive dressing and with readings performed typically between 48 and 96 h.

R. Steele (✉)
Section of Rheumatology, Allergy and Immunology, Yale School of Medicine, New Haven, CT, USA
e-mail: r.steele@yale.edu

Case 1

A 46-year-old male presents with a 2-year history of a diffuse erythematous and pruritic rash over his entire body. He has a history of intermittent eczema and seasonal rhinitis, but denies asthma, food, or drug allergies. The only thing that has improved his rash has been repeated injections of depomedrol. The patient had a biopsy from dermatology showing a spongiosis with intraepidermal vesicles and parakeratosis.

He has also seen an allergist as part of his workup for this dermatitis and had extensive skin testing for multiple and environmental and food allergens that were all negative except for a tree sensitization. He is adamant that foods are worsening his symptoms as he switched to a vegetarian diet several months before his rash first appeared.

Question 1

What is the most likely allergen responsible for this patient's symptoms?

A. Soybean
B. Tree nuts
C. Nickel
D. Asparagus
E. Balsam of Peru

Answer and Explanation

Answer: C—Nickel

To correctly answer this question, the type of allergic reaction occurring must first be identified. The lack of urticaria, angioedema, gastrointestinal, and respiratory symptoms as well as the patient's negative skin testing make an immediate hypersensitivity reaction less likely. The presence

M. Mahmoudi (ed.), *Absolute Allergy and Immunology Board Review*, https://doi.org/10.1007/978-3-031-12867-7_15

of a diffuse rash and pruritus correlated with food ingestion would suggest a possible diagnosis of systemic contact dermatitis. This variable presentation of ACD may occur with an ingested or possibly topically encountered allergen—notably nickel or balsam of Peru.

Soybean is a possible cause of nickel exposure in this patient, as dermatitis from dietary nickel has been demonstrated in many studies in nickel allergic patients. However, as the underlying allergen is the nickel present in the food and not the soybean itself this answer is not correct. Similarly, tree nuts, and asparagus are a significant source of dietary nickel, but again the allergy is not to the food itself. While balsam of Peru, like nickel, is a potential source of systemic contact dermatitis it is more commonly found in citrus fruits, spices, and chocolate.

2 Presentations of Allergic Contact Dermatitis

Acute ACD typically presents as pruritic eczematous papules or plaques and may be accompanied by erythema, vesicles, bullae, crusting, and edema. Chronic ACD lesions can be lichenified, hyperpigmented, and fissured. Atypical reactions such as purpuric and pustular skin lesions can also occur. ACD can affect any part of the body, but skin findings are typically located where allergen contact had occurred, although ectopic presentations can occur through allergen transfer via the nails for instance to the face, and injection/ingestion or transcutaneous exposure of certain allergens can lead to a diffuse rash causing systemic contact dermatitis (Table 1).

This systemic contact dermatitis may present as a diffuse rash and/or pruritus. There should be a corresponding positive patch test result to the suspected allergen, in this case, nickel. An accompanying hand dermatitis and further involvement of the extensor surfaces such as the elbow may further point to the possibility of a systemic contact dermatitis; however, the absence of this should not necessarily point away from this diagnosis. With balsam of Peru as another potential cause of systemic contact dermatitis, patients may experience stomatitis or cheilitis if it is encountered as a flavoring. Ingestion may also lead to generalized pruritus and/or dermatitis localized to the perianal region possibly from unabsorbed allergen present in feces. Additional symptoms that may occur are rhinitis, conjunctivitis, and plantar dermatitis.

Table 1 Dietary sources of some allergens associated with systemic contact dermatitis

Allergen	Common dietary sources	Avoidance strategies
Nickel	Wheat, soy products, beans, asparagus, broccoli, cauliflower, tree nuts, seeds, chocolate, canned foods, shellfish, stainless steel cookware with acidic foods	Minimize nickel-containing foods in the diet, avoid canned foods and vitamins containing nickel, eat high vitamin C and iron-containing foods, run tap water before cooking or ingesting it, avoid stainless steel cookware with acidic foods
Balsam of Peru	Coffee, flavored tea, alcohol such as vermouth and bitters, soft drinks, juice, citrus, tomatoes, spices, pickled vegetables, and sweets	Avoid foods that contain balsam of Peru including products containing citrus fruits, spices such as cinnamon and vanilla flavorings like candy and gum, avoid products with flavorings such as toothpaste/cough drops/mints, chocolate, vermouth, and cola

Whether the presentation of ACD is a systemic contact dermatitis, local dermatitis, or an ectopic dermatitis, it is a T cell mediated, DTH reaction resulting from exposure to a previously encountered substance. Immunologic events include a sensitization phase in which potential allergens via a hapten-induced activation of antigen-presenting cells such as Langerhans cells drain to regional lymph nodes and are recognized by hapten-specific T cells which then expand and generate memory cells. Further contact with the offending allergen results in an elicitation phase recruiting elements of both the adaptive and innate immune systems. Once an eczematous reaction is induced by an allergen it can take anywhere from 18 to 48 h to reach its maximum response. Although people are more likely to develop reactions to new exposures, reactions to chronically used allergens can also occur.

Question 2

Which of the following would be most supportive of a diagnosis of ACD in this patient?

A. History of a rash from a metal watch band
B. Local redness and itching at the site of a nickel containing
C. Ingestion of foods known to have a significant nickel content earring
D. A positive patch test result to nickel

Answer and Explanation

Answer: D—A positive patch test result to nickel

Although all the other answers are suggestive of a diagnosis of ACD, the gold standard remains patch testing. Patch testing should be performed whenever the history and physical exam are suggestive of an external trigger causing dermatitis. The procedure involves the cutaneous placement of a small number of selected allergens in an appropriate concentration and vehicle. Adhesive patches have chambers either loaded with a substance dissolved in petrolatum or for aqueous-based allergens, small filter papers are placed in the chambers. The test allergens are applied with the adhesive

Table 2 The International Contact Dermatitis Research Group grading scale

Reaction	Definition	Example
–	Negative reaction	
+	Weak positive reaction—Erythema, possible papules, and infiltration	
++	Strong positive reaction—Erythema, papules, vesicles, infiltration	
+++	Extreme positive reaction—Infiltration, coalescing vesicles, and intense erythema	
+?	Doubtful reaction—Faint macular erythema	
IR	Irritant reaction—May be more difficult to interpret as may be faint erythema or may even present as a bullous reaction	

patches to the patient's back for 48 h. Patients are instructed not to bathe, shower, or engage in exercise or activity that may cause excessive sweating. Two readings are generally recommended, the first after the removal of the patches on day 2 or 3, and the second on days 2–5 later. Some allergens, depending on the allergen tested, can cause late positive patch test reactions. The International Contact Dermatitis Research Group (ICDRG) has established an internationally accepted recording system for reading patch tests. Table 2 shows grades of patch test reactions, including questionable reactions, and weak and strongly positive and irritant reactions. Frequent difficulties are encountered in separating weak positive reactions from irritant reactions, and the patient's clinical history as well as physical exam must be considered.

3 Diagnosis of Allergic Contact Dermatitis

Diagnosis of ACD with other skin conditions by appearance alone can be difficult. Both ACD and ICD may present with a similar appearance of a demarcated rash with edema/erythema in the acute phase. The appearance of blistering, weepy, and vesicular rashes may be mistaken for skin infections and chronic ACD may also be difficult to distinguish from other forms of eczema. Histologic findings in ACD can include spongiosis and intraepidermal vesicles, and focal parakeratosis. Therefore, patch testing is the preferred diagnostic standard for ACD. However, allergen selection is key to a successful diagnosis.

The allergens chosen should cover the potential allergens based on the clinical history. Patch tests can be custom-made or built on top of a baseline series. One common baseline series in the United States is the Thin-layer Rapid Use Epicutaneous (T.R.U.E) Test which is FDA approved for patients over the age of 6. The T.R.U.E test covers 35 of the most common contact allergens and a negative control; however, this series cannot be considered a comprehensive diagnostic series by itself. Thus, expanded personalized panels in addition to the T.R.U.E test can improve its clinical utility. The North American Series and the American Contact Dermatitis Society Core Allergen Series are two comprehensive allergen series that may add to the sensitivity of patch testing. Additional series based on occupational and healthcare exposures are also available, as is the ability to test with the patient's personal products. Some common allergen groupings include:

- *Metals*

  Nickel has consistently been found to be one of the most common contact allergen. It is found in many products such as costume jewelry, piercings, toys, belt buckles, and fasteners in clothing. It is the fifth most common element. Its ubiquity has been suggested as a potential cause for its increasing prevalence as a contact allergen in children and adults. Reducing Nickel exposure in common products may reduce the development of contact allergy to Nickel. Commonly affected areas are eyelids, face, neck, wrists, hands, periumbilical area, and thighs. Reactions can range from mild dermatitis to systemic hypersensitivity responses and can possibly occur within the first 30 min of exposure if previously sensitized.

  Cobalt is another common contact allergen. It is often compounded with Nickel which explains its prevalence and can be found in jewelry, belt buckles, zippers, blue/green pigment in crayons, paints, pottery, hair dye, vitamins, foods, deodorants, and leather products.

- *Topical Antibiotics*

  Neomycin and bacitracin are also frequently reported as contact allergens. They are often used together in topical antibiotic products such as Neosporin that contain neomycin, bacitracin, and Polymyxin B. Neomycin can also be found in some corticosteroid preparations. Patients who have contact allergy to neomycin may also have reactions to gentamicin and tobramycin due to cross-reactivity of these antibiotics.
- *Fragrances/Preservatives and Other Common Contact Allergens*

  Fragrances are found in perfumes, cosmetics, personal care products, essential oils, detergents, shampoos, and cleansing wipes. Patch testing studies have shown Fragrance Mix I and II, and Balsam of Peru are the most commonly positive allergens. Fragrance Mix I contains a mixture of eight fragrances and Mix II contains a mixture of six fragrances. Balsam of Peru comes from the bark of the tree *Myroxolon balsamum* which is found in El

Salvador. In addition to its use in fragrances, it is also used as a flavoring agent and as an ingredient in some medical products for its healing properties. Exposure to these fragrances can occur secondhand through the use of these products by partners and caretakers and should be considered in ectopic sources of allergens in contact dermatitis. Consideration must be given when label reading products marketed as fragrance free and unscented, as a masking fragrance may be used in unscented products and essential oils/botanicals may not be listed as fragrances.

Formaldehyde and formaldehyde-releasing preservatives are found in many products and textiles as well as methylisothiazolinone/methylchloroisothiazolinone (MI/MCI). Lanolin, an emollient used in lotions/ointments/ and lip balms, is also commonly found to be a contact allergen. Lanolin sensitization is found to be higher in children with atopic dermatitis compared with nonatopic children, especially those with hand and foot eczema. This may be due to higher emollient exposure in children with atopic dermatitis.

The American Contact Dermatitis Society (ACDS) selects a "contact allergen of the year" to highlight common unrecognized contact allergens. The allergen for 2021 was acetophenone azine, commonly found in footwear and shin guards that contain the foam elastomer ethyl vinyl acetate. It is thought that this allergen is created during the manufacturing process by reactions between other additives.

Question 3

Which statement is most accurate in reference to the patient in the case presented:

A. The patient's atopic dermatitis history makes him more likely to have a positive patch test finding
B. The patient's atopic dermatitis history makes him less likely to have a positive patch test finding
C. The relationship between eczema and allergic contact dermatitis is unclear

Answer and Explanation

Answer: C—The relationship between eczema and allergic contact dermatitis is unclear

The question of whether atopic dermatitis (AD) predisposes individual to ACD or is protective has been a source of some controversy with mixed results in the literature. While some studies have shown a positive relationship between AD and ACD, other studies had found similar rates of ACD between healthy individuals and individuals with AD and some have even found patients with AD are at a lower risk of developing ACD.

This complex interplay between the two diseases is highlighted in retrospective case review of patients in the Pediatric Contact Dermatitis Registry. Patients with AD had a different reaction profile than those without AD with differences noted in some contact allergens frequently found in topical medicaments and skin care products used in the treatment of AD. Cocamidopropyl betaine, wool alcohol, lanolin, tixocortol pivalate, and parthenolide were found to be significantly greater in AD patients compared to patients without AD. Disease severity in AD appears to play a role in ACD prevalence as AD leads to an impaired skin barrier and allows for easier allergen sensitization that can result in induction of ACD. The association of more severe AD and longer duration of treatment with topical agents may explain this finding.

Question 4

In addition to a positive patch test result to nickel, the patient in the above case had positive patch test results to fragrance mix and lanolin. He is now asking for the best way to avoid future exposure to these allergens, you advise him to:

A. Read the ingredient labels for products
B. Look for products that do not contain these allergens in a standardized database
C. Perform a use test to see if the products elicit a reaction
D. All of the above

Answer and Explanation

Answer: D—All of the above

A comprehensive exposure history and patch testing are the mainstay of diagnosis for ACD. Management of ACD mainly involves minimizing exposure risk and treatment of accidental exposures. Complete avoidance of the allergen is the most effective intervention in ACD. A study found that avoidance resulted in complete remission in 90% of patients with contact dermatitis. One limitation of avoidance is that patients may not remember or be able to easily identify the allergens they are sensitized to. Studies suggest that patients tend to have poorer recall of their triggers the more time has passed since their evaluation. For those that live in areas with limited access to patch testing, a preemptive avoidance strategy (PEAS) can be offered. In this strategy, patients are given a safe list of products that avoid 25 of the most common allergens.

Once contact allergens have been identified, the next step is educating parents about what products are safe and which are to be avoided. Decision support tools have been created for this purpose. The Contact Allergen Management Program

Table 3 Contact allergen databases

Database name	Access	Approximate number of products in database
Contact allergen management program (CAMP)	Free to patients; physician ACDS membership	5000
SkinSAFE	Free to patients; physician membership	43,000
Consumer product information database	Free	8000

(CAMP), by the American Contact Dermatitis Society, assists providers in developing personalized safe lists for patients. A similar service called SkinSAFE developed with Mayo Clinic by HER inc is also available as well as the Consumer Product Information Database (Table 3).

For treatment, topical steroids are considered first-line therapies for localized ACD. Alternative therapy for localized ACD is topical calcineurin inhibitors. Although there is a lack of quality studies comparing these agents in ACD, a study did show 0.1% tacrolimus applied twice daily was more effective than placebo in treating ACD. For ACD involving >20% of body area or areas where fast resolution is needed, systemic corticosteroids have been used, as in the case of poison ivy treatment. For those with AD and ACD, there is some evidence that dupilumab may be beneficial.

Question 5

The patient in the above case, tests negative to all allergens including nickel at 96 h, what would you advise as a next step?

A. Stop your investigation for allergic causes for his symptoms
B. Advise empiric avoidance of major contact allergens
C. Have the patient return for a 7 day delayed patch test read
D. Patch test the patient again

Answer and Explanation

Answer: C—Have the patient return for a 7 day delayed patch test read

Although the optimal time for a final patch test read is usually around 96 h, certain allergens including metals, antibiotics and some topical corticosteroids may necessitate a delayed read. In this case, stopping the investigation for an allergic cause for the patient's symptoms would likely have resulted in a missed diagnosis. Empiric avoidance of major contact allergens may result in nickel avoidance but would also significantly impact the patient's quality of life and place an undue burden on them to avoid common potential triggers. Patch testing the patient again, depending on the timing and allergens tested without a delayed read may result in another false negative test.

Case 2

A 23-year-old female presents to the emergency room 1 week after she was treated at urgent care for a poison ivy rash on her leg. She was given a Clobetasol propionate 0.05% cream and her rash improved rapidly over the course of 3 days, but she then noticed that the leg started to become red and painful with blisters forming. In the emergency room, she was treated for presumed cellulitis and given a course of cephalexin and discharged home with instructions to take the antibiotic. She continued to use both the antibiotic and the topical steroid without improvement.

Question 1

What is the most likely cause of the patient's worsening rash?

A. Bacterial resistance to cephalexin
B. Allergy to clobetasol propionate
C. Inadequate topical steroid potency
D. Lack of sufficient antibiotic duration

Answer and Explanation

Answer: B—Allergy to clobetasol propionate

Whenever there is an initial improvement with use of a topical medicament, such as an antibiotic or steroid, followed by subsequent worsening of symptoms, this should raise a suspicion for ACD. This can be difficult to diagnose as the therapeutic action of the medication itself may initially improve the insult such as improving an infection in the case of a topical antibiotic or an inflammatory dermatitis in the case of a topical corticosteroid.

While it is possible that there is superimposed infection, the lack of fever/chills or other signs/symptoms of infection make bacterial resistance or lack of sufficient antibiotic duration less likely. Lack of steroid potency is another possible explanation for the lack of responsiveness; however, Clobetasol propionate 0.05% cream is a group 1 (super-high potency) steroid making this less likely as an answer choice (Table 4).

Question 2

How is this steroid allergy best diagnosed?

A. An oral challenge to corticosteroids
B. Skin prick testing to corticosteroids
C. Patch testing to corticosteroids
D. Intradermal testing to corticosteroids

Table 4 Potencies of selected topical corticosteroids

Potency Group	Corticosteroid
Super-high potency (Group 1)	Clobetasol propionate 0.05% cream/ointment, fluocinonide 0.1% cream
High potency (Group 2)	Amcinonide 0.1% ointment, halcinonide 0.1% cream/ointment/solution
High potency (Group 3)	Betamethasone valerate 0.1% ointment, fluticasone propionate 0.005% ointment
Medium potency (Group 4)	Clocortolone pivalate 0.1% cream, hydrocortisone valerate 0.2% ointment
Low-mid potency (Group 5)	Desonide ointment 0.05%, triamcinolone acetonide ointment 0.025%
Low potency (Group 6)	Alclometasone dipropionate 0.05% cream/ointment, fluocinolone acetonide 0.01% cream
Lowest potency (Group 7)	Hydrocortisone 1% ointment/cream

Answer and Explanation

Answer: C—Patch testing to corticosteroids

The reaction presented in this case, the development of rash with blisters over the course of several days, is more consistent with a delayed-type hypersensitivity reaction consistent with ACD and is best diagnosed via patch testing.

An oral challenge and skin prick testing to corticosteroids would be the appropriate diagnostic testing for a suspected immediate hypersensitivity reaction (IgE mediated) but not for a delayed reaction as occurred in this case. Intradermal testing may be used to diagnose immediate hypersensitivity reactions, and with a delayed read they may have a role in diagnosing delayed-type hypersensitivity reactions. However, the preferred testing modality is patch testing.

Question 3

The patient in the above case stops her original corticosteroid and would like an alternative treatment. What would be the best option to help improve her symptoms?

A. Topical mometasone furoate
B. Topical diphenhydramine cream
C. No intervention
D. Topical desoximetasone

Answer and Explanation

Answer: D—Topical desoximetasone

In this case, when there is an allergy to clobetasol propionate (a Group D1 corticosteroid) there is little to no cross-reactivity with desoximetasone (Group C), making this the best choice. Corticosteroids are grouped not only by potency but also by structure (Table 5).

Topical mometasone furoate is a Group D1 corticosteroid and would not be a good choice for this patient with an allergy to a topical steroid in the same class. Topical diphenhydramine may help with pruritus, but this drug also has the potential to develop into a contact allergen. While no intervention would allow the reaction to follow its natural history likely to resolution, the question specifies what would help improve the patient's symptoms.

Table 5 Classes of corticosteroids

Classes	Corticosteroids	Test allergen
Class A	Hydrocortisone (acetate) Methylprednisolone Prednisolone Tixocortol pivalate Oral cloprednol Oral fludrocortisone acetate Oral methylprednisolone Oral prednisolone	Tixocortol pivalate
Class B	Amcinonide Desonide Fluocinolone acetonide Halcinonide Triamcinolone (acetonide, diacetate) Oral budesonide Oral triamcinolone	Budesonide and triamcinolone
Class C	Clocortolone pivalate Desoximetasone Oral betamethasone Oral dexamethasone	Low potential for allergic sensitization
Class D1	Alclometasone dipropionate Betamethasone dipropionate Betamethasone valerate Clobetasol propionate Clobetasone butyrate Diflorasone diacetate Fluticasone propionate Mometasone furoate Oral/IM betamethasone	Clobetasol-17-propionate
Class D2	Hydrocortisone buteprate Hydrocortisone butyrate Hydrocortisone valerate Prednicarbate Hydrocortisone valerate	Hydrocortisone-17-butyrate
Cross-reactivities	Class A cross-reacts with Class D2. Budesonide cross-reacts with Class D2	

Question 4

The patient is concerned about using not only topical corticosteroids after this reaction, but also personal care products as well. It is suggested that they undergo patch testing, but they opt to delay the procedure at this time. What can you suggest as a strategy to help to determine if a product is a potential contact allergen?

A. Perform a repeat open application test
B. Avoid products containing clobetasol propionate
C. Avoid products containing all Class D1 steroids
D. All of the above

Answer and Explanation

Answer: D—All of the above

A repeat open application test (ROAT) is a simple method of testing for potential allergic contact dermatitis that can be performed at home. The product in question is applied to a small area of skin on several occasions. The tested area is monitored for signs and symptoms consistent with contact dermatitis. The ROAT is intended to help identify delayed-type allergic reactions (type-4 hypersensitivity reactions) and not immediate hypersensitivity reactions. As such the products most suitable for testing are leave-on products such as cosmetics, moisturizers, and topical medicaments. Less amenable to the ROAT are wash-off products such as shampoos and potential causes of irritant contact dermatitis such as detergents. For the patient in the case, it is also prudent to avoid the likely contact allergen of clobetasol propionate and related steroids.

Question 5
In order to avoid future occurrences of rashes from botanicals, the patient is asking about patch testing to poison ivy and related plants. You explain that:

A. Testing to poison ivy and related botanicals would be advised
B. The rash from poison ivy was not an allergic contact dermatitis and therefore patch testing is not indicated
C. Patch testing to poison ivy is not generally advised
D. Lab testing for poison ivy IgE

Answer and Explanation

Answer: C—Patch testing to poison ivy is not generally advised

ACD to plants such as poison ivy, poison oak, and poison sumac are common. However, patch testing is not recommended as it may cause large bullous reactions or sensitization. These plants are from the *Toxicodendron* genus and contact with them results in oil deposition from leaves, with some individuals developing an ACD.

Although it is an ACD, and thus a delayed hypersensitivity reaction, patch testing would not be the appropriate answer for the reasons previously discussed. By the same reasoning, the second answer is also incorrect. IgE testing for poison ivy would not be indicated as ACD is not an immediate hypersensitivity reaction (Type 1 hypersensitivity).

4 Summary

Contact dermatitis has a negative impact on patients' quality of life. For patients with concomitant skin conditions such as AD, ACD can complicate effective management. To best manage the impact of ACD, identification of contact allergens is essential. Patch testing is the gold standard in the diagnosis of ACD. Increased utilization of patch testing may lead to better management of patients suffering from dermatitis through identification of allergen triggers and education on avoidance measures.

Further Reading

1. Alinaghi F, Bennike NH, Egeberg A, Thyssen JP, Johansen JD. Prevalence of contact allergy in the general population: a systematic review and meta-analysis. Contact Dermatitis. 2019;80(2):77–85.
2. Belhadjali H, Mohamed M, Youssef M, Mandhouj S, Chakroun M, Zili J. Contact sensitization in atopic dermatitis: results of a prospective study of 89 cases in Tunisia. Contact Dermatitis. 2008;58(3):188–9.
3. Brar KK. A review of contact dermatitis. Ann Allergy Asthma Immunol. 2021;126(1):32–9.
4. Camacho-Halili M, Axelrod S, Michelis MA, Lighvani S, Khan F, Leon S, et al. A multi-center, retrospective review of patch testing for contact dermatitis in allergy practices. Ann Allergy Asthma Immunol. 2011;107(6):487–92.
5. Chipalkatti N, Lee N, Zancanaro P, Dumont N, Kachuk C, Rosmarin D. A retrospective review of dupilumab for atopic dermatitis patients with allergic contact dermatitis. J Am Acad Dermatol. 2019;80(4):1166–7.
6. Dumycz K, Osinka K, Feleszko W. Contact allergens in topical corticosteroid vehicles: analysis of product composition. Contact Dermatitis. 2017;76(4):254–5.
7. Fonacier L, Bernstein DI, Pacheco K, Holness DL, Blessing-Moore J, Khan D, et al. Contact dermatitis: a practice parameter-update 2015. J Allergy Clin Immunol Pract. 2015;3(3 Suppl):S1–39.
8. Goldenberg A, Jacob SE. Demographics of US pediatric contact dermatitis registry providers. Dermatitis. 2015;26(4):184–8.
9. Jacob SE, McGowan M, Silverberg NB, Pelletier JL, Fonacier L, Mousdicas N, et al. Pediatric contact dermatitis registry data on contact allergy in children with atopic dermatitis. JAMA Dermatol. 2017;153(8):765–70.
10. Malajian D, Belsito DV. Cutaneous delayed-type hypersensitivity in patients with atopic dermatitis. J Am Acad Dermatol. 2013;69(2):232–7.
11. Militello M, Hu S, Laughter M, Dunnick CA. American contact dermatitis society allergens of the year 2000 to 2020. Dermatol Clin. 2020;38(3):309–20.
12. Raison-Peyron N, Sasseville D. Acetophenone azine. Dermatitis. 2021;32(1):5–9.
13. Saito M, Arakaki R, Yamada A, Tsunematsu T, Kudo Y, Ishimaru N. Molecular mechanisms of nickel allergy. Int J Mol Sci. 2016;17(2):202.
14. Simonsen AB, Johansen JD, Deleuran M, Mortz CG, Sommerlund M. Contact allergy in children with atopic dermatitis: a systematic review. Br J Dermatol. 2017;177(2):395–405.
15. Thyssen JP. Nickel and cobalt allergy before and after nickel regulation – evaluation of a public health intervention. Contact Dermatitis. 2011;65(s1):1–68.
16. Zhu TH, Suresh R, Warshaw E, Scheinman P, Mowad C, Botto N, et al. The medical necessity of comprehensive patch testing. Dermatitis. 2018;29(3):107–11.

Insect Allergy

Donya Salmasinia Imanirad and Dennis Ledford

Abbreviations

AAAAI	American Academy of Allergy, Asthma, and Immunology
ACAI	American College of Allergy and Immunology
ACEi	Angiotensin-converting enzyme inhibitor
ARB	Angiotensin receptor blocker
BAT	Basophil activation test
HIT	Hymenoptera identification test
IFA	Imported fire ant
IgE	Immunoglobulin E
WBE	Whole-body extract

Case 1

A 10-year-old boy who lives in Florida presents to his pediatrician after experiencing a suspected wasp sting. He was playing in the backyard when it stung his forearm. He experienced immediate pain, erythema, and edema at the site of the sting. Within 4 hours (h), he developed progressive, localized edema and erythema of the forearm which expanded past the elbow.

He did not develop dyspnea, angioedema, nausea, rapid heartbeat, or light-headedness. His pediatrician examined him and reassured the patient and his family of the benign nature of the reaction and the lack of need for antibiotics. He prescribed topical 0.1% triamcinolone cream to apply to the inflamed skin every 8 h for 3 days. The patient had a prior episode of a wasp sting with less severe burning and swelling localized to the site of the sting. The pediatrician referred the patient to the allergist/immunologist because of the increased size of the most recent reaction. "We may need to investigate the development of allergy before it becomes serious," the pediatrician advised.

D. S. Imanirad · D. Ledford (✉)
Division of Allergy, Asthma, and Immunology, Department of Internal Medicine, Morsani College of Medicine, University of South Florida and James Haley VA Hospital, Florida, USA
e-mail: donya@usf.edu; dledford@usf.edu

Question 1

Following an insect sting, which clinical outcome is most likely in a patient with a history of increasing size of local reactions from prior insect stings?

A. Anaphylaxis
B. Generalized urticaria
C. Large local reaction
D. Serum sickness

Answer and Explanation

Answer: C

In patients without systemic reactions with prior insect stings, 10–20% may experience mild systemic reactions with subsequent stings. In contrast to the increased risk of systemic reactions in individuals with prior systemic reactions, those with large local reactions are not at increased risk for future serious systemic sting reactions.

The chance that a subsequent sting will cause a systemic hypersensitivity reaction depends on the history and presence of specific IgE. Generally, risk factors that predict a serious future systemic reaction to an insect sting include:

- Severity of prior reactions.
- Honeybee allergy.
- Lack of urticaria or angioedema with prior systemic sting reactions, usually with hypotension, as occurs with mast cell disorders.
- Limited time, months to a few years, since the last systemic insect sting reaction.
- Angiotensin-converting enzyme (ACE) inhibitor, beta blocker, or angiotensin receptor blocker (ARB) therapy.
- Multiple stings.
- Increased age.
- Male gender (this likely reflects exposure rather than a specific risk factor).

M. Mahmoudi (ed.), *Absolute Allergy and Immunology Board Review*, https://doi.org/10.1007/978-3-031-12867-7_16

In patients with a previous systemic allergic reaction to insect venom, about 45% experience significant, systemic reactions to future stings. Although the greatest risk is within 10 years of a prior systemic reaction, adults who had moderate to severe systemic reactions in childhood have a 30% risk of systemic sting reactions decades later. Honeybee allergy is associated with a greater risk for reoccurrence and severity of systemic reactions, probably due to the larger volume of venom in honeybee stings.

Question 2

Select the Hymenoptera species associated with the nest in the picture.

A. Domestic honeybee
B. Paper wasp
C. Africanized honeybee
D. Yellow jacket

Answer and Explanation

Answer: D

The nest shown in the picture is characteristic of yellow jacket nest which does not have an open face and can be found in woodpiles, decaying logs, underground or suspended from overhead, in this case on rafters as shown.

Important stinging Hymenoptera species are divided into three families: Apidae, Vespidae, and Formicidae. Figure 1 summarizes insect characteristics and nesting habits of common Hymenoptera species.

The Apidae family includes honeybee (*Apis mellifera*), bumblebee (*Bombus spp.*), and sweat bee (*Halictus spp.*, *Dialictus spp.*). Wild honeybees naturally form their nests in felled trees or other natural hollows. The sting apparatus of honeybee breaks away from the body, and thus the insect usually dies following a sting. Africanized honeybee, or "killer bees," may swarm with minimal provocation and sting in large numbers, potentially resulting in toxic, even fatal, reactions. The venoms of native and Africanized honeybees do not differ. Native honeybees and bumblebees are not aggressive, and systemic reactions to their stings usually occur in areas of high exposure, such as greenhouses for bumblebees and gardening among pollinating plants or walking barefoot in flowering clover for honeybees. Honeybee nests are generally found within enclosed spaces,

Fig. 1 Various flying insects within the Hymenoptera family and their habitat. (**a**) Honeybee. (**b**) Honeybee comb and nest. (**c**) Brown paper wasp. (**d**) Wasp nest. (**e**) Yellow jacket. (**f**) Yellow jacket nest. (**g**) New world hornet or aerial yellow jacket. (**h**) New world hornet nest

such as a hollow tree or in manmade beehives, and are composed of waxy, open-faced cells in one or more vertical levels. Each cell contains stored honey or developing larvae.

The Vespidae family includes paper wasps, yellow jackets, and new world hornets *(Vespa)*. Wasps are a commonly encountered stinging insect in the southern US, particularly the Gulf Coast, and the Mediterranean coastal areas in Europe; whereas yellow jackets are encountered more commonly in northern and central North America and central Europe. Generally, the insects within the Vespidae subfamilies are slim, elongated in shape, and hairless. Among these, wasps are black, red, or brown with yellow stripes on their abdomen. Their nests generally have round paper-like combs of cells that have an open face on a single level. Wasps usually form nests in woodpiles, porches, windowsills of homes, and railing of decks; thus, human contact frequently occurs in domestic settings.

Yellow jackets can be found in variable climates. Their name reflects the typical black and yellow color pattern that encircles their abdomen. Some may have white or red markings as well. They are distinguished from honeybees by their slim waist and more segmented body parts. Their nests are made of paper-like material, and they usually form their multilevel nests in the ground and are enclosed without an open face. These nests can sometimes be found in woodpiles and within walls of dwellings.

New World hornets, also called aerial yellow jackets, tend to be brown or black, with either white, orange, or yellow stripes and tend to be larger in size than ground-dwelling yellow jackets. They form paper-like multilevel nests that are oval and enclosed, very similar to the yellow jacket nests. In contrast to the propensity for yellow jackets to build subterranean nests, hornets prefer their nests to be above ground in tree hollows, branches, and shrubs.

The Formicidae family (ants) of the Hymenoptera insects are unique, as they lack wings, except during the mating of the queen in preparation for a new colony, and their venom contains a predominance of alkaloids. Additionally, they contain one or two pairs of antennae. Several other stinging ants can cause allergic reactions, such as Jack Jumper ants in Australia and green head ants in Asia; but in the United States, IFAs are the major types of ants associated with hypersensitivity reactions, particularly in the South and Gulf coast (Table 1).

Table 1 Insect characteristics and nesting habits

Insect	Nesting habits and locations	Insect characteristics
Honeybee	Waxy nests in trees and natural hollows, commercially in wooden beehives	Docile except for African honeybee which are aggressive
Wasps	Single level paper nest opens downwards and is usually located in bushes and around woodpiles, roof eaves, and porches	Aggressive when nest is disturbed
Yellow jacket	Multilevel paper nests in ground or wooded areas	Aggressive and readily swarm
Yellow and white face hornets (aerial yellow jackets)	Multilevel paper nest with surrounding papier-mâché usually above ground in trees and shrubs	Swarm aggressively when nest is disturbed; sensitive to vibration
Imported fire ants	Form earth mounds in sandy and moist soil	Aggressive when nest disturbed

The main allergenic proteins in vespid venoms include phospholipase A1, antigen 5, and hyaluronidase. Antigen 5 is the major allergenic component of all vespid venoms, whereas phospholipase A1 is the second most common vespid allergen, comprising up to 14% of the venom dry weight.

The most important honeybee venom allergen is phospholipase A2, which comprises about 12–15% of the bee venom and functions as a cytotoxin. Honeybee venom also contains hyaluronidase, which partially resembles the enzyme from the vespid venom. Additionally, melittin is a major component of bee venom, comprising about 50% of its protein. However, only 26% of patients with allergic reactions to honeybee venom show specific IgE against melittin, and the significance of melittin-specific IgE is unproven. Therefore, melittin is not generally identified as a honeybee allergen. Table 2 lists the common allergenic components of the three main Hymenoptera families.

Question 3

The development of a large local reaction to a *Polistes* (wasp) sting is most likely associated with:

A. Specific IgG4 to antigen 5
B. Specific IgG4 to hyaluronidase
C. Specific IgE to antigen 5
D. Specific IgE to phospholipase A2

Answer and Explanation

Answer: C

Question 4

Which of the following is responsible for the pustular lesions with fire ant (IFA) stings?

A. *Staphylococcus auerus* bacterial superinfection
B. Antigen 5 protein
C. Alkaloids
D. Ant bites associated with stings

Answer and Explanation

Answer: C

Imported fire ants (IFA) are in the family Formicidae from the Hymenoptera order and are associated with various allergic reactions, ranging from large local to life-threatening systemic reactions. IFA venom however differs from other Hymenoptera insect venoms in that it contains a very low amount of protein and a high concentration of toxic alkaloids. The alkaloids comprise 95% by volume of the venom

Table 2 Insects from the Hymenoptera group with their allergenic components

Apidae		Vespidae			Formicidae
Honeybee	Bumblebee	Wasp (*Polistes*)	Yellow jacket (Vespula)	New World hornet	Fire ant
Phospholipase A2	Phospholipase A2				
		Phospholipase A1	Phospholipase A1	Phospholipase A1	Phospholipase A1
Hyaluronidase	Hyaluronidase		Hyaluronidase	Hyaluronidase	
		Antigen 5	Antigen 5	Antigen 5	Antigen 5
Protease	Protease	Serine protease			
Acid phosphatase	Acid phosphatase				
Vitellogenin			Vitellogenin		
Dipeptidyl peptidase IV			Dipeptidyl peptidase IV		
Major royal jelly protein					
Protease inhibitor					
Carboxylesterase					
Melittin					

but are not allergenic. The alkaloids have a pH of approximately 12 and are responsible for most hemolytic, cytotoxic, and bactericidal IFA venom properties. These alkaloids cause the sterile pustules following IFA stings. The protein components in the IFA venom are the major allergens. Some of these proteins are homologous with the allergens in other Hymenoptera and specific IgE may cross-react among these proteins. The venom is delivered by a modified abdominal ovipositor, like the other Hymenoptera. IFAs do bite to maintain close contact with the sting victim, but these bites do not cause symptoms.

The whole-body extracts (WBE) from IFAs contain sufficient allergenic proteins to be used for testing and treatment of IFA allergy. This is not true for other Hymenoptera species, likely due to the larger body size of the other insects resulting in dilution or degradation of the venom allergens in WBEs.

Question 5
Which of the following insects is associated with nocturnal anaphylaxis?

A. Bedbugs
B. Tsetse flies
C. Triatoma bugs
D. Mosquitos

Answer and Explanation
Answer: C

Due to their tendency to bite at night, Triatoma or kissing bugs are associated with nocturnal anaphylaxis. These insects are from the subfamily Triatominae (*Triatoma protracta*) and typically are found in Central America and the southern USA. They are small and feed on small mammals but also on humans. They are more common in homes with organic construction materials such as thatched roofs. Triatoma bug bites are a cause of severe, systemic hypersensitivity reactions or anaphylaxis. In terms of allergenicity from other groups of insects, rare systemic allergic reactions occur from various biting insects, including mosquitos, fleas, kissing bugs, midges, and flies. Unlike the Hymenoptera insects with venom, several allergens in these mostly blood-feeding insects are within the salivary gland proteins. The more common biting or hematophagous insects that cause allergic reactions belong to the order Diptera, that includes mosquitos and flies. Among these, tsetse bugs, kissing bugs, and bedbugs result in reactions. Mosquito allergy typically causes large local reactions in children ("Skeeter syndrome"), which almost always resolves with time. Bedbugs bite almost exclusively at night and often cause urticaria or bullous lesions. Bedbug allergy may cause a presentation suggesting idiopathic or spontaneous urticaria, although individual wheals typically persist longer than 24 h. Histologically, bedbug hives may manifest features of a necrotizing, eosinophilic vasculitis. The most likely cross-reactive allergen in common with blood-feeding insects and Hymenoptera is hyaluronidase. It is found in the saliva of horse flies and biting midges as well as the salivary gland of other blood-feeding insects. There is up to 60% sequence homology among hyaluronidases from biting insects, wasp (Ves v 2), and honeybee (Api m 2). The importance of this observation in human allergy is not known. Current diagnostic tools for the diagnosis of biting insect allergy are limited and not validated. Commercial reagents for immunotherapy are not available or not validated, so the treatment focus is on avoidance measures and availability of epinephrine autoinjectors, as well as premedication with antihistamines before likely exposure and symptomatic treatment including topical corticosteroids.

Case 2
A 23-year-old male with a history of anxiety disorder and obesity presented for evaluation of an insect sting by an unidentified insect 3 weeks ago. He was standing in a garden with a can of soda when he was stung by an insect which was on the rim of his soda can. Within 5 minutes (min) of the sting, he developed pruritis and flushing and shortly after-

ward lost consciousness. The event was witnessed by friends, and he eventually awakened without treatment. Within a day he developed large edema of his lip and periorbital area. He endorsed shortness of breath at the time of the event but did not have trouble swallowing or breathing when he regained consciousness. There is no documentation of his blood pressure at the time of the event. A month after his syncopal episode, specific IgE blood and skin testing for all Hymenoptera venoms were negative.

Question 1
What is the next best test for evaluation of this patient?

A. Basophil activation testing
B. Baseline tryptase
C. Echocardiogram
D. Tilt table testing

Answer and Explanation
Answer: B
The presenting symptoms, in this case, suggest the possibility of mast cell activation disorder. Mast cell activation disorders frequently manifest as an immediate sting reaction involving at least two organ systems without urticaria. A systemic allergic reaction involves one or more organ systems, but concurrent involvement of two or more organ systems with cardiovascular or significant respiratory manifestations generally is accepted as anaphylaxis. Flushing, typically without urticaria, followed by hypotension and loss of consciousness is associated with insect sting reactions among people with mastocytosis and other mast cell disorders. Therefore, a baseline tryptase level in all subjects who experience severe hypersensitivity reactions to Hymenoptera venom should be considered, particularly those with hypotension but without acute urticaria. Patients with syncope or hypotension following Hymenoptera stings, even with baseline tryptase levels in the high normal range (7–11.5 ng/mL), should be considered for evaluation of the c-*KIT* mutation, D816V, in the peripheral blood. If positive, then a bone marrow biopsy is also recommended to evaluate the number, distribution, and morphology of the mast cells. Some systemic sting reactions are non-IgE-mediated, and all testing, including skin and blood testing for specific IgE, will yield negative results. This also is a reason to consider a mast cell disorder diagnosis. The absence of specific IgE when suspected should prompt consideration of repeat IgE testing in 3 months or use of alternative testing modalities if available. Alternative methods, not generally used in the United States but discussed in the literature, include sting challenge and the basophil activation test (BAT). The latter involves in vitro mixing of the test subject's blood with venom and measuring the expression of basophil activation markers by flow cytometry. Basophil activation tests may detect specific IgE when other testing is negative. Repeat tryptase measurements during acute symptoms would help confirm that symptoms are mast cell or basophil related even with negative specific IgE testing. During systemic symptoms, an acute increase of serum tryptase by 20% plus 2 ng/ml or 30% from the baseline tryptase suggests mast cell activation.

Question 2
What is the recommended indication for allergy testing in this patient if he were stung by IFAs in the future?

A. Progressive large local sting reactions
B. Generalized urticaria 10 min after one sting
C. Headache after 4 stings
D. Delayed anaphylaxis following ingestion of pork sausage

Answer and Explanation
Answer: B
The management for reactions to IFA stings is slightly different from Hymenoptera insects in the case of systemic cutaneous reactions in that the natural history of IFA hypersensitivity is less characterized, avoidance is difficult, and anxiety from the reactions may be significant. Thus, whole-body extract (WBE) immunotherapy may be a consideration in patients with a history of a systemic cutaneous reaction to IFA sting. This differs from the other Hymenoptera for which data show cutaneous systemic reactions are not predictive of more life-threatening reactions. Hymenoptera venom stings can cause several types of allergic reactions. Identifying historical points that correlate with a greater risk of venom anaphylaxis is crucial in the evaluation of suspected insect allergy. Cutaneous reactions are categorized as large local reactions or systemic reactions according to the size and distribution of the manifestations. Large local reactions are defined by edema surrounding the site of an insect sting that exceed 10 cm in diameter and increase in size within 24–48 h after the sting and last more than 72 h. Individuals with large local sting reactions have only 2–7% chance of experiencing future, mild systemic sting reactions. Similarly, systemic cutaneous sting reactions with urticaria do not predict a significant risk of subsequent clinically significant anaphylaxis. Thus, large local reactions and systemic cutaneous reactions are not indications for insect testing, as such reactions are usually not treated with immunotherapy due to very low risk of having a significant systemic reaction or anaphylaxis. However, insect testing and treatment are a consideration if the discomfort of large local reactions is significant and there is frequent, unavoidable exposure. Large local reactions are generally tolerable. They can be symptomatically treated with oral antihistamines and topical corticosteroids. During the assessment of venom allergy, it is important to consider conditions such as heart disease and clonal mast cell disorders

Table 3 Factors that suggest a need for prolonged or lifelong immunotherapy

Anaphylaxis to honeybee, especially in those at high risk of exposure to repeated stings (beekeepers)
Systemic reaction to insect immunotherapy
Systemic reaction to a sting during maintenance insect immunotherapy
High risk of a serious systemic reaction, which is greater in those with prior severe reactions, cardiovascular disease and use of angiotensin-converting enzyme inhibitors or Beta-blocking agents, older age
Elevated baseline tryptase or mast cell disorders including mastocytosis and mast cell activation disorders
Patient preference due to anxiety and fear of unexpected stings

that may increase the morbidity or severity of future reactions. Table 3 summarizes the recommended management of various types of reactions following an insect sting or stings. Anaphylaxis following ingestion of pork sausage is suggestive of sensitivity to alpha-galactose-alpha-1,3-galactose and not venom allergy. This condition may be related to tick bites which induce sensitivity to alphagalactose-alpha-1,3-galactose found in mammalian meat such as beef, pork and mutton. Non-specific symptoms such as headache are not an indication for evaluation of systemic insect allergy.

Figure 2 provides guidance for diagnostic steps and management of patients following Hymenoptera venom sting.

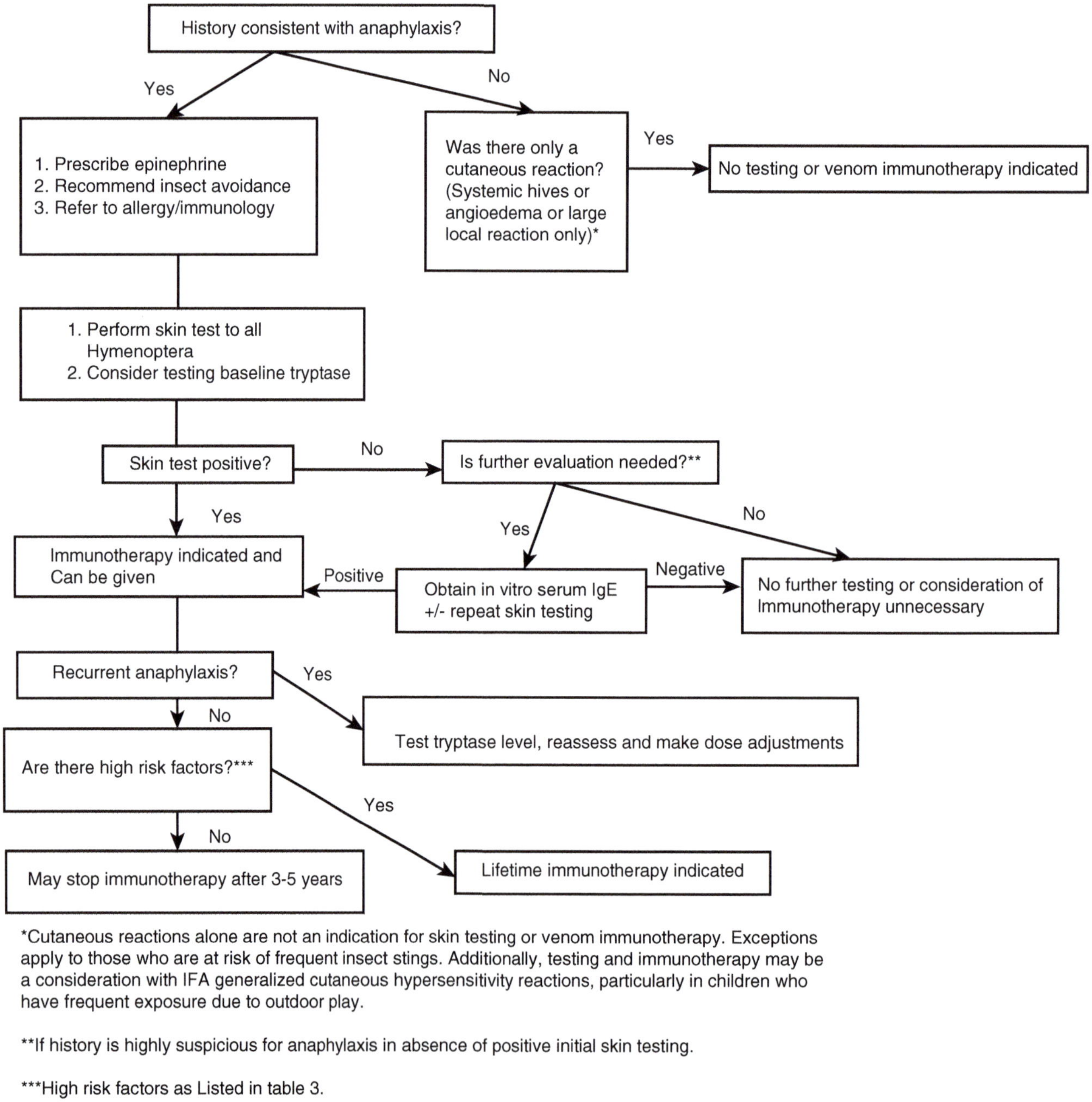

Fig. 2 Diagnostics and management algorithm following Hymenoptera venom sting

Question 3

On another occasion, the patient in this case reported lightheadedness, nausea, and generalized urticaria following an insect sting while trimming shrubbery. Symptoms improved after receiving oral diphenhydramine and lying down for approximately 30 min. He brought with him a crushed yellow and brown winged insect. The patient's father observed an open-faced, paper nest in the nearby shrubbery with similarly colored insects in proximity. Another allergy specialist performed skin testing for all Hymenoptera which were negative. What is the best next step?

A. Repeat skin testing to *Polistes*
B. Request in vitro specific IgG testing for *Polistes*
C. Order a basophil activation testing to all Hymenoptera
D. Repeat skin testing to Hymenoptera venom in 12–16 weeks

Answer and Explanation

Answer: D

Hymenoptera venom allergy testing generally is not performed within 2–6 weeks of a sting reaction due to the potential occurrence of false negative results. This refractory period is not well substantiated, and early testing is a consideration in high-risk situations. Repeat testing is recommended in 3–6 months, regardless of the timing of the first testing, in subjects with negative results and high pretest probability based upon a convincing history. In vitro allergy blood tests for specific-IgE against culprit insects could also be considered if skin testing is initially negative. Approximately 4–6% of the people who experience systemic allergic reactions to insect stings have negative venom skin tests and serum specific-IgE for the suspected culprit insect. There are several reasons for negative test results: (1) The test was performed for the wrong culprit insect; (2) Spontaneous loss of sensitization occurred between the insect exposure and testing; (3) The tests were done within the refractory period, 1–6 weeks, following the insect sting; (4) The reaction was not immunologic, for example, a vasovagal reaction or toxic effect of the venom and not an allergic reaction; and (5) Mast cell activation disorder in which negative insect allergy tests occur in up 15% of this group of patients.

Skin tests are more sensitive for insect venom allergy compared to in vitro, blood testing. However, if skin testing is negative, blood testing should be performed promptly, as this testing modality may be positive with negative skin tests in a minority of sensitive subjects. There is no value of venom specific-IgG in the assessment of potential sting allergy. When there is a convincing history of insect sting anaphylaxis, but initial specific IgE testing is negative, the skin testing should be repeated to all Hymenoptera after 3–6 months. The recommendation for testing for all Hymenoptera insects is supported by a multicenter study with a Hymenoptera identification test (HIT). This study demonstrated that most people are unable to identify the culprit insect, except for the honeybee.

Both skin testing with venom (or whole body in the case of IFA) extracts and serologic testing are options in patients with a history of sting anaphylaxis. However, skin testing is the preferable test or gold standard for Hymenoptera allergy due to greater sensitivity than serologic testing. Positive insect skin test results require a wheal of at least 3 mm or more in diameter within 15–20 min compared to the negative control. According to the AAAAI and ACAAI Joint Task Force practice parameter in 2016, it is also recommended to perform intradermal testing if the skin prick test is negative at a concentration of 100 mcg/mL of purified venom extract. In general, a positive intradermal skin test response to insect venom at a concentration of less than or equal to 1.0 mcg/mL indicates clinically significant specific IgE antibodies. In case of fire ant whole-body extract, a positive response to concentration of 1:100 wt/vol or less by skin prick method or 1:1000 wt/vol or less by the intradermal method is indicative of clinically significant specific IgE antibodies. The combination of skin prick and intradermal testing increases sensitivity to greater than 90%. Intradermal skin testing with venom concentration >1.0 mcg/ml results in an excessive number of false positive results.

Question 4

False positive results with insect testing with a non-culprit insect is most likely due to:

A. Identical protein structures among insects.
B. Enzyme activity of venom proteins.
C. Carbohydrate determinants common to different insects.
D. Specific IgG to venom antigens.

Answer and Explanation

Answer: C

The most effective way of evaluating insect allergy is with skin testing, although false positive and negative results do occur. Specific-IgE antibodies for common carbohydrate epitopes shared among insects may result in multiple positive tests that are not clinically relevant. Many of the insects within the same taxonomy family will have venom antigen similarities, for example, between yellow jackets and hornets, but the venom components are not identical. Proteins such as hyaluronidase are common to multiple families, albeit the hyaluronidases do differ. It is preferable to distinguish between true sensitivity and cross-reactivity and treat with the relevant insect reagent. Thus, in some countries sting challenge is performed to verify the importance of a specific insect. However, the negative impact on patient comfort, patient and physician anxiety, liability, and logistics (for example, access to stinging insects) limit the use of this strategy in clinical practice in the US. Another approach, not commercially accessible in the US but available in some other countries, is component resolved diagnosis (CRD) in cases of difficult interpretation of sensitization-based *Vespula* and *Polistes* specific IgE to whole venom. For example, testing with the purified component Ves v 5 and Pol d 5 may identify which is the culprit insect. Another test used to distinguish true and false positive results is in vitro inhibition of specific IgE or CAP-inhibition test (ImmunoCap®). It is a quantitative assay that measures inhibition of allergen-specific IgE in serum based upon the efficiency of one venom, mixed with the patient's serum prior to the assay, to inhibit reactivity to another venom. Antibodies with greater avidity for a specific inhibiting venom suggest the inhibiting venom is from the culprit insect. IgG antibodies do not result in positive venom allergy testing. The enzymatic activity of venom components may be responsible for toxic reactions but not allergy.

Question 5
Which of the following is an indication for lifelong venom immunotherapy?

A. Wasp allergy
B. Systemic reaction during venom immunotherapy
C. Asthma
D. Calcium channel blocker therapy of coexisting hypertension

Answer and Explanation
Answer: B

There are risk factors for post immunotherapy occurrence of sting anaphylaxis. These factors include a life-threatening reaction prior to initiating immunotherapy, systemic reactions to insect immunotherapy or to field stings during immunotherapy, anaphylaxis to honeybee stings, and mast cell disorders (Table 3). Aside from these exceptions, 3–5 years of venom immunotherapy would be sufficient to reduce the risk of life-threatening anaphylaxis from specific Hymenoptera venom. Insect sting immunotherapy is an efficient method of treating people with severe immediate systemic reactions. The treatment gradually diminishes the specific IgE to the treatment allergens via multiple mechanisms. Immunotherapy is performed with purified venoms from most insects of interest and whole-body extract from IFAs. Insect sting immunotherapy is indicated if there is a compelling history of anaphylaxis following the sting and specific IgE identified to the culprit insect. Insect immunotherapy will reduce the size of local reactions and may be considered in subjects with incapacitating large local reactions and inability to avoid stings, for example, in a work-related exposure. Approximately 3 years of venom immunotherapy reduces the risk of significant anaphylaxis by more than 80%, and this protection endures for 3–5 years or longer following discontinuation of immunotherapy. Despite the persistence of a positive skin test result while on venom immunotherapy, about 80–90% of patients will not experience a systemic allergic reaction to a subsequent insect sting after 3–5 years of immunotherapy. Repeat skin testing is not required for consideration of stopping immunotherapy. Relapse is less likely with 5 years than 3 years of venom immunotherapy. The duration is not yet defined for immunotherapy with IFA whole-body extract. More favorable immunotherapy responses generally occur in children, those with mild or moderate systemic reactions and those with history of *Vespula* allergy.

A shared decision-making discussion with individuals at greater risk of reoccurrence of sensitivity associated with anaphylaxis should be documented concerning lifetime immunotherapy. Honeybee sting anaphylaxis relapse is more likely, possibly due to the higher amount of venom injected by honeybees during a sting compared to other insects, such as wasps. Another reason may be that some honeybee allergens (e.g., Api m 3 and Api m 10) are of insufficient concentration in commercially purified venoms.

1 Summary

Systemic venom reactions following insect sting can be mild or life-threatening anaphylaxis, with involvement of two or more organ systems. Testing and venom immunotherapy is only indicated in case of a severe allergic reaction or anaphylaxis or risk factors for potentially fatal reactions with future insect stings. The American Academy of Allergy and Immunology (AAAAI) and American College of Allergy, Asthma, and Immunology (ACAAI) 2016 Joint Task Force insect sting allergy practice parameter recommends that testing should be repeated in 3–6 months if initially negative in patients with a high pretest probability of insect sting allergy. Baseline serum tryptase level should be considered in cases of severe allergic reactions, especially with hypotension without urticaria or with a convincing history and no detectable specific IgE. An elevated baseline tryptase may increase the risk of sting reactions. Insect sting testing is performed with commercially available purified venoms, except whole-body extracts are used for IFA. Long-term management of systemic insect allergy includes prescription of epinephrine autoinjectors, instruction in the timing and technique of epinephrine use, venom immunotherapy if relevant specific IgE is detected, instructions as to when to seek emergency assistance, and education about measures to avoid future stings.

Bibliography

1. Alfaya Arias T, Soriano Gómis V, Soto Mera T, Vega Castro A, Vega Gutiérrez JM, Alonso Llamazares A, et al. Key issues in hymenoptera venom allergy: an update. J Investig Allergol Clin Immunol. 2017;27(1):19–31.
2. Baker TW, Forester JP, Johnson ML, Stolfi A, Stahl MC. The HIT study: Hymenoptera identification test: how accurate are people at identifying flying insects? Ann Allergy Asthma Immunol. 2014;113(267)
3. Bilò MB, Tontini C, Martini M, Corsi A, Agolini S, Antonicelli L. Clinical aspects of hymenoptera venom allergy and venom immunotherapy. Eur Ann Allergy Clin Immunol. 2019;51(6):244–58.
4. Casale TB, Burks AW. Clinical practice. Hymenoptera-sting hypersensitivity. N Engl J Med. 2014;370(15):1432–9.
5. Castells M. Diagnosis and management of anaphylaxis in precision medicine. J Allergy Clin Immunol. 2017;140(2):321–33.
6. Golden DBK, Tabatabaian F, Müller-Gierok U, Lockey RF. Indications for and preparing and administering Hymenoptera vaccines. In: Allergens and allergen immunotherapy. 6th ed. Boca Raton, FL: CRC Press/Taylor and Francis Group; 2020. p. 423–35.
7. Golden DBK. Insect allergy. In: Burks AW, Holgate ST, O'hehir R, Broide DH, Bacharier LB, Hershey GKK, Peebles RS, Middleton E, editors. Middleton's allergy: principles and practice. Amsterdam: Elsevier; 2020.
8. Golden DBK, Demain J, Freeman T, Graft D, Tankersley M, Tracy J, et al. Stinging insect hypersensitivity: a practice parameter update 2016. Ann Allergy Asthma Immunol. 2017;118(1):28–54.
9. Hemmer W, Wantke F. Insect hypersensitivity beyond bee and wasp venom allergy. Allergol Select. 2020;4(01):97–104.
10. Niedoszytko M, Bonadonna P, Oude Elberink JNG, Golden DBK. Epidemiology, diagnosis, and treatment of Hymenoptera venom allergy in mastocytosis patients. Immunol Allergy Clin N Am. 2014;34(2):365–81.
11. Ochfeld EN, Greenberger PA. Stinging insect allergy and venom immunotherapy. Allergy Asthma Proc. 2019;40(6):372–5.
12. Tankersley MS, Ledford DK. Stinging insect allergy: state of the art 2015. J Allergy Clin Immunol Pract. 2015;3(3):315–22.

Latex Allergy

Abeer Siddiqi and Anu Mallapaty

1 Introduction

In this chapter, we will use a case-based approach to discuss the clinical manifestations, diagnosis, management, and clinical pearls of latex allergy. We will discuss the common presenting signs and symptoms followed by a discussion of the major allergens implicated in latex allergy. We will also review methods of diagnosis, describe latex-containing products, and discuss allergy to latex-related foods.

2 Cases

Case 1

A 35-year-old male, who works as a technician in a catheter-producing plant, presents to your clinic with complaints of dyspnea, chest tightness, wheezing, and perennial rhinitis. Symptoms are worse on Fridays and improve over the weekend. Vitals include temperature of 99F, respiratory rate of 13 breaths per minute, heart rate of 76 beats per minute, blood pressure of 120/80 mm Hg, and oxygen saturation of 96% on room air. On physical examination, you note bilateral clear lung fields with no clubbing. Chest X-ray is unremarkable. You send the patient to obtain pulmonary function testing. Results are significant for an FEV_1/FVC below 80% predicted, with a 20% improvement in FEV_1 post albuterol.

A. Siddiqi · A. Mallapaty (✉)
Division of Allergy, Immunology and Rheumatology, Department of Pediatrics, Columbia University Irving Medical Center, New York, NY, USA

Division of Pediatric Pulmonology, Allergy, Immunology & Sleep Medicine, Boston Children's Health Physicians, Valhalla, NY, USA
e-mail: Anu_Mallapaty@bchphysicians.org

Question 1

Which of the following types of allergen is likely implicated in this disorder?

A. Flour
B. Metal
C. Varnish
D. Latex
E. Wood dust

Explanation

Answer: D

While all the answers may be implicated in occupational asthma, it is essential to make the correlation with likely exposure at the workplace. Of these choices, latex is the most likely allergen in a catheter-producing plant.

Latex or natural rubber latex (NRL) is derived from the tropical lactiferous rubber tree, *Hevea braziliensis* (of the botanical family Euphorbiaceae). Current reports indicate that every year over 12 million tons of NRL is produced. It is extracted as a fluid, with a milk-like consistency, by tapping the tree bark and is then used for the manufacturing of a wide assortment of medical and nonmedical materials and equipment. Only a limited number of these products are regulated by the FDA and several of these do not contain precautionary labeling.

In the late 1980s, there were unexplained IgE-mediated immediate hypersensitivity reactions reported which included mild urticaria to angioedema and even life-threatening anaphylaxis. Later, these reactions were found to be likely attributed to trace amounts of NRL proteins present in a variety of materials. A decade later, in the 1990s, NRL proteins were also identified as potential airborne allergens causing rhinitis and occupational asthma through their binding to glove powder particles.

Natural rubber latex (NRL) allergy reached an epidemic pattern during the late 1900s owing to the increased production and manufacturing methods of various NRL-containing

M. Mahmoudi (ed.), *Absolute Allergy and Immunology Board Review*, https://doi.org/10.1007/978-3-031-12867-7_17

products, such as powdered gloves, condoms, and catheters. This condition was noted to be particularly prevalent in individuals with mucosal exposure to natural rubber latex (NRL), such as in those with spina bifida requiring repeated ureteral catheterizations with latex catheters, and in health care workers using natural rubber gloves. Health care workers developed occupational rhinitis and occupational asthma due to sensitization to NRL allergens. NRL-induced Occupational Asthma (OA) and Occupational Rhinitis (OR) were also seen in various other professions including hairdressers, food processors, and those who worked in greenhouses, glove manufacturing, laboratories, and the pharmaceutical industry.

The reported data suggests that the prevalence of latex allergy in the general population is around 4.3%, with a higher prevalence seen in patients that are susceptible (e.g., 7.2% in patients with spina bifida and 9.7% in health care workers). Health care workers often present with symptoms of OA and OR as aforementioned, although there is a scarcity of reported literature correlating the incidence of asthma due to latex sensitization.

Question 2

Which of the following allergens is the dominant sensitizer noted in this syndrome?

A. Hev b 1
B. Hev b 2
C. Hev b 3
D. Hev b 6
E. Hev b 8

Explanation

Answer: D

Within highly purified or recombinant NRL, more than 15 allergens have been identified at a molecular level. Exposure to these proteins arises through airborne, percutaneous, or parenteral routes and is prevalent among individuals in frequent contact with natural rubber latex materials. In health care workers, airborne exposure is prominent and results from the attachment of latex proteins to dry lubricants that are used to facilitate donning of gloves, although sensitization rates have decreased with new manufacturing techniques. For patients, exposure is often parenteral and caused by leaching from latex materials used surgically. The pattern of IgE reactivity is dependent on the route of sensitization. Hev b 4, 5, 6.01, and 6.02 are the predominant sensitizers in health care workers. Hev b1, b3, and b5 are the immunodominant allergens for patients with spina bifida. The remaining allergens are considered to be minor.

The following table delineates these antigens.

Allergen	Frequency of reactivity (%)	Biochemical name
Hev b 1	50–82	Rubber elongation factor
Hev b 2	20–61	Endo-1,3-beta-glucosidase
Hev b 3	79	Small rubber particle protein (SRPP)
Hev b 4	25–89	Lecithinase homologue
Hev b 5	56–92	Acidic protein
Hev b 6	83	Hevein precursor
Hev b 7	8–49	Patatin like protein
Hev b 8	24	Profilin
Hev b 9	15	Enolase
Hev b 10	4	Superoxide dismutase
Hev b 11	3	Class I chitinase (PR-3)
Hev b 12	24	Nonspecific lipid transfer protein type 1 (nsLTP1)
Hev b 13	63	Esterase
Hev b 14	3–67	Hevamine
Hev b 15	Unknown	Serine protease inhibitor

Question 3

How would you best approach diagnosing this patient?

A. Latex skin testing with commercially available extracts
B. A relevant medical history with Immunocap testing
C. Patch Testing to Latex
D. Strict avoidance of all potential workplace allergens
E. Latex catheter challenge

Explanation

Answer: B

There are various methods that may be employed in the evaluation of a patient with a likely allergic reaction to latex.

In vitro assays for serum specific IgE antibody to latex are commercially available (e.g. Immunocap Assay). A relevant medical history and a positive Immunocap to latex at a level above 0.7 kU/L has a >98% specificity. However, it should be noted that 30% and 9% of asymptomatic atopic and non-atopic individuals respectively have positive ImmunoCap latex results.

There are currently no commercially available latex skin testing extracts in the United States. Some allergists prepare their own extracts of unknown potencies by using latex gloves. One method involves soaking 2 fingers of a latex glove or a toy balloon in 5 mL of normal saline or buffer, and prick testing with the solution. This testing modality is not standardized and can potentially expose the patient to high levels of latex protein.

Patch testing to latex is not standard practice for diagnosing an IgE mediated latex allergy. Avoidance of all potential workplace allergens is unnecessary. A latex catheter challenge would be risky given history, and testing prior would be most prudent.

Question 4
When counseling the patient regarding products to be avoided in the future, which of the following statements is accurate:

A. Synthetic rubber should be avoided
B. No need to strictly avoid latex
C. Powdered natural rubber latex gloves may be used
D. Avoid all products that may contain latex in the future
E. Common household items such as balls do not contain latex anymore

Explanation
Answer: D
Currently, the mainstay of therapy includes avoidance, which may significantly impact quality of life and employment for individuals. Patients should be counseled to decrease future exposure (e.g. avoidance of latex containing products in common household and personal care items including balloons, balls, condoms and gloves). In a latex-allergic patient who needs surgery, every provision should be made to obtain a latex-free operating suite to decrease the risk of anaphylaxis.

Given the significant public health concern posed by NRL allergy, various policies have been put into place, including the recommendation made by the American Society for Testing and Materials to limit the amount of allergen protein to be less than 200 μg/m^2 (approximately 50 μg/g of glove) for medical gloves, and substitution of powdered NRL gloves by gloves with a low content of protein and powder.

Synthetic rubber can be safely used in a latex allergic patient.

Question 5
What is the natural prognosis/treatment of this disorder?

A. Patients will grow out of their latex allergy with continued exposure
B. Unknown
C. Patients maintain their sensitization to latex lifelong
D. Sublingual immunotherapy is available for latex
E. Subcutaneous immunotherapy is the gold standard for treatment of latex allergy

Explanation
Answer: B
The prognosis of this condition is still unclear. However, it is postulated that latex allergy may follow the same pattern as other drug allergens and result in decreased sensitivity over time as long as no further exposure.

There are a few studies investigating specific immunotherapy for latex using latex allergen extract. Limited studies have shown that immunotherapy may reduce cutaneous and respiratory symptoms on re-exposure to natural rubber latex in IgE sensitized allergic individuals. However, the risk of systemic reactions remains high. Further studies in this area of research are required, and it is currently not routinely available.

Case 2
A 4-year-old girl presents to the clinic with her grandmother. Grandmother reports that within 10 min of eating a spoonful of Kung Pao chicken from a new Chinese restaurant, her granddaughter complained of an itchy mouth and throat. No systemic involvement was noted and the symptoms resolved with one dose of an over-the-counter antihistamine. After reviewing the ingredients, chestnuts were the only item that the patient has not eaten and tolerated previously. You order specific serum IgE to chestnut and it is reported to be elevated. You suspect an oral allergy syndrome.

Question 1
Which of the following allergens is known to be cross-reactive with chestnuts?

A. Latex
B. Ragweed
C. Birch
D. Timothy grass
E. Orchard grass

Explanation
Answer: A
Latex-fruit syndrome (LFS) is noted in approximately 30–50% of individuals that are sensitized to NRL allergens. An increasing number of plant sources have been associated with this syndrome including a variety of fruits, with the most commonly reported being avocado, banana, and kiwi. Potato, tomato, chestnut, olive, fig, herbs, and carrot have also been implicated. Clinical reactions to these foods vary from limited oral symptoms to life-threatening anaphylaxis. The diagnosis is based on clinical history. The current management includes avoidance of cross-reactive foods to which one has demonstrated clinical symptoms.

Chestnut has not been reported to be cross-reactive with pollen protein.

Question 2
Which of the following are considered major pan-allergens responsible for Latex-fruit syndrome.

A. Class 1 chitinases
B. Class 2 chitinases
C. Class 3 chitinases
D. Class 5 chitinases
E. Class 4 chitinases

Explanation
Answer: A

There are 14 families of Pathogenesis Related (PR) proteins, of which 6 account for most of the cross-reactivity. Class 1 chitinases, which belong to the PR-3 and PR-4 families, have been identified as major pan-allergens responsible for latex-fruit syndrome. These are cross-reactive with hevein (Hev), namely Hev b2, Hev b7, Hev b8, and Hev b12, which are major allergens in latex. Allergens of PR Protein Group 2 (β-1,3-glucanase proteins) are responsible for the latex-fruit cross-reactivity seen between latex (Hev b2) and avocado, banana, chestnut, fig, and kiwi.

Class 2/3/4/5 chitinases are not major pan-allergens implicated in latex-fruit syndrome.

Question 3
Which of the following allergen protein in Chestnut is cross-reactive with Latex allergen Hev b 6.02.

A. Bos d 5
B. Gal d 1
C. Can f5
D. Ara h2
E. Cas s 5

Explanation
Answer: E

Latex allergen Hev b 6.02 is similar to PR-3 type proteins in chestnut (Cas s 5) and avocado (Pers a 1), and PR-4 type proteins (wound-induced proteins) found in tomato and potato (Win 1 and Win 2).

Below table is a summary of the cross-reactive allergens present in Latex and fruits:

	Cross-reactive Proteins
Latex Hev b2	β-1,3 glucanase
Latex Hev b6.02	Cas s 5, pers a 1, win 1, win 2
Latex Hev b2, Hev b7, Hev b8, Hev b12	Class 1 chitinases

Bos d 5 is a cow's milk component, gal d 1 is a heat stable egg component, Can f 5 is a cat component, Ara h 2 is a peanut component.

Question 4
What is the next best step:

A. Start latex immunotherapy for this patient
B. Obtain a history inquiring about latex exposure and potential clinical reactivity
C. Perform latex skin testing
D. Add latex to patient's allergy list
E. Instruct patient to continue eating chestnut even if symptom provoking

Explanation
Answer: B

Currently, there are no longitudinal studies determining the likelihood of development of latex allergy in patients with food reactions that have been implicated in this syndrome or vice versa. Theoretically, there may be an increased risk of an allergic reaction with latex exposure. However, there is no strong recommendation for or against the evaluation of latex allergy in people with oral symptoms to commonly implicated foods without a history of reaction to latex. Thus, obtaining a detailed history of latex exposure and reactivity is the best next step.

Question 5
The patient is due for vaccinations and develops localized hives and facial swelling within 30 min of an injection. Which of the following vaccines is known to contain latex.

A. Influenza, Fluarix Quadrivalent
B. Pneumovax 23
C. Hepatitis B, Engerix-B
D. Zoster, Shingrix
E. HPV, Gardasil-9

Explanation
Answer: C

Of these choices, Hepatitis B, Engerix-B is the only one that is latex containing, specifically the tip cap of the prefilled syringe.

Dry natural rubber latex can be present in tip caps, vial stoppers, and syringe plungers. In 2003, a review of over 167,000 reported reactions in the VAERS database revealed 28 immediate-type allergic reactions to a vaccine in patients also reported to be latex-allergic. Although rare, there have been a few case reports of allergic reactions presumed to be related to latex contamination by medications stored in containers with rubber stopper vials. In theory, storage of liquid in a vial containing dry natural rubber latex may result in contact and release of latex allergens into the solution. The US Food and Drug Administration (FDA) has regulations regarding the labeling of natural rubber-containing medical devices, but these rules have not been applied to medication vial stoppers. The Center of Disease Control has an updated list of latex-containing vaccine packaging on their website, which health care practitioners should be encouraged to refer to. Vaccine vial stoppers and syringe plungers made of synthetic rubber do not pose a risk to latex-allergic persons.

3 Conclusion

Over the last decade, multiple interventions including the substitution of powdered latex gloves with either low-protein latex gloves or non-latex gloves have dramatically decreased the incidence of sensitization to latex, resulting in a decline in sensitization by more than 50%.

Key Points

- Latex allergy continues to be a significant public health concern, particularly in high-risk individuals and health care workers.
- Patients with suspected latex allergy should be evaluated based on clinical history and demonstration of clinical sensitization on skin prick testing or serum Immunocap testing.
- The primary treatment modality remains avoidance and counseling.
- Latex-fruit syndrome is noted in 30–50% of patients sensitized to latex and is due to cross-reactivity between plant and latex allergens.

Acknowledgements We wish to confirm that there are no known conflicts of interest associated with this publication and there has been no significant financial support for this work that could have influenced its outcome.

All authors listed have contributed sufficiently to the project to be included as authors, and all those who are qualified to be authors are listed in the author byline.

Bibliography

1. Archambault S, Malo J, Infante-Rivard C, et al. Incidence of sensitization, symptoms and probable occupational rhinoconjunctivitis and asthma in apprentices starting exposure to latex. J Allergy Clin Immunol. 2001;107:921–3.
2. Kogevinas M, Zock JP, Jarvis D, et al. Exposure to substances in the workplace and new-onset asthma: an international prospective population-based study (ECRHS-II). Lancet. 2007;370:336–41.
3. Vandenplas O, Froidure A, Meurer U, et al. The role of allergen components for the diagnosis of latex-induced occupational asthma. Allergy. 2016;71:840–9.
4. Vandenplas O, Raulf M. Occupational latex allergy: the current state of affairs. Curr Allergy Asthma Rep. 2017;17:14.
5. Bousquet J, Flahault A, Vandenplas O, et al. Natural rubber latex allergy among health care workers: a systematic review of the evidence. J Allergy Clin Immunol. 2006;118:447–54.
6. Siracusa A, Folletti I, Gerth van Wijk R, et al. Occupational anaphylaxis–an EAACI task force consensus statement. Allergy. 2015;70:141–52.
7. Hamborsky J, Kroger A, Wolfe S. Epidemiology and prevention of vaccine-preventable diseases. In: Centers for Disease Control and Prevention. Washington, DC: Public Health Foundation; 2015.
8. Primeau M-N, Adkinson NF, Hamilton RG. Natural rubber pharmaceutical vial closures release latex allergens that produce skin reactions. J Allergy Clin Immunol. 2001;107:958–62.
9. Russell M, Pool V, Kelso JM, et al. Vaccination of persons allergic to latex: a review of safety data in the vaccine adverse event reporting system (VAERS). Vaccine. 2004;23:664–7.
10. Slater JE. Latex allergy. J Allergy Clin Immunol. 1994;94:139–49.
11. Matricardi PM, Kleine-Tebbe J, Hoffmann HJ, et al. EAACI molecular allergology user's guide. Pediatr Allergy Immunol. 2016;27:1–250.
12. Makatsori M, Durham S, Calderon MA. Specific immunotherapy for latex allergy. Cochrane Database Syst Rev. 2017;2017(4):CD009240. https://doi.org/10.1002/14651858.CD009240.pub2. PMCID: PMC6478135
13. Nucera E, Mezzacappa S, Buonomo A, Centrone M, Rizzi A, Manicone PF, Patriarca G, Aruanno A, Schiavino D. Latex immunotherapy: evidence of effectiveness. Postepy Dermatol Alergol. 2018;35(2):145–50. https://doi.org/10.5114/ada.2018.75235. Epub 2018 Apr 24. PMID: 29760613; PMCID: PMC5949543

Drug Allergy

Schuman Tam

Management of drug allergy is commonly encountered in the specialty of Allergy and Immunology. Primary care practitioners frequently request assistance from the allergist and immunologist in managing patients with drug allergy. It is essential that a trained allergy and immunology specialist be knowledgeable in the condition to assist our colleagues in caring for the patients. The following two clinical cases demonstrate the essential clinical principles one has to master to be a board-certified allergist and immunologist.

Case 1

A 67-year-old businessman presented to your clinic with a history of allergic reactions following flu shot and a steroid shot. In 1995, he received a steroid shot (believed to be Kenalog) into his shoulder joint and developed hives within half an hour without breathing problem; since then, he has not had any steroid shot. In 2017, he received Fluzone quadrivalent vaccine at a healthcare facility. Within minutes, he developed hives and then dyspnea and was given epinephrine for relief; since then, he has not had a flu shot. Past medical history included sleep apnea, hypercholesteremia, benign prostatic hyperplasia, anxiety, mild intermittent asthma, shellfish-induced anaphylaxis, and mild reaction to certain fresh fruits. Current medications at the time of presentation included Atorvastatin, Tamsulosin, Albuterol HHN, and Cetirizine. Prior drug reaction included Typhoid vaccine (injectable) induced urticaria, IV dye contrast-induced flushing, injectable Fluzone Quadrivalent induced anaphylaxis, and injectable triamcinolone induced urticaria. His son has allergic rhinitis and his daughter has shellfish allergy. His social history, surgical history, and review of system were noncontributory. Physical examination was normal.

He presented to your clinic in the morning requesting your advice on whether he should receive the COVID-19 mRNA vaccine scheduled for the same afternoon. His primary doctor said that he might be allergic to an inactive ingredient present in Triamcinolone Acetonide. Since this inactive ingredient was present in the COVID mRNA vaccine, he sought your advice.

You investigated the inactive ingredients present in Fluzone Quadrivalent, Triamcinolone acetonide, and Typhoid vaccine. Your research reviewed that Fluzone Quadrivalent contains egg protein, octylphenol ethoxylate (Triton X-100), sodium phosphate-buffered isotonic sodium chloride solution, thimerosal, and sucrose. Triamcinolone acetonide contains sodium chloride, benzyl alcohol, carboxymethylcellulose, and polysorbate 80. Typhoid vaccine contains Vi polysaccharide, sodium chloride, disodium phosphate, and monosodium phosphate.

Question 1

You informed the patient that polysorbate 80 which is present in the triamcinolone acetonide can cross-react with polyethylene glycol in the COVID 19 mRNA vaccine. The polysorbate 80 is also present in the Janssen COVID 19 adenovirus vector vaccine. You further told the patient the following (choose the best answer):

A. The Quadrivalent Fluzone contains no polyethylene glycol and no polyethylene glycol derivative
B. The Quadrivalent Fluzone contains polyethylene glycol or its derivative

Answer: B

Triton X-100, present in the Quadrivalent Fluzone, is a poly(ethylene glycol) derivative that is poly(ethylene glycol) in which one of the terminal hydroxy groups has been converted into the corresponding p-(2,4,4-trimethylpentan-3-yl) phenyl ether (Fig. 1).

S. Tam (✉)
University of California, San Francisco, CA, USA

Asthma and Allergy Clinic of Marin and San Francisco, Inc., Greenbrae, CA, USA
e-mail: schuman.tam@ucsf.edu

M. Mahmoudi (ed.), *Absolute Allergy and Immunology Board Review*, https://doi.org/10.1007/978-3-031-12867-7_18

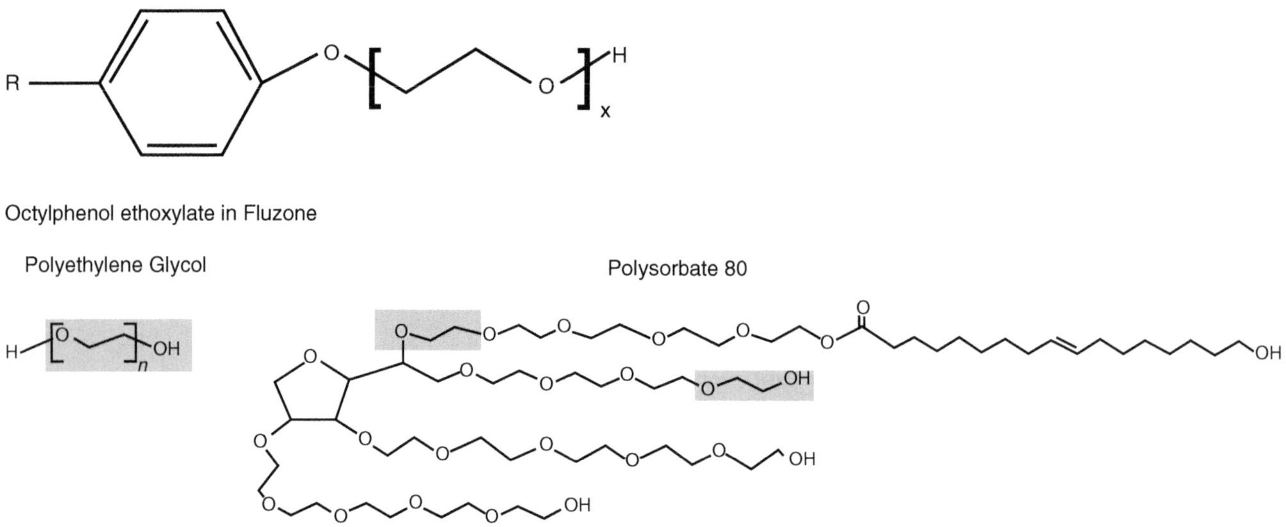

Fig. 1 Chemical structures of OctylPhenol ethoxylate (Triton X-100), polyethylene glycol, and polysorbate 80. All three chemicals contain polyether which contains multiple (-CH2CH2-O-) in repeat sequences. Octylphenol ethoxylate contains p-(2,4,4-trimethylpentan-3-yl)phenyl-ether at one of its terminals and H on the other end. Polyethylene glycol contains polyether (multiple-CH2CH2O- in sequence) terminating with H on one end and OH on the other end. Polysorbate 80 contains polyether with 1 big chemical structure, which contains polyether, on one end and another big chemical structure, which also contains polyether groups on the other end

It is important in managing drug allergy that if one encounters an unknown chemical, it is critical to look up the chemical structure. Pubchem is a good sauce to look up unfamiliar chemicals. For looking up inactive drug ingredients, a reliable sauce is DailyMed. In this case, Fluzone contains octylphenol ethoxylate (Triton X-100). Octylphenol ethoxylate is a derivative of polyethylene glycol and is structurally similar to polyethylene glycol. If the patient's prior allergic reaction to Fluzone was secondary to a reaction to Triton x-100, he may react to the COVID 19 mRNA vaccine since it contains polyethylene glycol. He may also react to the Jansen adenovirus COVID 19 vaccine since Triton X-100 is also structurally similar to polysorbate 80.

Question 2

You discussed with the patient the risk and benefits of getting the COVID-19 mRNA vaccine. With his reliable history of Type I allergy (hives and dyspnea) to Fluzone Quadrivalent (containing a derivative of polyethylene Glycol) and Triamcinolone acetonide (containing polysorbate 80 which is similar to polyethylene glycol), you recommended:

A. Routine vaccination with 15 min observation as the benefit of getting the COVID-19 mRNA vaccination is higher than the risk of potential allergic reaction
B. Routine vaccination with 30 min observation because of medium risk of allergic reaction to COVID-19 mRNA vaccine
C. Avoid COVID-19 vaccination even though the patient insists on getting the COVID-19 vaccination
D. Expanded skin testing using medications containing polyethylene glycol and polysorbate

Answer: D

Expanded skin testing using medications containing polyethylene glycol and polysorbate. The benefit of getting the COVID-19 vaccine is to reduce the risk of mortality if he acquires the COVID-19 infection. In public health standpoint, the vaccination may help to prevent the spread of COVID-19 infection in the community. The risk, based on history, is high in that he had a typical immediate allergic reaction to the Fluzone quadrivalent vaccine containing polyethylene glycol derivative and triamcinolone acetonide containing polysorbate 80. In this situation, skin testing may be helpful as a negative response may indicate that the patient may be able to tolerate the COVID-19 vaccine. Although skin testing using medications containing polyethylene glycol and polysorbate is not standardized, a positive result using a recommended concentration of the drugs suggests the presence of polyethylene glycol or polysorbate IgE antibodies. A negative skin test result does not rule out an allergy. Care should be taken when administering the COVID vaccine with 30 min of observation and be prepared to treat anaphylaxis. A positive skin test response, together with a history of prior allergic reaction to the drug, suggests a true allergic reaction and the COVID vaccine should not be given in the usual manner.

The skin test was performed urgently this morning before the patient's scheduled COVID-19 mRNA vaccination in the afternoon. The skin test showed a negative response to both the polyethylene glycol and polysorbate 80. You recommended the patient to proceed to get the vaccine with 30 min observation. The vaccine center in your hospital however declined to have the patient vaccinated in your clinic because of logistic reasons in getting the vaccine to you. You sent the patient to the vaccination center for administration.

Question 3

You received a call from the vaccination center that the patient developed flushing, diffuse urticaria, and shortness of breath. You rushed to the vaccination center. The patient was in supine position as he was feeling dizzy. He was able to speak with you but with a respiratory rate of 32. You could not detect a radial pulse but his femoral pulse was present. You would administer:

A. 0.3 mg Epinephrine subcutaneously into patient's right upper arm
B. 0.3 mg Epinephrine intramuscularly into the patient deltoid muscle
C. 0.3 mg Epinephrine IV push
D. 0.3 mg Epinephrine intramuscularly into patient's anterior-lateral mid-thigh
E. Either 0.3 mg Epinephrine intramuscularly into patient's anterior-lateral mid-thigh or 0.3 mg intramuscularly into patient's deltoid muscle, depending on how the patient was dressed

Answer: D

Intramuscularly into the patient's anterior-lateral mid-thigh. Table 1 shows that peak plasma epinephrine concentration is significantly higher after epinephrine is injected intramuscularly into the thigh than after epinephrine was injected intramuscularly or subcutaneously into the upper arm. Although injection of epinephrine into the deltoid muscle and into the vastus lateralis muscle (anterior-lateral mid-thigh) are both intramuscular, the injection to the later site produces a higher and earlier peak in plasma epinephrine level than the earlier site. The greater absorption of epinephrine from the vastus lateralis muscle in comparison with the deltoid muscle is most likely due to the greater blood flow in the vastus lateralis. When epinephrine is injected into the vastus lateralis, peak serum epinephrine can be achieved in 10 min and the peak plasma epinephrine level is five times higher than when injected intramuscularly into the deltoid muscle and is three times higher than that achieved by injecting subcutaneously into the upper arm. In anaphylaxis due to medication or other causes, the appropriate site of injection for epinephrine is intramuscularly into the anterior-lateral mid-thigh. Intramuscular injection is also a preferred route compared to subcutaneous route in pediatric patients with anaphylaxis. Epinephrine IV push should not be given to a patient who is not having cardiac arrest.

After appropriate administration of epinephrine, the patient improved and was transferred to the Emergency Department for further observation and treatment. The patient also received antihistamine and systemic steroid. Subsequently, he was discharged home.

Question 4

The patient returned to see you 2 months later. He said that his hospitalist obtained a serum tryptase about 2 h after the onset of his anaphylaxis to the COVID vaccine; the value was 8.2 ng/ml. His primary doctor obtained a serum tryptase a month after his anaphylaxis and the number was 5.1 ng/ml. You explained to the patient that:

A. Since both numbers were < 11.4 ng/ml, which was the normal cut-off reference value for the specific laboratory, the tests did not support the diagnosis of anaphylaxis secondary to his COVID vaccination. However, these numbers could be false negative finding for his anaphylactic episode

Table 1 Mean maximum plasma epinephrine concentration and time to achieve peak epinephrine concentration after its administration. The information is extracted based on a study performed by Simons et al. 2001. Commercial EpiPen was injected intramuscularly into mid anterior-lateral thigh. In addition, epinephrine (in syringe) was injected intramuscularly into mid anterior-lateral thigh, intramuscularly into deltoid, and subcutaneously into upper arm. Saline control was injected intramuscularly into deltoid and subcutaneously into the upper arm. Serum epinephrine levels were measured serially for 180 min after injection with epinephrine. Peak serum epinephrine levels were determined and the corresponding times reaching the peak were recorded. The peak serum epinephrine in pg/ml and the time required to reach the peak were as listed in the table. Among the methods used for injecting epinephrine, intramuscular injection into the mid anterior-lateral thigh achieved the highest peak in epinephrine concentration in plasma. The time required to reach that peak was the quickest (about 10 min after injecting epinephrine) for intramuscular injection into the mid anterior-lateral thigh

Injection route	EpiPen IM	Epinephrine IM	Epinephrine IM	Epinephrine SC	Saline IM	Saline SC
Injection site	Thigh	Thigh	Arm	Arm	Arm	Arm
C(max) mean in pg/ml	12,222	9722	1821	2877	1458	1495
Time in minutes from injection to peak serum level (Approximate)	10	10	180	120	N/A	N/A

IM intramuscular, *SC* subcutaneous, *C(max)* peak plasma epinephrine concentration, *N/A* not applicable

B. The tryptase value should be over 20 ng/ml within 3 h of his anaphylactic episode if he indeed had an anaphylaxis secondary to the COVID vaccine
C. The elevation of baseline tryptase of 5.1 ng/ml to 8.2 ng/ml supported the diagnosis of anaphylaxis after his recent COVID vaccination
D. Measuring serum tryptase is not helpful to confirm the diagnosis of anaphylaxis

Answer: C

Elevated serum tryptase level indicates IgE-mediated and non-IgE-mediated mast cell activation. Levels over 11.4 ng/ml are considered abnormal. If baseline tryptase is elevated to a level over 20 ng/ml, the practitioner needs to consider a possible diagnosis of systemic mastocytosis. Anaphylaxis can occur with normal acute serum tryptase. Comparing changes in a patient's tryptase level at baseline is important using the formula as shown in Fig. 2.

The patient worried that without the booster COVID-19 vaccine, he would not have sufficient protection against the virus. Although at that time, the number of community's COVID-19 infections had reduced, the pandemic had not been over. You ordered COVID-19 IgG and the patient had no detectable COVID-19 antibody, probably because of a systemic steroid given when he had the anaphylaxis following the first vaccination. At the present time, there were sufficient COVID vaccines available for the community. You discussed with the patient the possibility of COVID-19 vaccine desensitization although there had not been a published study on how to desensitize patient with COVID-19 vaccine induced anaphylactic shock. The patient would like to be desensitized to the COVID-19 vaccine so that he could have sufficient immunity against the virus. You admitted the patient to the intensive care unit for rapid desensitization using COVID-19 Moderna Vaccine 0.5 cc.

Question 5

After discussing with your patient, the risks and benefits of the desensitization, measuring patient's vital signs, performing a physical examination, and establishing an intravenous access, you will perform the following first (choose the best answer):

A. Since there are sufficient COVID vaccines at the present time for the whole community, you perform a skin test using the COVID vaccine before its administration: full dose will be given if the skin test is negative; desensitization protocol if the skin test is positive
B. Administer COVID 19 Moderna Vaccine 0.05 cc; if no reaction by 30 min, administer 0.45 cc of the same vaccine
C. Dilute the COVID vaccine using a diluent supplied by the manufacturer by 1000-fold; you will then administer 0.05 cc of the diluted vaccine
D. Dilute the COVID vaccine using a diluent supplied by the manufacturer by ten-fold; you will then administer 0.05 cc of the diluted vaccine

Answer: C

Starting dose for drug desensitization is typically 1/10,000 of the full dose and the dose is doubled every 15 min. Since the patient had anaphylactic shock following the COVID vaccine, starting the dose at 1/10,000 will be most appropriate. Since the patient's history and laboratory findings confirm the diagnosis of COVID 19 mRNA vaccine-induced anaphylactic shock, answer A is incorrect as a negative skin test using the COVID vaccine can be false negative since the sensitivity and specificity of skin test using COVID vaccine has not been established. Besides, you already performed the appropriate skin test at presentation and the result showed no sensitivity (thus false negative finding). The available clinical data suggests true COVID vaccine allergy. Answer B is a typical drug challenge protocol. This patient has a high probability of true severe allergic reaction to the COVID vaccine in question. Drug challenge with such a high dose is dangerous for the patient. The starting dose for Answer D is 100-fold. Desensitization starting from 1/100 of the full dose Moderna COVID-19 vaccine has been described for two patients who have documented mild (not anaphylactic shock) Type I allergy to the vaccine. Starting desensitization from 1/100 of the full dose may not be as safe as Answer C, for this patient who had anaphylactic shock using the full dose. After administering 1/10,000 of the normal dose, the administration of vaccine can be given every 15 min while monitoring the patient carefully for allergic reaction. Write your

Minimum elevated tryptase: (Baseline tryptase in ng/ml) x 1.2 + 2 ng/ml

Elevation of 20% of baseline tryptase plus 2 ng/ml suggests mast cell degranulation

Fig. 2 Formula for calculating MinimalSerum tryptase elevation to be considered having an active mast cell degranulation. Baseline tryptase is typically measured at about 24 h or more after the anaphylactic episode, or before the anaphylactic episode. By multiplying baseline tryptase in ng/ml by 1.2 (20% above baseline) and adding 2 ng/ml, one can obtain the minimal tryptase level to be considered as having evidence for mast cell activation. Acute tryptase is usually measured about 30 min to 2 h after the anaphylactic episode. If the acute tryptase number is more than or equal to the minimal calculated tryptase level, the patient likely has mast cell degranulation as the mechanism for the anaphylactic episode. Minimum elevated tryptase: (Baseline tryptase in ng/ml) × 1.2 + 2 ng/ml. Elevation of 20% of baseline tryptase plus 2 ng/ml suggests mast cell degranulation

own protocol based on desensitization principle and compare it to Tables 2 and 3.

Case 2

A 55-year-old female university nursing professor, referred by her primary doctor and gastroenterologist, to see you because of drug allergy and possible food allergy. In the early 1990s, during her training in a hospital in Spain, she had dermatitis of the hands when using latex gloves. She was told by an allergist at that time to avoid latex. In her early 20s, she developed dyspnea, angioedema of her face and hands, and diffuse urticaria within 30 min after taking her first dose of oral Amoxicillin prescribed for sinusitis. Since then, she has not had any penicillin-type antibiotics including cephalosporin. At age 41, she developed hives, dyspnea, and dizziness during a course of Ciprofloxacin given for her urinary tract infection. At age 45, she developed diffuse skin rash with blisters involving her trunk, mouth, and conjunctiva about 1 week after being started on Allopurinol for her gout. Since age 50, she has developed dysphagia when eating foods like chicken and beef and the frequency of her dysphagia has increased. Her other past medical history included borderline hypertension. Current medication included Metoprolol 25 mg every morning (she did not take the blood pressure medication yesterday and this morning). Drug allergy included latex, penicillin, ciprofloxacin, and allopurinol. Social history and family history were both noncontributory. Examination showed that her heart rate was 80 and regular. Blood pressure was 145/88. HEENT exam was normal. Cardiorespiratory exam was normal. Neurological exam was also grossly unremarkable.

Per patient, her primary doctor wanted to rule out patients penicillin allergy. Her gastroenterologist would like you to help with possible food allergy management if she would be confirmed to have eosinophilic esophagitis based on endoscopic examination and biopsy scheduled for the next day. Her gastroenterologist would also like to assure that patient was not allergic to latex before her endoscopy the next day since the gastroenterologist was not certain whether there would be latex present in the endoscopy suite and in the medication vial stoppers.

Since the patient had not taken any beta blocker in the past 1.5 days, you feel that it was safe to perform latex testing. However, workup for penicillin testing would be deferred until after her endoscopy since the patient would not be given any antibiotics during the procedure. You performed prick skin test using freshly extracted latex gloves

Table 2 COVID-19 moderna vaccine solution preparation for desensitization. COVID 19 Moderna vaccine consists of 100 mcg nucleoside modified mRNA, encoding the COVID-19 spike protein, in each 0.5 cc dose. Concentration of the vaccine is 200 mcg mRNA per cc solution. Recommended full dose is 0.5 cc of the 200 mcg mRNA per cc solution or 100 mcg mRNA. This table listed 4 solutions with different concentrations: 200 mcg/cc (actual vaccine concentration), 20 mcg/cc, 2 mcg/cc, and 0.2 mcg/cc. The prepared solutions are being utilized for desensitization as shown in Table 3

Moderna COVID mRNA	Volume (ml or cc)	Concentration	Total amount (mcg) nucleoside modified mRNA
Solution 1	0.5 cc	200 mcg/cc	100 mcg
Solution 2	1.0 cc	20 mcg/cc	20 mcg
Solution 3	1.0 cc	2 mcg/cc	2 mcg
Solution 4	1.0 cc	0.2 mcg/cc	0.2 mcg

Table 3 Desensitization protocol for Administration of COVID 19 Moderna Vaccine. Starting from the lowest concentration, 0.2 mcg/cc, one will administer an incrementally higher volume of the same solution every 15 min. When top volume was given, one will switch to a higher concentration and repeat the same procedures. It will take about 225 min to complete the desensitization while the patient is under close observation with examination before each dose

Step no.	Solution no.	Concentration of vaccine (mcg/cc)	Time (min)	Volume of IM dose (ml)	Administered dose (mcg)	Cumulative dose (mcg)
1	4	0.2	0	0.05	0.01	0.01
2	4	0.2	15	0.1	0.02	0.03
3	4	0.2	30	0.2	0.04	0.07
4	4	0.2	45	0.4	0.08	0.15
5	3	2.0	60	0.05	0.1	0.25
6	3	2.0	75	0.1	0.2	0.45
7	3	2.0	90	0.2	0.4	0.85
8	3	2.0	105	0.4	0.8	1.65
9	2	20.0	120	0.05	1.0	2.65
10	2	20.0	135	0.1	2.0	4.65
11	2	20.0	150	0.2	4.0	8.65
12	2	20.0	165	0.4	8.0	16.65
13	1	200.0	180	0.05	10	26.65
14	1	200.0	195	0.1	20	46.65
15	1	200.0	210	0.15	30	76.65
16[a]	1	200.0	225	0.2	40	116.65

[a] Final cumulative dose will be 116.65 mcg mRNA

and the result was normally negative. Oral mucosal challenge by touching the patient's oral cavity with a latex glove revealed no allergic reaction. You did not order blood test to search for latex-specific IgE as the result would take a week to come back. You felt confident that the patient was not allergic to latex based on your testing and advised her to proceed with endoscopy the next day.

Question 6

On your way back to your clinic the next day, you received a call from the patient's gastroenterologist asking you to see the same patient urgently in the hospital. The gastroenterologist said that the patient developed swelling of the face and dyspnea shortly after the upper endoscopy procedure. When you saw the patient in the hospital, she complained of throat tightness sensation and shortness of breath. A brief quick visual examination revealed that her face was obviously swollen compared to the day before. Palpation of the patient's face showed crunching. She had no urticaria at that time. Respiratory rate was 28. Monitor at bedside showed heart rate was 110 and blood pressure was 80/50. Oxygen saturation was 92% with supplemental nasal cannula oxygen. You would:

A. Immediately administer 0.3 mg intramuscularly to patient's mid anterior-lateral thigh and simultaneously increase patient's intravenous fluid infusion rate
B. Urgently intubate the patient before complete upper airway blockage
C. Urgently administer Icatibant as if you suspected that the endoscopy induced an attack of underlying hereditary angioedema
D. Urgently order equipment for tube thoracostomy while examining patient's chest
E. Urgently ask for hospitalist to assist in managing patient's anaphylaxis while you quickly review the anesthesia record in order to determine the time course of the reaction relative to medications given as you were certain that the allergic reaction was not caused by latex

Answer: D

Urgently order equipment for tube thoracostomy after determining the side of the chest that caused the tension pneumothorax. The patient did not have angioedema. Palpation of the swollen face revealed crepitation, suggesting subcutaneous emphysema. Given the history of endoscopy with esophageal biopsy, the patient likely developed pneumothorax with air getting into the subcutaneous tissue. She likely had tension pneumothorax causing breathing problem and hypotension.

Although an allergist frequently evaluates drug reaction after it has occurred, sometimes you may see a case when it is occurring. In this situation, history and examination are important to make the correct diagnosis and treatment. Sometimes, a case that appears like an allergic reaction to a drug may not be a true drug allergy. Choice A is incorrect as the patient does not have anaphylaxis. Choice B and choice C are appropriate if the patient has hereditary angioedema. However, crepitation of the swollen face rules out typical angioedema. Choice E is probably appropriate if the patient is hemodynamically stable and you miss the diagnosis. This question emphasizes the importance in considering the differential diagnosis of drug allergy.

After assuring that you are dealing with a drug reaction and not another cause like in this case, a thorough history will help to identify the cause of the potential drug reaction. Figure 3 shows the classification of drug reactions. It is important to know the pharmacological property of the drug in question in order to distinguish between Type A and Type B reaction. Type A reaction is dose-dependent and predictable whereas Type B reaction is dose-independent and unpredictable. An example of Type A reaction is bleeding while on Coumadin. Type B reaction can be intolerance, idiosyncrasy, and true drug allergy. Next, it will be helpful to know if the patient has taken the medication in the past. If he or she takes the medication in the past and now has reaction, he or she may have been sensitized to the drug previously causing the current immunological IgE-mediated allergic reaction. Next, it is important to know the clinical manifestation of the timing of the reaction. Does the patient have typical IgE-mediated reactions like urticaria, angioedema, hypotension, or bronchospasm? If you have a chance to examine the patient when he or she has an allergic reaction, it will be helpful. Sometimes, one has to rely on a record from another physician who examined the patient at the time of the reaction. Immediate Type I IgE-mediated allergic reaction occurs within minutes to hours after the administration of the drug. Delayed Type IV T cell-mediated drug reaction occurs days to weeks after the administration of the drug and may present with maculopapular, bullous, and pustular exanthem. It is not only important to know when a reaction occurs relative to the timing of the administration of the drug one also has to consider other concurrent administered drugs and renders a decision whether the reaction is caused by another drug. Therefore, temporal association between initiation of drug therapy and onset of allergic reaction will be essential. In conclusion, typical workup for potential drug allergy includes history, examination, and review of prior records to assure that you are dealing with drug reaction and not other causes that mimic allergic drug reaction. After that, the algorithm as outlined in Fig. 3 can assist you in making the correct diagnosis.

Question 7

Five days later, you received a call from an infectious disease specialist who was caring for your patient in the hospital.

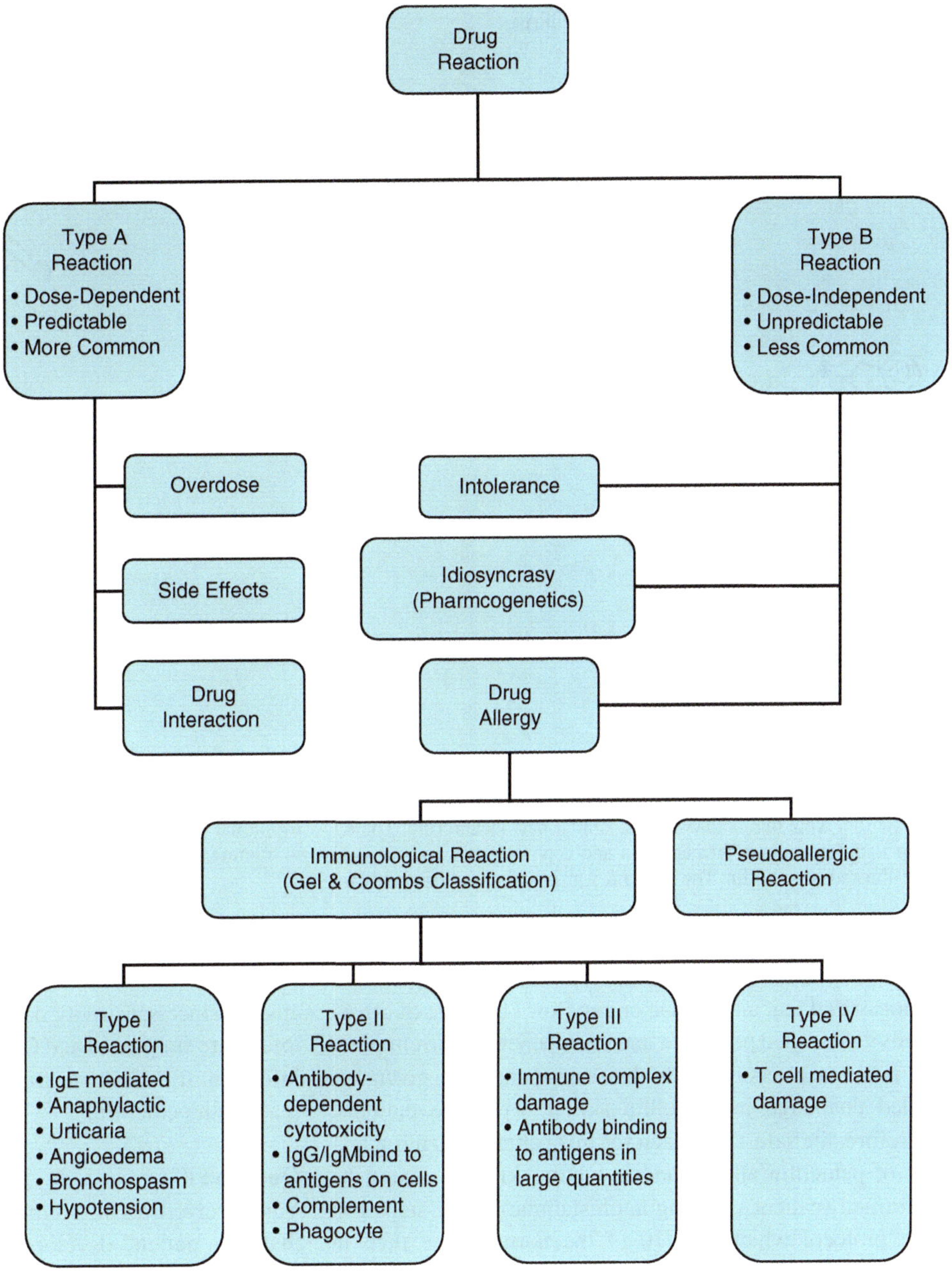

Fig. 3 Classification of drug reaction. Drug reaction can be either Type A or Type B. The table indicates what is the definition of each of the reaction and examples of the respective type of reaction. Drug allergy is a Type B drug reaction. Drug allergy is divided into immunological type or pseudo-allergic type. Within the immunological type, it is divided, based on gel and coombs classification, into Type I, Type II, Type III, and Type IV immunological reactions

Your patient developed pseudomonas infection of the lung and would require antibiotics. Medication given to the patient in the hospital included Metoprolol daily and diphenhydramine given at bedtime for insomnia. Preliminary drug sensitivity showed that Ciprofloxacin, Cefepime, and Imipenem have excellent antibacterial activity against the bacteria. Aztreonam had borderline activity against the bacteria. The infectious disease specialist asked you which antibiotics listed below should be used based on the patient's drug allergic history

A. Ciprofloxacin, under 2 step drug challenge protocol, since the patient's possible allergic reaction to the drug was more than 10 years ago
B. Imipenem, under 2 step drug challenge (test dose) protocol
C. Aztreonam, since it does not cross-react with penicillin and is the safest antibiotic to give
D. Cefepime, under 2 step challenge (test dose) protocol
E. B and D

Answer: E

Using either Imipenem or Cefepime is appropriate. Imipenem, which is a carbapenem, shares a common beta

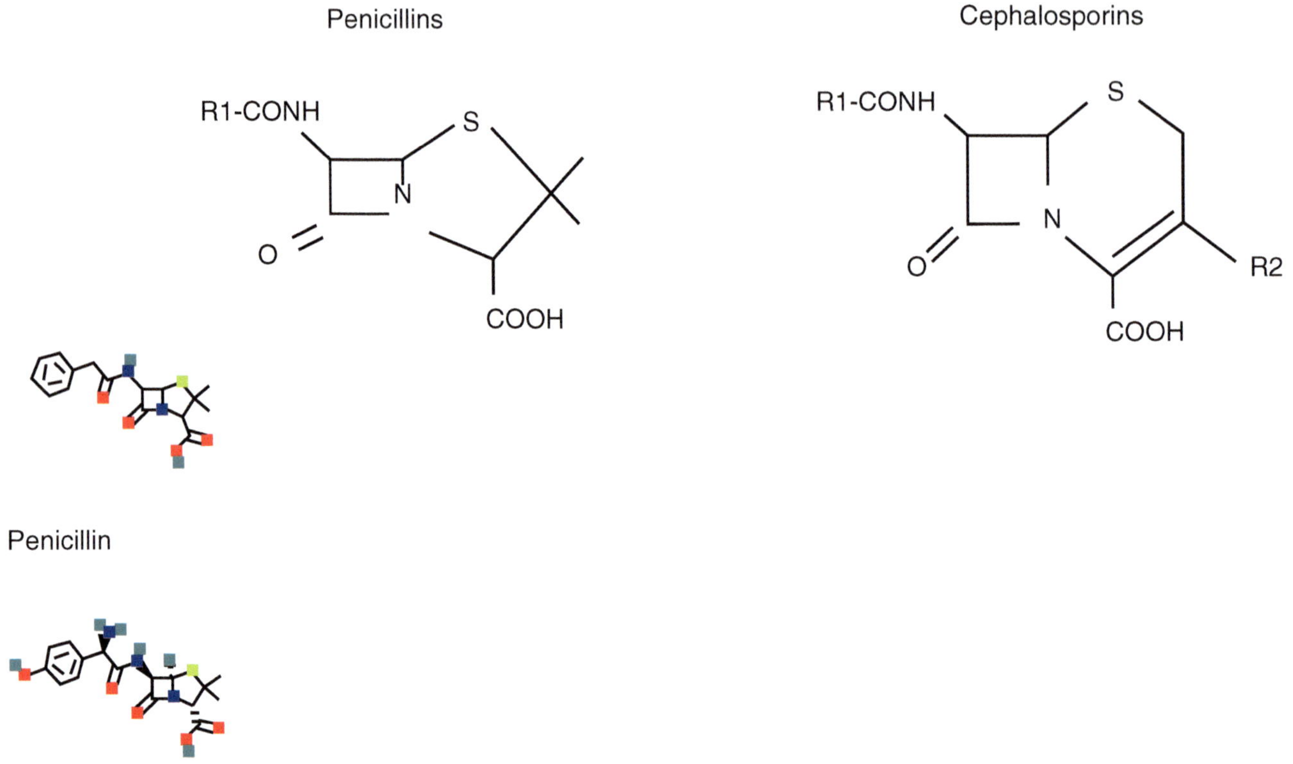

Fig. 4 Chemical structures of Penicillin, Cephalosporins, and Amoxicillin. Note that both contain a similar beta lactam ring. The R1 group is different between penicillin and cephalosporin and between penicillin and amoxicillin. The bigger R 1 group structure especially for the third- and fourth-generation cephalosporins makes cephalosporin dissimilar to penicillin and thus penicillin-allergic patient may be able to tolerate the cephalosporin

lactam ring with penicillin. However, clinical studies have demonstrated that an absence or very low (1%) rate of cross-reactivity between penicillin and carbapenems. Current practice parameter recommends that carbapenem be given via graded challenge in penicillin skin test-positive patients. Therefore, the safest approach for this patient who has a history of penicillin allergy and to whom skin test cannot be performed (patient is taking antihistamine) is a 2 steps test dose protocol which is 1/10 of the normal dose then in 30–60 min, administers full dose.

Cephalosporin allergy is not as common as penicillin allergy. IgE-mediated allergy to cephalosporins is likely directed mostly at the R1 group side chain (Fig. 4) rather than the core beta lactam ring; this is especially true for the third- and fourth-generation cephalosporin with a bigger side chain structure. Overall, about 3% of penicillin skin test-positive patients react to cephalosporins. Therefore, the safest approach for this patient who has a history of penicillin allergy and to whom skin test cannot be performed (patient is taking antihistamine) is a 2 steps test dose protocol which is 1/10 of the normal dose then in 30–60 min, administers full dose.

Answer A will not be appropriate since the patient has a history of Ciprofloxacin allergy. It will be appropriate if it is given in a desensitization protocol. Answer C is not the best answer since sensitivity data shows that using Aztreonam is inferior to Imipenem and Cefepime against the pseudomonas infection. It is however true that Aztreonam does not cross-react with penicillin.

Question 8

You received a call from the infectious disease specialist that the patient developed hives and hypotension after receiving Imipenem. Two doses of Epinephrine have been given intramuscularly into patient's mid anterior-lateral thigh without sufficient response. What will you recommend next:

A. Glucagon 1 mg intravenously bolus, followed by 5–10 mcg/min titrated against the clinical response
B. Epinephrine IV drip
C. Benadryl intravenously
D. Solumedrol intravenously

Answer: A

Patients taking beta blocker may resist standard treatment including epinephrine. Glucagon has both positive inotropic and chronotropic effects on the heart. The inotropic effect does not depend on catecholamine receptors and therefore not affected by being on beta blocker.

Your patient was subsequently treated with Cefepime for her pseudomonas lung infection and recovered. She presented to your office for follow-up skin test to rule penicillin allergy.

Question 9

Select from the following the best combination of prick skin test and intradermal skin test before oral challenge in this patient:

A. 0.00006 mol/L penicilloyl-polylysine (PPL, Pre-Pen), 10,000 Unites/ml penicillin G by prick skin test followed the same concentration for both ingredients utilizing intradermal method if the initial prick skin tests are negative
B. 0.00006 mol/L penicilloyl-polylysine (PPL, Pre-Pen), 10,000 Unites/ml penicillin G by prick skin test followed ten-fold dilution the for both ingredients via intradermal method if the initial prick skin tests are negative
C. 5 mg/cc Amoxicillin by prick then if negative, intradermal method
D. 6 mol/L penicilloyl-polylysine (PPL, Pre-Pen), 10,000 Unites/ml penicillin G by prick skin test followed the same concentration for both ingredients utilizing intradermal method if the initial prick skin tests are negative
E. A and C
F. B and C
G. C and D

Answer: E

The most commonly performed drug skin test is for penicillin as the testing is validated. The beta lactam ring of penicillin (Fig. 4) is unstable and degrades after administration to reactive intermediates that act as haptens and covalently bind to self-protein. The most important intermediate to induce allergenic response is penicilloyl (the major antigenic determinant for penicillin). Skin test using the major determinant, penicilloyl, and the minor determinant penicillin G has a negative predictive value of >95%. The concentration of penicilloyl-polylysine, commercially available as Pre-Pen, utilized for both prick and intradermal skin test, is 6 x 10 -5 M (or 0.00006 M or 0.00006 mol/L). Therefore, B and D are incorrect. The standard concentration utilized for penicillin G prick and intradermal skin testing is 10,000 units per ml.

Since the patient gave a history of allergic reaction to Amoxicillin, amoxicillin should be included in the skin testing panel. If this patient is selectively sensitive to amoxicillin and not penicillin (specific IgE antibody is directed against the R1 side chain (see Fig. 4), skin test can be positive for Amoxicillin and negative for Pre-Pen and penicillin G. In this situation, the patient is allergic to the Amoxicillin, but is able to tolerate penicillin. It is important to point out that the skin test for amoxicillin has not been validated. Therefore, oral challenge using amoxicillin is required if skin test is negative for amoxicillin, Pre-Pen, and penicillin G, in order to rule out amoxicillin allergy.

The result of the skin test showed that the patient is sensitive to Amoxicillin by prick skin test but negative by both prick and intradermal methods for Pre-Pen and penicillin G. You challenged the patient orally with penicillin VK and showed tolerance. You determined that the patient is likely allergic to amoxicillin based on her past history of reaction and current result of skin test.

Question 10

You advised the patient and her primary physician that the patient can take penicillin and should avoid amoxicillin. If she requires cephalosporin urgently in the future and skin test is not readily available, graded test dose administration of the appropriate cephalosporin can be given except the following cephalosporin:

A. Ceftazidime, ceftriaxone, cefuroxime
B. Cefadroxil, Cefprozil, Cefatrizine
C. Cefaclor, cephalexin, cephradine
D. A and B
E. B and C

Answer: E

Cefadroxil, cefprozil, and cefatrizine share an identical R group (Fig. 4) with Amoxicillin. Cefaclor, cephalexin, and cephradine share a similar R group with Amoxicillin. Therefore, these antibiotics may have a higher risk of causing an allergic reaction in the patient; thus, test dosing protocol is not appropriate for this patient. However, the R groups of ceftazidime, ceftriaxone, and cefuroxime are dissimilar to the R group of Amoxicillin; test dosing protocol is appropriate as the risk of reaction is low.

You informed the patient that her reaction to Allopurinol was likely Steven Johnson Syndrome or Toxic Epidermal Necrolysis. Testing for this reaction is not available and desensitization does not work for this drug and is considered dangerous. You were seeing this patient with a medical resident.

Question 11

You asked your medical resident which genetic risk factor is associated with the development of Steven Johnson syndrome after taking Allopurinol:

A. HLA-B*13:01
B. HLA-B*15:02
C. HLA-B*58:01
D. HLA-B*57:01

Table 4 Human leukocyte antigen association with serious drug allergy syndrome. The table lists the five drugs that have HLA association with different types of serious drug allergy syndromes

Drug	HLA allele	Drug allergy syndrome
Abacavir	B*57:01	Abacavir hypersensitivity syndrome
Allopurinol	B*58:01	SJS/TEN/DRESS
Carbamazepine	B*15:02; A*31:01	SJS/TEN/DRESS
Dapsone	B*13:01	DRESS
Vancomycin	A*32:01	DRESS

Abacavir hypersensitivity syndrome: fever, malaise, nausea, diarrhea, skin rash
DRESS drug reaction with eosinophilia and systemic symptoms, *HLA* human leukocyte antigen, *SJS* Stevens-Johnson syndrome, *TEN* toxic epidermal necrolysis

E. HLA-A*32:01

Answer: C

HLA-B*58:01 is a risk allele associated with Allopurinol induced Steven Johnson syndrome, TEN (Toxic epidermal necrolysis), and DRESS (drug reaction with eosinophilia and systemic symptoms). HLA-B*13:01 is associated with Dapsone-induced DRESS. HLA-B*15:02 is associated with carbamazepine-induced Steven Johnson Syndrome, TEN, and DRESS. HLA-B*57:01 is associated with Abacavir hypersensitivity syndrome. HLA-A*32:01 is associated with Vancomycin-induced DRESS. One may want to commit Table 4 into memory for the board examination.

Bibliography

1. Banerji A, Wickner PG, Saff R, Stone CA, Robinson LB, Long AA, Wolfson AR, et al. mRNA Vaccines to Prevent COVID-19 disease and reported allergic reactions current evidence and suggested approach. J Allergy Clin Immunol. 2020;9(4):1423–37.
2. Broyles AD, Banerji A, Barmettler S, Biggs CM, Blumenthal K, Brennan PJ, Breslow RG, et al. Practical guidance for the evaluation and management of drug hypersensitivity: specific drugs. J Allergy Clin Immunol. 2020;8(9S):S16–S122.
3. Broyles AD, Banerji A, Castells M. Practical guidance for the evaluation and management of drug hypersensitivity: general concepts. J Allergy Clin Immunol. 2020;8(9S):S3–S15.
4. DailyMed. NIH National Library of Medicine. https://dailymed.nlm.nih.gov/dailymed/. Accessed 7 Mar 2021.
5. Dreskin SC, Stitt JM. Anaphylaxis. In: Burks AW, Holgate ST, O'Hehir RE, Broide DH, Bacharier LB, Hershey GKK, Peebles RS, editors. Middleton's allergy principles and practice. Philadelphia, PA: Elsevier Inc.; 2020. p. 1228–46.
6. Konvinse KC, Trubiano JA, Pavlos R, James I, Shaffer CM, et al. HLA-A*32:01 is strongly associated with vancomycin-induced drug reaction with eosinophilia, and systemic symptoms. J Allergy Clin Immunol. 2019;144:183–92.
7. Mustafa SS, Ransey A, Staicu ML. Administration of a second dose of the moderna COVID-19 vaccine after an immediate hypersensitivity reaction with the first dose: two case reports. Ann Intern Med. 2021;174(8):1177–8. https://doi.org/10.7326/L21-0104. Online ahead of print
8. PubChem. NIH National Library of Medicine. National Center for Biotechnology Information. https://pubchem.ncbi.nlm.nih.gov/#query=. Accessed 7 Mar 2021.
9. Simons FE, Roberts JR, Gu X, Simons KJ. Epinephrine absorption in children with a history of anaphylaxis. J Allergy Clin Immunol. 1998;101:33–7.
10. Simons FE, Gu X, Simons KJ. Epinephrine absorption in adults: intramuscular versus subcutaneous injection. J Allergy Clin Immunol. 2001;108:871–3.
11. Solensky R, Phillips EJ. Drug allergy. In: Burks AW, Holgate ST, O'Hehir RE, Broide DH, Bacharier LB, Hershey GKK, Peebles RS, editors. Middleton's allergy principles and practice. Elsevier Inc.; 2020. p. 1261–82.
12. Tam S. Chemotherapy and biologic drug allergy. In: Mahmoudi M, editor. Allergy and asthma. Cham: Springer; 2019. https://doi.org/10.1007/978-3-030-05147-1_24. Print ISBN: 978-3-030-05146-4. Online ISBN: 978-3-030-05147-1.
13. Tam S. Drug allergy. In: Mahmoudi M, editor. Allergy and asthma, practical diagnosis and management. Switzerland: Springer International Publishing; 2016. p. 407–26.

Autoimmunity

Jack Jeskey, Lauren Fill, Madiha Huq, Sandeep Sarkaria, Remie Saab, and Robert Hostoffer

Abbreviations

ACLE	Acute cutaneous lupus erythematous
ANA	Anti-neutrophilic antibody
BMI	Body mass index
BP	Bullous Pemphigoid
BP180	Bullous Pemphigoid Antigen 180
BP230	Bullous Pemphigoid Antigen 230
CCLE	Chronic cutaneous lupus erythematosus
CNS	Central nervous system
DIF	Direct Immunofluorescence
DLE	Discoid lupus erythematous
EM	Erythema Multiforme
IgE	Immunoglobulin E
IIF	Indirect Immunofluorescence
IVIG	Intravenous immunoglobulins
LE	Lupus erythematous
LP	Lichen planus
MAC	Membranolytic attack complex
PMLE	Polymorphic light eruption
PV	Pemphigus Vulgaris
SCLE	Subacute cutaneous lupus erythematous
SLE	Systemic lupus erythematous

J. Jeskey · L. Fill · M. Huq · R. Saab · R. Hostoffer
Division of Pulmonary, Critical Care and Sleep Medicine, UH Cleveland Medical Center, University Hospital's Pediatric and Adult Allergy/Immunology Fellowship, Cleveland, OH, USA

S. Sarkaria (✉)
Division of Pulmonary, Critical Care and Sleep Medicine, UH Cleveland Medical Center, University Hospital's Pediatric and Adult Allergy/Immunology Fellowship, Cleveland, OH, USA

Allergy/Immunology Associates, Inc., Mayfield Heights, OH, USA

1 Introduction

Autoimmune disorders arise from a failure of self-tolerance, or immunologic unresponsiveness to an individual's own antigens, in genetically susceptible individuals. The mechanisms that result in tissue damage in autoimmune diseases parallel the normal responses of adaptive immunity and may include autoantibodies, immune complexes and/or autoreacting T lymphocytes. The clinical manifestations of autoimmune disorders are extremely varied but tend to be long-lasting and progressive. Immune responses may be directed against a single tissue resulting in organ specific disease such as bullous pemphigoid (BP). In contrast, some autoimmune disorders target a widespread antigens resulting in diseases like systemic lupus erythematous (SLE). We aim to introduce the reader to a better understanding of some common autoimmune diseases, as well as appropriate treatment approaches.

Case 1

An 80-year-old male with a past medical history of hypertension and atrial fibrillation presented to the outpatient clinic for evaluation of a new rash. His rash had started 3 months prior, with small punctate lesions scattered over his body. Initially, the patient went to the emergency department and was given a course of oral prednisone, which did improve his symptoms for a brief time. He tried topical over-the-counter moisturizers which did not help. He saw dermatology 1 month after his rash started and was given a topical steroid cream which did not improve his symptoms. At the time of presentation, he had blisters extending from his neck to his toes. He reports that the blisters would burst and ooze a clear liquid. He denied any new household or personal products (Fig. 1).

Further subjective history was negative for any autoimmune conditions to the patient's knowledge. Family history was also unremarkable. The patient was a nonsmoker.

His vitals identified a healthy body mass index (BMI) (23.7, normal range [NR] 18.5–24.9 kg/m^2). The physical

M. Mahmoudi (ed.), *Absolute Allergy and Immunology Board Review*, https://doi.org/10.1007/978-3-031-12867-7_19

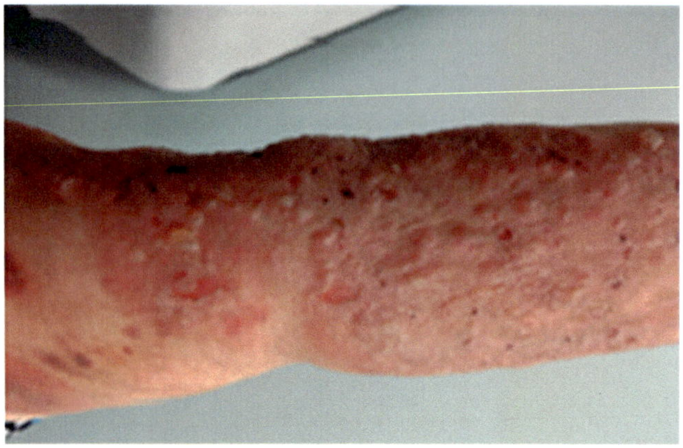

Fig. 1 Picture of the patient's skin lesion

examination revealed an erythematous, macular rash with thin-walled blisters. No evidence of oral mucosal lesions or ocular findings.

At his evaluation, the patient had a skin biopsy revealing subepidermal vesicles containing a dense infiltrate of inflammatory cells including eosinophils. Direct and indirect immunofluorescence (DIF and IIF) studies revealed linear deposition of IgG and C3 along the basement membrane.

Question 1

Based on the biopsy results the most likely diagnosis would be:

A. Erythema multiforme
B. Pemphigus vulgaris
C. Mucous membrane pemphigoid
D. Bullous pemphigoid
E. Lichen planus

Answer: D

The patient in this case is presenting with bullous pemphigoid, an autoimmune disorder characterized by blistering (Table 1). It presents with urticarial and eczematous lesions on the trunk and upper legs that progress to tense bullae. The most common age of onset for BP is 80-years old; however, drug-induced pemphigus can also be seen in the pediatric population, and young adults. BP is the most frequently encountered autoimmune blistering disease and is caused by autoantibodies directed against the hemidesmosomal proteins, BP antigen 180 (BP180) and BP antigen 230 (BP230). The diagnosis can be made with a perilesional biopsy with staining and direct immunofluorescence. A skin biopsy demonstrates subepithelial vesicles with mixed inflammatory infiltrate containing eosinophils. DIF microscopy shows linear IgG and/or C3 staining along the basement membrane zone.

Erythema Multiforme (EM) is an acute inflammatory reaction of the skin, with rare involvement of the mucous membranes, that presents as erythematous macules that classically evolve into targetoid lesions on the extensor extremities. It represents a cell-mediated immune reaction associated with infections (mainly Herpes Simplex Virus), drugs, vaccinations, and autoimmune disorders. The histological features in EM are nonspecific with perivascular inflammation, interface dermatitis, and epidermal necrosis. Unlike BP, DIF is nonspecific and IIF would be negative, making this answer choice incorrect.

Table 1 Characteristics of common blistering skin disorders

Disease	Clinical features	Histology	Autoantibody target	Immunofluorescence
Erythema Multiforme	Sudden onset. Variable lesions: Erythematous macules, target lesions, vesicles. Located on extensor extremities → spread centripetally. Mucous membranes are sometimes involved	Nonspecific. Perivascular inflammation, interface dermatitis, epidermal necrosis with intra and subepithelial vesicular formation	Nonspecific	Nonspecific
Pemphigus vulgaris	No sudden onset. Painful, flaccid bullae with erosions on skin and mucosa. Nikolsky sign is positive	Intraepithelial vesicles with acantholytic keratinocytes, intact basal layer	Epidermal desmosomes (Desmoglein-1 and Desmoglein-3)	Intercellular, reticulated IgG and C3
Mucous membrane pemphigoid	No sudden onset. Vesiculoulcerative lesions involving the mucosa only. May lead to scarring	Perivascular inflammation with subepithelial vesicles	BP230, BP180, α6-integrin, Laminin 5, Laminin 6, b4-integrin, type 7 collagen	Linear IgG and IgA along basement membrane
Bullouspemphigoid	No sudden onset. Pruritic, tense bullae with rare mucosal involvement	Perivascular infiltrate with eosinophils and subepithelial vesicles	BP230, BP180	Linear IgG and C3 along basement membrane
Lichen planus	No sudden onset. Pruritic, flat topped papules and plaques with involvement of the skin (flexor extremities), hair, nails, and mucous membranes	Hyperkeratosis, hypergranulosis, interface dermatitis, saw-toothed epithelial vacuolation with apoptosis	None/unknown	Fibrin deposits along the basement membrane

Pemphigus Vulgaris (PV) is a chronic, autoimmune blistering disease of the skin and mucous membranes. In contrast to BP, PV presents at an earlier age (40–60 years old) with painful, flaccid bullae that rupture easily to form erosions. In addition, the mucosa is almost always involved compared to BP. The primary defect in PV is IgG autoantibodies directed against keratinocyte adhesion proteins resulting in loss of cell adhesion or acantholysis. In turn, DIF demonstrates intercellular deposits of IgG in a reticular pattern around keratinocytes making this an incorrect answer choice.

Mucous Membrane Pemphigoid is a rare, antibody-mediated blistering disorder that only affects the mucous membranes, making this answer choice incorrect. In addition, although DIF would demonstrate linear bands of IgG deposited along the basement membrane, IIF is usually negative.

Lichen planus (LP) is an idiopathic inflammatory disorder of the skin. The skin lesions of LP are characterized by the "6P's" including purple, pruritic, planar, polygonal, papules, and plaques with an overlying reticulated, fine scale known as Wickham striae. The wrist and ankles are common sites of involvement, but any location may be affected including the hair, nails, and mucous membranes. Characteristic histological features of LP include a band-like lymphocytic infiltrate along the dermal-epidermal junction, saw-toothed epithelial vacuolation, and apoptosis with a thickened granular cell layer and stratum corneum making this an incorrect answer choice. In addition, LP will have a negative IIF.

Question 2

What are possible complications of bullous pemphigoid?

A. Neurologic disorders
B. Malignancy
C. Thrombosis
D. A and B
E. All the above

Answer: E

There have been studies that have demonstrated an association between BP and neurologic disorders. A recent systematic review with meta-analysis evaluated 14 studies. Results of the analysis indicated that individuals with BP were five times more likely to develop neurologic disorders such as dementia, epilepsy, multiple sclerosis, Parkinson's disease, and stroke. This review also found that the neurologic disorder typically precedes the onset of BP by 5.5 years. The most common associated neurologic disorder is multiple sclerosis with a 5–12 time risk of development of BP. Unfortunately, the pathogenesis linking BP and neurologic disorders is still not completely understood.

Another study found that bullous pemphigoid antigen (BP180 and BP230) are expressed both in the central nervous system (CNS) and skin. This may be the common feature which connects the clinical manifestations of BP to the CNS. It is postulated that any insult to the CNS can trigger increased levels of anti-BP180. Increases in these levels have been found to correlate with the severity of dementia in patients with Alzheimer's disease.

There is conflicting data regarding BP association with malignancy. Two Japanese studies found higher rates of malignancy in their BP subjects when compared to age-matched controls. One study demonstrated a rate of 5.8% of their BP subjects with malignancies including lymphoma, gastric, colorectal, prostate, lung, and uterine cancers.

In addition to its association with neurologic disorders and malignancy, BP also has been associated with an increased thrombotic risk. BP promotes a dysregulated immune response mediated by Th1 and Th2 cells resulting in an increased synthesis of IL-1B, TNF-α, IL-5, IL-6, IL-8, IL-10, and IL-15. The production of these pro-inflammatory cytokines upregulates vascular endothelial growth factor and E-selectin which results in endothelial cell activation. These patients also have been found to have increased circulating levels of prothrombin and D-dimer as well as overexpression of tissue factor in lesional skin. These levels have been shown to return to normal with disease control. Together this evidence suggests a prothrombotic state exists in the BP patient. This may lead to a higher risk of thromboembolic events, including pulmonary embolism and stroke in comparison to age-matched controls.

Question 3

Which of the following would be the first-line treatment in a patient diagnosed with mild to moderate BP?

A. Doxycycline
B. Low potency topical corticosteroid
C. High potency topical corticosteroid
D. Oral prednisolone
E. Methotrexate

Answer: C

BP is a chronic disease that can persist for many years with a risk of relapse. The main purpose of treatment is to promote healing of the skin lesions as well as to decrease itching and prevent recurrence to improve patients' quality of life.

The first-line treatment in BP depends on the disease severity and spread. For localized and moderate forms of BP the current first-line treatment consists of super potent topical corticosteroids (e.g., clobetasol propionate). The effectiveness of topical clobetasol propionate cream was proven in extensive BP with less mortality and side effects when compared with oral prednisone in a randomized control trial. Limitations of extensive topical steroid use include skin atrophy and difficult application for the elderly person.

In the case of extensive and advanced BP, systemic corticosteroids are considered the first line of treatment. Other alternative therapies may be considered to counteract the systemic effects of oral steroids. For patients who fail to respond to corticosteroids or who develop side effects, other therapeutic options are available.

Doxycycline has proven effective as an alternative to steroids. In a randomized control trial of 253 patients with BP, Doxycycline was noninferior to oral prednisolone for short-term blister control.

The patient was started on an oral prednisone taper and empiric mycophenolic acid (Cellcept) 750 mg twice daily for treatment for BP.

Despite starting mycophenolic acid, the patient had persistent symptoms and reported increased pruritus. He completed two 10-day prednisone tapers and started hydroxyzine 10 mg three times daily as needed for pruritus. There was minimal improvement in his rash and his blisters were in various stages of healing.

After a 26-day trial of mycophenolic acid, the patient elected to stop treatment due to persistence of his rash. Alternative treatment options were considered and discussed with the patient.

Question 4

Which of the following would be the next best step in management for a patient who has failed corticosteroids and corticosteroid-sparing therapy?

A. Intravenous immunoglobulins (IVIG)
B. Cyclosporine
C. Methotrexate
D. Doxycycline
E. Topical urea

Answer: A

Addition of an oral immunosuppressive agent (cyclosporine, methotrexate, mycophenolic acid. Etc.) can be considered with severe disease. However, in cases refractory to corticosteroids and corticosteroid-sparing therapy, biological therapies should be considered. Various biologic therapies can be used including rituximab, omalizumab, dupilumab, and IVIG (Table 2). Therefore, the next most appropriate choice in treatment in this case would be IVIG. Multiple treatment cycles are typically needed for disease improvement. Studies have shown that IVIG leads to a decline in serum levels of BP180 and BP230 antibodies.

The patient was started on 70 g of IVIG daily for 2 consecutive days to treat his resistant disease state. Over the next 6 weeks, he had near complete resolution of his rash and associated symptoms. Monthly IVIG infusions were continued for disease maintenance.

Table 2 Biologic therapies available for refractory BP

Biologic agents	Target	Average response rate based on case series (%)
Rituximab	CD20	85
Omalizumab	Immunoglobulin E (IgE)	84
Dupilumab	IL-4 and IL-13	92
Intravenous immunoglobulin (IVIG)	BP180 and BP230 antibodies	86

Question 5

By what mechanism does IVIG affect autoimmune disorders in general?

A. Modulation of pathogenic autoantibodies
B. Inhibition of complement activation and interception of membranolytic attack complex (MAC) formation, an action relevant to the complement-mediated mechanisms
C. Modulation of the inhibitory or activation Fc receptors on macrophages invading targeted tissues
D. Downregulation of pathogenic cytokines and adhesion molecules
E. All the above

Answer: E

Many mechanisms have been suggested to account for the beneficial action of IVIG in autoimmune and inflammatory disorders including blockade of Fc receptors on cells of the reticuloendothelial system, the neutralization of pathogenic autoantibodies, and the attenuation of complement-mediated tissue damage. The major role of IVIG in these disorders appears to be through its effects on Fc receptors. FcRn blockade leads to an accelerated neutralization and clearance of autoantibodies. The blockade of activating FcγRIII results in reduced opsonization of antigens and decreased pro-inflammatory responses from innate effector cells. Furthermore, sialylated IgG exerts anti-inflammatory effects by upregulated inhibitory FcγRIIB on macrophages. Still other mechanisms of IVIG may include the downregulation of pro-inflammatory genes in macrophages, modulation of dendritic cell maturation and function, and inhibition of lymphocyte autoreactivity. The pleiotropic role and effect of IVIG on autoimmune disease such as BP has yet to be completely elucidated.

Case 2

A 35-year-old female presents to the outpatient clinic for evaluation of a new skin rash. The rash started 4 months prior with scaling, erythematous lesions on the upper trunk. She was seen at a local urgent care clinic and was given an oral

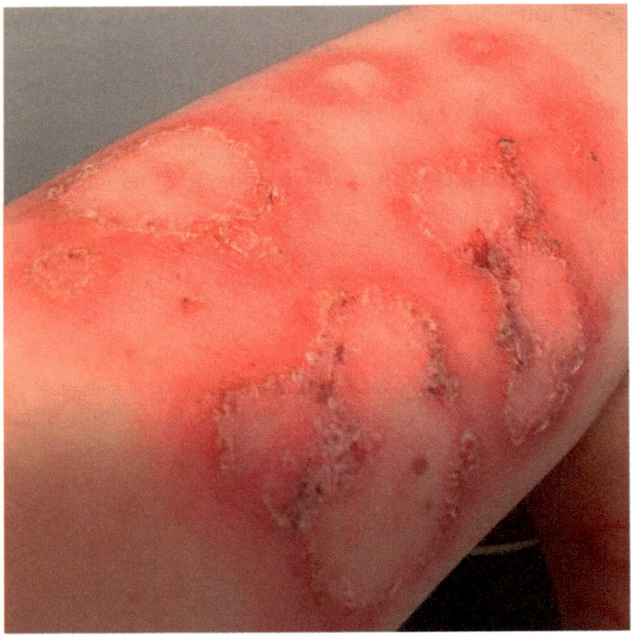

Fig. 2 Picture of the patient's skin lesion

prednisone taper with temporary resolution of her symptoms. The lesions have since progressed with involvement of the extremities. She has been working as a lifeguard for the past year and notes the lesions are exacerbated by the sun. She denies any symptoms other than occasional pruritis. Her medical history is significant for Hashimoto's thyroiditis for which she takes levothyroxine. She has no known allergies. She is a nonsmoker without any recent travel history or sick contacts. Her father has a history of psoriasis limited to the scalp.

Her vital signs were stable. Skin examination revealed erythematous, annular plaques with central clearing forming polycyclic patterns. There are variable amounts of scale-crust at the margins as shown in Fig. 2. There are similar lesions symmetrically distributed over the neck and trunk; however, the face is spared. There are no mucosal lesions or joint swellings present. The remainder of the physical exam was unremarkable.

Laboratory results at the time of evaluation revealed a positive anti-neutrophilic antibody (ANA), but normal blood cell counts and urinalysis. A skin biopsy was performed and demonstrated lymphocytic infiltrate along the dermal-epidermal junction with vacuolar basal cell degeneration. DIF was performed and revealed granular deposits of IgG and C3 along the dermal-epidermal junction.

Question 1

Which of the following is the most likely diagnosis:

A. Discoid lupus erythematous
B. Subacute cutaneous lupus erythematosus
C. Psoriasis
D. Lichen planus
E. Polymorphic light eruption

Answer: B

Lupus erythematous (LE) is a multisystemic autoimmune disorder characterized by the formation of autoantibodies that cause tissue injury mainly through the deposition of immune complexes as well as binding of antibodies directly to cells. Cutaneous manifestations are frequently the initial presentation of LE and can occur with or without systemic disease. LE-specific skin disease includes acute cutaneous lupus erythematous (ACLE), subacute cutaneous lupus erythematous (SCLE), and chronic cutaneous lupus erythematosus (CCLE). Although all types share similar histologic features like interface dermatitis and deposits of immunoglobulin and complement, they differ in clinical presentation, association with systemic disease, and long-term complications.

The clinical vignette above describes a case of SCLE. SCLE occurs most common in middle aged females. Two morphologic variants of SCLE exist: annular and papulosqamous. In the annular variant, scaling, erythematous, plaques with central clearing coalesce to form polycyclic configurations. The papulosqamous variant resembles plaque psoriasis. SCLE is a photosensitive rash and exacerbations occur with sun exposure. In turn, the most affected areas are sun exposed skin like the neck, upper trunk, and extensor portion of the upper extremities; however, the face is usually spared. SCLE heals without scarring or atrophy. Post inflammatory hypopigmentation may result, but usually resolves overtime.

Discoid lupus erythematous (DLE) is the most common type of CCLE. It presents as well-defined, round, erythematous plaques with an adherent scale that involves the hair follicles. Unlike SCLE, the lesions scar as they evolve into atrophic plaques with central hypopigmentation and peripheral hyperpigmentation, making this an incorrect answer choice. In addition, the head and neck are common sites of involvement. Lesions of the scalp can result in permanent alopecia.

Psoriasis is a chronic inflammatory disorder of the skin characterized by well-defined erythematous plaques with overlying silvery scales. It commonly affects the scalp, elbows, knees, and intergluteal cleft but any site may be affected. The nails and joints may also be involved. Psoriasis is not usually exacerbated by UV-radiation, but instead may be beneficial. In fact, phototherapy may be used as a treatment modality in severe disease. Laboratory evaluation in psoriatic patients would demonstrate a negative ANA. The histological findings of psoriasis are characterized by parakeratosis, a decreased granular layer, and a neutrophilic infiltrate in the dermis. DIF would also be negative.

Lichen planus is an idiopathic inflammatory disorder of the skin. The lesions manifest as pruritic, flat-topped papules and plaques with an overlying reticulated scale known as Wickham striae. The wrist and ankles are common sites of involvement, but any location may be affected. The rash is typically not photosensitive. Characteristic histological features of LP are described in Table 1. Both ANA and DIF would be negative.

Polymorphic light eruption (PMLE) is an idiopathic inflammatory disorder of the skin. The lesions can take many morphologies including erythematous macules, papules, plaques, or vesicles that occur after sun exposure. PMLE is a diagnosis of exclusion that may resemble the cutaneous features of lupus erythematous; however, DIF would be negative in PMLE, making this answer choice incorrect.

Question 2

Which of the following is most likely to be positive in this patient:

A. Anti-SSA/Ro antibody
B. Anti-dsDNA antibody
C. Anti-histone antibody
D. Anti-Jo-1 antibody
E. Anticentromere antibody

Answer: A

Detection of serum autoantibodies can aid in the diagnosis of autoimmune diseases. SCLE is strongly associated with Anti-SSA/Ro antibodies. Studies have shown that 70% of patients are positive for anti-SSA/Ro and 70–80% are positive for ANA. In addition, half of the patients with SCLE will be positive for Anti-SSB/La. In contrast to systemic lupus erythematous, only 5% of patients are positive for anti-dsDNA.

Anti-histone antibodies are associated with drug-induced lupus which presents with acute onset fever, arthralgias, and serositis after starting the causative drug. Common culprits are hydralazine, procainamide, isoniazid, and TNF-alpha inhibitors. Unlike SLE, Drug-induced lupus is less likely to have a cutaneous, central nervous system, or renal involvement.

Anti-Jo-1 antibodies are specific for inflammatory myositis, which presents as progressive proximal muscle weakness and atrophy. Dermatomyositis can also present with similar skin manifestations as SLE.

Anticentromere antibodies are associated with disease limited systemic sclerosis or CREST syndrome, which usually presents with fibrosis of the skin limited to fingers and face, calcinosis cutis, Raynaud phenomenon, esophageal dysmotility, sclerodactyly, and telangiectasias.

The patient was started on an oral prednisone taper supplemented with 0.1% tacrolimus cream. She was advised to apply broad spectrum sunscreen (SPF of at least 50) 20–30 min before sun exposure. The rash resolved without scarring and minimal dyspigmentation.

One year later the rash returned despite extensive photoprotection and continued topical maintenance therapy. In addition to the recurrence of the annular erythematous plaques, the patient also now complains of fatigue, joint pains, and oral ulcers.

The laboratory evaluation revealed an elevated ANA, hemoglobin of 13, and platelet count of 120,000. Urinalysis was normal.

Question 3

What percentage of patients will progress to systemic lupus erythematous?

A. 5%
B. 10%
C. 25%
D. 50%
E. 90%

Answer: D

Cutaneous lupus erythematosus is a common initial manifestation of LE and can occur with or without systemic disease. Fifty percent of patients with SCLE will eventually meet the criteria for systemic disease outlined by the American College of Rheumatology. However, systemic symptoms are mild and most commonly include a photosensitive rash, oral ulcers, arthritis, and positive serology. Most patients do not have central nervous involvement or lupus nephritis. The progression of chronic cutaneous lupus erythematosus to systemic disease is even rarer. In contrast, acute cutaneous lupus erythematosus is almost always associated with systemic disease.

Question 4

Which of the following is the best next step in treatment for this patient?

A. Hydroxychloroquine
B. Methotrexate
C. Mycophenolate mofetil
D. Rituximab
E. Acitretin

Answer: A

The first-line treatment for cutaneous lupus erythematous includes photoprotection, topical or oral steroids (depending on the extend of the disease), and hydroxychloroquine. For localized disease, topical steroids or topical calcineurin inhibitors can be used. For disease that is widespread, scarring or refractory to topicals, hydroxychloroquine is considered the drug of choice. Methotrexate, mycophenolate mofetil, rituximab, and acitretin may be considered in cases refractory to hydroxychloroquine.

Question 5

Which of the following should be monitored while taking this medication?

A. Liver function tests
B. Visual acuity
C. Blood cell counts
D. Thyroid function tests
E. Lipid panel

Answer: B

Hydroxychloroquine is the drug of choice for systemic lupus erythematosus as well as cutaneous lupus erythematous that is widespread, severe or refractory to other first-line treatments. Although hydroxychloroquine is normally well tolerated, there is an increased risk of retinopathy after 5–7 years of use. The American Academy of Ophthalmology recommends a baseline examination for patients starting hydroxychloroquine with annual eye examinations after 5 years.

Bibliography

1. Kumar V, Abbas AK, Aster JC. Robbins and Cotran pathologic basis of disease. 9th ed. Philadelphia, PA: Elsevier/Saunders; 2015. p. 211–7.
2. Miyamoto D, Santi CG, Aoki V, Maruta CW. Bullous pemphigoid. An Bras Dermatol. 2019;94(2):133–46.
3. Pratasava V, Sahni VN, Suresh A, Huang S, Are A, Hsu S, Motaparthi K. Bullous pemphigoid and other pemphigoid dermatoses. Medicina. 2021;57(10):1061.
4. Schmidt E, Della Torre R, Borradori L. Clinical features and practical diagnosis of bullous pemphigoid. Immunol Allergy Clin N Am. 2012;32(2):217–32.
5. Memis I, Andreadis D, Apessos I, Georgakopoulou E, Poulopoulos A. Paraneoplastic autoimmune multi-organ syndrome and oral mucosa involvement: an intriguing disorder. Cancer Res Front. 2015;1(3):268–79.
6. Li N, Culton D, Diaz LA, Liu Z. Modes of action of intravenous immunoglobulin in bullous pemphigoid. J Invest Dermatol. 2018;138(6):1249–51.
7. Ahmed AR. Intravenous immunoglobulin therapy for patients with bullous pemphigoid unresponsive to conventional immunosuppressive treatment. J Am Acad Dermatol. 2001;45(6):825–35.
8. Santoro FA, Stoopler ET, Werth VP. Pemphigus. Dent Clin N Am. 2013;57(4):597–610.
9. Trayes KP, Love G, Studdiford JS. Erythema multiforme: recognition and management. Am Family Physician. 2019;100(2):82–8.
10. Xu HH, Werth VP, Parisi E, Sollecito TP. Mucous membrane pemphigoid. Dent Clin N Am. 2013;57(4):611–30.
11. Lehman JS, Tollefson MM, Gibson LE. Lichen planus. Int J Dermatol. 2009;48(7):682–94.
12. Joly P, Roujeau JC, Benichou J, et al. A comparison of oral and topical corticosteroids in patients with bullous pemphigoid. N Engl J Med. 2002;346(5):321–7.
13. Williams HC, Wojnarowska F, Kirtschig G, et al. Doxycycline versus prednisolone as an initial treatment strategy for bullous pemphigoid: a pragmatic, non-inferiority, randomised controlled trial. Lancet. 2017;389(10079):1630–8.
14. Czernik A, Toosi S, Bystryn JC, Grando SA. Intravenous immunoglobulin in the treatment of autoimmune bullous dermatoses: an update. Autoimmunity. 2012;45(1):111–8.
15. Sami N, Ali S, Bhol KC, Ahmed AR. Influence of intravenous immunoglobulin therapy on autoantibody titres to BP AG1 and BP AG2 in patients with bullous pemphigoid. J Eur Acad Dermatol Venereol. 2003;17(6):641–5.
16. Kremer N, Snast I, Cohen ES, et al. Rituximab and omalizumab for the treatment of bullous pemphigoid: a systematic review of the literature. Am J Clin Dermatol. 2019;20(2):209–16.
17. Abdat R, Waldman RA, De Bedout V, et al. Dupilumab as a novel therapy for bullous pemphigoid: a multicenter case series. J Am Acad Dermatol. 2020;83(1):46–52.
18. Galeotti C, Kaveri SV, Bayry J. IVIG-mediated effector functions in autoimmune and inflammatory diseases. Int Immunol. 2017;29(11):491–8.
19. Grönhagen CM, Nyberg F. Cutaneous lupus erythematosus: an update. Indian Dermatol Online J. 2017;5(1):7–13.
20. Gruber-Wackernagel A, Byrne SN, Wolf P. Pathogenic mechanisms of polymorphic light eruption. Front Biosci (Elite Ed). 2009;1:341–54.
21. Trayes KP, Savage K, Studdiford JS. Annular lesions: diagnosis and treatment. Am Fam Physician. 2018;98(5):283–91.
22. Okon LG, Werth VP. Cutaneous lupus erythematosus: diagnosis and treatment. Best Pract Res Clin Rheumatol. 2013;27(3):391–404.
23. Marmor MF, Kellner U, Lai TY, et al. Revised recommendations on screening for chloroquine and hydroxychloroquine retinopathy. Ophthalmology. 2011;118(2):415–22.
24. Wieczorek IT, Propert KJ, Okawa J, Werth VP. Systemic symptoms in the progression of cutaneous to systemic lupus erythematosus. JAMA Dermatol. 2014;150(3):291–6.

Vasculitis

Nikita Patel and Stratos Christianakis

Abbreviations

AAV	ANCA-associated vasculitis
ANCA	Antineutrophil cytoplasmic antibodies
CDC	Centers for disease control
CT	Computed tomography
DIF	Direct immunofluorescence
EM	Electron microscopy
EPGA	Eosinophilic granulomatosis with polyangiitis
FDG	Fluoro-deoxyglucose
GN	Glomerulonephritis
GPA	Granulomatosis with polyangiitis
HBV	Hepatitis B virus
HCV	Hepatitis C virus
HIV	Human immunodeficiency virus
IBD	Inflammatory bowel disease
IV	Intravenous
MPA	Microscopic polyangiitis
MPO	myeloperoxidase
MRI	Magnetic resonance imaging
PAN	Polyarteritis nodosa
PCR	polymerase chain reaction
PET	Positron emission tomography
PR 3	Proteinase 3
PUK	Peripheral ulcerative keratitis
RA	Rheumatoid arthritis
SAGN	Streptococcal-associated glomerulonephritis
TB	Tuberculosis

N. Patel (✉) · S. Christianakis
LAC-USC Division of Rheumatology, Los Angeles, CA, USA

1 Introduction

Vasculitis is a broad topic and the clinical presentation can vary depending on the type of vessel and organ system affected. Large vessel vasculitis includes Giant Cell Arteritis and Takayasu Arteritis. Medium vessel vasculitis includes Polyarteritis nodosa and Kawasaki Disease. Small vessel vasculitis can be categorized as immune complex-mediated or pauci-immune such as ANCA-associated vasculitis. Immune complex small vessel vasculitis includes cryoglobulinemic vasculitis, IgA vasculitis, and hypocomplementemic urticarial vasculitis. ANCA-associated small vessel vasculitis includes MPA, GPA, and EPGA.

For this chapter, due to the diversity of types of vasculitis, we will focus on ANCA-associated vasculitis and large vessel vasculitis. The patient cases and questions will highlight the complexity of organ involvement and how to recognize and diagnose vasculitis through clinical scenarios. The focus will be on laboratory, imaging, and pathology findings that would be useful in aiding the diagnosis. We will not focus on treatment, as this topic is more detailed and not pertinent to this chapter. The goal of this chapter is to make the non-Rheumatologist clinician aware of when to be concerned about vasculitis.

Case 1

A 57-year-old man with a history of prior treatment for pulmonary tuberculosis (TB) presents with worsening right eye pain and acute on chronic renal failure.

Patient's social history includes tobacco use with smoking 5 cigarettes per day for 25 years. He denied any current alcohol use. He works as a mechanic. He lives with his wife and son. He is originally from Mexico, immigrated to the United States approximately 10 years ago.

He was treated for pulmonary TB approximately 1 year prior to presentation with rifampin, isoniazid, ethambutol, pyrazinamide, and pyridoxin. At that time, he had symptoms of congestion, cough, shortness of breath, and fevers. He also noted some eye irritation however was given eye

M. Mahmoudi (ed.), *Absolute Allergy and Immunology Board Review*, https://doi.org/10.1007/978-3-031-12867-7_20

drops for possible allergies or dryness. A preliminary chest radiograph showed left lung pneumonia with concerning features, so it was followed up with chest computed tomography (CT) (Fig. 1).

Question 1

Which ANCA-associated vasculitis is most likely to have this type of pulmonary involvement?

A. Eosinophilic Granulomatosis with Polyangiitis (EGPA)
B. Microscopic Polyantiitis (MPA)
C. Granulomatosis with Polyangiitis (GPA)
D. Polyarteritis Nodosa (PAN)

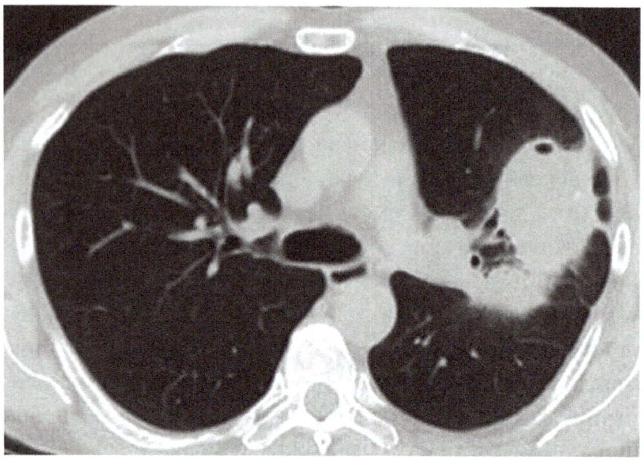

Fig. 1 CT of the chest without contrast showing left lung with large consolidative mass in the upper lobe with cavitation in the anterior, inferior portion, and surrounding smaller nodules in bilateral lungs

Answer: C

Discussion: There are many different causes of cavitary lung lesions that must be considered in the differential (see Table 1 and Fig. 2). Among autoimmune causes of cavitary lung lesions, ANCA-associated vasculitis is high on the differential. The main types of lung lesions in AAV include necrotizing granulomatous or cavitary lesions (lung nodules), tracheobronchial inflammation, pulmonary capillaritis manifesting as diffuse alveolar hemorrhage (DAH), interstitial lung disease (ILD), and asthma.

ANCA-associated vasculitis variants include MPA, GPA, EGPA, and single organ limited AAV. GPA is characterized by necrotizing granulomatous inflammation. GPA can involve both the upper and lower respiratory tracts.

Table 1 Causes of cavitary lung lesions

Cancer	Bronchogenic carcinoma Metastasis
Autoimmune	Granulomatosis with polyangiitis Rheumatoid arthritis
Vascular	Pulmonary emboli
Infection	Tuberculosis Pulmonary abscess
Trauma	Traumatic pulmonary pseudocyst
Youth	Congenital pulmonary airway malformation Pulmonary sequestration Bronchogenic cyst

C	Cancer	Bronchogenic carcinoma (most common SCC) Metastasis (most common SCC)
A	Autoimmune	Granulomatosis with polyangiitis Rheumatoid arthritis
V	Vascular	Pulmonary emboli
I	Infection	Tuberculosis pulmonary abscess
I	Trauma	Traumatic pulmonary pseudocyst
Y	Youth	Congenital pulmonary airway malformation pulmonary sequestration Bronchogenic cyst

SCC, squmous cell carcinoma.

Fig. 2 Differential diagnosis for cavitary lung lesions with pneumonic "CAVITY"

Upper tract involvement tends to be more symptomatic with hoarseness, dyspnea, stridor, or wheezing. Lower respiratory tract involvement in GPA can present as any of the lung lesions listed above; however, lung nodules can be found in up to 50% of the patients with GPA. Cavities are usually thick-walled, with irregular inner margins and absence of calcification. They are often in relation to vessels but do not necessarily have a predilection for specific lung zones. Depending on the degree of lung involvement patients may experience cough or development of hemoptysis. Chest X-ray must be followed up by CT of the chest if abnormalities are noted.

Interstitial lung disease (ILD) in AAV mostly occurs in the context of MPA and in patients with anti-myeloperoxydase (MPO)-ANCA. The most frequent high-resolution computed tomography (HRCT) findings of ILD associated with MPA include ground-glass and reticular opacities, interlobular septal thickening, parenchymal consolidations, and honeycombing.

Question 2

If this gentleman were to have a history of adult-onset asthma, which of the following vasculitides should be suspected?

A. Eosinophilic Granulomatosis with Polyangiitis (EGPA)
B. Microscopic Polyangiitis (MPA)
C. Granulomatosis with Polyangiitis (GPA)
D. Polyarteritis Nodosa (PAN)

Answer: A

Asthma is the main clinical manifestation of EGPA and occurs in 95–100% of patients. Asthma in EGPA is also associated with chronic rhinosinusitis, atopic background, blood eosinophilia, and the presence of ANCA (though can be absent in up to 60% of patients) most often directed against MPO antigen. Radiological abnormalities like peripheral ground-glass opacities, consolidation, bronchial thickening, or pleural effusions can be found in the most severe cases.

Transbronchial or surgical biopsies may be considered to rule out differential diagnoses (mainly tumors or infections, especially fungal caused by histoplasmosis, coccidioidomycosis, blastomycosis, or nocardia) or to confirm the diagnosis of vasculitis in cases of isolated lung nodules without other organ involvement.

Further workup along with infectious considerations should include inflammatory markers (ESR, CRP), ANCA with MPO and PR3 titers, as well as ruling out other possible inflammatory causes such as RA and Sarcoidosis in the correct clinical scenario. Lung nodules usually respond to immunosuppressive therapies.

Case Continued

The patient continued to complain of eye irritation and 2 months after completing TB treatment presented to the Emergency Department (ED) with bilateral eye redness, pain, and blurring of the vision. He had two subsequent visits to the ED with cough, rhinorrhea, and chest discomfort. Each time he was treated with supportive care for presumed viral etiologies with CXR showing the previously seen findings on the left lung field. Antineutrophil cytoplasmic antibodies were sent which returned with results below (Table 2 ANCA testing).

Question 3

Which of the following is the most likely diagnosis for this patient?

A. Episcleritis
B. Anterior scleritis
C. Scleromalacia perforans
D. Nodular scleritis

Answer: B

Discussion: The sclera is the outer type I collage of the eyes with scleritis representing inflammation of the sclera. The differential diagnosis of scleritis includes infection, malignancy, inflammatory arthritides, systemic autoimmune disease, and primary systemic vasculitides. Clinically scleritis can present with erythema, decreased visual acuity, pain, and photophobia. Ocular pain typically helps differentiate episcleritis from scleritis. Scleritis can be characterized based on anatomical distribution, as anterior or posterior to recti muscle insertion site. The subtypes of anterior scleritis include diffuse, nodular, and necrotizing. Scleromalacia perforans is a type of necrotizing scleritis in which patients do not experience pain. An Ophthalmologic emergency can lead to blindness if not quickly treated. Necrotizing scleritis has a poor prognosis and the highest association with systemic disease. After ruling out infectious, malignant, and drug-induced causes of scleritis, autoimmune etiologies must be further investigated. The most common autoimmune associations in order of prevalence include rheumatoid arthritis, GPA, relapsing polychondritis, inflammatory bowel disease, and systemic lupus erythematosus. Ocular involvement in GPA occurs in approximately 38% of patients with scleritis being the common manifestation.

Table 2 ANCA testing

Variable	Patient value
ANCA screen	C-ANCA positive
C-ANCA titer	1:160
P-ANCA titer	Not detected
Proteinase 3 antibody (antibody index)	42.1
Myeloperoxidase antibody (antibody index)	<1.0

Necrotizing scleritis and corneal involvement are more frequent and anterior, diffuse scleritis is most common.

Given the potential for overlap between infectious and autoimmune etiology, infection must be ruled out. In a patient from an area with higher endemic rates of tuberculosis this is especially true. Treatments with immunosuppression may also be needed thus treating any superimposed infection on top of an autoimmune process is especially important.

Case Continued

Unfortunately, the patient developed worsening renal function. The Indomethacin was discontinued, and a discussion was had about starting patient on Prednisone. Referral to Rheumatology clinic had been made. Prior to the Rheumatology appointment, he had worsening ocular symptoms and found to have peripheral ulcerative keratitis (PUK) with corneal thinning and high concern for impending corneal perforation. He was admitted to the hospital for expedited treatment and workup.

Question 4

Which laboratory and pathologic features would help distinguish GPA from MPA?

A. MPO+ with granulomatous changes on pathology
B. MPO+ without granulomatous changes on pathology
C. PR3+ with granulomatous changes on pathology
D. PR3+ without granulomatous changes on pathology

Answer: C

Discussion: The vessels involved in AAV are typically capillaries, but small arteries and veins may also be affected. ANCAs are unique markers that support the classification and diagnosis of GPA, MPA, and EGPA. Antineutrophil cytoplasmic antibodies are to PR3 (PR3-ANCA) or MPO (MPO-ANCA); PR3 and MPO are found primarily in neutrophils. GPA and MPA can involve small blood vessels in any organ or tissue but commonly affect the upper and lower respiratory tract and the kidneys. GPA is predominantly associated with PR3-ANCA and its clinical features typically include sino-nasal disease, lower respiratory tract involvement with pulmonary hemorrhage and granulomatous inflammation, and glomerulonephritis. MPA is usually associated with MPO-ANCA and clinical features include more severe renal disease and some of the manifestations of GPA but without granulomatous inflammation.

The indirect immunofluorescence test is commonly the initial screening test for ANCA but high-quality immunoassays are preferred. ANCA specificity also predicts differences in long-term prognosis: PR3-ANCA+ patients are at higher risk of relapse than MPO-ANCA+ patients.

Kidney, lung, skin, or other tissue biopsy is often important in establishing the diagnosis along with serologies. The different forms of AAV show histopathological differences. Fibrinoid necrosis and inflammation of small vessels, sometimes accompanied by thrombosis, is the hallmark of acute injury in all forms of AAV. In GPA there are granulomatous changes often adjacent to the vessels, in EGPA there may be eosinophils in addition to the necrotizing small vessel vasculitis, and in MPA these features are absent.

Case Continued

He was treated with intravenous (IV) methylprednisolone 1000 milligrams (mg) for 3 days. Rheumatology and Nephrology were consulted. Pertinent laboratory data obtained are shown in Table 3. In addition, anti-glomerular

Table 3 Laboratory data

Variable	Reference range	Hospital admission	Prior to admission
Blood			
White blood cell count (10^3 cells per cubic milliliter)	4.5–10	9.2	7.8
Hemoglobin (g/dL)	13.5–16.5	10.5	11.5
Hematocrit (%)	40–49	32.1	35.1
Platelet (10^3 cells per cubic milliliter)	160–360	381	382
Sodium (mmol/L)	135–145	141	137
Potassium (mmol/L)	3.5–5.1	4.3	4.2
Chloride (mmol/L)	100–110	101	105
Carbon dioxide (mmol/L)	20–30	25	22
Urea nitrogen (mg/dL)	8–22	41	52
Creatinine (mg/dL)	0.5–1.3	3.31	4.05
Glucose (mg/dL)	65–99	100	101
Calcium (mg/dL)	8.5–10.3	9.2	9.1
Total protein (g/dL)	6.0–8.0	7.5	7.9
Albumin (g/dL)	3.5–5.0	4.3	4.0
Alanine aminotransferase (U/L)	10–50	20	14
Aspartate aminotransferase (U/L)	10–50	18	12
Alkaline phosphatase (U/L)	40–129	87	70
Total bilirubin (mg/dL)	<1.0	0.3	0.7
C-reactive protein (mg/L)	<4.9	58.3	
Urine			
Color	Yellow	Straw	Light yellow
Clarity	Clear	Clear	Clear
pH	5.0–8.0	6.0	6.0
Specific gravity	1.005–1.030	1.017	1.003
Glucose (mg/dL)	Negative	Negative	Negative
Ketones (mg/dL)	Negative	Negative	Negative
Leukocyte esterase	Negative	Negative	Negative
Nitrite	Negative	Negative	Negative
Blood	Negative	Large	Small
Protein	Negative	>300	50
Erythrocytes (per high power field)	0–2	11–25	4–5
Leukocytes (per high power field)	0–3	11–30	11–30

basement membrane antibody test was negative. A renal ultrasound showed normal-sized kidneys with no other abnormalities. A chest radiograph is shown in Fig. 3. A renal biopsy was obtained (Fig. 4). The patient received Rituximab and oral Prednisone taper. He had improvement in his scleritis following hospital stay. His renal function remained stable and he did not need to receive dialysis since he was still able to produce urine and serum electrolyte values remained within normal ranges.

Question 5

If this patient was treated as outlined above but later became febrile with blood cultures positive for Methicillin-Resistant *Staphylococcus aureus* (MRSA), which of the following pathology reports would give concern for an alternative diagnosis to ANCA-associated vasculitis?

A. Acute tubular necrosis with tubular vasculopathy. Diffuse proliferative and exudative glomerulonephritis with prominent endocapillary proliferation and numerous intracapillary neutrophils
B. Interstitial nephritis associated with vasculitis in the vasa recta, Necrotizing glomerulonephritis
C. Crescentic glomerulonephritis, both pauci immune and with anti-tubular basement membrane antibodies present

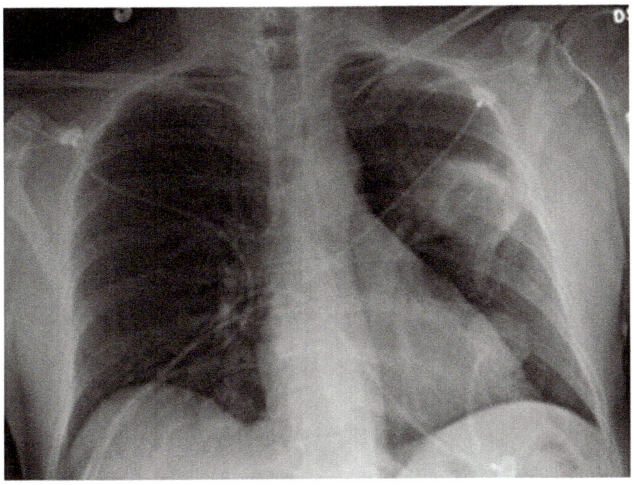

Fig. 3 Chest radiograph. CXR with interval increase in cavitation in the left lung superior lesion

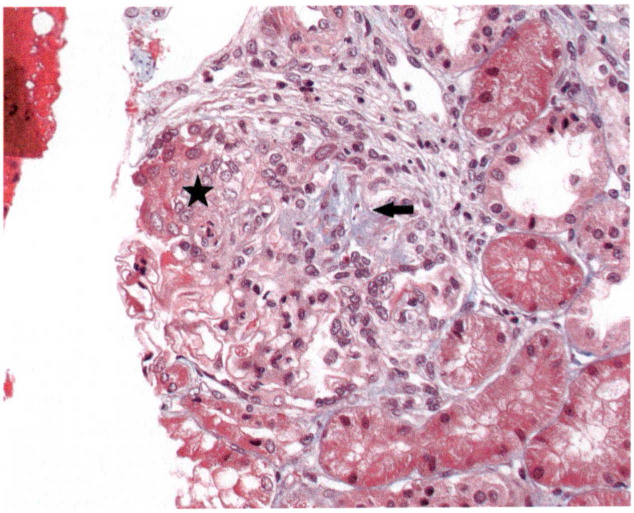

Fig. 4 Renal biopsy. Glomerulus with segmental sclerosis (star) in setting of small fibrous crescent (arrow); *(Masson trichrome stain; original magnification 400x)*. (Credit: William Wallace, MD, Department of Pathology and Laboratory Medicine Keck Hospital of USC)

Answer: A

It is important to discuss causes other than AAV which can give rise to a positive ANCA. ANCA can be seen following environmental exposure such as silica in association with certain drugs or during the course of certain disease processes. The disease processes can be divided into three categories (Table 4). All of these conditions may elevate inflammatory markers. Infection and inflammatory processes may cause hematuria and/or nephrotic range proteinuria. A positive ANCA in noninfectious inflammatory conditions other than AAV could be triggered by secondary infections. Severe manifestations mediated by autoantibodies triggered by infection may require specific treatment. The exact histological diagnosis of renal disease in patients with infections and multiplicity of autoantibodies is important for understanding the biology of the process.

Indications for a renal biopsy include hematuria (in the absence of primary renal infection such as pyelonephritis or cystitis and no concern for primary structural urological causes) in association with proteinuria; HTN and positive serum biomarkers; significant proteinuria greater than 1 g/day; systemic disease; unexplained renal involvement. A renal biopsy in AAV is considered the gold standard for diagnosis and prognosis. An adequate biopsy should contain at least 10 glomeruli but greater than 20 may be needed to diagnose focal glomerular disease.

The characteristic lesion in AAV is segmental necrosis of glomerular capillary loops, with little or no deposition of immunoglobulin or complement, termed "pauci-immune"

Table 4 Causes of positive ANCA

Chronic inflammatory	Neoplasm	Infection
• Rheumatoid arthritis • Systemic lupus erythematosus • Mixed connected tissue disease Autoimmune hepatitis • Inflammatory bowel disease	• Solid tumors • Atrial myxoma • Lymphoma • Myeloma • Myelodyscrasias • Hematopoietic stem cell transplant	• Viral • Bacterial • Fungal

focal necrotizing and crescentic glomerulonephritis. Crescents may be fibrous or fibrocellular with the absence of endocapillary hypercellularity. The necrotizing arterial lesions on direct immunofluorescence (DIF) show a pauci-immune pattern and no immune complexes seen on electron microscopy (EM). In streptococcal-associated GN (SAGN) endocapillary hypercellularity is seen in non-crescentic glomeruli, there is no necrotizing vasculitis, complement 3 can be seen on DIF and dome-shaped subepithelial deposits are seen on EM.

Glomerular lesions are used to stage renal disease in AAV by a histopathological classification, where the dominant lesion is linked to outcomes. There are four.

patterns of injury, namely sclerotic (≥50% globally sclerosed glomeruli, worst outcome), focal (≥50% normal glomeruli, best outcome), crescentic (≥50% cellular crescents, intermediate outcome), and mixed (no single dominant type of lesion, outcome better than the sclerotic but worse than the crescentic class) (Table 4).

Case 2

A 38-year-old female presents with squeezing retrosternal chest pain of sudden onset with radiation to the back. She has associated nausea and light-headedness. Her chest pain improved with sublingual nitroglycerin. Her electrocardiogram did not show any abnormalities.

The patient's past medical history included diabetes mellitus type 2 requiring insulin, hypertension, pituitary tumor followed by neurosurgery known to be stable, and heavy menstrual cycles.

She denied any tobacco or alcohol use. She works as a caregiver for the elderly. She had immigrated to the United States from the Philippines at age 15. She has four children who are healthy.

Upon admission to the hospital, cardiac catheterization was done which showed nonobstructive coronary arteries and diffuse aneurysms (Fig. 5). A transthoracic echocardiogram showed a left ventricle ejection fraction of 60%, aortic root enlargement, atrioventricular sinus enlargement, elevated right ventricle systolic pressure of 31, and no findings of pericardial effusion (Fig. 6).

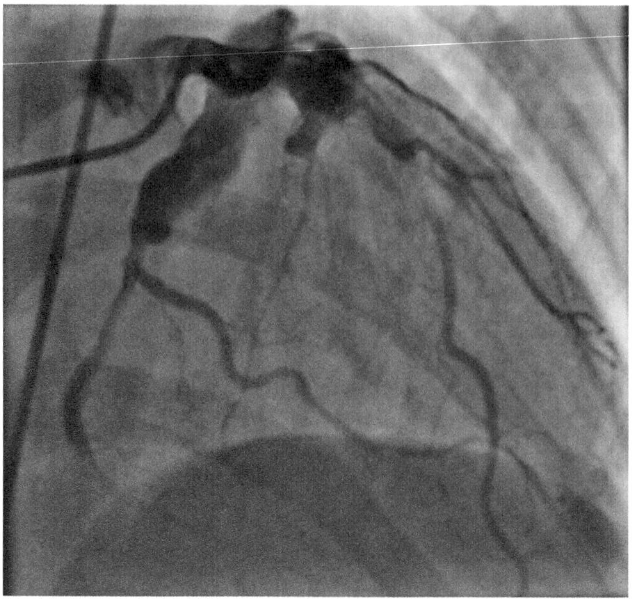

Fig. 5 Coronary catheterization showing diffuse aneurysms

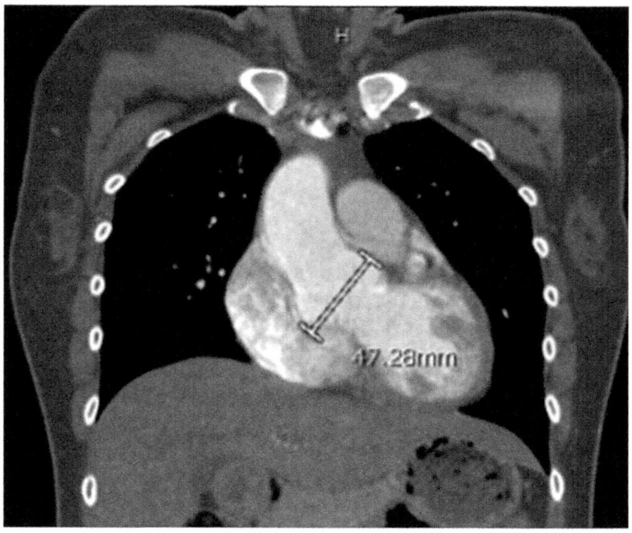

Fig. 6 Computed tomography angiogram of the chest. Computed tomography angiogram of the chest showing an aneurysmal dilatation of the aortic root measuring 4.7 cm

Question 1

A biopsy is needed to diagnose Takayasu arteritis?

A. True
B. False

Answer: B

Discussion: Large vessel vasculitides includes Giant Cell Arteritis and Takayasu Arteritis. Histopathologically, they are indistinguishable; however, clinically the presentations differ. TAK generally affects those aged 50 and younger, while GCA affects patients aged 50 and older. Different diagnostic and classification criteria exist for TAK with the Ishikawa Diagnostic criteria modified by Sharma as an updated diagnostic criteria (we refer the reader to De Souza et al.). Thus, it is important to note that a biopsy is not needed for diagnosis. Some of the signs and symptoms to look for include: pulselessness, differences in pulses and/or blood pressure in the arms, unobtainable blood pressure, easy limb fatigability or pain, unexplained fever, transient amaurosis, blurred vision, syncope, dyspnea, palpitations, hypertension or aortic regurgitation.

Large vessel vasculitis affects the aorta and its branches including the coronary arteries. Coronary artery aneurysms can also be associated with atherosclerotic disease, other autoimmune inflammatory diseases, infection, connective tissue disease, drugs, and trauma. Ten to twelve percent of TAK patients present with coronary artery aneurysms. Treatment is often complex and multidisciplinary as it can involve the typical medical management of aneurysms along with immunosuppressive agents and surgical interventions.

Case Continued

Rheumatology consultation was placed to help evaluate for potential causes of the patient's presentation. Further history was obtained and the patient was noted to have 4 years of fatigue and intermittent exertional substernal left chest pain. She had been seen in the ED for chest pain previously, the first time 3 years ago, at which time she was diagnosed with hypertension. She otherwise denied any other associated symptoms.

Some of the pertinent physical exam findings at the time of hospitalization:

Right upper extremity blood pressure: 90/79 mmHg; Left upper extremity blood pressure: 108/82.

Cardiovascular exam: normal heart rate and rhythm. Non-palpable bilateral radial pulses, 1+ dorsalis pedis, and posterior tibial pulses bilaterally.

Laboratory findings are noted in Table 5.

In addition, patient had negative ANA, ANCA, Beta2 glycoprotein, anti-cardiolipin antibody, lupus anticoagulant, hepatitis B serology, hepatitis c serology, and HIV was nonreactive.

Table 5 Laboratory data

Variable	Reference range	Patient value
Blood		
White blood cell count (10^3 cells per cubic milliliter)	4.5–10	14.2
Hemoglobin (g/dL)	13.5–16.5	9.9
Hematocrit (%)	40–49	3.8
Platelet (10^3 cells per cubic milliliter)	160–360	321
Sodium (mmol/L)	135–145	136
Potassium (mmol/L)	3.5–5.1	4.1
Chloride (mmol/L)	100–110	98
Carbon dioxide (mmol/L)	20–30	25
Urea nitrogen (mg/dL)	8–22	17
Creatinine (mg/dL)	0.5–1.3	0.63
Glucose (mg/dL)	65–99	107
Calcium (mg/dL)	8.5–10.3	8.8
Total protein (g/dL)	6.0–8.0	7.4
Albumin (g/dL)	3.5–5.0	3.8
Alanine aminotransferase (U/L)	10–50	40
Aspartate aminotransferase (U/L)	10–50	21
Alkaline phosphatase (U/L)	40–129	38
Total bilirubin (mg/dL)	< 1.0	0.2
C-reactive protein (mg/L)	< 4.9	38.8
Erythrocyte sedimentation rate (mm/h)	0–29	77
Urine		
Color	Yellow	Straw
Clarity	Clear	Cloudy
pH	5.0–8.0	6.5
Specific gravity	1.005–1.030	1.016
Glucose (mg/dL)	Negative	Negative
Ketones (mg/dL)	Negative	Negative
Leukocyte esterase	Negative	Negative
Nitrite	Negative	Negative
Blood	Negative	Small
Protein	Negative	Negative
Erythrocytes (per high power field)	0–2	4–5
Leukocytes (per high power field)	0–3	11–30

Question 2

When suspecting Takayasu's Arteritis, which initial imaging modality is generally the preferred choice of diagnosis given risks and benefits to the patient?

A. Magnetic resonance angiography
B. Conventional angiogram
C. FDG-PET/CT from skull base to mid thighs
D. Computed tomography angiography

Answer: A

Discussion: Advanced Imaging has come to replace the more invasive method of traditional angiography. Early imaging in a patient with clinical features of an LVV is recommended. The earliest detectable abnormality in TAK is thickening of the vessel wall from inflammation. Other characteristic signs of vasculitis on imaging include narrowing, stenosis, or aneurysm. MR imaging is generally recommended as the initial imaging modality and contrast-enhanced MR imaging allows better soft-tissue differentiation and can depict other signs of inflammation including mural edema and increased mural vascularity as well as wall thickening. In addition, MR imaging spares the patient from excess radiation iodinated contrast exposure.

Question 3

Which infectious etiologies should be considered in this patient as a trigger for her vasculitis?

A. Staphylococcus aureus infection
B. Mycobacterium tuberculosis
C. Hepatitis B
D. Hepatitis C

Answer: B

Discussion: The cause of most vasculitides has not been fully elucidated, but infections have been suggested to contribute to the induction and reactivation of some forms. Takayasu arteritis has been linked to Mycobacterium tuberculosis infection, perhaps via cross-reactivity against vascular peptides that mimic the antigens of M. tuberculosis. In granulomatosis with polyangiitis (GPA), chronic nasal carriage of Staphylococcus aureus is related to endonasal activity and disease relapses which can be reduced with antibacterial treatment; current research aims to identify the staphylococcal traits that are implicated in disease onset and relapses.

PAN, one of the medium-sized vessel vasculitides, has been associated with HBV, HCV, and HIV infection.

Hepatitis B virus (HBV)-associated and hepatitis C virus (HCV)-associated polyarteritis nodosa (PAN) have a clinical presentation different from nonviral PAN and require antiviral treatment.

Question 4

Emerging data suggest which auto-inflammatory disease coexists more frequently with Takayasu's Arteritis by virtue of HLA marker?

A. Adamantides-Behcet's syndrome
B. ANCA-Associated vasculitis
C. Giant Cell arteritis
D. Ulcerative Colitis

Answer: D.

Discussion: Genome-wide association studies of TAK and Ulcerative colitis (UC) patients have found the HLA B52:01 gene to be associated with both diseases. Prevalence of inflammatory bowel disease (IBD) is approximately 6% in TAK patients, much higher than the prevalence in the general population. A diagnosis of IBD can precede the TAK diagnosis. Both diseases also have granulomatous histopathology. TAK should be considered in IBD patients presenting with concerning vascular symptoms.

Case Continued

After patient was stabilized during hospital stay and appropriate referrals were made, conservative medical therapies of antiplatelet agents, statin, beta-blocker, and ace inhibitor were initiated. She was also started on Prednisone with plan to taper and oral Methotrexate.

The patient subsequently presented months later with similar chest pain symptoms. At that time computed tomography angiography revealed again the known LAD artery aneurysm measuring 13 mm now with a thrombus and small secular component. The left circumflex artery aneurysm was the same size as prior and an increased aneurysmal dilation of the aortic root was noted to be 5.3 cm, splenic artery aneurysm measuring 12 mm, along with aneurysms of the superior mesenteric artery and uterine vessels.

A Bentall procedure was performed by cardiothoracic surgery to replace the aortic valve and repair the aortic root. Pathology specimens from the aorta showed intramural fibrosis and myxomatous change.

Question 5

In addition to lab interpretation, the following imaging would suggest further evidence of vascular inflammation if MRA was negative:

A. Ultrasound subclavian artery
B. Conventional Angiogram of left upper extremity
C. FDG-PET/CT from skull base to mid thighs
D. CT Angiogram of left upper extremity

Answer: C

Discussion: Assessing disease activity in TAK can be difficult. Inflammatory markers do not always correlate with disease activity and can be normal even in the presence of symptoms or radiographic progression of the disease. Guidelines continue to be updated as imaging modalities and expertise becomes more advanced.

Positron emission tomography (PET) is a noninvasive metabolic imaging based on regional distribution of the glucose analog fluoro-deoxyglucose (FDG) imaging. It detects areas of active metabolism such as inflammatory cells or malignant cells. It should be noted to assess inflammation in cranial vessels MR imaging is recommended due to brain metabolic activity.

Vasculitis may be active and these ongoing changes can be seen on PET-FDG; however, the use of this imaging in the absence of symptoms is not fully understood at this point.

Bibliography

1. Alibaz-Oner F, Dede F, Ones T, Turoglu HT, Direskeneli H. Patients with Takayasu's arteritis having persistent acute-phase response usually have an increased major vessel uptake by 18F-FDG-PET/CT. Mod Rheumatol. 2015;25(5):752–5.
2. Dejaco C, Ramiro S, Duftner C, Besson FL, Bley TA, Blockmans D, et al. EULAR recommendations for the use of imaging in large vessel vasculitis in clinical practice. Ann Rheum Dis. 2018;77(5):636–43.
3. De Souza AWS, de Carvalho JF. Diagnostic and classification criteria of Takayasu arteritis. J Autoimmun. 2014;48–49:79–83.
4. Dhaun N, Bellamy CO, Cattran DC, Kluth DC. Utility of renal biopsy in the clinical management of renal disease. Kidney Int. 2014;85(5):1039–48.
5. Ford J, Monach P. Disease heterogeneity in antineutrophil cytoplasmic antibody-associated vasculitis: implications for therapeutic approaches. Lancet Rheumatol. 2019;1:e247–56.
6. Gómez-Puerta JA, Hernández-Rodríguez J, López-Soto A, Bosch X. Antineutrophil cytoplasmic antibody-associated vasculitides and respiratory disease. Chest. 2009;136(4):1101–11.

7. Guggenberger KV, Bley TA. Imaging in vasculitis. Curr Rheumatol Rep. 2020;22(8):34.
8. Jennette JC, Falk RJ, Bacon PA, Basu N, Cid MC, Ferrario F, et al. 2012 revised international Chapel Hill consensus conference nomenclature of vasculitides. Arthritis Rheum. 2013;65(1):1–11.
9. Kitching AR, Anders H-J, Basu N, Brouwer E, Gordon J, Jayne DR, et al. ANCA-associated vasculitis. Nat Rev Dis Primers. 2020;6(1):71.
10. Konstantinov KN, Emil SN, Barry M, Kellie S, Tzamaloukas AH. Glomerular disease in patients with infectious processes developing antineutrophil cytoplasmic antibodies. ISRN Nephrol. 2013 Feb;19(2013):1–18.
11. Nevares A, Raut R, Libman B, Hajj-Ali R. Noninfectious autoimmune scleritis: recognition, systemic associations, and therapy. Curr Rheumatol Rep. 2020;22(4):11.
12. Pham V, de Hemptinne Q, Grinda J-M, Duboc D, Varenne O, Picard F. Giant coronary aneurysms, from diagnosis to treatment: a literature review. Arch Cardiovasc Dis. 2020;113(1):59–69.
13. Sacato G, Boukhal S, Specks U, Flores-Suarez LF, Cornec D. Lung involvement in ANCA-associated vasculitis. Presse Med. 2020;49:104039.
14. Satoskar AA, Parikh SV, Nadasdy T. Epidemiology, pathogenesis, treatment and outcomes of infection-associated glomerulonephritis. Nat Rev Nephrol. 2020;16(1):32–50.
15. Seyahi E. Takayasu arteritis: an update. Curr Opin Rheumatol. 2017;29(1):51–6.
16. Terao C, Matsumura T, Yoshifuji H, Kirino Y, Maejima Y, Nakaoka Y, et al. Brief report: Takayasu arteritis and ulcerative colitis: high rate of co-occurrence and genetic overlap: co-occurrence and genetic overlap in TAK and UC. Arthritis Rheumatol. 2015;67(8):2226–32.
17. Valenzuela F, Perez V. Scleritis- infectious versus inflammatory. US Ophthalmic Rev. 2016;2:92–6.

B Cell Deficiency

Aishwarya Navalpakam and Pavadee Poowuttikul

Abbreviations

ADEM	Acute disseminated encephalomyelitis
AID	Activation-induced cytidine deaminase
ARA	Autosomal recessive agammaglobulinemia
BACH2	BTB (Broad-Complex, Tramtrack and Bric a brac) domain and CNC (Cap'n'collar) homolog 2
BAFFR	B cell activating factor of the TNF family receptor
BCR	B cell receptor
BLK	B lymphocyte tyrosine kinase
BLNK	B cell linker protein
Btk	Bruton's tyrosine kinase
CD20 (MS4A1)	Membrane spanning 4-domains A1
CD21 (CR2)	Complement receptor type 2
CD81	Tetraspanin family
CTLA4	Cytotoxic T lymphocyte associated antigen 4
CVID	Common Variable Immunodeficiency
HRCT	High resolution computed tomography
HSCT	Hematopoietic stem cell transplant
IBD	Inflammatory bowel disease
ICOS	Inducible T cell co-stimulator
IKZF1 (IKAROS)	Ikaros family zinc finger 1
IRF2BP2	Interferon regulatory factor 2 binding protein 2
IVIG	Intravenous Immunoglobulin
LRBA	Lipopolysaccharide-responsive beige-like anchor protein
NFKB1	Nuclear factor kappa B 1
NFKB2	Nuclear factor kappa B 2
PIK3CD	Phosphatidylinositol-4,5-bisphosphate 3-kinase catalytic subunit p110δ
PIK3R1	Phosphoinositide-3-kinase regulatory subunit 1
PJP	*Pneumocystis jirovecii* pneumonia
PLCG2	Phospholipase C gamma 2
PRKCD	Protein kinase C delta
RAC2	Rac family small GTPase 2
SLP-65	SH2 (Src homolog 2) binding leukocyte phosphoprotein of 65 kD
STAT1	Signal transducer and activator of transcription 1
STAT3	Signal transducer and activator of transcription 3
TACI	Transmembrane activator and calcium modulator and cyclophilin ligand interactor
THI	Transient hypogammaglobulinemia of infancy
TNFRSF13B	Tumor necrosis factor receptor superfamily member 13B
TNFRSF13C	Tumor necrosis factor receptor superfamily member 13C
TNFRSF7 (CD27)	Tumor necrosis factor receptor superfamily member 7
TNFSF12	Tumor necrosis factor receptor superfamily member 12
TWEAK	TNF related WEAK inducer of apoptosis
VAV1	VAV guanine nucleotide exchange factor 1
XLA	X-linked agammaglobulinemia

A. Navalpakam · P. Poowuttikul (✉)
Division of Allergy, Immunology and Rheumatology, Department of Pediatrics, Children's Hospital of Michigan, Detroit, MI, USA

Central Michigan University College of Medicine, Mt. Pleasant, MI, USA
e-mail: anavalpa@dmc.org; ppoowutt@dmc.org

M. Mahmoudi (ed.), *Absolute Allergy and Immunology Board Review*, https://doi.org/10.1007/978-3-031-12867-7_21

1 Introduction

Humoral immunodeficiency disorders are characterized by B cell abnormalities which can lead to a decrease in the level and/or function of B cells and antibodies. Patients with humoral immunodeficiencies generally present with recurrent bacterial sinopulmonary or gastrointestinal infections and have a propensity for autoimmune diseases. Bacterial infections are caused by encapsulated pyogenic bacteria such as *Haemophilus influenzae*, *Moraxella catarrhalis*, *Streptococcus pneumoniae*, and *Pseudomonas aeruginosa.* Patients classically present after 4–6 months of age when maternally acquired IgG decreases, when it is expected that individuals produce their own antibodies.

Laboratory values to evaluate for humoral immunodeficiencies generally include measuring serum immunoglobulin levels of IgG, IgA, and IgM; antibody responses to protein and polysaccharide vaccines; and B cell flow cytometry.

Examples of humoral immunodeficiencies include agammaglobulinemia, common variable immunodeficiency, hyper IgM syndrome, selective IgA deficiency, and specific antibody deficiency. In this chapter, we use case-based discussion of common B cell immunodeficiency disorders focusing on X-linked agammaglobulinemia (XLA) and Common Variable Immunodeficiency (CVID).

XLA is caused by a mutation in the *BTK* gene and is characterized by absence of mature B cells and serum immunoglobulins while CVID is a group of disorders with no unifying genetic etiology. CVID is characterized by hypogammaglobulinemia and decreased specific antibody production. We will further discuss in detail regarding the etiology, pathophysiology, clinical manifestations, diagnosis and differential diagnosis, management, and prognosis of these diseases.

B cell deficiencies or humoral immunodeficiencies are primarily antibody deficiencies leading to recurrent sinopulmonary infections. Certain B cell deficiencies, including XLA and CVID, are associated with comorbidities such as bronchiectasis and chronic lung disease. Management of humoral deficiencies includes treating infections aggressively, immunoglobulin replacement therapy, and addressing complications including chronic lung disease, autoimmune disease, and malignancy. In this chapter, we will use case-based discussion to review common humoral immunodeficiencies, including their pathophysiology, clinical manifestation, diagnosis, treatment, and prognosis.

Case 1

A 11-month-old boy had four episodes of acute otitis media and two episodes of pneumonia starting from the age of 6 months. Family history is pertinent for a maternal uncle who passed away due to a severe infection.

Question 1

The most likely affected gene in this patient leads to

A. Absence of immunoglobulin heavy chain class switching
B. Arrest of B lymphocyte development from Pre-B to Pro-B stage
C. Arrest of B lymphocyte development from Pro-B to Pre-B stage
D. Decrease in maturation of B lymphocyte to follicular B cell
E. Defect in enzyme needed for B cell affinity maturation

Answer and Explanation

Answer: C

The most likely diagnosis is X-linked agammaglobulinemia (XLA). XLA is caused by a mutation in the *BTK* gene which is involved in the development of B cells from the Pro-B stage to the Pre-B stage. Therefore, the correct answer is C and not B. Figure 1 represents major developmental stages of B cell and the important cell surface markers. This figure will be further described in Sect. 2.1.

Answer choice A refers to hyper IgM syndrome. The most common cause is due to CD40L deficiency and it is also an X-linked disorder. Patients with CD40L deficiency have a combined immunodeficiency and generally present with sinopulmonary infections and opportunistic infections, such as *Pneumocystis jirovecii* pneumonia (PJP).

Answer choice D refers to Common Variable Immunodeficiency (CVID). Patients with CVID generally present in their teenage years or in early adult life. Diagnosis is typically made after 4 years of age. CVID patients can present with recurrent sinopulmonary infections and tend to have autoimmune manifestations.

Answer choice E refers to defect in enzyme AID (activation-induced cytidine deaminase) which is the autosomal recessive cause of the second most common form of hyper IgM syndrome. AID is an enzyme involved in class switch recombination and somatic hypermutation. These two processes are important for the selection and proliferation of B cells expressing B cell receptors with high affinity to antigen. Patients with AID deficiency generally present with recurrent sinopulmonary infections, have an increased likelihood for autoimmune disorders, and have characteristic lymphoid hyperplasia with giant germinal centers.

2 X-Linked Agammaglobulinemia

X-linked agammaglobulinemia (XLA) is a primary immunodeficiency disorder caused by a mutation in *BTK* gene and leads to the inability of the B cells to mature and produce

	Stem Cell	Pro- B cell	Large Pre-B cell	Small Pre-B cells	Immature B cell	Mature B cell
Ig Expression	None	None	Pre- B receptor • μ heavy chain • Surrogate light chains (VpreB and λ5) • Igα and Igβ signaling molecules	Cytoplasmic μ	Membrane IgM	Membrane IgM and IgD
Ig DNA, RNA	Germline DNA	D_H to J_H	V_H to DJ recombination, μ mRNA	Recombined H chain gene (VDJ), V to Jk recombination	μ and κ/λ RNA	Alternative splicing of VDJ- C RNA to form Cμ and Cδ mRNA
Surface Markers	CD43+	CD43+ CD19+ CD10+	B220^{lo} CD43+	B220^{lo} CD43+	IgM+ CD43–	IgM^{hi} IgD+

Fig. 1 B cell development

immunoglobulins. Prevalence has been estimated to be one in every 379,000 births. Many patients (41%) with XLA have a male family member with the disorder. The X-linked form of agammaglobulinemia accounts for 85% of the cases, and the autosomal recessive form accounts for 15%.

2.1 Etiology and Pathophysiology of XLA

B cell development begins in the bone marrow with pro-B stage. VDJ recombination occurs within the Ig heavy chain, and a μ heavy chain protein is generated. This forms the pre-B cell with a pre-BCR (B cell receptor) complex. The pre-BCR complex is characterized by a μ heavy chain, surrogate light chains (VpreB and λ5), Igα and Igβ signaling molecules. See Fig. 1. It is important to know that expression of this pre-BCR is the first checkpoint in B cell maturation.

Bruton's tyrosine kinase (Btk) is a TEC-family kinase. It is activated through the signaling of pre-BCR which leads to survival and differentiation of these cells. *BTK* gene is found on the X chromosome. Mutation in *BTK* gene leads to X-linked agammaglobulinemia. This can be de novo or familial. This defect leads to failure of B cell maturation and arrests B lymphocyte development from Pro-B to Pre-B stage.

Immature B cells form, and VJ recombination occurs to produce Ig light chains. Heavy and light chain molecules are assembled to form IgM which is then expressed on the surface of Immature B cells as shown in Fig. 1. The cells that are not self-reactive survive (negative selection) exit the bone marrow. It is important to know that B cell negative selection process occurs at the immature B cell stage when the process called receptor editing occurs. It is also important to know the surface markers and receptors on each stage of B cell development as depicted in Fig. 1.

Immature B cells complete their development into mature naïve B cells in the peripheral lymphoid organs. Mature naïve B cells express IgM and IgD on their cell surface. Mature naïve B cells are then activated by their specific antigen and helper T cells to become plasma cells or memory cells in the peripheral lymphoid tissue.

As stated earlier, if the expression of pre-BCR is the first checkpoint in B cell maturation, then deficiency in either μ heavy chain, surrogate light chains (VpreB and λ5), Igα, or Igβ signaling molecules will also lead to the arrest in B cell development similar to XLA. Another adaptor protein known as SLP-65 (SH2-binding leukocyte phosphoprotein of 65 kD), also called BLNK (B cell linker protein), is closely linked to Btk enzyme. BLNK is important for B cell signaling pathway for B cell activation and differentiation (not shown in Fig. 1).

Although rare, there are autosomal recessive forms of agammaglobulinemia (15%) caused by a μ heavy chain

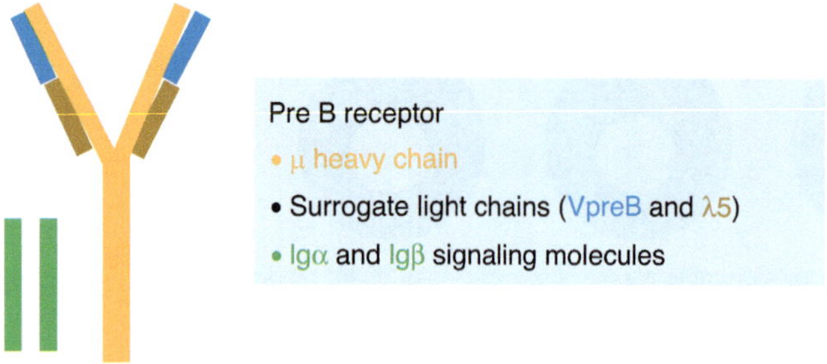

Fig. 2 Structure of Pre-B cell receptor

deficiency, λ5 deficiency, Igα deficiency, Igβ deficiency, and BLNK deficiency. There are still other unknown genetic mutations that cause autosomal recessive agammaglobulinemia. Figure 2 represents the structure of a pre-B cell receptor.

Question 2
Which physical exam finding helps support the diagnosis of XLA?

A. Absence of tonsils and adenoids
B. Eczema
C. Splenomegaly
D. Lymphoid hypertrophy
E. Retained primary teeth

Answer and Explanation
Answer: A

Absence of tonsils, adenoids, and palpable lymph nodes are characteristic features of XLA as there is a lack of mature B cells in these lymphoid tissues.

Severe eczema can be associated with several immunodeficiency disorders including Wiskott–Aldrich Syndrome, Hyper IgE syndrome, and Omenn syndrome. Patients with Hyper IgE syndrome also have retained primary teeth. Lymphoid hypertrophy is seen in patients with Hyper IgM syndrome and CVID. Splenomegaly is also observed in some CVID patients.

2.2 Clinical Manifestation of XLA

XLA should be suspected in male infants aged 6–12-month-old with recurrent sinopulmonary infections, a positive family history, and physical exam findings of absence of lymphoid tissue.

Most patients with XLA present after 4–6 months old when maternal IgG that is transferred through the placenta during pregnancy begins to disappear. With this reason, XLA is unlikely to present prior to 3 months of age. Generally, at 4–6 months of age, infants begin to develop their own antibodies. As XLA infants cannot produce their own antibodies, this is when they become susceptible to infections.

Sometimes the process of production of IgG can be delayed and this can lead to hypogammaglobulinemia. This is defined as Transient Hypogammaglobulinemia of Infancy (THI), when the IgG, with or without IgA and IgM, is less than 2 standard deviations below the mean for age. Although hypogammaglobulinemia is present, these children do respond to vaccine antigens. They generally do not have severe infections but have an increased likelihood of having atopic diseases. In THI, antibody levels correct by mean age of 27 months and most are normalized by 59 months. In XLA, there is no recovery of antibody production.

Infectious presentation seen in patients with XLA includes recurrent otitis media, conjunctivitis, respiratory tract infections, septic arthritis, and/or gastrointestinal infections. We will discuss infections in further detail later in this chapter.

Classic physical exam finding of decreased or absence of tonsils, adenoids, and lymph nodes is pathognomonic of XLA. There is generally normal growth and development in these patients, however, in cases of delayed diagnosis, failure to thrive can occur due to chronic infections.

Question 3
Prior to the IVIG treatment, septic arthritis in patients with XLA is mainly caused by

A. *Kingella kingae*
B. *Pseudomonas aeruginosa*
C. *Staphylococcus aureus*
D. *Streptococcus pyogenes*
E. *Streptococcus pneumoniae*

Answer and Explanation
Answer: E

Streptococcus pneumoniae along with *Haemophilus influenzae* type B are mainly the cause of septic arthritis in patients with XLA before intravenous immunoglobulin (IVIG) treatment. *Streptococcus pyogenes*, *Pseudomonas aeruginosa*, and *Staphylococcus aureus* can cause septic arthritis in XLA patients, but not as common as *S. pneumoniae* or *H. influenzae* type B. *Pseudomonas* species are reported to be the most common cause of septicemia in XLA patients. *Kingella kingae* is one of the common causes of septic arthritis in immunocompetent children aged 3 months to 3 years.

2.3 Infections in XLA

XLA patients have recurrent infections of the respiratory tract including chronic sinusitis, bronchitis, and pneumonias. The most common presenting infection in patients with XLA in a cohort of 201 studied in the United States was otitis media (69%), pneumonia (53%), sinusitis (37%), and diarrhea (14%). Generally, the etiology of the pneumonia is not identified but when isolated pneumococcus is the most common cause followed by *H. influenzae* type B, *Haemophilus parainfluenza*, *Pseudomonas* spp., and *Staphylococcus* spp. Recurrent lower respiratory tract infections can lead to chronic lung disease and bronchiectasis despite treatment with IVIG. Moreover, respiratory tract infections are common in patients before and after treatment with IVIG.

Patients with XLA are susceptible to infections caused by encapsulated pyogenic bacteria, specifically *Streptococcus pneumoniae*, *Haemophilus influenzae*, and *Streptococcus pyogenes*. These bacteria are generally the cause of infections prior to the initiation of immunoglobulin therapy. *Pseudomonas* spp. are most commonly implicated in septicemia in XLA patients, followed by *H. influenzae*, *S. pneumoniae*, and *S. aureus*.

H. influenzae type b and *S. pneumoniae* are mainly the cause of septic arthritis prior to IVIG treatment, while viral cause of septic arthritis is more commonly seen after IVIG treatment. Meningitis caused by these encapsulated pyogenic bacteria is also observed in patients with XLA prior to IVIG treatment.

Mycoplasma spp. can cause infections of the respiratory tract, urogenital tract, and joint. Sometimes mycoplasma infection in XLA patients can be prolonged and severe as usual cultures may not identify the organism. *Mycoplasma* spp. can cause severe disease when superimposed with other bacterial infections. *Mycoplasma* and *Ureaplasma* spp. cause some cases of septic arthritis and osteomyelitis typically in XLA patients who are receiving inadequate IVIG. It is important to remember that serology testing for infections is not helpful in patients with XLA due to the inability to produce antibodies in these patients.

Question 4

Later on, the patient missed several doses of IVIG and developed progressive neurological symptoms with ataxia and loss of developmental skills. Peripheral edema and erythematous rash were noted on physical examination. CSF cultures and PCR studies were negative. What is the most likely cause of the patient's meningoencephalitis?

A. Coxsackie virus
B. *Cryptococcus neoformans*
C. Epstein–Barr virus
D. *Mycobacterium avium*
E. *Neisseria meningitidis*

Answer and Explanation

Answer: A

Chronic enteroviral meningoencephalitis is a cause of progressive neurological deterioration and a dermatomyositis-like syndrome in patients with untreated XLA. Examples of enterovirus include poliovirus, echovirus, and coxsackie virus. Although *Cryptococcus neoformans* and *Neisseria meningitidis* can cause meningoencephalitis, they are not associated with the chronicity of symptoms or the dermatomyositis-like syndrome seen in enteroviral infections. Mycobacterial infections are opportunistic and seen rarely in patients with humoral immunodeficiencies. Epstein–Barr virus can cause viral meningitis and acute disseminated encephalomyelitis (ADEM), also known as postinfectious encephalomyelitis, described in immunocompetent children. However, dermatomyositis is not observed in these patients.

Enterovirus infections, such as poliovirus, echovirus, and coxsackie virus, can cause infections in patients with XLA. Chronic enteroviral meningoencephalitis develops slowly with progressive loss of motor and cognitive skills. The disease can culminate in coma and death. Dermatomyositis has also been associated with enteroviral infection. It is difficult to diagnose and requires a high index of suspicion considering that the virus is not always isolated in the CSF. Often, it requires brain biopsy for diagnosis. Enteroviral meningoencephalitis is managed with high dose IVIG, intrathecal immunoglobulin, and other medications such as interferon gamma and antiviral drugs.

Although rare, some XLA patients are also susceptible to *Pneumocystis jirovecii* pneumonia (PJP). PJP is generally associated with T cell mediated defects but B cells and opsonization by antibodies play a role in combatting *Pneumocystis jirovecii*. It appears that poor nutritional status is a risk factor for acquiring PJP. Some PJP cases are considered to be secondary to another severe infection.

Patients with XLA are also susceptible to gastrointestinal tract infections. When XLA patients present with persistent and chronic diarrhea and malabsorption, *Giardia lamblia* must be considered. Diarrhea that presents along with fever and skin manifestations is associated with *Campylobacter jejuni*. Skin manifestations include an erysipelas-like rash and in other cases are maculopapular with small vesicles or crusts. *Salmonella* spp. are also a cause of diarrhea and gastrointestinal infections in XLA patients.

2.4 Autoimmune Disorders in XLA

Although not as common as patients with CVID, patients with XLA have been described to have autoimmune conditions, specifically arthritis (16%), hypothyroidism (5%), and inflammatory bowel disease (IBD) (0.7%). In another cohort of XLA patients, the diagnosis of IBD was found in 10%.

Although there is no consensus recommendation, screening for autoimmune disease symptoms and referring to specialists if concerned are appropriate.

2.5 Malignancies in XLA

About 4% of primary immunodeficiency related malignancies are reported in XLA versus 24% seen in patients with CVID in one study. These malignancies associated with XLA include lymphomas, colon cancer, gastric adenocarcinoma, and squamous cell lung carcinoma. While other studies examining patients with XLA found no malignancies. Risk of malignancy is not established in patients with XLA, and there is a lack of data for specific cancer surveillance recommendation. Therefore, as with other primary immunodeficiencies, age-appropriate cancer surveillance would be prudent.

2.6 Diagnosis of XLA

Laboratory findings in patients with XLA are notable for low to undetectable immunoglobulins, absence or near absence of B lymphocytes (CD19+) in the peripheral blood, and low antibody response to antigens or immunizations. B lymphocytes are also absent in the lymphoid tissues. Lymphoid follicles and germinal centers are absent. Isohemagglutinins, which are IgM antibodies against ABO blood group antigens, are also absent or remarkably decreased in patients with XLA. T lymphocyte function appears to be normal in XLA patients, however the CD3+ T cell numbers and CD4+/CD8+ ratio can be variable.

Table 1 summarizes the diagnostic criteria for XLA.

These criteria are based on the Joint Task Force of the American Academy of Allergy, Asthma and Immunology and the American College of Allergy, Asthma and Immunology on the Practice parameter for the diagnosis and management of primary immunodeficiency and the Diagnostic criteria for Primary Immunodeficiencies established by the Pan-American Group for Immunodeficiency (PAGID), and European Society for Immunodeficiency (ESID).

Table 1 Diagnostic criteria for XLA

Definitive diagnosis[a]
Male patient with less than 2% CD19+ B lymphocytes *and* at least *one* of the following: – Mutation in *BTK* – Absence of Btk mRNA on northern blot analysis of neutrophils or monocytes – Absence of Btk protein in monocytes or platelets – Maternal cousins, uncles, or nephews with less than 2% CD19+ B lymphocytes
Probable diagnosis[a]
Male patient with less than 2% CD19+ B lymphocytes in whom *all* of the following are positive: – Onset of recurrent bacterial infections in the first 5 years of life – Serum IgG, IgM, and IgA less than 2 SD below normal for age – Absence of isohemagglutinins and/or poor antibody response to vaccines – Other causes of hypogammaglobulinemia have been excluded
Possible diagnosis
Male patient with less than 2% CD19+ B lymphocytes in whom other causes of hypogammaglobulinemia have been excluded *and* at least *one* of the following is positive: – Onset of recurrent bacterial infections in the first 5 years of life – Serum IgG, IgM, and IgA less than 2 SD below normal for age – Absence of isohemagglutinins

[a] Patients with definitive or probable XLA diagnosis are assumed to have greater than 98% and 85% probability, respectively, that in 20 years they will have the same diagnosis

2.7 Differential Diagnosis

The differential diagnosis for XLA includes the following:

2.7.1 Transient Hypogammaglobulinemia of Infancy (THI)

THI is a delayed production of immunoglobulins that improves over time. Typically, these patients are able to produce specific antibodies to vaccines and isohemagglutinins against ABO blood type. By definition of THI, serum immunoglobulins and response to vaccines should be normalized by 4 years of age. Patients with THI generally do not have severe infections. However, upper and lower respiratory tract infections are common (40–90% of patients with THI).Other infections observed include otitis media, urinary, and gastrointestinal tract infections. In rare cases, bacterial sepsis and meningitis are reported. Interestingly, atopic diseases, particularly eczema and food allergies, are seen in up to 30–50% of the patients. Immunoglobulin levels correct by mean age of 27 months and most are normalized by 59 months. Recurrent infections typically resolve by 9–15 months old.

2.7.2 Common Variable Immunodeficiency (CVID)

CVID is characterized by hypogammaglobulinemia and decreased specific antibody production. CVID is seen in both males and females, unlike XLA. In the majority of patients, CVID presents in adolescents and young adults. Many patients present after 20 years of age. Autoimmune disorders and lymphoproliferative disorders are commonly seen in patients with CVID. We will discuss CVID in further detail later in this chapter.

2.7.3 Autosomal Recessive Agammaglobulinemia (ARA)

As discussed previously, patients with ARA present similarly to XLA. However, unlike XLA which is X-linked and found in males, ARA is found in both males and females. ARA accounts for about 15% of cases with agammaglobulinemia. Defects that cause ARA include μ heavy chain deficiency, λ5 deficiency, Igα deficiency, Igβ deficiency, and BLNK deficiency.

Agammaglobulinemia can also be found in combined T and B cell immunodeficiency.

2.8 Management of XLA

The mainstay of treatment for X-linked agammaglobulinemia is immunoglobulin replacement and antimicrobial treatment for infections.

Immunoglobulin can be administered intravenously or subcutaneously. The recommended dose is 400–500 mg/kg every 3–4 weeks. IgG trough level of greater than 500 mg/dL is known to prevent acute bacterial infections such as pneumonias, septicemia, dermatitis and cellulitis. However, IgG trough levels greater than 800 mg/dL may be needed to prevent complications such as bronchiectasis, chronic sinusitis, and non-bacterial infections (meningoencephalitis, exudative enteropathy, and aseptic arthritis).

Enteroviral meningoencephalitis, as discussed previously, should be treated with high dose IVIG to maintain IgG trough level of greater than 1000 mg/dL. IVIG with specific antibodies to the serotype of the infecting enterovirus will have more efficacy in treating this disease.

When any bacterial infection is suspected in patients with XLA, there should be a low threshold to obtain culture and begin empirical antibiotic treatment as soon as possible. Antibiotics should cover encapsulated pyogenic bacteria as XLA patients are often susceptible to these organisms. They should also cover for *Pseudomonas* spp. and *Mycoplasma* spp. if applicable. It is often beneficial to involve infectious disease in infection management. And in some cases, prolonged antimicrobial therapy may be required.

Live viral vaccines are contraindicated in patients with XLA because it can cause infections secondary to vaccination. Although patients with XLA show poor antibody responses to antigens, clinicians should still consider giving inactivated viral or bacterial vaccines, including the influenza vaccine and the COVID-19 vaccine. These vaccines can stimulate T cell-mediated immune responses and can potentially protect against infections. Patients with XLA also acquire passive immunization via immunoglobulin replacement therapy. As influenza strains change each year, and immunoglobulin products may not contain antibody against each year influenza strains, it should be emphasized that XLA patients should get inactivated influenza vaccine yearly.

Although hematopoietic stem cell transplant (HSCT) can offer a cure for patients with XLA, it is not generally considered due to the risks associated with the procedure. Moreover, with immunoglobulin replacement therapy, XLA patient's antibody deficits are generally addressed. Cases of XLA patients undergoing successful HSCT have been reported for patients who have significant episodes of infections despite being on immunoglobulin replacement therapy.

Lentiviral vector for gene therapy for XLA has been studied in murine models but have not been studied in humans at this time.

Question 5
What is the most likely cause of mortality in patients with X-linked agammaglobulinemia?

A. Chronic lung disease
B. Fulminant infectious mononucleosis
C. Malignancy
D. Nodular regenerative hyperplasia
E. Severe enteropathy

Answer and Explanation
Answer: A

The most likely cause of mortality in patients with XLA is complications related to chronic lung disease. Malignancy is rare in patients with XLA and when compared to chronic lung diseases, it is not a significant cause of mortality. Hepatic nodular regenerative hyperplasia is more described in CVID, though it has reported quite frequently in one XLA cohort. The disease has high mortality rate; however, it is not as common and not well described as chronic lung disease in XLA patients.

Diseases that are associated with radiation hypersensitivity including Nijmegen breakage syndrome, DNA ligase I deficiency, DNA ligase IV deficiency, and Bloom syndrome are at risk for mortality associated with malignancies. Fulminant infectious mononucleosis is a cause of mortality in X-linked lymphoproliferative disease. Enteropathy and malignancies, such as non-Hodgkin lymphoma and gastric cancer, are common in CVID and increase the disease morbidity.

2.9 Complications Related to XLA

Chronic lung disease and bronchiectasis are common in patients with XLA, and related respiratory failure is a strong cause of mortality in these patients. Prevalence of bronchiectasis varies (17–36%) and happens despite treatment with immunoglobulins. These patients can have chronic cough and wheezing with frequent infectious exacerbations or may be asymptomatic. It is important to note that chronic lung disease and bronchiectasis are a marker of poor prognosis in XLA. Therefore, it is recommended that these patient's pulmonary status is monitored with routine pulmonary function tests. High resolution computed tomography (HRCT) is useful and sensitive for diagnosing and monitoring chronic lung disease. There is no clear recommendation if clinicians should obtain HRCT based on clinical presentation or routinely to monitor for pulmonary changes.

Immunoglobulin therapy is essential in preventing respiratory tract infections which lead to chronic lung disease. However, as mentioned previously, chronic lung disease can develop despite high dose immunoglobulin replacement therapy. Management of bronchiectasis and chronic lung disease includes early and aggressive treatment of infections with appropriate antimicrobials, which may require a prolonged course up to 10–14 days to prevent recurrence.

In patients with cystic fibrosis who tend to have recurrent infections leading to bronchiectasis, the use of macrolide antibiotics such as azithromycin can have both anti-inflammatory and antimicrobial function. A double-blind, placebo controlled, randomized clinical trial was designed to evaluate the benefits of oral azithromycin 250 mg for 3 days a week in patients with primary antibody deficiencies and chronic infection related pulmonary disease. The investigators found that there were reduced infectious respiratory exacerbations, decreased courses of antibiotics, and decreased risk of hospitalization in the treatment group. Therefore, macrolides are recommended for XLA patients with bronchiectasis who have an infectious exacerbation. Pulmonary physical therapy to move secretions out of the airways is also recommended. Patients who develop severe bronchiectasis and chronic lung disease may need lung transplantation.

Hepatic nodular regenerative hyperplasia is reported as one of the late-onset complications of XLA. It causes noncirrhotic portal hypertension characterized by the diffuse transformation of normal hepatic parenchyma into small, regenerative nodules. In one U.S. XLA cohort, the authors reported that nodular regenerative hyperplasia is seen in 29% of their patients. The disease is highly associated with increased mortality (>80%). Nodular regenerative hyperplasia is seen more in XLA patients who have persistent thrombocytopenia, elevated ALP, hepatomegaly, or splenomegaly.

Other complications such as malignancies and autoimmune disorders associated with XLA are previously discussed in clinical manifestations of XLA.

2.10 Prognosis of XLA

A recent long-term follow-up study of XLA patients revealed that the risk of developing chronic lung disease in almost half of XLA patients when they reach the age of 40–50. This occurs despite immunoglobulin therapy. The overall survival rate in this cohort of patients with XLA was 93% at 43 years of age. However, when compared to healthy individuals of the same age, life expectancy is reduced.

Case 2

A 7-year-old girl was evaluated for chronic abdominal pain and diarrhea. Due to concern for celiac disease, a screen was done and noted that her IgA was undetectable. However, her IgG and IgM were both within normal limits.

Question 1

Mutation in which gene can put this patient at risk for developing Common Variable Immunodeficiency (CVID)?

A. *BLNK*
B. *ICOS*
C. *PIK3CD*
D. *TNFRSF13B*
E. *TNFRSF13C*

Answer and Explanation

Answer: D

Genetic defect in the tumor necrosis factor receptor superfamily member 13B (*TNFRSF13B*) or transmembrane activator and calcium modulator and cyclophilin ligand interactor (*TACI*) is seen in both patients with selective IgA deficiency and common variable immunodeficiency (CVID). This receptor is important for B cell activation and differentiation. It is suggested that some patients with selective IgA deficiency may develop into CVID. Mutation in *TACI* appears to induce susceptibility to CVID.

Selective IgA deficiency is defined by serum IgA less than 0.07 g/L with normal serum IgG and IgM levels in patients greater than 4 years of age. Most of these patients are asymptomatic with no significant infectious history. Selective IgA deficiency patients with mutation in *TACI* are more likely to develop CVID and have autoimmune conditions.

Mutations in B cell activating factor of the TNF family receptor (*BAFFR*, also known as *TNFRSF13C*) lead to impaired B cell development/survival and are observed in CVID patients. The interaction between BAFF and BAFFR is important for transitional B cell differentiation into marginal zone and follicular B cells. Patients with BAFFR defect have B cell lymphopenia, relatively increased transitional B cells, and preserved IgA levels. *BARFFR* mutations are not associated with selective IgA deficiency.

Inducible T cell co-stimulator (ICOS) interacts with ICOS ligand to aid in the formation of germinal centers. It is important for B cell differentiation, immunoglobulin class switching, and B cell tolerance. Patients with ICOS deficiency have low to absent memory B cells and a loss of plasma cells. This is thought to be secondary to poor germinal center reactions with loss of ICOS signaling. ICOS deficient patients can also have T cell defects and can present with viral and opportunistic infections. Patients with ICOS deficiency have low IgG, low or normal IgM, and low or normal IgA. Patients with ICOS deficiency present with recurrent respiratory infections and tend to have autoimmune disease.

Gain of function mutation in phosphatidylinositol-4,5-bisphosphate 3-kinase catalytic subunit p110δ (PIKC3D) can lead to lymphoproliferative disease by stimulating cellular division. It is also known as activated PI3Kδ syndrome (APDS). ARPDS causes defects in B and T cell signaling leading to recurrent infections and T cell senescence. These patients present with recurrent infections with encapsulated organisms and severe herpes virus family infections. Patients with APDS also have autoimmune diseases, lymphoproliferative diseases, and lymphomas. APDS has been treated with rapamycin which is an antiproliferative medication.

ICOS and PIKC3D mutations are described in CVID but they are not associated with selective IgA deficiency. Other monogenic defects associated with CVID and their effects on B cell are listed in Table 2 below.

Table 2 Biologic effects of CVID-associated monogenic gene defects

Impaired B cell development and survival	
IKZF1 (*IKAROS*)	Ikaros family zinc finger 1
TNFRSF13C or BAFFR	Tumor necrosis factor receptor superfamily member 13C or B cell activating factor receptor
TNFSF12 or TWEAK	Tumor necrosis factor receptor superfamily member 12 or TNF related WEAK inducer of apoptosis
NFKB2	Nuclear factor kappa B 2
IRF2BP2	Interferon regulatory factor 2 binding protein 2
TNFRSF7 (*CD27*)	Tumor necrosis factor receptor superfamily member 7
STAT1 GOF	Signal transducer and activator of transcription 1, gain of function
Impaired class switch recombination/somatic hypermutation	
IL21	
IL21R	
BACH2	BTB domain and CNC homolog 2
Excessive lymphoproliferation	
LRBA	Lipopolysaccharide-responsive beige-like anchor protein
CTLA4	Cytotoxic T lymphocyte associated antigen 4
PIK3CD	Phosphatidylinositol-4,5-bisphosphate 3-kinase catalytic subunit p110δ
PIK3R1	Phosphoinositide-3-kinase regulatory subunit 1
STAT3 GOF	Signal transducer and activator of transcription 3, gain of function
Impaired B cell activation and tolerance	
NFKB1	Nuclear factor kappa B 1
TNFRSF13B or *TACI*	Tumor necrosis factor receptor superfamily member 13B or transmembrane activator and calcium modulator and cyclophilin ligand interactor
CD19	
CD81	Tetraspanin family
CD21 (*CR2*)	Complement receptor type 2
CD20 (*MS4A1*)	Membrane spanning 4-domains A1
ICOS	Inducible T cell costimulatory
BLK	B lymphocyte tyrosine kinase
PLCG2	Phospholipase C gamma 2
Other	
VAV1	VAV guanine nucleotide exchange factor 1
RAC2	Rac family small GTPase 2
PRKCD	Protein kinase C delta

B cell linker protein (*BLNK*) mutation is associated with autosomal recessive agammaglobulinemia described previously in this chapter, not CVID.

Question 2

The patient is lost to follow up and returns to immunology clinic after 13 years. She is now 20 years old and has had 3 pneumonias and 4 episodes of sinus infections in the past 2 years. Her primary care provider recommended that she see immunologist again. What immunologic findings would be supportive of the diagnosis of CVID?

	Absolute CD19 count (B lymphocyte) [95–515] cells/μL	IgG [768–1728] mg/dL	IgA [99–396] mg/dL	IgM [38–266] mg/dL	Responses to pneumococcal polysaccharide vaccine
A	0	10	<7	10	Low
B	172	232	20	16	Low
C	189	554	117	192	Low
D	225	805	<7	115	Normal
E	315	950	110	256	Low

Answer and Explanation

Answer: B

Classically, patients with CVID have normal CD19+ B lymphocytes, low IgG and low IgA and/or IgM with a poor antibody response to polysaccharide vaccines. Some patients with CVID can have low CD19+ B cell counts; however, they are usually not absent.

Profile A is seen in patients with X-linked agammaglobulinemia or autosomal recessive agammaglobulinemia with absence of CD19+ B lymphocytes, low immunoglobulins of all types, and poor response to pneumococcal vaccine.

Profile C is seen in patients with IgG isotype deficiency with isolated low IgG, while IgA and IgM levels are normal. Patients with IgG isotype deficiency can have loss of antibody function. These patients do not fit criteria for CVID diagnosis but sometimes are treated like CVID. Patients with CVID, unlike IgG isotype deficiency, tend to have low class switched memory B cells and increased CD19+CD21^{low} B cells. CVID patients, unlike patients with IgG isotype deficiency, are more likely to have bronchiectasis, lower baseline serum IgG, and worse vaccine response.

Profile D is seen in patients with selective IgA deficiency with normal CD19+ B lymphocytes, low or absence of IgA but other types of immunoglobulin are normal, and normal response to pneumococcal vaccine.

Profile E is seen in patients with specific antibody deficiency with normal CD19+ B lymphocytes, normal immunoglobulins, and poor response to pneumococcal vaccine.

3 Common Variable Immunodeficiency

Common variable immunodeficiency (CVID) is a group of disorders characterized by hypogammaglobulinemia, decreased specific antibody production, and recurrent infections. Prevalence of CVID has been estimated to be one in every 25,000 people.

3.1 Etiology and Pathophysiology of CVID

Most cases of CVID are sporadic. About 10% are familial with 80% of those cases being autosomal dominant. About 15% of patients with CVID have a relative with either IgA deficiency, CVID, or specific antibody deficiency. *TACI* gene (*TNFRSF13B*) mutation has been identified in both patients with IgA deficiency and CVID. In patients with selective IgA deficiency, those with autoimmune disorders and severe infections are more likely to develop CVID.

Generally, CVID is characterized by impaired B cell differentiation into plasma cells that secrete immunoglobulins. There are reduced levels of immunoglobulins, reduced isohemagglutinins, and abnormal response to protein and polysaccharide antigens.

The pathophysiology of CVID is not completely understood and there are many genetic defects that can result in the clinical presentation of CVID. Due to high-throughput gene sequencing, monogenic defects impairing B cell activation are identified in approximately 20–30% of patients with CVID. Most CVID patients can have multiple gene defects. These defects along with epigenetic and environmental factors can play a role in disease manifestation and lead to diverse CVID phenotypes.

Patients with CVID who have monogenic defects are more likely to have lymphoproliferative disease, autoimmune disease, B cell lymphopenia, and family history of CVID as key features of their disease. Lymphoproliferative disease includes lymphadenopathy, splenomegaly, and/or granulomatous lymphoinfiltrative lung disease.

Table 2 includes the genes implicated in monogenic CVID categorized by their biologic effects.

In B cell maturation, TACI, BAFFR, and BCMA (B cell maturational agent) through APRIL (a proliferation inducing ligand) and BAFF (B cell activating factor, also called BLys or B lymphocyte stimulator) allow for downstream signaling of B cells. This leads to class switch recombination, differentiation and survival of plasma cells and B cells, and T cell independent responses to antigens. See Fig. 3.

Mutations in the genes encoding for these receptors (TACI, BAFFR, and BCMA), therefore, are seen in patients with CVID.

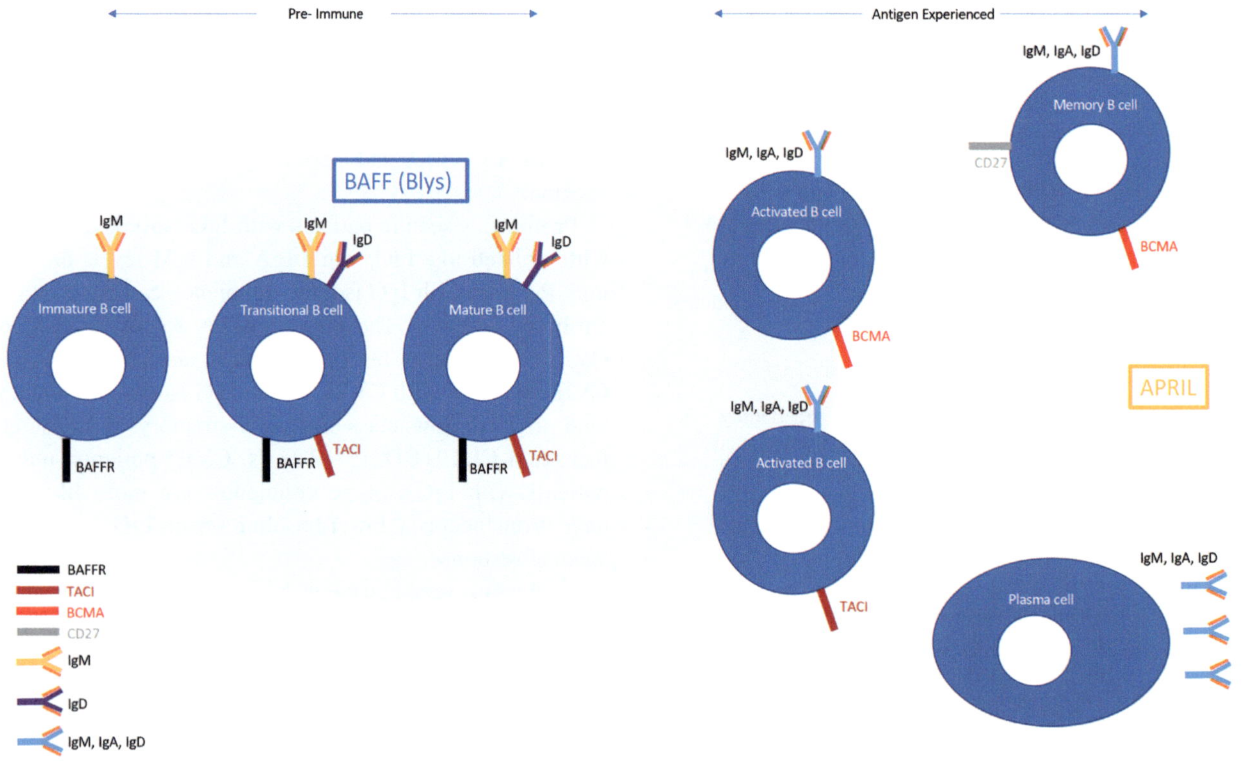

Fig. 3 B cell maturation

Table 3 Markers for peripheral B cell subsets

Types of B cells	Surface markers
Total mature B cells	CD45+CD19+CD20+
Transitional B cells	CD19+CD24^{hi}CD38^{hi}CD10+CD27−CD21$^{low/int}$IgM+++
Naïve B cells	CD19+ IgM+ IgD+CD27−CD38−CD21+++
Memory B cells	CD19+CD27+
Switched memory B cells	CD19+CD27+IgM−IgD− (IgG+ or IgA+ or IgE+)
Marginal zone B cells	CD19+CD27+IgM+IgD+
IgM-only memory B cells	CD19+CD27+IgM+IgD−
IgD-only memory B cells	CD19+CD27+IgM−IgD+
Plasmablasts	CD19+CD20−IgD−IgM−CD38++CD27+

B lymphocytes develop from stem cells in the bone marrow. They express CD19 on their surface. Immature B cells express IgM on their cell surface. Mature B cells express IgM and IgD. When B cells are activated by antigen and other signals, they become antibody secreting B cells. They then complete their path with either becoming memory B cells or plasma cells. Memory B cells express CD27. When memory B cells encounter their specific antigen again, they differentiate into plasma cells.

In CVID, there are maturation defects and the B cells do not differentiate into follicular B cells, plasma cells, or memory cells. Furthermore, monogenic CVID associated genes have been found to have a defect in B cell development and survival; impaired class switch recombination and somatic hypermutation; excessive lymphoproliferation; and defect in B cell activation and tolerance as described in Table 2.

Question 3

Detailed B cell subset analysis is performed in the patient and shows that there is a decrease in CD19+CD27+IgD−IgM− percentage and absolute cell counts. The decrease in CD19+CD27+IgD−IgM− cells is associated with many complications of CVID, **except** for

A. Autoimmune hemolytic anemia
B. Chronic lung disease
C. Granulomatous disease
D. Inflammatory bowel disease
E. Splenomegaly

Answer and Explanation

Answer: D

Patients with CVID who have low switched memory B cells (CD19+CD27+IgD−IGM−) have an increased likelihood of autoimmune diseases, in particular autoimmune hemolytic anemia and autoimmune thrombocytopenia.

Low switched memory B cells and increased CD19+CD21^{low} B cells are also associated with granulomatous disease in patients with CVID. CVID patients with low switched memory B cell are at increased risk for splenomegaly, lymphadenopathy, and lymphoproliferative disorders.

Some studies have found that reduced switched memory B cells and non-switched IgM memory B cells are associated with chronic lung disease.

Other autoimmune diseases have been observed in CVID, including inflammatory bowel disease, systemic lupus erythematosus, rheumatoid arthritis, and Sjogren's syndrome. However, the relationships of B cell phenotypes and these autoimmune diseases have not been clarified. Table 3 demonstrates markers of B cell subsets.

B cell numbers can be normal or low in CVID. However, lower B cell numbers may be associated with a worse prognosis.

Several studies have aimed to classify CVID based on memory B cells. There is an association between decreased class switched memory B cells and increased CD19+CD21^{low} B cells with autoimmune diseases, splenomegaly, lymphadenopathy, and granulomatous disease.

Some studies have also found that reduced class switched memory B cells and non-switched IgM memory B cells are associated with chronic lung disease. Low class switched memory B cells is also associated with bronchiectasis and lymphocytic pneumonitis. Decreased IgG level is associated with increased likelihood of pulmonary complications including bronchiectasis. Patients with CVID with reduced IgM memory B cells have less IgM antibodies to pneumococcal polysaccharide antigens and therefore, they have recurrent lower respiratory tract infections. These recurrent lower respiratory tract infections could lead to bronchiectasis and chronic lung disease.

About 50% of CVID patients may have reduced T cell number and decreased lymphocyte proliferation to mitogens and antigens. A subset of CVID patients, about 25–35%, have increased number of CD8+ T cells, decreased CD4+ T cells, with a reduced CD4+/CD8+ T cell ratio. These patients present with opportunistic infections and are considered to have late-onset combined immunodeficiency (LOCID). These LOCID subtype of CVID patients have a high consanguinity rate in parents and a higher likelihood of having lymphoproliferative disorder, splenomegaly, and infectious gastrointestinal disorders.

3.2 Clinical Manifestation of CVID

CVID is frequently diagnosed after puberty with a majority of patients diagnosed between ages of 20 and 45. Patients with CVID have both acute and chronic recurrent infections. Commonly, they have pneumonia, sinusitis, and otitis media secondary to encapsulated bacteria such as *Streptococcus pneumoniae* and *Haemophilus influenzae*. They also tend to have septic arthritis, meningitis, and gastrointestinal infections. Gastrointestinal infections can occur secondary to *Giardia lamblia*, *Salmonella* spp., *Campylobacter* spp., *Clostridium difficile*, and *Yersinia enterocolitica*.

CVID patients who have T cell immunodeficiency can develop opportunistic infections from fungi, parasites, viruses, and mycobacteria including *Toxoplasma* spp., *Aspergillus* spp., *Pneumocystis jirovecii*, cytomegalovirus, and adenovirus.

Physical examination findings in patients with CVID can be normal. However, there can be signs of complications associated with CVID including lymphadenopathy, splenomegaly, arthritis, and skin changes associated with autoimmune diseases such as vitiligo. Lung exam in patients with chronic lung disease may reveal wheezing, crackles, and use of accessory muscles for respiration. Patients with chronic lung disease may also have digital clubbing.

Failure to thrive, developmental delay, depression has been more commonly reported in pediatric patients with CVID than adult patients. Whereas adults with CVID are more likely to have symptoms of chronic illness such as fatigues, aches, and arthralgia.

Question 4
The patient later develops a disease which is associated with worse prognosis of CVID and a higher rate of lymphoproliferative disease. What is the most likely diagnosis?

A. Autoimmune cytopenia
B. Bronchiectasis
C. Granulomatous–lymphocytic interstitial lung disease
D. Inflammatory bowel disease
E. Pernicious anemia

Answer and Explanation
Answer: C
Granulomatous–lymphocytic interstitial lung disease (GLILD) defines as granulomatous infiltrations along with lymphoid interstitial disease of the lungs. It is associated with worse prognosis and higher rate of lymphoproliferative disease in CVID.

Chronic lung disease with bronchiectasis is secondary to recurrent infections and is a common complication of CVID. Inflammatory bowel disease and autoimmune cytopenias are also common complications of CVID. Autoimmune gastrointestinal diseases in CVID also include pernicious anemia, which is caused by antibodies to intrinsic factor in the gastric parietal cells leading to vitamin B12 malabsorption and atrophic gastritis. However, answer choices A, B, D, and E are not associated with higher rate of lymphoproliferative diseases.

In a cohort of CVID patients who were studied for their morbidity and mortality, the authors found that CVID patients with certain noninfectious complications have reduced survival when compared to CVID patients without these complications. These noninfectious complications include gastrointestinal disease, lymphoma, and chronic lung disease. Importantly, the study found that lung disease, specifically granulomatous lung disease, was the main cause of death (36.5%), followed by lymphoma (18%), other cancers (10.7%), liver disease (8.6%), and other infections (5.3%). Inflammatory bowel disease and autoimmune disease were not found to be a cause of death in this cohort of patients.

Question 5
The patient also develops bronchiectasis due to her history of recurrent pneumonias. Which treatment is recommended?

A. Interferon gamma injection
B. Immunoglobulin replacement with IgG trough levels of 400–700 mg/dL
C. Immunoglobulin replacement with IgG trough levels of 800–1000 mg/dL
D. Systemic corticosteroid
E. TNF alpha inhibitor

Answer and Explanation
Answer: C
Patients with CVID with lung complications should maintain a higher trough level of IgG. Interferon gamma therapy is used in patients with chronic granulomatous diseases, mendelian susceptibility to mycobacterial diseases, and in select cases of hyper IgE syndrome. Autoinflammatory syndromes are treated with anti-inflammatory agents such as TNF alpha inhibitors and corticosteroids. CVID patients with GLILD can be resistant to steroid and immunoglobulin replacement even with maintaining higher IgG trough levels. TNF inhibitors rituximab and azathioprine have been used in these cases with various improvement.

3.2.1 Pulmonary Disease

Chronic lung disease is common in patients with CVID and is observed in approximately 20% of the patients. It is a cause of morbidity and mortality in CVID patients. Recurrent infections lead to inflammation and damage of lung tissues, which results in bronchiectasis and interstitial lung disease. The CVID subgroup with low IgM memory B cells and reduced levels of anti-pneumococcal polysaccharide IgM antibodies is at increased risk for chronic lung disease.

Obstructive lung disease such as asthma is seen in 9–15% of CVID patients.

Various forms of inflammatory lung diseases occur in 30–60% of CVID patients and are diagnosed by high resolution computed tomography (HRCT). Interstitial lung disease contributes to significant morbidity and mortality in patients with CVID. It is important to know that interstitial lung disease is commonly associated with previous autoimmune cytopenia, lymphadenopathy, and splenomegaly. Interstitial lung disease has benign lymphoproliferative pathology, including follicular bronchiolitis, lymphocytic interstitial pneumonitis, and nodular lymphoid hyperplasia. Some patients (5–22%) also develop non-caseating granulomatous infiltrations of the lung, lymphoid tissues, or liver. These granulomas can be difficult to differentiate from those found in sarcoidosis.

When granulomatous infiltrations along with lymphoid interstitial disease are present, this disease entity is called granulomatous-lymphocytic interstitial lung disease (GLILD). GLILD occurs despite immunoglobulin treatment and is difficult to treat. Patients with hypersplenism and polyarthritis are at increased risk for GLILD. Patients with GLILD are at increased risk for lymphoproliferative diseases and tend to have a worse prognosis. GLILD is a significant cause of morbidity and mortality in CVID patients.

HRCT of the chest is more sensitive than a chest radiograph in detecting interstitial lung disease, GLILD, or bronchiectasis. It is preferred in patients with CVID who have parenchymal lung defects. HRCT of the chest should be obtained during the initial evaluation if clinically suspected, and for monitoring of disease progression. Since radiation exposure should be limited, HRCT is not recommended to be done frequently or annually. Pulmonary function tests with spirometry should also be performed at the time of diagnosis and at regular intervals to monitor for changes in respiratory symptoms. Pulmonary function test with diffuse capacity for carbon monoxide can also help evaluate lung impairment. Open lung or video-assisted thoracoscopic biopsy may be required to identify the type of cellular infiltrate.

GLILD as discussed above is challenging to treat. It is resistant immunoglobulin replacement therapy. Low doses of systemic steroid are used to treat acutely, however, they have limited long-term benefit. Steroid sparing agents and TNF inhibitors have had some success but data is lacking. In some patients, cyclophosphamide, cyclosporine, or a combination of rituximab and azathioprine has been of benefit.

As discussed with XLA previously, immunoglobulin therapy is essential to preventing respiratory tract infections which lead to chronic lung disease. However, chronic lung disease can develop despite high dose immunoglobulin replacement therapy. Higher trough IgG levels are recommended for CVID patients with pulmonary diseases. IgG trough levels of 800–1000 mg/dL have been suggested to reduce infectious lung disease in CVID patients.

Management of bronchiectasis and chronic lung disease includes early and aggressive treatment of infections with appropriate antimicrobials, which may require a prolonged course up to 10–14 days to prevent recurrence. Macrolide antibiotics are recommended in patients with antibody deficiencies with bronchiectasis who have an infectious exacerbation. Azithromycin has shown benefit in reducing respiratory exacerbation and chronic infection-related pulmonary disease in CVID patients. Bronchiectasis is managed by mobilizing pulmonary secretions through chest physiotherapy and use of albuterol. Patients who develop severe bronchiectasis and chronic lung disease may need lung transplantation.

3.2.2 Autoimmune Disease

About 20–25% of patients with CVID develop autoimmune conditions. Pathogenesis of autoimmune disease is associated with B cell tolerance defect and the inability to counter select self-reactive B cell clones.

Autoimmune cytopenia which includes immune thrombocytopenia and autoimmune hemolytic anemia occurs in 4–20% of patients with CVID. Autoimmune cytopenia is the most common autoimmune disorder observed in CVID patients. The diagnosis relies on persistently abnormal complete blood counts with review of peripheral blood smear. Treatment is the same as in patients without CVID. Referral to hematologist is recommended. Management generally includes IV steroids followed by moderate doses of oral steroids or high dose IVIG. Patients with recurrent cytopenias may need chronic immunoglobulin therapy to maintain consistent IgG trough level of 700 mg/dL or more. The use of thrombopoietin receptor agonists, rituximab, azathioprine, or mycophenolate mofetil is reported to treat refractory episodes, especially to avoid splenectomy.

Autoimmune rheumatological diseases, most frequently rheumatoid arthritis, are reported in patients with CVID. Systemic lupus erythematosus, vasculitis, Sjögren's syndrome, and myositis are less commonly reported. Having other inflammatory complications, malignancy, higher baseline IgG levels, formation of autoantibodies, or family history of autoimmune diseases appear to be associated with autoimmune rheumatologic diseases in CVID patients. Screening for rheumatological disease through a thorough review of systems and referral to a rheumatologist as needed is recommended. Detailed joint and skin exam along with relevant imaging (X-ray or magnetic resonance imaging) are utilized to make diagnosis. Treatment for rheumatological diseases is the same as in patients without CVID. This includes immunosuppressive medications such as methotrexate or cyclophosphamide, and anti-TNF agents or anti-CD20 (rituximab) are considered.

Although rare, autoimmune dermatological conditions such as alopecia, psoriasis, and vitiligo are reported in patients with CVID. A thorough skin exam and prompt referral to dermatology when concerned is recommended.

Autoimmune gastrointestinal disease in CVID is discussed next in gastrointestinal disease.

3.2.3 Gastrointestinal Disease

Inflammatory bowel disease (IBD) like colitis is found in about 10% of CVID patients with severe enteropathy. It is suggested that this is due to defects in cellular immunity. These patients present with chronic diarrhea, weight loss, malabsorption, blood loss, and abdominal pain. CVID associated colitis is challenging to treat and sometimes not responsive to standard treatment of inflammatory bowel disease. As with many other complications associated with CVID, IBD-like colitis can develop despite immunoglobulin replacement therapy. Management includes referral to gastroenterology, nutritional support, antibiotics for bacterial overgrowth, immunosuppressants, and corticosteroids. Anti-inflammatory biological therapies such as anti-tumor necrosis factor alpha, anti-IL12, and anti-IL 23 agents have been used.

Enteropathy in CVID tends to occur with autoimmunity, lymphadenopathy, and splenomegaly. Chronic enteropathy can further cause loss of serum IgG and low serum albumin. This CVID-related enteropathy appears similar to the enteropathy seen in celiac disease with short villi, crypt hyperplasia, and intraepithelial lymphocytes. These patients present with malnutrition, weight loss, and chronic diarrhea which leads to increased IgG requirement and replacement.

Prolonged diarrhea can cause deficiencies in fat soluble vitamins (Vitamin A, D, E, and K), iron, calcium, and zinc. Nutritional supplementation may be required. Consultation with a clinical dietician may be beneficial. In patients with severe enteropathy or colitis, total parenteral nutrition may be considered.

Liver dysfunction is seen in 10% of patients with CVID. When compared to those without liver disease, CVID patients with liver disease have reduced survival and increased mortality.

Nodular regenerative hyperplasia (NRH) secondary to intra-hepatic vasculopathy is the most common form of liver disease in patients with CVID found in 5–12%. NRH leads to hepatocyte injury and regeneration creating perisinusoidal fibrosis. Patients with NRH also have intrasinusoidal inflammatory CD8+ T cell mediated infiltrates leading to non-necrotizing granuloma formation. NRH has been associated with autoimmune cytopenia, lymphoproliferation, and granulomatous disease. Patients with NRH present with jaundice, hepatomegaly, pruritis, ascites, and esophageal varices. They are found to have elevated alkaline phosphatase and gamma-glutamyl-transpeptidases. Liver biopsy is needed for diagnosis. There is no medical treatment to halt the progression of NRH and only liver transplant can allow for long-term survival. However, risk associated with long-term immunosuppression with liver transplant in CVID patients needs to be considered.

Infectious hepatitis B and C secondary to iatrogenic contamination of immunoglobulin preparations have been reported in the past. There is no definitive evidence to suggest that CVID patients with iatrogenically infected HBV or HCV are at increased risk for mortality when compared to the general population. With improved de-contamination of blood products, more recent studies show a lower prevalence of viral hepatitis in CVID patients.

In the general population, *Helicobacter pylori* infections and pernicious anemia are associated with increased risk for gastric cancer. Autoimmune gastrointestinal diseases in CVID also include pernicious anemia, which is caused by antibodies to intrinsic factor in the gastric parietal cells leading to vitamin B12 malabsorption and atrophic gastritis. Atrophic gastritis can lead to achlorhydria which is found to be more severe in CVID patients placing them at higher risk for dysplasia and carcinoma. Management of pernicious anemia includes Vitamin B12 replacement, monitoring for *Helicobacter pylori*, and evaluating for malignant gastric mucosal changes.

H. pylori can cause chronic active gastritis in patients with CVID. When compared to the general population, *H. pylori* infection rates in CVID patients are not increased. *H. pylori* is a gram-negative bacterium and is associated with development of gastritis, ulcers, gastric carcinoma, and lymphoma. Patients with *H. pylori* infection may present with acute or chronic dyspepsia, iron deficiency anemia, immune thrombocytopenia, or pernicious anemia secondary to atrophic gastritis. *H. pylori* infection can also be asymptomatic. Patients with CVID may lack the specific antibodies needed to kill *H. pylori* placing them at increased risk for gastritis and gastric dysplasia leading to gastric carcinoma.

Diagnosis of *H. pylori* is made by urea breath test (gold standard), fecal antigen immunoassay, or endoscopic biopsy. It is recommended by certain experts outside of the United States that all patients with CVID should be tested for *H. pylori* with urea breath test (UBT) at diagnosis and screen for signs of dyspepsia, weight loss, iron deficiency anemia, and vitamin B12 deficiency at follow up. If positive UBT, then treat *H. pylori* infection with triple therapy of proton pump inhibitor, amoxicillin and clarithromycin for 14 days.

3.2.4 Lymphoma and Cancers

Patients with CVID are susceptible to lymphoma and gastric cancer. A Scandinavian study notes association with higher IgM with development of lymphoma and lymphoid hyperplasia. One-third of patients with CVID develop a lymphoproliferative disorder that is characterized by lymphadenopathy, splenomegaly, or intestinal lymphoid hyperplasia. This lymphoproliferative disorder is associated with increased likelihood of developing B cell malignancies and poor prognosis.

There is also a relationship between non-Hodgkin lymphoma (NHL) and congenital immunodeficiency. Patients with CVID have a 438-fold increased likelihood of developing NHL and ten-fold increased likelihood of developing gastric cancer than the general public.

A recent study showed incidence of gastric cancer in CVID to be 0.6%, lower than previously thought. The mechanism is not completely understood but as discussed previously *H. pylori* infection and pernicious anemia place patients with CVID at increased risk for gastric cancer. This is due to gastritis leading to gastric dysplasia and thereby carcinoma. Regular esophageal duodenoscopy (EGD) is recommended in patients with concern for gastritis or gastric dysplasia along with close follow-up with gastroenterology.

Prevalence of non-Hodgkin lymphoma in patients with CVID was found to be about 1.8–8.2%. Patients with NHL present with fever, weight loss, night sweats, lymphadenopathy, hepatomegaly, or splenomegaly. Lymphomas in CVID are generally extranodal and of B cell origin. They are well differentiated, secrete antibodies, and found predominantly in females. Frequent cases of mucosa-associated lymphoid tissue (MALT) lymphoma have been reported. In patients with CVID, new enlarged lymph nodes or hepatosplenomegaly should be evaluated for lymphoma. Treatment is the same as in patients without CVID with chemotherapy and rituximab. In a cohort of CVID patients followed over four decades found that lymphoma was the second most common cause of death. Furthermore, a history of lymphoma is associated with reduced survival rates in patients with CVID when compared to age and sex matched controls.

3.3 Diagnosis of CVID

CVID is characterized by hypogammaglobulinemia and poor specific antibody response. Therefore, measurement of immunoglobulin levels and response to vaccinations are evaluated.

In patients with CVID, serum IgG is below two standard deviations of the mean (the lower limit of normal) and at least less than 450–500 mg/dL. IgA and/or IgM are also below the lower limit of normal and at least less than the 5th percentile. A recent review of monogenic etiology of CVID suggests the utility of measuring serum IgE as patients with CVID are more likely to have undetectable IgE. If serum IgE is high, then it is unlikely to be CVID and alternative diagnoses should be considered.

It is also valuable to remeasure serum immunoglobulin levels as they can fluctuate over time and vary between individuals. Immunoglobulin levels can also vary with various medication use, infections, or chronic illnesses. In CVID, hypogammaglobulinemia is persistent without any secondary etiologies. In some patients whose serum IgG and/or IgA and IgM levels are just outside of the reference range, this antibody deficiency can progress or resolve. Therefore, it is reasonable to monitor these patient's antibody levels and function. Isohemagglutinins, which are IgM antibodies against ABO blood group antigens, are absent or decreased in patients with CVID.

In patients with CVID, response to protein and polysaccharide vaccines is decreased. Responsiveness is not affected by steroid therapy or illness. Generally, IgG antibodies to tetanus, diphtheria, and either the 14 or 23 serotypes of polysaccharide pneumococcal vaccine are measured.

Flow cytometry to measure memory B cells and switched memory B cells may be useful in characterizing CVID as discussed previously.

Genetic studies may be useful in patients with CVID as 10% of cases are familial, and 20–30% are associated with monogenic defects, particularly in those with lymphoproliferative disease, autoimmune disease, and B cell lymphopenia. Genetic testing may be useful for disease prognosis, targeted treatment, family planning, and excluding other causes of immunodeficiencies.

Of note, it is difficult to diagnose CVID in children younger than 4 years of age as low immunoglobulin levels can be from transient hypogammaglobulinemia of infancy (THI). In both THI and CVID, there is hypogammaglobulinemia. However, children with THI can produce specific antibodies but patients with CVID cannot. Normal children less than 2 years of age may have poor response to polysaccharide vaccinations making it difficult to differentiate between CVID and THI. THI resolves with age but CVID does not. Therefore, monitoring patients less than 4 years of age with concern for CVID and considering immunoglobulin replacement therapy when appropriate is essential.

There is no universal definition of CVID but various criteria have been proposed that slightly differ. Table 4 lists the diagnostic criteria based on the Joint Task Force of the American Academy of Allergy, Asthma and Immunology and the American College of Allergy, Asthma and Immunology on the Practice parameter for the diagnosis and management of primary immunodeficiency.

Table 4 Diagnostic criteria for CVID

Diagnostic criteria for CVID
– Patients older than 4 years of age
– Decreased serum IgG and decreased serum IgA *or* IgM
– Poor antibody response to immunizations
– Exclude primary or secondary immunodeficiencies that cause antibody deficiency

3.3.1 Differential Diagnosis of CVID

The differential diagnosis of CVID includes the following:

3.3.2 Secondary Hypogammaglobulinemia

Hypogammaglobulinemia secondary to other disorders can occur due to several etiologies as listed below:

- Decreased production of immunoglobulins secondary to immunosuppressant medications, such as glucocorticoids, anti-CD20 therapy (rituximab), antiepileptics.
- Decreased production of immunoglobulins due to malignancy involving bone marrow.
- Increased loss of immunoglobulins secondary to protein losing enteropathy, nephrotic syndrome, burns, chylothorax/lung effusions, and other causes of fluid losses including trauma.

3.3.3 Primary Hypogammaglobulinemia

As discussed previously in Sect. 2, primary hypogammaglobulinemia can occur due to X-linked agammaglobulinemia, autosomal recessive agammaglobulinemia, and transient hypogammaglobulinemia of infancy.

3.3.4 Specific Antibody Deficiency

Similar to CVID, specific antibody deficiency is characterized by recurrent sinopulmonary infections and decreased response to polysaccharide and/or protein antigens. However, unlike CVID, in specific antibody deficiency patients have normal immunoglobulin levels.

3.3.5 IgG Subclass Deficiency

Specifically, IgG1 which comprises 75% of serum IgG can cause hypogammaglobulinemia. These patients with IgG1 deficiency can have poor vaccine response.

3.3.6 Hyperimmunoglobulin M Syndrome (HIGM)

HIGM can occur secondary to various defects. Patients with HIGM can have recurrent and severe sinopulmonary infections, hypogammaglobulinemia with low IgG and IgA, normal to elevated IgM, and poor response to vaccine antigens. The X-linked form of the disease is caused by a deficiency in CD40 ligand which prevents isotype switching from IgM to IgA or IgG. Other causes are due defects in activation-induced cytidine deaminase (AID), uracil nucleoside glycosylase (UNG), or CD40 deficiency which also cause defects in class switching.

3.3.7 Combined Immunodeficiencies

Combined immunodeficiencies can be difficult to differentiate from CVID. There are mild forms of adenosine deaminase deficiency (ADA) and recombination activating gene (RAG) mutations, and Artemis defects that can present similarly to CVID.

As discussed previously, there are a subset of CVID patients who have decreased CD4+ T cells, present with opportunistic infections, and are considered to have late-onset combined immunodeficiency (LOCID).

3.4 Management of CVID

The mainstay of treatment for CVID is with immunoglobulin replacement and antimicrobials for infections.

Immunoglobulin can be administered intravenously or subcutaneously. The recommended dose is 400–600 mg/kg every 3–4 weeks to maintain IgG trough levels of >500 mg/dL and as high as 800–1000 mg/dL for patients with lung damage/chronic lung disease. Higher doses are also recommended for refractory infections, enteropathy, and autoimmune cytopenias. It is to be noted that clinicians should focus on optimal doses of IgG to reduce infections rather than for certain trough levels.

With the multitude of clinical complications associated with CVID, close monitoring is necessary as discussed individually in previous sections. For pulmonary complications, routine spirometry and lung imaging should be obtained as clinically indicated. If gastrointestinal symptoms such as diarrhea, weight loss, and abdominal pain arise, then a referral to a gastroenterologist for endoscopy and imaging should be considered. Routine laboratory for CVID should include monitoring for autoimmune cytopenias. Age-appropriate cancer screening is vital.

3.5 Prognosis

Mortality in patients with CVID has improved with the introduction of immunoglobulin replacement therapy. A large cohort of CVID patients followed for over 40 years revealed mortality rate to be 19.6% with reduced survival compared with age and sex matched controls. Patients with any noninfectious complications, including gastrointestinal disease, lymphoma, and chronic lung disease, had decreased survival rate. However, these reduced survival rates were not seen in CVID patients with a history of autoimmunity. Lung disease (36.5%) appears to be the main cause of death followed by lymphoma (18%), then other cancers (10.7%), liver disease (8.6%), and other infections (5.3%).

4 Summary

Common B cell immunodeficiency disorders that were discussed in this chapter include X-linked agammaglobulinemia (XLA) and Common Variable Immunodeficiency (CVID).

X-linked agammaglobulinemia is caused by a mutation in the *BTK* gene and is characterized by absence of mature B cells and serum immunoglobulins. XLA should be suspected in male infants 6–12 months of age with recurrent sinopulmonary infections, a positive family history, and physical exam finding of absence of lymphoid tissues. Laboratory findings in patients with XLA are notable for low to undetectable immunoglobulins, absence of B lymphocytes (CD19+) in the peripheral blood, and low antibody response to antigens or immunizations. Management for XLA is immunoglobulin replacement therapy (400–500 mg/kg every 3–4 weeks to maintain an IgG trough level of greater than 500 mg/dL) and antimicrobial treatments for infections. Live viral vaccinations are contraindicated in patients with XLA. Chronic lung disease and bronchiectasis are common in patients with XLA and a strong cause of mortality in these patients. Pulmonary status is monitored with routine pulmonary function tests and high resolution chest tomography. Immunoglobulin therapy is essential to preventing respiratory tract infections that could lead to chronic lung disease. However, chronic lung disease can develop despite high dose immunoglobulin therapy. Macrolides and pulmonary physical therapy are used for management for acute respiratory exacerbations in XLA patients with bronchiectasis.

Common variable immunodeficiency is a group of disorders with no unifying genetic etiology and is characterized by hypogammaglobulinemia and decreased specific antibody production. CVID should be suspected in patients greater than 4 years of age who have recurrent sinopulmonary and gastrointestinal infections. CVID is characterized by impaired B cell differentiation into plasma cells that secrete immunoglobulins. There are reduced levels of immunoglobulins, reduced isohemagglutinins, and abnormal response to protein and polysaccharide antigens. CVID can be divided into subgroups based on their B cell subset characterization. Patients with CVID who have low class switched memory B cells are more likely to have autoimmune diseases, in particular autoimmune hemolytic anemia and autoimmune thrombocytopenia, splenomegaly, lymphadenopathy, and lymphoproliferative disorders. Low class switched memory B cells and increased $CD19+CD21^{low}$ B cells are also associated with granulomatous disease in patients with CVID. Some studies have found that reduced class switched memory B cells and non-switched IgM memory B cells are associated with chronic lung disease. Patients with CVID are at risk for pulmonary complications with lymphoid interstitial pneumonitis, non-caseating granulomatous infiltrations of the lung, and granulomatous lymphocytic interstitial lung disease. Patients with CVID can develop autoimmune conditions including autoimmune hemolytic anemia, immune thrombocytopenia, rheumatoid arthritis, and vitiligo. CVID patients commonly have gastrointestinal disease complications including inflammatory bowel disease like colitis, enteropathy, and nodular regenerative hyperplasia of the liver. CVID patients are at increased risk for gastrointestinal malignancies due to their susceptibility for pernicious anemia and *Helicobacter pylori* gastritis. They also have a 438-fold increased likelihood of developing non-Hodgkin lymphoma. Management of CVID is with immunoglobulin replacement therapy (400–600 mg/kg every 3–4 weeks to maintain IgG trough levels of >500 mg/dL and as high as 800–1000 mg/dL for patients with lung damage/chronic lung disease) and antimicrobials for infections. Monitoring for multitude of clinical complications associated with CVID is recommended.

Bibliography

1. Winkelstein JA, Marino MC, Lederman HM, Jones SM, Sullivan K, Burks AW, et al. X-linked agammaglobulinemia: report on a United States registry of 201 patients. Medicine (Baltimore). 2006;85(4):193–202.
2. Abbas AK, Abbas AK, Lichtman AH, Pillai S. Cellular and molecular immunology. Philadelphia: Elsevier; 2018.
3. Aghamohammadi A, Notarangelo LD, Rezaei N, editors. Primary immunodeficiency diseases: definition, diagnosis, and management. 2nd ed. Berlin: Springer; 2017.
4. Ochs HD, Smith CI. X-linked agammaglobulinemia. A clinical and molecular analysis. Medicine (Baltimore). 1996;75(6):287–99.
5. Conley ME, Howard V. Clinical findings leading to the diagnosis of X-linked agammaglobulinemia. J Pediatr. 2002;141(4):566–71.
6. Conley ME, Rohrer J, Minegishi Y. X-linked agammaglobulinemia. Clin Rev Allergy Immunol. 2000;19(2):183–204.
7. Burks AW, Holgate ST, O'Hehir RE, Broide DH, Bacharier LB, Hershey GKK, et al. Middleton's allergy: principles and practice. 9th ed. Edinburgh: Elsevier; 2020.
8. Dittrich A-M, Schulze I, Magdorf K, Wahn V, Wahn U. X-linked agammaglobulinaemia and Pneumocystis carinii pneumonia—an unusual coincidence? Eur J Pediatr. 2003;162(6):432–3.
9. Autenrieth IB, Schuster V, Ewald J, Harmsen D, Kreth HW. An unusual case of refractory Campylobacter jejuni infection in a patient with X-linked agammaglobulinemia: successful combined therapy with maternal plasma and ciprofloxacin. Clin Infect Dis. 1996;23(3):526–31.
10. Shapiro RS. Malignancies in the setting of primary immunodeficiency: implications for hematologists/oncologists. Am J Hematol. 2011;86(1):48–55.
11. Lavilla P, Gil A, Rodríguez MC, Dupla ML, Pintado V, Fontán G. X-linked agammaglobulinemia and gastric adenocarcinoma. Cancer. 1993;72(5):1528–31.
12. Bachmeyer C, Monge M, Cazier A, Le Deist F, de Saint BG, Durandy A, et al. Gastric adenocarcinoma in a patient with X-linked agammaglobulinaemia. Eur J Gastroenterol Hepatol. 2000;12(9):1033–5.
13. Echave-Sustaeta JM, Villena V, Verdugo M, López-Encuentra A, de Agustín P, Alberti N. X-linked agammaglobulinaemia and squamous lung cancer. Eur Respir J. 2001;17(3):570–2.
14. Plebani A, Soresina A, Rondelli R, Amato GM, Azzari C, Cardinale F, et al. Clinical, immunological, and molecular analysis in a large cohort of patients with X-linked agammaglobulinemia: an Italian multicenter study. Clin Immunol. 2002;104(3):221–30.
15. Hermaszewski RA, Webster AD. Primary hypogammaglobulinaemia: a survey of clinical manifestations and complications. Q J Med. 1993;86(1):31–42.

16. Hernandez-Trujillo VP, Scalchunes C, Cunningham-Rundles C, Ochs HD, Bonilla FA, Paris K, et al. Autoimmunity and inflammation in X-linked agammaglobulinemia. J Clin Immunol. 2014;34(6):627–32.
17. Barmettler S, Otani IM, Minhas J, Abraham RS, Chang Y, Dorsey MJ, et al. Gastrointestinal manifestations in x-linked agammaglobulinemia. J Clin Immunol. 2017;37(3):287–94.
18. Plebani A, Fischer MB, Meini A, Duse M, Thon V, Eibl MM. T cell activity and cytokine production in X-linked agammaglobulinemia: implications for vaccination strategies. Int Arch Allergy Immunol. 1997;114(1):90–3.
19. Pituch-Noworolska A, Zwonarz K, Błaut-Szlósarczyk A, Szaflarska A, Kowalczyk D, Siedlar M. T lymphocytes and NK cells in X-linked agammaglobulinemia. Przegl Lek. 2013;70(12):1048–50.
20. Bonilla FA, Khan DA, Ballas ZK, Chinen J, Frank MM, Hsu JT, et al. Practice parameter for the diagnosis and management of primary immunodeficiency. J Allergy Clin Immunol. 2015;136(5):1186–1205.e1–78.
21. Conley ME, Notarangelo LD, Etzioni A. Diagnostic criteria for primary immunodeficiencies. Representing PAGID (Pan-American group for immunodeficiency) and ESID (European society for immunodeficiencies). Clin Immunol. 1999;93(3):190–7.
22. Quartier P, Debré M, De Blic J, de Sauverzac R, Sayegh N, Jabado N, et al. Early and prolonged intravenous immunoglobulin replacement therapy in childhood agammaglobulinemia: a retrospective survey of 31 patients. J Pediatr. 1999;134(5):589–96.
23. Bearden D, Collett M, Quan PL, Costa-Carvalho BT, Sullivan KE. Enteroviruses in x-linked agammaglobulinemia: update on epidemiology and therapy. J Allergy Clin Immunol Pract. 2016;4(6):1059–65.
24. Ikegame K, Imai K, Yamashita M, Hoshino A, Kanegane H, Morio T, et al. Allogeneic stem cell transplantation for X-linked agammaglobulinemia using reduced intensity conditioning as a model of the reconstitution of humoral immunity. J Hematol Oncol. 2016;9:9.
25. Kerns HM, Ryu BY, Stirling BV, Sather BD, Astrakhan A, Humblet-Baron S, et al. B cell-specific lentiviral gene therapy leads to sustained B-cell functional recovery in a murine model of X-linked agammaglobulinemia. Blood. 2010;115(11):2146–55.
26. Abolhassani H, Hirbod-Mobarakeh A, Shahinpour S, Panahi M, Mohammadinejad P, Mirminachi B, et al. Mortality and morbidity in patients with X-linked agammaglobulinaemia. Allergol Immunopathol (Madr). 2015;43(1):62–6.
27. Tarzi MD, Grigoriadou S, Carr SB, Kuitert LM, Longhurst HJ. Clinical immunology review series: an approach to the management of pulmonary disease in primary antibody deficiency. Clin Exp Immunol. 2009;155(2):147–55.
28. Milito C, Pulvirenti F, Cinetto F, Lougaris V, Soresina A, Pecoraro A, et al. Double-blind, placebo-controlled, randomized trial on low-dose azithromycin prophylaxis in patients with primary antibody deficiencies. J Allergy Clin Immunol. 2019;144(2):584–593.e7.
29. Lougaris V, Soresina A, Baronio M, Montin D, Martino S, Signa S, et al. Long-term follow-up of 168 patients with X-linked agammaglobulinemia reveals increased morbidity and mortality. J Allergy Clin Immunol. 2020;146(2):429–37.
30. Gathmann B, Mahlaoui N, CEREDIH, Gérard L, Oksenhendler E, Warnatz K, et al. Clinical picture and treatment of 2212 patients with common variable immunodeficiency. J Allergy Clin Immunol. 2014;134(1):116–26.
31. Romberg N, Lawrence MG. Birds of a feather: common variable immune deficiencies. Ann Allergy Asthma Immunol. 2019;123(5):461–7.
32. Bogaert DJA, Dullaers M, Lambrecht BN, Vermaelen KY, De Baere E, Haerynck F. Genes associated with common variable immunodeficiency: one diagnosis to rule them all? J Med Genet. 2016;53(9):575–90.
33. Ahn S, Cunningham-Rundles C. Role of B cells in common variable immune deficiency. Expert Rev Clin Immunol. 2009;5(5):557–64.
34. Tam JS, Routes JM. Common variable immunodeficiency. Am J Rhinol Allergy. 2013;27(4):260–5.
35. Mannina A, Chung JH, Swigris JJ, Solomon JJ, Huie TJ, Yunt ZX, et al. Clinical predictors of a diagnosis of common variable immunodeficiency-related granulomatous-lymphocytic interstitial lung disease. Ann Am Thorac Soc. 2016;13(7):1042–9.
36. Warnatz K, Denz A, Dräger R, Braun M, Groth C, Wolff-Vorbeck G, et al. Severe deficiency of switched memory B cells (Cd27(+) igm(−)igd(−)) in subgroups of patients with common variable immunodeficiency: a new approach to classify a heterogeneous disease. Blood. 2002;99(5):1544–51.
37. Piqueras B, Lavenu-Bombled C, Galicier L, Bergeron-van der Cruyssen F, Mouthon L, Chevret S, et al. Common variable immunodeficiency patient classification based on impaired B cell memory differentiation correlates with clinical aspects. J Clin Immunol. 2003;23(5):385–400.
38. Wehr C, Kivioja T, Schmitt C, Ferry B, Witte T, Eren E, et al. The EUROclass trial: defining subgroups in common variable immunodeficiency. Blood. 2008;111(1):77–85.
39. Sánchez-Ramón S, Radigan L, Yu JE, Bard S, Cunningham-Rundles C. Memory B cells in common variable immunodeficiency: clinical associations and sex differences. Clin Immunol. 2008;128(3):314–21.
40. Detková D, de Gracia J, Lopes-da-Silva S, Vendrell M, Alvarez A, Guarner L, et al. Common variable immunodeficiency: association between memory B cells and lung diseases. Chest. 2007;131(6):1883–9.
41. Filion CA, Taylor-Black S, Maglione PJ, Radigan L, Cunningham-Rundles C. Differentiation of common variable immunodeficiency from igg deficiency. J Allergy Clin Immunol Pract. 2019;7(4):1277–84.
42. Malphettes M, Gérard L, Carmagnat M, Mouillot G, Vince N, Boutboul D, et al. Late-onset combined immune deficiency: a subset of common variable immunodeficiency with severe T cell defect. Clin Infect Dis. 2009;49(9):1329–38.
43. Oksenhendler E, Gérard L, Fieschi C, Malphettes M, Mouillot G, Jaussaud R, et al. Infections in 252 patients with common variable immunodeficiency. Clin Infect Dis. 2008;46(10):1547–54.
44. Sanchez LA, Maggadottir SM, Pantell MS, Lugar P, Rundles CC, Sullivan KE, et al. Two sides of the same coin: pediatric-onset and adult-onset common variable immune deficiency. J Clin Immunol. 2017;37(6):592–602.
45. Cunningham-Rundles C, Bodian C. Common variable immunodeficiency: clinical and immunological features of 248 patients. Clin Immunol. 1999;92(1):34–48.
46. Agarwal S, Cunningham-Rundles C. Autoimmunity in common variable immunodeficiency. Ann Allergy Asthma Immunol. 2019;123(5):454–60.
47. Uzzan M, Ko HM, Mehandru S, Cunningham-Rundles C. Gastrointestinal disorders associated with common variable immune deficiency (Cvid) and chronic granulomatous disease (Cgd). Curr Gastroenterol Rep. 2016;18(4):17.
48. Pecoraro A, Crescenzi L, Varricchi G, Marone G, Spadaro G. Heterogeneity of liver disease in common variable immunodeficiency disorders. Front Immunol. 2020;11:338.
49. Dhalla F, da Silva SP, Lucas M, Travis S, Chapel H. Review of gastric cancer risk factors in patients with common variable immunodeficiency disorders, resulting in a proposal for a surveillance programme. Clin Exp Immunol. 2011;165(1):1–7.
50. Cunningham-Rundles C. How I treat common variable immune deficiency. Blood. 2010;116(1):7–15.

51. Chapel H, Lucas M, Lee M, Bjorkander J, Webster D, Grimbacher B, et al. Common variable immunodeficiency disorders: division into distinct clinical phenotypes. Blood. 2008;112(2):277–86.
52. Lucas M, Lee M, Lortan J, Lopez-Granados E, Misbah S, Chapel H. Infection outcomes in patients with common variable immunodeficiency disorders: relationship to immunoglobulin therapy over 22 years. J Allergy Clin Immunol. 2010;125(6):1354–1360.e4.
53. Resnick ES, Moshier EL, Godbold JH, Cunningham-Rundles C. Morbidity and mortality in common variable immune deficiency over 4 decades. Blood. 2012;119(7):1650–7.
54. Quinti I, Soresina A, Spadaro G, Martino S, Donnanno S, Agostini C, et al. Long-term follow-up and outcome of a large cohort of patients with common variable immunodeficiency. J Clin Immunol. 2007;27(3):308–16.

T Cell Deficiency

Divya Seth and Pavadee Poowuttikul

Abbreviations

aCGH	Array comparative genomic hybridization
ADA	Adenosine deaminase
ARDS	Acute respiratory distress syndrome
BCG	Bacillus Calmette–Guerin
CHARGE syndrome	Coloboma, Heart anomaly, Choanal atresia, Retardation, Genital and ear anomalies
CID	Combined immunodeficiency diseases
CMCC	Chronic mucocutaneous candidiasis
CMV	Cytomegalovirus
COMT	Catechol-*O*-methyltransferase
CRAC	Calcium release activated component
CT	Computerized tomography
CXR	Chest X-ray
DGS	DiGeorge syndrome
DOCK-8	Dedicator of cytokinesis 8
ED	Emergency Department
FISH	Fluorescence in situ hybridization
GVHD	Graft versus host disease
HIV	Human immune deficiency virus
HLA	Human leukocyte antigen
HSCT	Hematopoietic stem cell transplant
IL2RG	Interleukin 2 receptor gamma
IL7R	Interleukin 7 receptor
IVIG	Intravenous Immunoglobulins
JAK3	Janus kinase 3
MLPA	Multiplex ligation-dependent probe amplification
NBS	Newborn screening
NK cells	Natural killer cells
OS	Omenn syndrome
PHA	Phytohemagglutinin
PID	Primary immune deficiency
PJP	*Pneumocystis jirovecii* pneumonia
PNP	Purine nucleoside phosphorylase
RAC-2	Ras-related C3 botulinum toxin substrate 2
RAG	Recombinase activating genes
RSV	Respiratory syncytial virus
RT-PCR	Real time polymerase chain reaction
SCID	Severe combined immune deficiency
SNP	Single nucleotide polymorphism
TCL	T cell lymphopenia
TCR	T cell receptor
TREC	T cell receptor circles
WAS	Wiskott Aldrich Syndrome

D. Seth (✉) · P. Poowuttikul
Department of Pediatrics, Children's Hospital of Michigan, Central Michigan University College of Medicine, Detroit, MI, USA
e-mail: dseth@dmc.org; ppoowutt@dmc.org

1 Introduction

T cell deficiency can be a Primary Immune Deficiency (PID) or secondary to medications, certain malignancies, severe infections, and nutritional deficiencies. T cell-mediated PIDs usually present in infancy or early childhood. They are associated with more grave clinical problems, including infections, than those seen in antibody deficiency diseases. In the absence of early diagnosis and treatment, survival is uncommon beyond infancy or childhood. There is a wide spectrum of T cell defects, ranging from severe combined immunodeficiency (SCID), which is associated with absent or low T cell function, to combined immunodeficiency diseases (CID) wherein there is inadequate T cell function. These diseases are diagnosed by reduced number of T cells and/or reduced proliferative response to polyclonal T cell activators such as phytohemagglutinin. SCID

M. Mahmoudi (ed.), *Absolute Allergy and Immunology Board Review*, https://doi.org/10.1007/978-3-031-12867-7_22

is the most severe form of PID and is characterized by severe T cell lymphopenia. SCID is the first PID that has been added to Newborn Screening (NBS) Panel to facilitate the disease early diagnosis and treatment. The most common treatment for SCID is hematopoietic stem cell transplant (HSCT). DiGeorge syndrome is a congenital disorder characterized by a triad of cellular immune deficiency, congenital cardiac anomalies, and hypocalcemia. DGS is associated with variable degree of T cell lymphopenia and proliferative responses. Complete DGS occurs due to thymic aplasia and is associated with severe T cell immunodeficiency similar to that seen in SCID. Thymus transplantation has shown to restore immune function in complete DGS. Partial DGS, on the other hand, occurs due to abnormal thymic migration. These patients have mild to moderate T cell lymphopenia.

Case Presentation 1

A 4-month-old, full-term boy presents to the emergency department (ED) with fever, failure to thrive and respiratory distress. The family has recently moved to the US. The patient was previously doing well per parents. There is no sick contact in the family. Birth history was non-significant. Mother received prenatal care. Human immune deficiency virus (HIV) test was negative at time of delivery. Chest X-ray obtained in the ED shows diffuse bilaterally symmetric interstitial opacification. Complete blood cell count shows severe lymphopenia. Laboratory tests are ordered for lymphocyte enumeration and mitogen proliferation. Results show normal absolute count of CD19+ cells, but nearly absent CD3+ cells and absent CD3−CD56+CD16+ cells. Mitogen proliferation response is also very low.

Question 1

What is the most likely diagnosis in this case?

1. DiGeorge syndrome
2. GATA2 deficiency
3. HIV infection
4. Idiopathic T cell lymphopenia
5. Severe combined immunodeficiency

Answer: E

Severe combined immunodeficiency (SCID) is a group of inherited immunodeficiency diseases which is characterized by markedly low number of T cells. T cell proliferation response to mitogens is absent or very low as seen in our patient here.

HIV infection can be associated with CD4+ T cell lymphopenia. Thus, HIV infection should be ruled out in all cases of T cell lymphopenia and suspected SCID. Maternal lab was negative for HIV in this case, thus making the diagnosis of HIV less likely.

Patients with partial DiGeorge syndrome have mild to moderate T cell lymphopenia and normal to low mitogen proliferation response, unlike this patient. Though patients with complete DGS can present like SCID, there are no other clinical characteristics that are supportive of DGS in this patient.

Patients with GATA-2 deficiency are susceptible to viral, fungal, mycobacterial infections, and malignancies. These patients may exhibit aplastic anemia and chronic neutropenia. T cell lymphopenia is not commonly encountered in GATA-2 deficiency patients, though B and Natural Killer (NK) cell lymphopenia may be observed. GATA-2 deficiency also presents in older children or adults compared to other classic immunodeficiencies.

Idiopathic T cell lymphopenia (TCL) is defined as infants without SCID, other genetic defects, or medical syndrome. TCL usually requires ongoing monitoring or treatment. In the absence of further testing, the diagnosis of idiopathic TCL cannot be established in this patient.

2 Severe Combined Immunodeficiency (SCID)

Severe combined immunodeficiency (SCID) is a group of inherited immunodeficiency diseases which is characterized by markedly low number of T cells. T cell proliferation response to mitogens is absent or very low. This results in complete absence of cell-mediated immunity and impaired humoral immunity. The prevalence can vary from 1 per 50,000 to 1 per 100,000 live births. As described in Case Presentation 1, patients born with SCID usually appear normal at birth. These patients are at increased risk of developing serious, life-threatening infections around 4–6 months of age after waning of transplacentally acquired maternal antibodies. In the absence of timely diagnosis and treatment, patients with SCID often succumb to life-threatening and opportunistic infections by 2 years of age. Thus, SCID can be considered a pediatric emergency.

Question 2

What is the most likely cause of the chest X-ray finding in this patient?

A. Acute respiratory distress syndrome (ARDS)
B. Graft versus host disease (GVHD)
C. *Mycobacterium tuberculosis* pneumonia
D. *Pneumocystis jirovecii* pneumonia (PJP)
E. Respiratory syncytial virus (RSV) pneumonia

Answer: D

Patients with SCID are prone to develop serious and fatal infections. Infections can be from fungi, especially from

Candida spp. and *Pneumocystis jirovecii*. PJP usually presents with respiratory difficulty in conjunction with other features such as failure to thrive, diarrhea, and thrush. Chest X-ray (CXR) in PJP shows bilateral homogeneous interstitial opacification as seen in the patient here. The opacifications can be reticular or granular. Chest computerized tomography (CT) scan can be done in case of normal CXR which can show extensive ground glass attenuation.

ARDS is defined as an acute, inflammatory lung injury that can result in respiratory failure. It may occur in patients with SCID related to a severe infection. However, ARDS usually presents with acute onset of respiratory distress with or without fever. Failure to thrive is unlikely related to ARDS given the acute onset of the event. In ARDS, CXR shows bilateral, somewhat asymmetrical consolidations, that are usually peripheral with air bronchograms. These findings are not described in this patient.

Presenting symptoms of GVHD, including skin rash, jaundice, and hepatitis, are not observed in this patient. GVHD is also less likely in this patient in the absence of prior history of transplant or blood transfusion. In GVHD, CXR findings can vary from mild perihilar or diffuse interstitial fibrosis to pulmonary cysts or nodules.

Mycobacterial infections are generally rare in SCID patients. However, fatal infections can occur in SCID patients related to Bacillus Calmette-Guerin (BCG), which is a live attenuated vaccine that is administered early in life. CXR findings in Mycobacterium pneumonia show infiltrates or consolidations with or without cavities in the upper lungs. This may be associated with mediastinal or hilar lymphadenopathy.

RSV infection in very young infants (less than 6 months old) may present as cough, apneic episodes, decreased activity, or appetite with or without fever. However, failure to thrive is unlikely in an acute setting of RSV infection. Furthermore, CXR findings in RSV pneumonia typically show air trapping, consolidation, or peribronchial thickening. CXR of the patient described here does not show these changes.

2.1 Clinical Manifestations of SCID

Patients with SCID often present with recurrent severe infections, failure to thrive, and chronic diarrhea. The most common infections reported in SCID are oral candidiasis (43%), viral infections (35.5%), and *Pneumocystis jirovecii* (26%) pneumonia. Viral infections include respiratory viruses such as parainfluenza, respiratory syncytial virus (RSV), adenovirus, measles, varicella, cytomegalovirus (both congenital and acquired), Epstein–Barr virus, herpes simplex virus, and rotavirus. Congenital cytomegalovirus (CMV) infection can initially be asymptomatic but if it not diagnosed and treated early on, it may result in serious and fatal disease in these patients.

Fatal infections can occur related to live attenuated vaccine such as rotavirus, oral poliovirus, measles, mumps, rubella, varicella as well as BCG vaccine. Live vaccines are thus contraindicated in patients with SCID.

On physical examination, patients often have signs of failure to thrive. Oral and/or genital candidiasis may be present. Lymphoid tissues such as tonsils and lymph nodes are absent. Some types of SCID are associated with skin rash which can particularly be severe as seen in Omenn syndrome. Skin rash may also be related to graft versus host disease (GVHD) which is caused by transplacental transfer of maternal T cells or donor lymphocytes present in blood products that are transfused. With this reason, SCID patients should receive only irradiated, leukocyte poor, and CMV negative blood products to minimize GVHD and/or CMV infection.

Question 3
What is the most likely deficiency in this patient?

A. ADA
B. IL2RG
C. IL7R
D. RAG1
E. JAK3

Answer: B

IL2RG is the most common genetic defect seen in SCID and is X-linked, unlike other genetic defects described in SCID. Our patient is a male infant with T−B+NK− phenotype which is consistent with IL2RG deficiency. The gamma chain of the IL 2 receptor (IL2R) is a common component of lymphocyte receptors for various cytokines including IL 2, 4, 7, 9, 15, and 21. Absence of IL 7 derived signals results in absence of T and NK cells. B cells number varies from normal to high.

JAK3 deficiency is associated with similar lymphocyte phenotype but is autosomal recessive in inheritance and thus can be seen in both males and females. JAK3 deficiency is also rare compared to IL2RG deficiency; therefore, the most likely deficiency in patient is IL2RG.

ADA defect is characterized by T−B−NK− phenotype. IL7R deficiency results in T−B+NK+ phenotype, while RAG deficiency leads to T−B−NK+ phenotype. Lymphocyte phenotypes of SCID are shown in Table 1.

2.2 Etiology of SCID

SCID is genetically heterogeneous. To date, mutations in more than 20 genes have been identified. Table 1 demon-

Table 1 Genetic defects in SCID and lymphocyte phenotypes

Type of SCID	Protein involved	Cell phenotype
X-linked SCID		
IL2RG deficiency (IL 2 receptor gamma)	Common γ-chain (γc) of receptors for IL 2, 4, 7, 9, 15 and 21	T−B+NK−
Autosomal recessive (AR) SCID		
JAK3 (Janus kinase) deficiency	JAK3	T−B+NK−
CD25 deficiency	IL2RA	T−B+NK−
ADA (adenosine deaminase) deficiency	ADA enzyme	T−B−NK−
IL7R deficiency	α-Chain of IL 7 receptor	T−B+NK+
RAG1 and RAG2 (recombinase activating genes) deficiency (leaky SCID)	RAG1 and RAG2 Antigen receptor gene rearrangement	T−B−NK+ Radiosensitive type
Cernunnos deficiency	NHEJ1	T−B−NK+ Radiosensitive type
Artemis deficiency	Artemis	T−B−NK+ Radiosensitive type
DNA ligase IV deficiency	DNA ligase IV	T−B−NK+ Radiosensitive type
Reticular dysgenesis	Adenylate kinase 2	T−B−NK−
TCR (T cell receptor) deficiency	TCR	T−B+NK+
CORO1a deficiency	CORO1a	T−B+NK+

strates types of SCID, involved genes and proteins, and lymphocyte phenotypes. All genetic mutations result in a marked decrease in circulating naïve T cells and subsequent impairment of cellular and humoral immunity. Involvement of B cells and NK cells is variable and is determined by underlying genetic defect. Most of SCID cases are X-linked disorder caused by deficiency in IL 2 receptor gamma (IL2RG). Since the initiation of newborn screening (NBS), a much higher incidence of SCID (1:58,000) is observed than previously reported (1:100,000). One study showed that only 19% of the infants had X-linked SCID as compared to the rate of 50% reported prior to the NBS. Autosomal recessive SCID is more prevalent in societies where consanguineous marriages are common.

It is important to know that only IL2RG deficiency causes X-linked SCID while other genetic defects lead to AR SCID. The lymphocyte phenotypes seen in each type of SCID is also of importance. Knowing SCID by the mechanism of defects will help understanding the lymphocyte phenotypes.

There are four classifications of SCID based on the mechanism of defects.

1. **Defect in Cytokine signaling—Absence of T and NK cells, normal to high B cells (T−B+NK− SCID)**
   (a) X-linked SCID

   Defects in the common gamma chain of the interleukin receptor (IL2RG, CD132) result in X-linked SCID. IL 2, 4, 7, 9, 15, and 21 share the common γ-chain (γc) receptor. IL 7 provides survival and proliferative signals for early T cell development while IL 15 is important for NK cell development. The defect in γc thus leads to the inhibition of both T and NK cell development which results T−NK− phenotype. B cells counts vary from normal to high. However, in the absence of T cell help, B cells function is impaired. As a result, both cell-mediated and humoral immunity are absent or markedly impaired in these patients.

   (b) Autosomal recessive SCID (JAK3 and CD25 deficiency)

   JAK3 is a signaling protein tyrosine kinase for γc and is essential for lymphoid cell development. Therefore, the lymphocyte phenotype observed in *JAK3* deficiency is like those seen in X-linked SCID (T−B+NK−). Mutation in *IL2Rα* (CD25) also results in T−B+NK− SCID.
2. **Defect in nucleotide salvage pathways—Absence of T, B, and NK cells (T−B−NK− SCID)**
   (a) ADA deficiency

   Mutation in adenosine deaminase (*ADA*) gene results in ADA-SCID which is a purine metabolism disorder. Premature lymphoid progenitor cells death occurs from toxicity due to accumulating toxic metabolites. This results in decrease in T, B, and NK cells. Lymphocyte cell counts may be normal at birth but there is a progressive decline in number of lymphocytes with time due to accumulation of toxic metabolites. ADA-SCID is one of the more common forms of SCID.

   (b) PNP deficiency

   Mutation in purine nucleoside phosphorylase (*PNP*) gene results in PNP-SCID which is also a purine metabolism disorder. This results in build up of toxic metabolites that kills all lymphoid progenitor cells similar to ADA-SCID. PNP-SCID is rare compared to ADA-SCID.

   (c) Reticular dysgenesis

   Reticular dysgenesis is a rare form of SCID that results from the defect in mitochondrial adenylate kinase 2. This also leads to accumulation of toxic metabolites that affect all hematopoietic stem cells, including both lymphoid and myeloid cells. The T−B−NK− lymphocyte phenotype is observed along with granulocytopenia and deafness.

3. **Defect in V(D)J recombination—Absence of T and B cells (T−B−NK+ SCID)**

   Since the defect is in lymphocyte antigen receptor rearrangement, this will affect T and B cells development but not the NK cells. Therefore, diseases in this group result in T−B−NK+ lymphocyte phenotype.

   (a) RAG 1 and RAG 2 deficiency

   RAG1 and RAG2 enzymes help to mediate double strand DNA breakage recombination sites. Deficiency in RAG1 or RAG2 results in cleavage defect during V(D)J recombination. Missense mutations in RAG1 or RAG2 genes cause Omenn syndrome (OS), a type of leaky SCID, wherein autologous oligoclonal T cells become activated and cause disease. Patients typically present with failure to thrive, erythroderma, diarrhea, enlarged lymph nodes, and hepatosplenomegaly.

   (b) Artemis, DNA ligase IV, and Cernunnos deficiency

   Mutations in *Artemis* gene lead to failure to resolve hairpins during V(D)J recombination. Similar phenotype is seen with mutations in DNA ligase IV and cernunnos/XLF. These genes also participate in double strand DNA break repair. As a result, SCID caused by these genetic defects is also associated with radiosensitivity.

4. **Specific inhibition of T cell development—Absence of T cells, while B cells are normal or reduced and NK cells are normal (T−B+NK+ SCID)**

   (a) IL7Rα (CD127) deficiency

   IL 7 provides survival and proliferative signals for early T cell development. Therefore, IL7α deficiency leads to SCID with T−B+NK+ phenotype. IL7Rα is one of the more common forms of SCID.

   (b) Defective pre-TCR (T cell receptor) checkpoint

   Mutations in CD45, CD3ε, CD3δ, Orai 1 (CRAC; Calcium release-activated channels) result in absence of T cells, while NK cells are normal. B cells can be normal or reduced.

   (c) *CORO1a* deficiency

   CORO1a gene defect leads to Coronin-1A deficiency. Coronin-1A is an actin regulating protein which involves in actin cytoskeleton regulation. It is essential for T cell egression from the thymus. *CORO1a* deficiency, therefore, leads to SCID with T−B+NK+ phenotype with the present of thymus.

2.3 Laboratory Findings in SCID

The laboratory abnormalities diagnostic of "Typical" SCID are as described in Box 1, while laboratory findings in "Partial" or "Leaky" SCID are described in Box 2.

Box 1: Laboratory Abnormalities Diagnostic of *"Typical"* SCID

1. Absent/low T cell numbers. An absolute CD3+ T cell count <300 cells/μL.
2. Absolute lymphocyte count <2500 cells/μL.
3. Absent/low T cell function. T cell proliferation response to mitogens is absent or low. In particular, response to phytohemagglutinin (PHA) is <10% compared with control value.

Laboratory results may also show hypogammaglobulinemia which could be due to lack of T cell help. Some phenotypes of SCID may also have absent/low number of B cells as well as functional abnormalities of B cells. This is described previously in the classification of SCID. IgA and IgM levels may be low or absent. IgG levels can be low or normal due to transplacental transfer of maternal IgG.

Box 2: Laboratory Abnormalities Diagnostic of *"Partial"* or *"Leaky"* SCID, Including Omenn Syndrome

1. T cell lymphopenia
   (a) <2 years-T cells <1000 cells/μL
   (b) 2–4 years-T cells <800 cells/μL
   (c) >4 years-T cells <600 cells/μL
2. Response to PHA is <30% of the control value.
3. Absence of maternal engraftment.
4. Hypomorphic or incomplete mutation in SCID genes.

Patients with partial or leaky SCID do not necessarily display typical clinical and laboratory features of SCID as shown in Box 2. They may have delayed onset of disease and less severe infectious complications. T cells have a skewed T-helper-2 (Th2) phenotype resulting in elevated Th2 cytokines, IL 4 and IL 5. This can lead to increased inflammatory markers, eosinophilia, and high IgE levels.

Definitive diagnosis for SCID requires genetic testing. Genetic testing is important for the preparation of hematopoietic stem cell transplant (HSCT). When family history is significant for SCID, prenatal diagnosis can be established by genetic studies.

Question 4

What condition is likely to be missed on newborn screening for TREC?

A. ADA deficiency
B. Complete DiGeorge syndrome
C. IL2RG deficiency
D. IL7R deficiency
E. JAK3 deficiency

Answer: A

Defects in nucleotide salvage pathways, such as ADA deficiency, can lead to the accumulation of toxic metabolites. T cell counts may be normal at birth, hence, normal TREC at birth, but there is a progressive decline in number of lymphocytes with time due to more accumulation of toxic metabolites. With similar reason, defects in DNA repair mechanisms, such as Artemis deficiency, can also lead to a progressive decline in number of lymphocytes with time, while TREC can be normal at birth.

IL2RG, IL7R, and JAK3 deficiency also cause SCID, which T cell lymphopenia is present at birth. Therefore, these conditions would lead to positive TREC screenings. Patients with complete DiGeorge syndrome also have severe T cell lymphopenia at birth, therefore, they will likely have a positive TREC screening.

Since ADA-SCID is one of the more common forms of SCID and can be missed on newborn screening based on TREC assay, this has led to some states to add ADA enzyme screening to NBS.

2.4 Newborn Screening for SCID and TREC Analysis

The advent of newborn screening (NBS) program for SCID has facilitated the diagnosis of SCID at birth. It has been implemented in all 50 states in the US since December 2018. NBS for SCID uses T cell receptor excision circles (TREC) assay which is a DNA biomarker of normal T cell development. This tends to remain stable in the dried blood spots (DBS) which are routinely obtained for newborn screening for other conditions as well. TRECs are measured in laboratories using a real time quantitative polymerase chain reaction (RT-PCR). Term newborns have a high rate of new T cell production such that TREC numbers account for almost 10% of total T cell numbers. Premature infants have lower TREC numbers at birth, which increase over time. Low TREC levels generally only imply T cell lymphopenia (TCL) which can be seen in SCID and also in many non-SCID conditions described in Box 3.

Box 3: Causes of T Cell Lymphopenia

SCID

Non-SCID syndromes

- DiGeorge syndrome (DGS)
- Jacobsen syndrome
- Trisomy 21
- CHARGE syndrome (coloboma, heart anomaly, choanal atresia, retardation, genital and ear anomalies)
- Ras-related C3 botulinum toxin substrate 2 (RAC2) deficiency
- Dedicator of cytokinesis 8 (DOCK8) deficiency
- Idiopathic CD4 lymphopenia (ICD4L)
- Premature infants
- Neonatal thymectomy during cardiac surgery
- Conditions associated with lymph loss (chylothorax and lymphangiectasia)
- Stress associated with severe illness or infection

Patients with Down syndrome can have T cell lymphopenia early in life, typically in the first 15 months of life. T cell lymphopenia in patients with Down syndrome tends to gradually normalize, while B cell counts remain low through childhood. Severe illness or infection in infants can cause neutrophilia, hence lead to lymphopenia, or they can cause leukopenia in general. These can result in T cell lymphopenia and low TRECs in NBS. In these cases, the repeat TREC assay or flow cytometry may be necessary.

Question 5

The genetic testing is confirmed for IL2RG deficiency. What would be the appropriate measures in this patient while waiting for bone marrow transplantation?

A. Avoiding live vaccines
B. Monthly Intravenous Immunoglobulins (IVIG) replacement
C. Receive only irradiated, leukocyte poor, and CMV negative blood products
D. Trimethoprim-sulfamethoxazole prophylaxis
E. All of the above

Answer: E

Initial management of SCID should include measures to protect patients against infections such as strict isolation, avoiding all live vaccines, use of antimicrobial prophylaxis, and IVIG administration.

2.5 Management of SCID

Initial management of SCID should include measures to protect patients against infections. Infants should avoid all live vaccines as they can cause severe infection. The use of antimicrobial agents is necessary to prevent certain opportunistic infections such as *Pneumocystis jirovecii*, *Mycobacterium avium*, *Candida* spp., and RSV. If hospitalized, infants should be in strict isolation. IVIG treatment is also necessary as in the absence of T cell help, B cells function is impaired. Box 4 summarizes initial management of SCID.

Box 4: Initial Management of Patients with SCID

Protective isolation at home/hospital
Avoidance of all live vaccines
Administration of IVIG
Antimicrobial prophylaxis 1. *Pneumocystis jirovecii* (trimethoprim-sulfamethoxazole or pentamidine alternatively) 2. *Mycobacterium avium* prophylaxis with azithromycin 3. Antifungal (*Candida* spp.) with fluconazole 4. Antiviral prophylaxis (a) RSV prophylaxis with palivizumab during RSV season (b) Ganciclovir and/or foscarnet if CMV infection is identified
CMV urine and saliva viral culture/CMV PCR testing
Avoid fresh breast milk from CMV seropositive mothers to reduce risk of CMV transmission
Blood products should be irradiated, leukocyte poor, and CMV negative
Consultation to pediatric bone marrow transplant team, pediatric immunology and infectious diseases specialist

Treatment should be initiated in case of low T cell (CD3) count (<300 cells/μL) and poor mitogen proliferative response (<10% of normal). The most definitive treatment for most types of SCID is HSCT. Human leukocyte antigen (HLA) identical or haplo-identical marrow, T cell depleted from a related donor is life saving in almost all forms of SCID particularly if performed early in infancy. HSCT has shown most success if it is performed in first 3.5 months of life. In most cases, immune reconstitution occurs. The Primary Immune Deficiency Treatment Consortium reports that when patients are treated with HSCT in the first 3.5 months of life, the survival rate is as high as 94%. The survival rate is 90% in older infants who had not developed any infections while the rate is lower at 82% in older infants with resolved infections. On the other hand, the survival rate is much lower (50%) when transplant was performed in older infants with active infection. Thus, early diagnosis and treatment of SCID significantly improve outcome.

Certain forms of SCID have been successfully treated with gene therapy such as ADA deficiency and X-linked SCID, especially when HSCT cannot be performed due to unavailability of HLA-identical donor. Gene therapy also results in immune reconstitution. The use of retroviral vectors for gene therapy in the past years was associated with leukemia, which limited its widespread utility. The risk is much lower with the use of newer viral vectors. Lentiviral vector gene therapy has been used successfully in patients with newly diagnosed X-linked SCID with few side effects.

Enzyme replacement therapy is yet another treatment modality that has been used to treat newly diagnosed ADA-SCID. This type of therapy restored immune function and helped to stabilize the patients who are still waiting for the definitive treatment.

Viral infections tend to be the most common cause of mortality in SCID patients before and after HSCT. These infections can be treated with adoptive immunotherapy with virus specific T cells as well as antiviral medications.

Case Presentation 2

A 6 days old, full-term boy is born via normal vaginal delivery to a healthy mother. He is breastfeeding well. Parents are notified that his NBS is positive for SCID and recommended to take him to an immunologist for further evaluation. Examination reveals a cardiac murmur. Cardiology consult is initiated and an echocardiogram is performed which shows congenital heart defect. In view of positive NBS, laboratory tests are ordered for lymphocyte enumeration and mitogen proliferation. Results show decreased but not absent CD3+ cells, normal absolute count of CD19+ cells and CD3−CD56+CD16+ cells. Mitogen proliferation response is low but not absent.

Question 6

What is the most likely gene involved?

A. *AIRE*
B. *IL7RG*
C. *ORAI1*
D. *TBX*
E. *WASP*

Answer: D

DiGeorge syndrome (DGS) is most likely the diagnosis in this case. DiGeorge syndrome is a congenital disorder characterized by T cell immune deficiency, congenital cardiac defects, hypocalcemia, dysmorphic facies, and learning disabilities. In majority of DGS patients, deletions occur in *TBX1* gene which encodes for a "T box" transcription factor. The presence of cardiac defect and T cell lymphopenia in this patient favor the diagnosis of DiGeorge syndrome.

Wiskott Aldrich syndrome (WAS) is an X-linked disorder characterized by defect in *WASP* gene. It is associated with immunodeficiency and infection, thrombocytopenia and bleeding, and eczema. There are no associated cardiac defects. However, aneurysm formation and vasculitis may occur.

ORAI1 gene encodes part of the calcium channel which is required for T cell activation. Mutations in *ORAI*1 gene result in an autosomal recessive type of SCID. Patients develop recurrent infections and congenital myopathy. T cell and B cell numbers are normal; however, their function is severely impaired. There are no known associated cardiac defects, though cardiomyopathy can coexist with skeletal muscle weakness.

Chronic mucocutaneous candidiasis (CMCC) is associated with mutation in *AIRE* gene. Patients have recurrent candida infection of skin and mucocutaneous tissues, endocrinopathies, and autoimmune cytopenias. T cell deficiency may be seen. CMCC is not associated with any cardiac defects though patients may have associated vasculitis.

Mutation in *IL7RG* gene results in SCID. The disease is associated with severe T cell lymphopenia but it is not usually associated with cardiac defects.

3 DiGeorge Syndrome

DGS occurs due to abnormal development of the pharyngeal arches and pouches. Thymus arises from the third and fourth arch structures of the pharyngeal apparatus during embryonic development, along with parathyroid gland. The incidence of DGS is 1 in 4000–6000 live births and affects both sexes equally. DGS is one of the most common genetic diseases.

3.1 Etiology of DiGeorge Syndrome

Majority of cases of DGS are associated with hemizygous deletion of chromosome 22q11.2. Most deletions arise de novo. Only 10% are inherited from the parents in an autosomal dominant fashion. Deletion in 22q11.2 is related to wide spectrum of disorders which have overlapping phenotypes. "22q11.2 deletion syndrome" defines patients who have the particular deletion, while the term "DGS" best describes patients who have the triad of T cell immunodeficiency, congenital cardiac defects, and hypocalcemia, with or without a demonstrable deletion.

In the majority of DGS patients, deletions occur in *TBX1* gene which encodes for a "T box" transcription factor. These genetic defects account for the characteristic features of DGS which are dysmorphic facial features, thymic hypoplasia, cardiac defects, parathyroid hypoplasia, and cleft palate defects.

Apart from *TBX1*, other genes may be affected in patients with DGS, including *Crkl* and *COMT* genes. *Crkl* encodes an adaptor protein that is expressed in the tissues derived from neural crest during development. Catechol-*O*-methyltransferase (COMT) is involved in catecholamines metabolism and has been linked to psychiatric disease in DGS patients. In a small subset of DGS patients, deletions on the short arm of chromosome 10p13-14 have been identified (1:200,000 live births). Patients with this deletion are more likely to develop growth and mental retardation, sensorineural hearing loss, and renal anomalies. Deletions on chromosomes 17p13 and 18q21 have also been identified in few patients with DGS.

Question 7

What clinical comorbidity is **not** commonly described in patients with DiGeorge syndrome?

A. Arthritis
B. Asthma
C. Autoimmune cytopenia
D. Eczema
E. Neonatal jaundice

Answer: E

DGS is characterized by the triad of immune deficiency, congenital cardiac defects, and hypocalcemia. Patients with DGS are at risk for developing autoimmune disorders, including autoimmune cytopenias, arthritis, and autoimmune endocrinopathy. An increased risk in atopic diseases, such as asthma and eczema, is also observed. Neonatal jaundice is not a described feature of DGS.

3.2 Clinical Characteristics of DiGeorge Syndrome

Immune deficiency in DGS is associated with thymic aplasia or hypoplasia and subsequent T cell lymphopenia (TCL). DGS is associated with variable degree of T cell lymphopenia and proliferative responses. Complete DGS accounts for less than 0.5% of DGS cases. It occurs due to thymic aplasia and is associated with severe T cell immunodeficiency similar to that seen in SCID. T cell receptor excision circles (TRECs) are markedly reduced in complete DGS patients. Defect in T cell-mediated immunity predisposes these patients to *Pneumocystis jirovecii* infection as well as viral and other fungal infections. These infections result in increased morbidity and mortality. Partial DGS, on the other hand, occurs due to abnormal thymic migration. These patients have mild to moderate T cell lymphopenia.

Patients with DGS are at increased risk for developing autoimmune diseases. This has been attributed to impaired thymic central tolerance, impaired development of T regulatory cells, and frequent infections. In a study of patients with DGS, the percentage of CD4+CD25+ T regulatory cells was significantly lower compared to normal control, particularly during infancy. There was, however, no difference in the percentage of CD4+CD25+ T regulatory cells in DGS patients with or without autoimmune disease. Autoimmune diseases are seen in 10% of patients with DGS. These include autoimmune cytopenias, juvenile rheumatoid arthritis-like polyarthritis, and autoimmune endocrinopathy.

Patients with DGS, similar to many other immune deficiency diseases, are more prone to develop atopic diseases such as asthma and eczema. Frequent respiratory tract infections earlier in life can damage the growing lungs, resulting in respiratory abnormalities. The infections can also alter immune regulation of the host. These contribute to the development of childhood asthma and atopy. Furthermore, it is postulated that frequent use of antibiotics early in life may predispose these patients to develop asthma.

Congenital cardiac defects are a common cause of increased morbidity and mortality in patients with DGS. Tetralogy of Fallot, interrupted aortic arch, right aortic arch, and aberrant right subclavian artery are some of the common cardiac defects seen in these patients. Cardiac defects are an important cause of increased mortality earlier in life.

Neonatal hypocalcemia is a characteristic feature of DGS and is related to hypoplasia of the parathyroid glands. Patients may present with tetany or seizures. However, as the parathyroid glands increase in size, hypocalcemia may resolve over the first year of life. Patients may also have latent hypoparathyroidism and may be unable to increase parathyroid hormone level during a hypocalcemic event.

Apart from these cardinal features, other organ systems may also be involved in DGS as outlined in Box 5.

Box 5: Clinical Features of DiGeorge Syndrome

- Cardiac defects (80% of patients)
- Hypocalcemia
- Thymic aplasia/hypoplasia resulting in immunodeficiency (T cell deficiency)
- Renal abnormalities (multicystic dysplastic kidney, duplicated collecting system, horseshoe kidney, or single kidney)
- Developmental delay (short stature, speech delay and learning disabilities)
- Cleft palate and feeding difficulties
- Distinctive facial features (short philtrum, small mouth, small palpebral fissures, telecanthus, low set ears with abnormal folding of the pinna and deafness)
- Psychiatric disorders (ADHD, autism spectrum disorders, schizophrenia, and other neurological abnormalities)

Question 8

What is the test of choice to diagnose DGS?

A. Chest X-ray
B. CBC
C. Echocardiography
D. FISH (fluorescence in situ hybridization)
E. Serum calcium levels

Answer: D

FISH analysis using probe to identify deletions is the diagnostic modality of choice. CBC may show low to normal lymphocyte count and is not diagnostic of DGS. Echocardiography is often employed to identify the cardiac defects. Some cardiac defects are more commonly associated with DGS such as outflow tract defects. However, echocardiography is not diagnostic of DGS. In DGS, thymic shadow might be absent on CXR but this is not reliable. Patients have hypocalcemia but it is not the test of choice to diagnose DGS.

3.3 Diagnosis of DGS

A positive FISH test that identifies chromosome 22q11.2 deletion is the diagnostic test of choice. In case of negative 22q11.2 FISH test but clinical phenotype is suggestive of DGS, a 10p deletion FISH should be requested. Dual probe FISH assays can help to diagnose deletions at 22q11.2 and at 10p13-14. It can diagnose up to 95% of DGS cases. In the remainder 5% of the cases, more specialized tests may be required, such as multiplex ligation-dependent probe amplification (MLPA), single nucleotide polymorphism (SNP) array, and array comparative genomic hybridization (aCGH), to make the diagnosis due to their higher resolution. Karyotype should be obtained to identify any chromosomal abnormalities.

Question 9

The patient CD3+ T cell is 120/μL. FISH is positive for 22q11del; however, the patient has no facial anomalies compatible with DGS and does not have hypocalcemia. This patient most likely has

A. Definitive complete DGS
B. Definitive partial DGS
C. Possible partial DGS
D. Probable complete DGS
E. Probable partial DGS

Answer: B

Diagnosis of DGS is based on number of T cells as well as associated clinical features, including cardiac features,

Table 2 Diagnosis of DGS

Type of DGS	CD3+ T cells	Associated features
Complete DGS		
Definitive	Reduced/ absent (<50/ μL)	Athymia, cardiac defect hypocalcemia
Partial DGS		
Definitive	<500/μL (first 3 years of life)	Cardiac defect and chromosome 22q11.2 deletion *with* or *without* hypocalcemia
Probable	<1500/μL (first 3 years of life)	Chromosome 22q11.2 deletion *without* other DGS clinical characteristics
Possible	<1500/μL (first 3 years of life)	Cardiac defect or hypocalcemia or dysmorphic facies/palatal abnormalities *without* chromosome 22q11.2 deletion

hypocalcemia, and the presence of chromosome 22q11.2 deletion, as shown below in Table 2.

T cells count in the patient presented here is 120/μL; therefore, he does not have complete DGS, which is defined as reduced or absent CD3+ T cell of less than 50/μL. With the CD3+ T cell of less than <500/μL, cardiac defect, and chromosome 22q11.2 deletion, even without other DGS characteristics including facial anomalies and hypocalcemia, this patient fits the diagnosis of definitive partial DGS.

Probable partial DGS is defined as patients with CD3+ T cell of less than <1500/μL *with* chromosome 22q11.2 deletion *without* other DGS clinical characteristics. While possible partial DGS is defined as patients with CD3+ T cell of less than <1500/μL *with* cardiac defect or hypocalcemia or characteristic facial/palatal abnormalities *without* chromosome 22q11.2 deletion.

DGS is characterized by thymic hypoplasia or aplasia leading to variable number of T cell counts and proliferative responses. In patients with complete DGS, laboratory tests show very low number of T cells (1–2%) in peripheral blood and absence of proliferative response to mitogens. Partial DGS patients, on the other hand, have mild to moderate T cell lymphopenia and low to normal proliferative responses. In some partial DGS patients, T cell numbers may be normal. Mitogen proliferation responses may help to assess T cell function more accurately compared to T cell counts and thus help to distinguish partial versus complete DGS.

In the absence of T cell help, B cell function is often dysregulated. Immunoglobulin levels are often normal in partial DGS, though some patients may also develop hypergammaglobulinemia due to dysregulated B cell function caused by T cell deficiency. In patients with severe partial DGS or complete DGS, hypogammaglobulinemia can be seen, commonly with poor antibody responses to polysaccharide vaccines. Other defects in humoral immunity observed in some DGS patients include antibody deficiency such as selective IgA, IgM, or IgG sub-class deficiency, or abnormal responses to protein vaccines (diphtheria and tetanus). Significant association has been seen between these defects in humoral immunity and recurrent infections.

The diagnosis of DGS and associated clinical features are shown in Table 2.

Question 10

What treatment modality is the most commonly used for immunological abnormality seen in patients with DGS?

A. Ganciclovir prophylaxis
B. Interferon-gamma injection
C. Intravenous Immunoglobulin (IVIG)
D. Itraconazole prophylaxis
E. Trimethoprim-sulfamethoxazole prophylaxis

Answer: E

Prophylactic antibiotics for *Pneumocystis jirovecii* pneumonia should be initiated for all patients with DGS due to risk of opportunistic infections unless immune competence is documented. For this purpose, trimethoprim-sulfamethoxazole or, alternatively, pentamidine can be used. IVIG is indicated only for hypogammaglobulinemia accompanied by poor responses to vaccines. It may be required in selective cases of DGS, particularly complete DGS. Interferon-gamma has no therapeutic role in patients with DGS. It is often used in patients with chronic granulomatous disease or patients with defects in IL12/Interferon-gamma pathway defects. Itraconazole and ganciclovir prophylaxis are not routinely used in DGS patients.

3.4 Management of DGS

In most patients with partial DGS, immune deficiency improves over time. On the other hand, complete DGS is associated with severe immune deficiency. Due to risk of opportunistic infections, prophylactic antibiotics for *Pneumocystis jirovecii* pneumonia should be initiated unless immune competence is documented.

Patients with complete DGS require transplant for immune reconstitution. Recently, trials have been conducted using postnatal thymic tissue transplant in infants with complete DGS. These trials showed that early thymic transplant prior to onset of infections successfully restores immune function. Mature functional T cells were present in peripheral circulation 4 months post-transplant with adequate mitogen proliferative responses. Survival rate was high (75%) amongst the infants who underwent transplant with post-transplantation follow-up as long as 13 years. All deaths occurred within 12 months of transplantation and were related to the congenital conditions associated with DGS. CMV infection, tracheostomy, and prolonged mechanical ventilation were recognized as main risk factors for death. Overall, thymic transplantation was well tolerated and resulted in stable immune reconstitution in these infants. Bone marrow and peripheral blood T cell transplantation have also been successfully used in some patients with complete DGS. The donor is usually an HLA-matched sibling.

4 Summary

T cell-mediated immune deficiency should be suspected in patients presenting with failure to thrive, chronic diarrhea, and recurrent infections usually caused by intracellular organisms such as viruses and fungi (including *P. jirovecii*). Patients often have T cell lymphopenia and absent or low proliferative responses to mitogens. TREC analysis as part of NBS aids in early identification of T cell lymphopenia, including severe T cell deficiency diseases such as SCID and complete DGS. However, some genetic defects, such as ADA-SCID, may not be identified through NBS TREC assay. Physicians should thus maintain a higher index of suspicion while evaluating patients presenting with life-threatening infections. HSCT and thymic transplant should be attempted early if needed, preferably prior to the onset of infections due to significantly higher survival rate.

Bibliography

1. Notarangelo LD. Primary immunodeficiencies. J Allergy Clin Immunol. 2010;125(2 Suppl 2):S182–94.
2. Vogel BH, Bonagura V, Weinberg GA, Ballow M, Isabelle J, DiAntonio L, Parker A, Young A, Cunningham-Rundles C, Fong CT, Celestin J, Lehman H, Rubinstein A, Siegel S, Weiner L, Saavedra-Matiz C, Kay DM, Caggana M. Newborn screening for SCID in New York state: experience from the first two years. J Clin Immunol. 2014;34:289–303.
3. Lipstein EA, Browning MF, Green NS, et al. Systematic evidence review of newborn screening and treatment of severe combined immunodeficiency. Pediatrics. 2010;125:1226–35.
4. Griffith LM, Cowan MJ, Notarangelo LD, et al. Improving cellular therapy for primary immune deficiency diseases: recognition, diagnosis, and management. J Allergy Clin Immunol. 2009;124(6):1152–60.
5. Patel NC, Hertel PM, Estes MK, et al. Vaccine-acquired rotavirus in infants with severe combined immunodeficiency. N Engl J Med. 2010;362(4):314–9.
6. Heimall J, Logan BR, Cowan MJ, et al. Immune reconstitution and survival of 100 SCID patients post-hematopoietic cell transplant: a PIDTC natural history study. Blood. 2017;130(25):2718–27.
7. Kwan A, Abraham RS, Currier R, Brower A, Andruszewski K, Abbott JK, et al. Newborn screening for severe combined immunodeficiency in 11 screening programs in the United States. JAMA. 2014;312(7):729–38.
8. Giblett ER, Anderson JE, Cohen F, Pollara B, Meuwissen HJ. Adenosine-deaminase deficiency in two patients with severely impaired cellular immunity. Lancet. 1972;2:1067–9.
9. Shinkai Y, Rathbun G, Lam KP, Oltz EM, Stewart V, Mendelsohn M, Charron J, Datta M, Young F, Stall AM, et al. RAG-2-deficient mice lack mature lymphocytes owing to inability to initiate V(D)J rearrangement. Cell. 1992;68:855–67.
10. Shearer WT, Dunn E, Notarangelo LD, et al. Establishing diagnostic criteria for severe combined immunodeficiency disease (SCID), leaky SCID, and Omenn syndrome: the Primary Immune Deficiency Treatment Consortium experience. J Allergy Clin Immunol. 2014;133(4):1092–8.
11. Sobacchi C, Marrella V, Rucci F, Vezzoni P, Villa A. RAG-dependent primary immunodeficiencies. Hum Mutat. 2006;2006(27):1174–84.
12. Chan K, Puck JM. Development of population-based newborn screening for severe combined immunodeficiency. J Allergy Clin Immunol. 2005;115(2):391–8.
13. Kwan A, Church JA, Cowan MJ, Agarwal R, Kapoor N, Kohn DB, et al. Newborn screening for severe combined immunodeficiency and T-cell lymphopenia in California: results of the first 2 years. J Allergy Clin Immunol. 2013;132(1):140–50.
14. Pai SY, Logan BR, Griffith LM, Buckley RH, et al. Transplantation outcomes for severe combined immunodeficiency, 2000-2009. N Engl J Med. 2014;371(5):434–46.
15. Kuo CY, Kohn DB. Gene therapy for the treatment of primary immune deficiencies. Curr Allergy Asthma Rep. 2016;16(39):1–8.
16. Kohn DB, Hershfield MS, Puck JM, Aiuti A, et al. Consensus approach for the management of severe combined immune deficiency caused by adenosine deaminase deficiency. J Allergy Clin Immunol. 2019;143(3):852–63.
17. Devriendt K, Fryns JP, Mortier G, van Thienen MN, Keymolen K. The annual incidence of DiGeorge/velocardiofacial syndrome. J Med Genet. 1998;35:789–90.
18. Baldini A. DiGeorge syndrome: the use of model organisms to dissect complex genetics. Hum Mol Genet. 2002;11:2363–9.
19. Haynes BF, Markert ML, Sempowski GD, Patel DD, Hale LP. The role of the thymus in immune reconstitution in aging, bone marrow transplantation, and HIV-1 infection. Ann Rev Immunol. 2000;18:529–60.
20. Marcinkowski M, Bauer K, Stoltenburg-Didinger G, Vogel M, Versmold H. Fatal aspergillosis with brain abscesses in a neonate with DiGeorge syndrome. Pediatr Infect Dis J. 2000;19:1214–6.
21. McLean-Tooke A, Spickett GP, Gennery AR. Immunodeficiency and autoimmunity in 22q11.2 deletion syndrome. Scand J Immunol. 2007;66:1–7.
22. Marino B, Digilio MC, Toscano A, Anaclerio S, Giannotti A, Feltri C, de Ioris MA, Angioni A, Dallapiccola B. Anatomic patterns of conotruncal defects associated with deletion 22q11. Genet Med. 2001;3(1):45–8.
23. Cheung EN, George SR, Costain GA, Andrade DM, Chow EW, Silversides CK, Bassett AS. Prevalence of hypocalcaemia and its associated features in 22q11·2 deletion syndrome. Clin Endocrinol (Oxf). 2014;81(2):190–6.
24. Botto LD, May K, Fernhoff PM, Correa A, et al. A population-based study of the 22q11.2 deletion: phenotype, incidence, and contribution to major birth defects in the population. Pediatrics. 2003;112:101–7.
25. Barrett DJ, Ammann AJ, Wara DW, Cowan MJ, Fisher TJ, Stiehm ER. Clinical and immunologic spectrum of the DiGeorge syndrome. J Clin Lab Immunol. 1981;6:1–6.
26. Junker AK, Driscoll DA. Humoral immunity in DiGeorge syndrome. J Pediatr. 1995;127:231–7.
27. Ryan AK, Goodship JA, Wilson DI, Philip N, Levy A, Seidel H, Schuffenhauer S, Oechsler H, Belohradsky B, Prieur M, Aurias A, Raymond FL, Clayton-Smith J, Hatchwell E, McKeown C, Beemer FA, Dallapiccola B, Novelli G, Hurst JA, Ignatius J, Green AJ, Winter RM, Brueton L, Brondum-Nielsen K, Scambler PJ, et al. Spectrum of clinical features associated with interstitial chromosome 22q11 deletions: a European collaborative study. J Med Genet. 1997;34:798–804.

28. Larson RS, Butler MG. Use of fluorescence in situ hybridization (FISH) in the diagnosis of DiGeorge sequence and related diseases. Diagn Mol Pathol. 1995;4(4):274–8.
29. Markert ML, Sarzotti M, Ozaki DA, Sempowski GD, et al. Thymus transplantation in complete DiGeorge syndrome: immunologic and safety evaluations in 12 patients. Blood. 2003;102:1121–30.
30. Matsumoto T, Amamoto N, Kondoh T, Nakayama M, et al. Complete-type DiGeorge syndrome treated by bone marrow transplantation. Bone Marrow Transplant. 1998;22:927–30.
31. Bensoussan D, Le Deist F, Latger-Cannard V, Gregoire MJ, et al. T-cell immune constitution after peripheral blood mononuclear cell transplantation in complete DiGeorge syndrome. Br J Haematol. 2002;117:899–906.

Combined Immunodeficiency Disorders

Jenny Huang and Pavadee Poowuttikul

Abbreviations

AICDA	Activation-induced cytidine deaminase
AID	Activation-induced cytidine deaminase
ALPS	Autoimmune lymphoproliferative syndrome
AP-1	Activate transcriptions factors activator protein-1
Arp2/3	Actin-related protein
BCR	B cell receptor
BTK	Bruton's tyrosine kinase
CD40LG	CD40 ligand gene
CID	Combined immunodeficiency disorders
CSR	Class switch recombination
CVID	Common variable immunodeficiency
EBV	Epstein–Barr virus
ESID	European Society for Immunodeficiencies
FASLG	Fas ligand
GV	Gammaretroviral vector
HIGM	Hyper-IgM
HSCT	Hematopoietic stem cell transplantation
Ig	Immunoglobulin
IPEX	Immune dysregulation, polyendocrinopathy, enteropathy, X-linked syndrome
IVIG	Intravenous immunoglobulin
LAD 3	Leukocyte adhesion defect type 3
LV	Lentiviral vector
NK	Natural killer
NPF	Nucleation-promoting factors
PCR	Polymerase chain reaction
PI3K-delta	Phosphoinositide 3-kinases delta
PIDD	Primary immunodeficiency disorders
PMS2	Postmeiotic segregation increased 2
RAG 1/2	Recombination-acting protein 1 and 2
SCID	Severe combined immunodeficiency
STAT	Activators of transcriptions
TCR	T cell receptor
TdT	Terminal deoxynucleotidyl transferase
TRAF	TNF receptor associated factors
Treg	T regulatory cells
UNG	Uracil-DNA-glycosylase
USIDNET	US immunodeficiency network
WAS	Wiskott-Aldrich syndrome
WASp	WAS protein
XHIGM	X-linked hyper-IgM
XLA	X-linked agammaglobulinemia
XLP	X-linked lymphoproliferative disease
XLT	X-linked thrombocytopenia

J. Huang · P. Poowuttikul (✉)
Division of Allergy/Immunology, Department of Pediatrics, Children's Hospital of Michigan, Detroit, MI, USA

Central Michigan University College of Medicine, Mt. Pleasant, MI, USA
e-mail: Jhuang2@dmc.org; Ppoowutt@dmc.org

Case 1

A 15-month-old, previously full-term, boy is admitted to the hospital with suspected *Pneumocystis jirovecii* pneumonia. He has a history of recurrent diarrhea with *Cryptosporidium* sp. since 12 months of age. His maternal uncle had recurrent infections, liver disease, and osteopenia. On physical exam, the infant is in the 1 percentile for weight and has no palpable lymphadenopathy or organomegaly and mild diaper dermatitis without evidence of other skin rashes. He has visible tonsils. His serum laboratory studies show a low absolute neutrophil count, low IgG level, and slightly high IgM level. Genetic testing confirms his diagnosis.

Question 1

What is the most likely gene defect associated with his disease?

A. *CD40LG*
B. *DOCK8*
C. *FASLG*

M. Mahmoudi (ed.), *Absolute Allergy and Immunology Board Review*, https://doi.org/10.1007/978-3-031-12867-7_23

D. *ICOS*
E. *LYST*

Answer and Explanation
Answer: A

The most likely diagnosis for this patient is X-linked hyper-IgM (XHIGM). XHIGM is caused by a mutation in the *CD40LG* gene encoding the CD40 ligand protein responsible for immunoglobulin isotype switching. Patients with XHIGM have combined immunodeficiency and can present with recurrent infections, opportunistic infections, and failure to thrive. Due to a class switching defect, low immunoglobulin (Ig)G and IgA levels are observed with normal-high or high IgM level, as seen in this patient.

The other choices can cause combined immunodeficiency phenotypes; however, the patient does not have clinical characteristics that fit each disease.

Answer choice B refers to autosomal recessive hyper-IgE syndrome from *DOCK8* mutation. Patients with *DOCK8* deficiency generally present with severe viral skin infections (human papillomavirus and herpes simplex), mucocutaneous candidiasis, and eczema.

Answer choice C refers to autoimmune lymphoproliferative syndrome (ALPS). Patients with ALPS typically present with nonmalignant lymphadenopathy, organomegaly, and autoimmune disease.

Answer choice D is associated with common variable immunodeficiency (CVID). A patient with CVID can have recurrent sinopulmonary and GI infections and at a higher risk of developing autoimmune diseases, similar to an XHIGM patient. However, they would be more likely to present later in life with low IgG and IgA or IgM and impaired post-immunization antibody response. CVID is generally considered as a humoral immunodeficiency disorder; however, T cell deficiency has been described in some patients. Typically, the diagnosis of CVID is not made until at least 4 years of age.

Answer choice E is associated with Chediak-Higashi syndrome, which is due to a defect in leukocyte granule formation causing oculocutaneous albinism, recurrent infections, and the tendency to bruise or bleed easily.

1 X-Linked Hyper-IgM

1.1 Classification, Epidemiology, and Genetics of the Hyper-IgM Syndromes

Patients with hyper-IgM (HIGM) syndrome typically present with recurrent infections. Many patients develop opportunistic infections or recurrent bacterial infections as a young infant such as diarrhea caused by *Cryptosporidium* sp. or pneumonia induced by *Pneumocystis jirovecii.* Laboratory studies may show elevated serum IgM with low or absent IgG, IgA, or IgE. However, serum IgM can be normal or only moderately elevated while other serum immunoglobulin levels are below normal but not absent.

HIGM is either acquired or primary. Acquired forms of HIGM can be a result of malignancy (multiple myeloma, chronic lymphocytic leukemia, and lymphomas), nephrotic syndrome, and autoimmune disease. Primary forms of HIGM are classified into types 1–5. The most common and well-recognized form of the HIGM syndromes is the X-linked form, or HIGM type 1, caused by mutations in the CD40 ligand gene (*CD40LG*). This gene is located on chromosome Xq26.3.

X-linked hyper-IgM (XHIGM) was first described in 1961 in two males with autoimmunity and recurrent sinopulmonary infections. They were also found to have dysgammaglobulinemia. In the late 1980s, the gene for the clinical phenotype was mapped to the long arm of the X chromosome. By the early 1990s, the responsible mutation was found in the *CD40LG* gene encoding the CD40 ligand protein expressed on T cells required for immunoglobulin isotype switching. CD40L also binds to CD40 on various other immune cells including B cells, macrophages, and dendritic cells to induce their activation. The loss of CD40L function results in a combined immunodeficiency clinical phenotype. XHIGM accounts for about 65–70% of all cases of HIGM syndrome. XHIGM in the US affects 2 in 1,000,000 males.

HIGM types 2–5 are all autosomal recessive forms caused by mutations in the genes involved in class switch recombination, somatic hypermutation, or DNA repair mechanisms. HIGM types 2–5 are caused by defects in activation-induced cytidine deaminase (AID), CD40, unknown defect, and uracil-DNA-glycosylase (UNG), respectively. From genetic testing studies to date, HIGM phenotypes have been discovered to also involve novel genes including *PI3K-delta*, *INO80*, *MSH6*, and postmeiotic segregation increased 2 (*PMS2*). Globally, all forms of HIGM make up about 0.3–2.9% of all primary immunodeficiency.

Question 2
The most common gene defect in HIGM affects what part of B cell activation and antibody production?

A. Affinity maturation
B. Class switch recombination
C. Combinatorial diversity
D. DNA repair
E. Receptor editing

Answer and Explanation
Answer: B

The most common form of HIGM is XHIGM. The defect in the *CD40LG* gene in patients with XHIGM produces an abnormal CD40L protein that affects immunoglobulin class switch recombination (CSR). CSR requires T cell and B cell engagement and downstream signals that results in the production of IgG, IgA, or IgE from a mature IgM/IgD producing B cell. CSR defects are also responsible for autosomal recessive forms of HIGM. These include CD40, AID, and UNG deficiency. CD40L on the T cell binds to CD40 on the B cell. This leads to B cell activation and heavy chain class switching. AID and UNG enzyme are also important in B cell heavy chain class switching and affinity maturation.

Affinity maturation is a process that leads to the selective survival of B cells to produce high-affinity antibodies. Impaired affinity maturation has been associated with CVID. AID and UNG enzyme, as stated earlier, are also important for B cell affinity maturation. The deficiency in both AID and UNG leads to autosomal recessive HIGM type 2 and 5, respectively. However, AID and UNG deficiency are not the most common type of HIGM.

Combinatorial diversity is a result of V, D, and J gene segment recombination. Several proteins including RAG1/2, DNA-dependent protein kinase complex and Artemis are important in VDJ recombination. Defects in these proteins lead to SCID.

The DNA repair process allows cells to identify and fix damage to the DNA in its genome. DNA repair defects result in radiosensitive severe combined immunodeficiency (SCID) caused by defects in the proteins such as Artemis, Cernunnos, and DNA ligase IV, which are important in the DNA repair process. Patients with ataxia-telangiectasia are also radiosensitive due to a mutation in the *ATM* gene, which is responsible for DNA double-stranded break repairs.

Receptor editing is an important process in central B lymphocyte tolerance. A defect in receptor editing causes the recognition of a self-antigen and leads to autoimmunity.

1.2 Pathogenesis of Hyper-IgM Syndromes

XHIGM arises from defects in immunoglobulin class switch recombination (CSR), primarily in membrane interactions between T cells and B cells. CSR is a process by which IgG, IgA, or IgE is produced after effective class switching from a mature IgM/IgD producing B cell. It requires DNA rearrangement and T-B cell engagement. Antibody-mediated immunity is important in the defense against extracellular microbes and viruses. The production of antibodies starts in the bone marrow with V(D)J recombination in B cells. This allows genomic rearrangement of the immunoglobulin heavy and light chain genes. IgM and IgD antibodies of low avidity are generated following this process. The proteins involved in V(D)J recombination include recombination-acting protein 1 and 2 (RAG1/2), Artemis, terminal deoxynucleotidyl transferase (TdT), and DNA ligase IV. Defects in these proteins lead to forms of SCID.

The now mature B cell with surface IgM and IgD migrates to secondary lymphoid organs. There, the mature B cells interact with CD4+ T cells, mainly through the interaction of CD40/CD40L. This interaction provides the signal for CSR. The binding of CD40 to CD40L is important in the induction of CSR. Mutation in the *CD40LG* gene, encoding CD40L, is responsible for the X-linked form of hyper-IgM syndrome. CD40L is a transmembrane protein predominately expressed on the surface of activated CD4+ T cells. CD40 is a transmembrane molecule found on the cell surface of antigen-presenting cells such as B lymphocytes, macrophages, dendritic cells as well as non-immune endothelial and neuronal cells. The engagement of CD40 by CD40L leads to receptor trimerization and signal transduction.

The downstream events activated following CD40/CD40L interaction can involve TNF receptor associated factors (TRAF) which recruit kinases such as phospholipase Cγ, nuclear factor (NF) kB inducing kinase, phosphoinositol 3 kinase, c-Jun-N-terminal kinase, and Bruton's tyrosine kinase (BTK) to activate transcription factors activator protein-1 (AP-1) and NF-kB. CD40 activation is also important in the expression of genes required for CSR including activation-induced cytidine deaminase (*AICDA*) gene, which encodes the protein activation-induced cytidine deaminase (AID).

Somatic hypermutation follows CSR and involves the introduction of point mutations to the variable region of the immunoglobulin genes to expand the antibody repertoire. Both CSR and somatic hypermutation require two crucial enzymes, AID and uracil-DNA-glycosylase (UNG). CSR and somatic hypermutation are interdependent processes. Defects in both of these events lead to HIGM syndromes. See Fig. 1 for the overview of events in immunoglobulin generation.

CD40/CD40L defects can also compromise dendritic cell activation and T cell priming which explains its effects on T lymphocyte function. Defects in both CD40L deficiency and CD40 deficiency cause a combined immunodeficiency phenotype and more severe than pure antibody deficiency.

HIGM type 3 is caused by mutations in the gene encoding the CD40 protein. This protein is constitutively expressed on the surface of B cells, dendritic cells, monocytes, and activated epithelial cells. The crosslinking of CD40 with CD40L on activated T cells leads to B cell signaling and increased expression of CD80/CD86 on B cells. CD80/86 binds to T cell surface molecules CD28, important to T cell priming. Patients with HIGM type 3 can present very similarly to XHIGM with opportunistic infections in early infancy, failure to thrive, and a combined immunodeficiency phenotype. CD40 deficiency should be suspected in any female present-

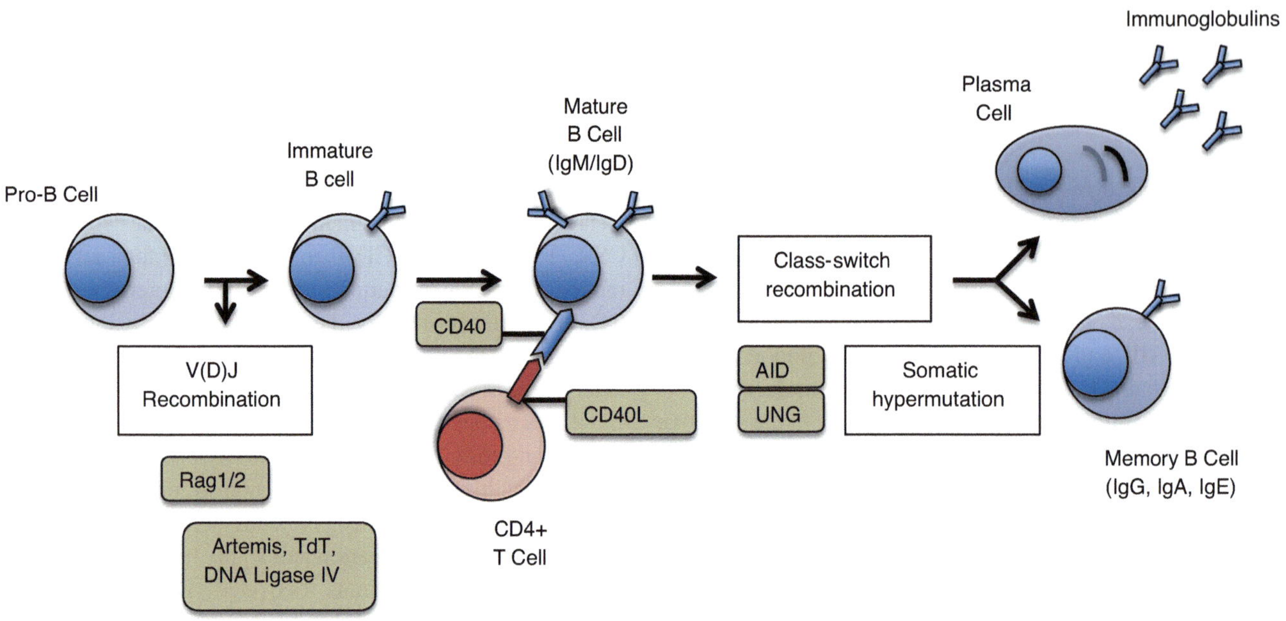

Fig. 1 Events in immunoglobulin generation

ing with an XHIGM clinical phenotype or any male with an XHIGM phenotype but normal expression of CD40L.

Both HIGM type 2 and type 5 are caused by defects in proteins required for CSR and somatic hypermutation. Somatic hypermutation relies on functional AID and UNG proteins to generate double-stranded breaks in the immunoglobulin variable switch regions. Upon CD40/CD40L interaction and B cell downstream signaling, AID is unregulated. AID converts cytosine to uracil residues to initiate somatic hypermutation, followed by the excision of uracil by UNG to create sites where DNA nicks can be made. AID and UNG defects cause HIGM type 2 and HIGM type 5, respectively. Patients affected by these defects generally have milder disease as compared to XHIGM and HIGM type 3. This is because AID and UNG defects are primarily antibody deficiency with intact T cell immunity. Patients can present in early childhood with recurrent bacterial infections affecting the sinopulmonary system and the gastrointestinal tract. Unlike CD40/CD40L deficiency, patients with AID/UNG deficiency are not susceptible to opportunistic infections. Lymphoid hyperplasia is a characteristic finding in patients with AID/UNG deficiency due to large germinal centers filled with proliferating B cells expressing IgM and IgD in the pathogenesis of the disease.

Question 3

What is the more common type of malignancy associated with XHIGM?

A. Cholangiocarcinoma
B. Chronic lymphocytic leukemia
C. Melanoma
D. Non-Hodgkin lymphoma
E. Squamous cell carcinoma

Answer and Explanation

Answer: A

Many primary immunodeficiency disorders (PIDD) are associated with an increased susceptibility to malignancy in part due to genetic predisposition and impaired immune functions. Patients with XHIGM are at higher risk of developing chronic sclerosing cholangitis, which can lead to cirrhosis and cholangiocarcinoma. Chronic lymphocytic leukemia is a secondary cause of hypogammaglobulinemia, which may benefit from immunoglobulin replacement therapy. Melanoma has been reported in CVID. Many PIDD are associated with non-Hodgkin's lymphoma including ALPS, CVID, SCID, ataxia-telangiectasia, hyper-IgE syndrome with *STAT3* mutation, Wiskott-Aldrich syndrome, and autosomal recessive HIGM caused by AID or UNG deficiency. Some patients with XHIGM can also develop lymphoma especially those who have lymphadenopathy. However, hepatobiliary carcinomas are more common in XHIGM. Patients with hyper-IgE syndrome with *STAT3* mutation and *DOCK8* mutation are at higher risk of developing squamous cell carcinoma.

1.3 Clinical Manifestations of X-Linked Hyper-IgM

Patients with XHIGM are susceptible to recurrent opportunistic infections, sinopulmonary infections, neutropenia, autoimmune disease, and cancers affecting the pancreas, liver, and biliary tree.

The inability to activate monocytes and dendritic cells via CD40/CD40L interaction confers susceptibility to opportunistic infections. Patients can present during childhood with *Pneumocystis jirovecii* pneumonia and *Cryptosporidium* sp. induced diarrhea. They can also present as an adolescent with parvovirus induced aplastic anemia. Some patients may not have significant symptoms until the fifth decade of life. The most common clinical manifestation is recurrent infection of the sinopulmonary system including viral upper respiratory infections, otitis media, sinusitis, and lower respiratory tract infections. The pathogens causing lower respiratory tract pneumonias include *Pneumocystis jirovecii*, *Mycobacterium bovis*, *Mycoplasma pneumoniae*, atypical mycobacterium, *Streptococcus* sp., cytomegalovirus, and adenovirus. *P. jirovecii* pneumonia can be the presenting feature in a patient with XHIGM. The classic chest-imaging finding shows a bilateral interstitial pattern with granular, reticular, or glass-ground pattern. Recurrent pneumonia in patients with all forms of HIGM predisposes them to developing bronchiectasis.

Extra-pulmonary symptoms involving the gastrointestinal, hematologic, and central nervous systems can also occur. Patients can present with more than one manifestation. See Table 1 for a list of clinical manifestations in patients with XHIGM.

Gastrointestinal findings typically involve the biliary system or the liver and are the second most common clinical manifestation in XHIGM patients. Biliary disease can lead to abnormal liver function and the development of sclerosing cholangitis. Cytomegalovirus infection has been implicated in some reports of chronic sclerosing cholangitis. The development of liver disease and sclerosing cholangitis is associated with higher mortality in patients with XHIGM. Chronic sclerosing cholangitis can lead to cirrhosis and increases the risk of cholangiocarcinoma.

Patients with AID and UNG deficiency are at increased risk of developing lymphomas due to the associated lymphoproliferation seen in these HIGM types. Although lymphadenopathy is not a typical finding in patients with XHIGM, the affected ones also have higher propensities of developing lymphomas. In XHIGM, hepatobiliary carcinomas such as hepatocellular carcinoma, cholangiocarcinoma, and biliary tree tumors are more common. GI malignancies account for approximately 25% of mortality in adolescent and young adult with XHIGM. Neuroendocrine carcinoma of the appendix and colon is rare but has been described in patients with CD40/CD40L deficiency.

Neutropenia has been reported to develop in about 50% of cases with either a transient or prolonged and persistent clinical course. Neutropenia contributes to patient's susceptibility to bacterial infections. The underlying mechanism for the development of neutropenia is not well understood. It is proposed that CD40/CD40L is important in the stimulation of myelopoiesis, and the lack of stimulation may play a role. Neutropenia is responsive to granulocyte colony-stimulating factor.

Autoimmune diseases are seen in all types of HIGM but occur more commonly in some types of HIGM. In general, the prevention of autoimmunity relies on intact peripheral and central B cell tolerance. CD40/CD40L is involved in peripheral B cell tolerance but not central B cell tolerance. Therefore, autoimmune complications have been described in some patients with XHIGM and HIGM type 3. In contrast, AID and UNG play a role in both central and peripheral B cell tolerance. The loss of both central and peripheral B cell tolerance places patients with AID and UNG defects at a higher risk for autoimmune disease as compared to XHIGM

Table 1 X-linked hyper-IGM clinical manifestations

Respiratory and ear/nose/throat system
Upper respiratory infection
Otitis media
Sinusitis
Pneumonia
Pneumocystis jirovecii, *Mycobacterium bovis*, *Mycoplasma pneumoniae*, atypical mycobacterium, *Streptococcus* sp., cytomegalovirus, and adenovirus
Histoplasmosis
Gastrointestinal system
Chronic diarrhea
Enteritis/colitis
Cryptosporidium parvum, *Giardia lamblia*, *Salmonella* sp., and *Entamoeba histolytica*
GI ulcers
Ischemic colitis
Oral ulcers
Esophagitis
Gastritis
GERD
Liver disease
Sclerosing cholangitis
Cryptosporidium parvum and cytomegalovirus
Inflammatory bowel disease
Malignancy
Adenocarcinoma of the GI tract (hepatocellular carcinoma, bile duct carcinoma, and biliary tree tumors)
Neuroendocrine carcinoma
Central nervous system
Meningoencephalitis
Toxoplasma gondii, *Cryptosporidium* sp., *Haemophilus influenzae*, herpes simplex virus, and enterovirus
Hematologic system
Aplastic anemia
Parvovirus
Thrombocytopenia
Neutropenia
Autoimmune hemolytic anemia
Bone
Osteopenia

and HIGM type 3. Autoimmune conditions described in HIGM syndromes include polyarthritis, inflammatory bowel disease, thrombocytopenia, autoimmune hemolytic anemia, autoimmune hepatitis, and autoimmune thyroiditis.

Osteopenia has also been described in XHIGM, presenting as spontaneous rib fractures without history of trauma. Patients with CD40L deficiency have lower bone mineral density due to increased osteoclastic activity from defective T cell interaction.

Female carriers of CD40L deficiency are typically asymptomatic. However, occasional symptomatic female carriers with lyonization have been reported.

Question 4

What laboratory abnormality is most likely associated with XHIGM?

	CD3+ counts	CD19+ counts	IgG	IgA	IgM	Antibody response to pneumococcal antigens	Mitogen proliferation assays
A	Normal	Normal	Normal	Low	Normal	Normal	Normal
B	Normal	Normal	Normal	Normal	Normal	Low	Normal
C	Normal	Normal	Low	Low	High	Low	Low
D	Normal	Low	Low	Low	Low	Low	Normal
E	Low	Normal	Normal	Normal	Normal	Normal	Low

Answer and Explanation

Answer: C

Patients with XHIGM typically have low serum IgG, IgA, IgE and normal or high IgM. Flow cytometry can show a normal number of CD19+ B cells and a low number of CD27+IgM−IgD− switched memory B cells. T cell number is typically normal; however, impaired T cell function is observed.

Answer choice A refers to selective IgA deficiency with low serum IgA but normal IgG, IgM, and vaccine titers. The majority of patients is asymptomatic or develops mild sinopulmonary infections. Patients with selective IgA deficiency are at risk for atopic diseases and autoimmune disease.

Answer choice B refers to specific antibody deficiency. Patients with specific antibody deficiency typically have normal serum IgG, IgA, and IgM but poor polysaccharide response to the pneumococcal vaccine. These patients primarily present with sinopulmonary infections.

Answer choice D refers to X-linked agammaglobulinemia (XLA). Patients with XLA present with very low immunoglobulins and low or absent CD19+ B cells. These patients have recurrent infections with sinopulmonary, GI infections and can develop severe viral encephalitis.

Answer choice E refers to DiGeorge syndrome. DiGeorge syndrome patients typically exhibit T cell immunodeficiency with T cell lymphopenia and impaired T cell function. Classically, patients with DiGeorge syndrome can present with cardiac defects, abnormal facies, thymic hypoplasia, cleft palate, and hypocalcemia.

1.4 Diagnosis and Laboratory Abnormalities in Hyper-IgM Syndrome

The diagnosis of HIGM syndrome should be suspected in patients with a convincing clinical history, family history, immunologic laboratory workup, and genomic sequencing. The European Society for Immunodeficiencies (ESID) developed criteria for the diagnosis of HIGM syndrome. See Table 2 for the ESID diagnostic criteria.

Table 2 European Society for Immunodeficiencies (ESID) criteria for high syndrome

At least one of the following:
Increased susceptibility to infections (recurrent and/or opportunistic, including *Cryptosporidium* sp.)
Immune dysregulation (autoimmunity, lymphoproliferation, and sclerosing cholangitis)
Cytopenia (neutropenia or autoimmune)
Malignancy (lymphoma)
Affected family member
AND both of the following laboratory abnormalities:
Marked decrease of IgG (measured at least twice)
Normal or elevated IgM (measured at least twice)
AND the exclusion of the following conditions:
Other defined causes of hypogammaglobulinemia
Profound T cell deficiency, defined as two out of three of the following:
- Absolute CD4 numbers/microliter:
  0–6 months <1000, 6 months–1 year <800, 1–2 years <500, 2–6 years <300, 6–12 years <250, >12 years <200
- % naive CD4:
  0–2 years <30%, 2–6 years <25%, 6–16 years <20%, >16 years 10%
- Absent T cell proliferation

Patients with HIGM typically have low serum IgG, IgA, IgE and normal or high IgM. Due to the defect in class switch recombination, all forms of HIGM syndrome will have markedly decreased serum levels of switched immunoglobulins. In the United States Immunodeficiency Network (USIDNET) Registry cohort of 145 patients, 80% of the patients with HIGM had normal IgM levels and 15% of the patients had normal serum IgA. This is interesting considering that the main pathogenic mechanism for HIGM syndrome is the loss of isotype switching. In these cases, it suggests that alternative mechanisms for isotype switching may be implicated.

In patients with HIGM of all types, flow cytometry typically shows normal numbers of total B cells. Reduced numbers of isotype switched memory B cells (CD27+IgM−IgD−) are seen in patients with XHIGM and CD40 deficiency. Abnormal CD40L or CD40 can also lead to defective T cell co-stimulation and proliferation, hence a combined cellular and humoral immunodeficiency phenotype is observed.

HIGM patients with AID and UNG deficiency typically present with normal numbers of total B cells, normal switched memory B cells, and normal T cell immunity. The generation of switched memory B cells requires CD40/CD40L interaction, explaining why low switched memory B cells are seen in CD40L and CD40 deficiency but not AID or UNG deficiency. Genetic analysis can be helpful to determine the gene defect in HIGM.

Question 5

The patient lives on a farm, drinks from well water, tends to livestock, and swims in a nearby pond in the summers. What preventative counseling should be given to this patient?

A. Drinking well water unfiltered is acceptable as long as it is freshly collected
B. No extra precautions are needed while traveling
C. Ponds are safe to swim in but not lakes
D. Regular contact with livestock does not pose a risk
E. Swimming in public pools should be limited

Answer and Explanation

Answer: E

Patients with XHIGM are susceptible to recurrent infection with *Cryptosporidium* sp. *Cryptosporidium* sp. are elevated in the summer months and can be chlorine-resistant. Patients should be counseled to avoid high-risk activities including swimming in public swimming pools, lakes, and ponds, drinking unfiltered water, and contact with livestock. Patients with XHIGM should also take extra precautions while traveling.

1.5 Treatment and Prognosis of X-Linked Hyper-IgM

The mainstay treatment of XHIGM is immunoglobulin replacement, antimicrobial prophylaxis, preventative measures to avoid *Cryptosporidium* sp. exposure and treatment of associated complications such as hematological and liver diseases.

The use of immunoglobulin replacement therapy has been helpful to decrease chronic infections in patients with XHIGM. For patients with chronic diarrhea and severe pulmonary complications, a high dose of immunoglobulin replacement is associated with better outcomes.

Oral trimethoprim-sulfamethoxazole or intravenous pentamidine can be used to prevent *P. jirovecii* infections. Azithromycin prophylaxis for *Cryptosporidium* sp. has been described but not well established. *Cryptosporidium* sp. is typically elevated in the summer months and can be chlorine-tolerant. Given this, prevention strategies to avoid *Cryptosporidium* sp. exposure are recommended for patients with XHIGM. This includes avoidance of public swimming pools, lakes, ponds, limiting contact with livestock, drinking only boiled or filtered water. Extra precautions while traveling, including hygienic measures and drinking bottled water, may be helpful to prevent infection with *Cryptosporidium* sp.

Careful monitoring for hematologic, autoimmune, and liver diseases is important. For severe chronic neutropenia and autoimmune disorders, recombinant granulocyte colony-stimulating factor and immunosuppressant therapy can be used, respectively. Careful monitoring of liver function can help prevent liver disease. Liver transplantation has been used in advanced liver disease such as sclerosing cholangitis, however relapse can occur.

For XHIGM patients with combined immunodeficiency, hematopoietic stem cell transplantation (HSCT) is the only corrective therapy. The first reported successful HSCT for a XHIGM patient was in 1995 and since then, there have been many more showing that HSCT can improve the well-being of XHIGM patients. A recent report showed that there is no significant difference in survival between patients treated with or without HSCT, however when adjusted for the year of transplant, a gradual survival benefit was noted for the transplant group after 1987 and significantly improved by 1993. This may be due to better HLA typing methods, antimicrobial prevention, and monitoring in recent decades. Earlier age at transplantation and absence of liver disease are associated with the best survival outcomes. HSCT should preferably be done before the age of 10 years and prior to the development of liver disease.

The leading causes of death for XHIGM patients include infectious complications, liver failure, and malignancy. In a study of a large cohort of HIGM patients, the overall median survival age is 25 years from the time of diagnosis. Genetic counseling should be offered to families affected by XHIGM.

As patients with HIGM syndrome can develop hematologic, pulmonary, and gastrointestinal complications, it is important for routine surveillance of these organ systems. At least annually, a complete blood count with differential should be performed to monitor for cytopenias. Similarly, liver function testing and abdominal ultrasound should also be completed. Pulmonary function testing and chest imaging may be indicated for patients who are at risk for developing bronchiectasis.

To monitor for progression of their immunodeficiency, immunoglobulin levels and lymphocyte subpopulations should be checked routinely. Polymerase chain reaction (PCR) based testing for stool *Cryptosporidium* sp. and *Microsporidium* sp. are helpful to detect early signs of infection. Clinicians should be aware that antibody-based testing is not reliable in these patients. If these organisms are detected, the patient should undergo treatment with azithromycin or nitazoxanide to prevent biliary tract and liver complications.

Case 2

A 5-year-old boy presents with excessive gum bleeding and easy bruising. He has a history of recurrent otitis media requiring antibiotic therapy at least six times in the past year. There is no family history of recurrent infections. On physical exam, the patient is in the 45 percentile for weight, he has severe eczema and several bruises over his trunk, but no palpable lymphadenopathy or organomegaly. He has visible tonsils. His serum laboratory studies show a platelet count of 20,000 kU/L, normal levels of IgG and IgM, poor antibody response to immunizations, and an abnormal T cell proliferation study. Blood smear shows small platelets. Genetic testing confirms the diagnosis.

Question 6

The gene defect in this disease results in the deficiency of a protein involved in which important cell function?

A. Actin polymerization
B. Class switch recombination
C. Cytokine activation of integrins
D. Somatic hypermutation
E. VDJ recombination

Answer and Explanation

Answer: A

The most likely diagnosis for this patient is Wiskott-Aldrich syndrome (WAS) presenting with the classic triad of recurrent infections, eczema, and microthrombocytopenia. This patient's thrombocytopenia causes him to bleed and bruise easily. WAS is caused by a mutation in the *WAS* gene which encodes WAS protein (WASp). WASp is involved in several different cellular processes including actin polymerization and cell signaling. Immunology laboratory studies can show normal serum IgG and IgM levels, high IgE levels, poor antibody response to immunizations, and low T cell proliferation to mitogen.

Answer choice B and D refer to hyper-IgM syndrome. The most common form of hyper-IgM syndrome is X-linked and caused by a defect in *CD40LG*. This gene encodes the CD40 ligand protein, which is important for class switch recombination in B cells. Patients with XHIGM can present with opportunistic infections of the respiratory and gastrointestinal tract, failure to thrive, autoimmunity and hepatobiliary disease. Autosomal recessive hyper-IgM can be due to AID and UNG defects, which are important for somatic hypermutation of B cells. Patients with AID and UNG deficiency suffer from recurrent sinopulmonary infection, gastrointestinal infections, lymphoid hyperplasia, and autoimmunity. Patients with HIGM syndrome would not be expected to have eczema or microthrombocytopenia. Neutropenia is more commonly seen in HIGM syndrome.

Answer choice C refers to leukocyte adhesion defect type 3 (LAD3). LAD3 is an autosomal recessive disorder caused by a defect in the Kindlin 3 protein, essential for integrins activation. Integrins are important for leukocytes and platelets to bind to the endothelium. Kindlin 3 protein deficiency can present with recurrent infections and impaired pus formation. Patients have poor wound healing and delayed detachment of the umbilical cord, but not eczema as seen in WAS. Patients with LAD3 can have inflammation of the gums and bleeding due to abnormal clotting. However, unlike in WAS, patients with LAD3 would not have microthrombocytopenia.

Choice E refers to autosomal recessive severe combined immunodeficiency (SCID) from *RAG1/2* gene defect leading to defects in V(D)J rearrangement of T and B cells. Patients with RAG1/2 deficiency present with a combined immunodeficiency phenotype. Those with Omenn syndrome typically have severe dermatitis along with severe and recurrent infections. Thrombocytopenia is not a feature.

2 Wiskott-Aldrich Syndrome

2.1 Genetics and Pathogenesis of Wiskott-Aldrich Syndrome

Wiskott-Aldrich syndrome (WAS) is an X-linked disease affecting multiple organ systems. Classically, it presents with the triad of thrombocytopenia with small platelets, eczema, and immunodeficiency. Patients with WAS have an increased risk of malignancy and autoimmunity. The estimated incidence is approximately 1–4 per 1 million males. The defect involved in WAS is the WAS protein (WASp) encoded by the *WAS* gene on the short arm of the X chromosome. WASp is a 502-amino acid intracellular protein found in non-erythroid hematopoietic cells. WASp is important in actin polymerization, cytoskeletal rearrangement, receptor engagement, and cell signaling. WASp is a member of the

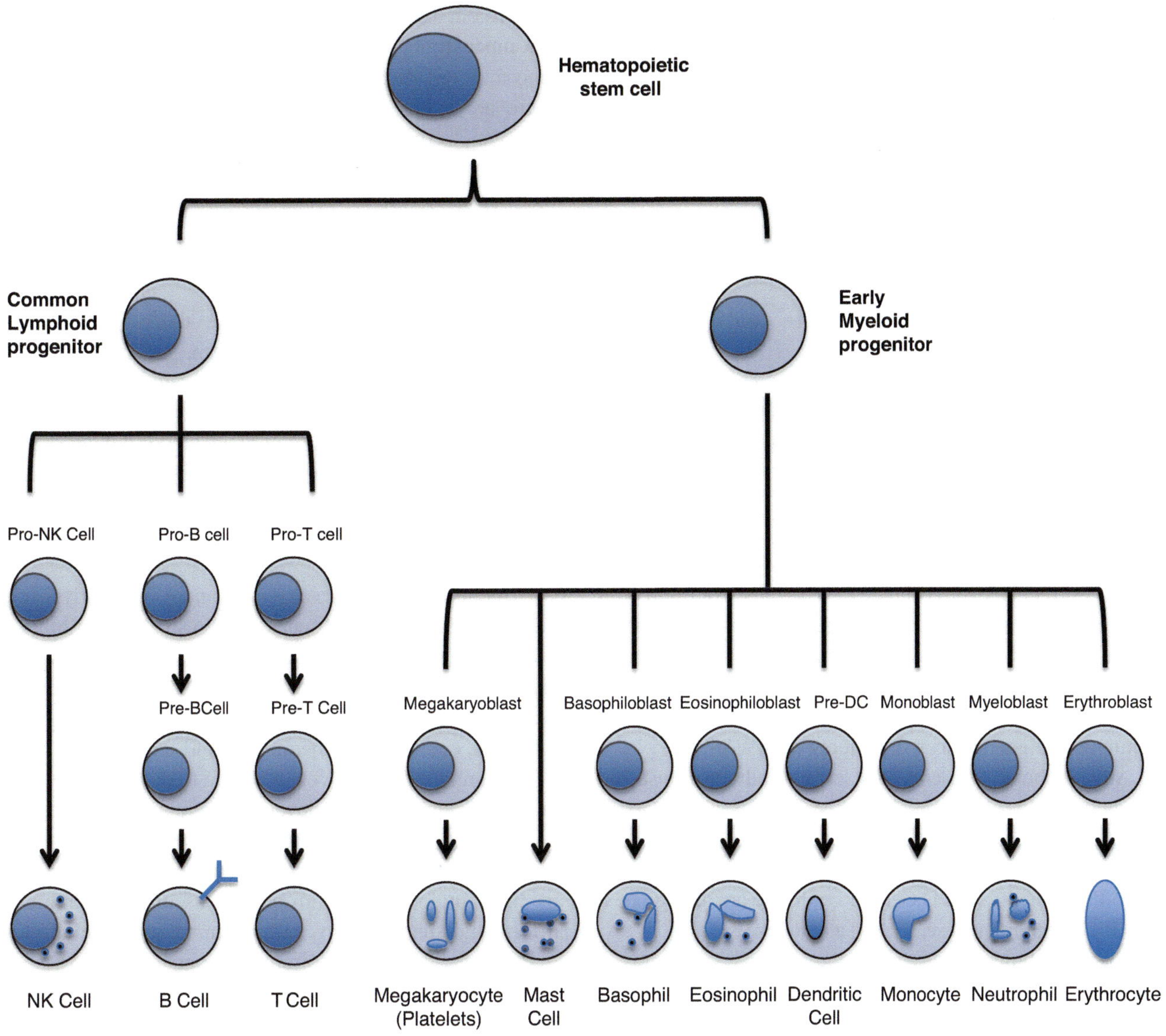

Fig. 2 Major lineages of the hematopoietic tree

family of actin nucleation-promoting factors (NPFs). At steady state, WASp is auto-inhibited. Upon activation, WASp is recruited to the site of membrane signaling and interacts with actin-related protein (Arp2/3) complex to initiate actin polymerization.

WASp involves in intrinsic functions of both lymphoid cells (T cells, B cells, and natural killer (NK) cells) and myeloid cells (monocytes, macrophages, dendritic cells, and neutrophils). See Fig. 2 for the hematopoiesis and the development of the major lineages of blood cells. Due to the important of WASp to the development of all non-erythroid hematopoietic cells, loss of function mutations in *WAS* can cause a combined immunodeficiency. Early progenitors of the lymphoid lineage can develop normally, however WASp is required for the survival of the terminal differentiation of these cells.

In B cells, WASp deficiency impairs the B cell receptor (BCR) and integrin signaling that disrupts proper immunological synapse formation and impacts B cell activation. Intrinsic B cell dysfunction may explain altered antibody production in WAS. In T cells, WASp is present in the nucleus and may play a role in expression of cytokines required for T helper 1 cell differentiation, leading to a skewing towards the T helper 2 cell type. This may contribute to the atopic phenotype and high IgE levels seen in patients with WAS.

WASp may also play a role in inducing apoptosis in T cells that respond to autoantigens. Under T cell receptor (TCR) stimulation and interaction with Fas ligand (FasL), activated T cells responding to autoantigens can undergo apoptosis. In a WASp deficient mouse model, there is an impaired TCR-induced FASL secretion in CD4+ T cells with increased autoantibody production. T regulatory (Treg) cell numbers are normal in patients with WAS, but the function of these cells is altered. Both abnormal apoptosis of autoreactive T cells and altered function of Treg cells may explain the development of autoimmunity in WAS.

Question 7

In addition to recurrent infections, what are the most common clinical manifestations in WAS?

A. Dysmorphic facies, intellectual disability, thymic hypoplasia
B. Lymphadenopathy, recurrent diarrhea, eczema
C. Lymphoma, autoimmune hemolytic anemia, eczema
D. Progressive neuronal loss, lymphoma, dysarthria
E. Retained primary dentition, scoliosis, eczema

Answer and Explanation

Answer: C

The correct answer is choice C. The most common clinical manifestations in WAS include microthrombocytopenia, eczema, and immunodeficiency. Patients with WAS can also have autoimmune diseases such as autoimmune hemolytic anemia and autoimmune neutropenia. Lymphoma is the most common malignancy seen in patients with WAS.

Answer choice A describes DiGeorge syndrome, most commonly from a chromosome 22 deletion. Patients with DiGeorge syndrome present with cardiac defects, dysmorphic facies, thymic hypoplasia, cleft palate, hypocalcemia, and intellectual disability. Patients with DiGeorge syndrome are also prone to develop autoimmune diseases and atopic diseases.

Answer choice B describes the clinical symptoms of Omenn syndrome, which can present with failure to thrive, chronic diarrhea, lymphadenopathy, hepatosplenomegaly, severe dermatitis, and recurrent infections.

Answer choice D describes ataxia-telangiectasia, a progressive disease characterized by neuronal loss, oculomotor apraxia, dysarthria, and telangiectasia.

Answer choice E describes autosomal dominant hyper-IgE syndrome caused by *STAT3* mutations. In addition to recurrent abscesses, fungal infections, and bacterial pneumonias, patients with *STAT3* mutations can have severe eczema, connective tissue disease, and skeletal abnormalities such as retained primary dentition and scoliosis. These patients can develop lymphomas, but not typically hematologic autoimmune diseases.

2.2 Clinical Manifestations and Diagnosis in Wiskott-Aldrich Syndrome

WAS gene mutation leads to a spectrum of disease severity. The clinical phenotype of WAS correlates with the level of WASp protein. Complete absence of WASp leads to defective protein function in multiple hematopoietic cell lines. Thrombocytopenia with small platelets, lymphopenia, and abnormal lymphoid and myeloid cell function are seen. There are approximately 300 mutations reported in *WAS* gene, which contributes to this wide spectrum of phenotypes. The most common symptoms include microthrombocytopenia, eczema, immunodeficiency, autoimmunity, and malignancy.

The most frequent finding in patients with WAS is microthrombocytopenia.

Microthrombocytopenia increases the risk of bleeding diathesis in the central nervous system, gastrointestinal system, and oral mucosal system. More than 80% of patients have frequent hemorrhages and severe bleeding which is a cause for mortality in 4–10% of patients. It is to be noted that WAS is associated with microthrombocytopenia, unlike the other more common bleeding disorder such as idiopathic thrombocytopenic purpura (ITP) which is associated with macrothrombocytopenia.

Difficult to treat severe eczema is more frequently associated with a complete absence of WASp. Susceptibility to infections results from defects in T cell, B cell, and NK cell function. Patients can develop serious bacterial, viral, and fungal infections including sinopulmonary infections, skin abscesses, enterocolitis, meningitis, and sepsis.

Autoimmunity in WAS can be due to defect in Treg cells and B cells leading to the development of autoantibodies and loss of immune regulation. Autoimmune diseases most frequently associated with WAS are autoimmune hemolytic anemia, autoimmune neutropenia, vasculitis, autoimmune arthritis, IgA nephropathy, and inflammatory bowel disease.

Patients who have autoimmune sequalae of WAS are at higher risk of developing malignancies. Abnormal NK function and cytotoxic T cell function play a role in the pathogenesis of malignancy in WAS patients. Malignancies affect older children with WAS compared to infants. Epstein–Barr virus (EBV) positive B cell lymphoma is the most frequently reported. Pre-B cell lymphoma, B cell lymphoma, glioma, reticulum cell sarcoma, lymphoblastic leukemia, testicular carcinoma, and acoustic neuroma have also been reported in patients with WAS.

X-linked thrombocytopenia (XLT) has generally a milder clinical phenotype compared to WAS. Mild decrease in the level of WASp protein expression typically leads to XLT, whereas complete loss of protein expression leads to WAS. XLT is characterized by thrombocytopenia and possible eczema and mild immunodeficiency. As the clinical symptoms overlap, it is helpful to be able to differentiate between XLT and WAS. A scoring system has been utilized to separate these two entities based solely on the severity of clinical symptoms. The scoring system focuses on the above-mentioned five main symptoms, which are thrombocytopenia, eczema, immunodeficiency, autoimmunity, and

Table 3 Scoring system for Wiskott-Aldrich syndrome and X-linked thrombocytopenia

	iXLT	XLT		WAS			
Clinical scores	<1	1	2	3	4	5A	5M
Thrombocytopenia	+/–	+	+	+	+	+	+
Eczema	–	–	+/–	+	++	++/–	++/–
Immunodeficiency	–	–	+/–	+	++	++/–	++/–
Autoimmunity	–	–	–	–	–	+	–
Malignancy	–	–	–	–	–	–	+

iXLT intermittent X-linked thrombocytopenia
Scoring system: + is moderately present, – is absent, +/– is mildly present or absent, ++ is severely present, ++/– is severely present or absent, 5A: score of 5 with autoimmunity, 5M: score of 5 with malignancy

malignancy. A score of 0.5–1 is assigned to each symptom domain based on the severity. See Table 3 for the components of the scoring system in detail. Patients who are initially diagnosed for XLT should be monitored closely and their clinical scores updated if they develop more profound immunodeficiency, autoimmunity, or malignancy. These patients may have been prematurely classified as XLT. However, with time other clinical manifestations may develop.

Question 8

What laboratory abnormality is most likely associated with WAS?

	T cell	B cell	NK cell	Immunoglobulins	Antibody response to vaccine	Other findings
A	Absent Th17 cell	Normal	Normal	Low IgG, normal IgM, very high IgE	Low	Eosinophilia
B	Low CD4+ counts and function	Normal	Low counts and function	Low IgG, normal IgM	Low	Anemia
C	Low counts and function	Normal	Normal counts, low function	Variable levels	Low	Thrombocytopenia
D	Impaired Treg	Normal	Normal	Normal IgG, normal IgM, high IgE	Normal	Anti-enterocyte antibodies
E	Normal	Very low	Normal	Very low	Low	Neutropenia

Answer and Explanation

Answer: C

Patients with WAS typically present with T cell lymphopenia, low T cell proliferation in response to mitogen stimulation, variable immunoglobulin numbers, poor response antibody response to immunizations, normal NK cell numbers, low NK cell function, and thrombocytopenia.

Answer choice A refers to autosomal dominant hyper-IgE syndrome from *STAT3* defects. Typical laboratory abnormalities seen in this disease include low IgG, very elevated IgE levels, poor antibody production following immunizations, normal numbers of B cells and NK cells, but absence of Th17 cells. Patients can also have high eosinophil counts. Clinically, patients can present with recurrent infections, severe eczema, connective tissue disease, and skeletal abnormalities. Although patients with WAS can have high IgE levels and normal IgM levels, they would not have absent Th17 cells.

Answer choice B can be seen in X-linked lymphoproliferative disease (XLP), most commonly due to a mutation in the *SH2D1A* gene. On laboratory studies, patients with XLP typically have low CD4+ T cell counts, impaired T cell function, normal B cell numbers, low serum IgG, impaired antibody response to immunizations, low NK cell number and function. Patients with XLP can present with fulminant EBV infections, hemophagocytic lymphohistiocytosis, aplastic anemia, and lymphomas induced by EBV. Although patients with WAS can have low NK cell function, their NK cell counts are typically normal. Thrombocytopenia is more common in WAS rather than anemia.

Answer choice D refers to immune dysregulation, polyendocrinopathy, enteropathy, X-linked (IPEX) syndrome. IPEX syndrome is a form of autoimmune enteropathy due to a defect in the FOXP3 protein resulting in the depletion of Tregs. Patients with IPEX syndrome typically have normal serum IgG and IgM, high IgE, normal antibody response to immunizations, normal number of NK cells and B cells, but impaired Treg cells. Patients also have detectable number of circulating autoimmune antibodies. Circulating anti-enterocyte antibodies can be detected in IPEX patients. Although patients with WAS can have high IgE, normal IgG and autoimmune diseases, anti-enterocyte antibodies are not associated with WAS.

Answer choice E refers to X-linked agammaglobulinemia (XLA), which can present with very low or absent CD19+ B cells, undetectable immunoglobulins, low post-immunization titers, but normal T cell counts and function. Patients are susceptible to infections with typical bacteria and recurrent sinopulmonary infections leading to chronic lung changes such as bronchiectasis. Patients with XLA can also have neutropenia, which places them at higher risk of infections. Although patients with WAS can have poor antibody response to immunizations, they should not have very low CD19+ B cells and very low immunoglobulin levels of all types.

2.3 Laboratory Abnormalities in Wiskott-Aldrich Syndrome

Due to the effects of a defective WASp on cytoskeletal rearrangement in hematopoietic cells, patients with WAS can show decreased lymphocyte counts, defects in T cell, B cell and NK cell function on laboratory evaluation. These combined immune defects place patients with WAS at risk for serious bacterial, viral, and fungal infections along with the development of malignancy and autoimmune diseases.

Abnormal T cell development and defect in the actin cytoskeleton leading to T cell lymphopenia are the major cause of immunodeficiency in WAS. In order to activate T cells, a stable interface must form between the T cell and the antigen-presenting cell known as the immune synapse. WASp is involved in the formation of this synapse, explaining the functional defect in T lymphocytes. Proliferation to mitogen stimulation can vary among patients due to the variability of T cell counts or the level of the WASp in each cell. Complete loss of WASp leads to T cell lymphopenia and abnormal proliferative responses to mitogens.

B cell development is not greatly affected; however, cytoskeletal defects can impact the migration of B cells. Patients with WAS can display elevated levels of serum IgG, IgE and variable levels of serum IgA and IgM. Elevated levels of serum IgE and the imbalance of Th1 to Th2 cytokine production are responsible for the development of atopic diseases. In fact, 80% of patients have eczema. The more severe forms of eczema persistent to adulthood are resistant to conventional therapies and prone to bacterial or herpes simplex viral super-infections.

NK cell levels may be normal or even elevated in WAS. However, function can be impaired due to the disorganized NK cell immune synapse from the loss of WASp. Abnormal function of NK cells may contribute to the underlying pathogenesis of malignancy in patients with WAS.

Complete blood cell count may reveal anemia and thrombocytopenia. Platelet numbers are affected in WAS due to the decreased production and increased clearance. The abnormal structure due to loss of WASp leads to trapping of platelets in the spleen and other organs along with increased opsonization by macrophages. 50% of patients with WAS will have severe thrombocytopenia with platelet count less than 20,000/μL, placing them at risk for serious bleeding complications.

Question 9

The patient undergoes HSCT for treatment of WAS. Following the procedure, he continues to have bleeding episodes and found to have chronic and refractory thrombocytopenia. What treatment is the best option to treat his thrombocytopenia?

A. Cyclophosphamide
B. Prednisone
C. Rituximab
D. Splenectomy
E. Trimethoprim-sulfamethoxazole

Answer and Explanation

Answer: D

Although all of the answer choices are potential therapeutic options for thrombocytopenia in patients with WAS, splenectomy is the best answer for the treatment of refractory thrombocytopenia, especially post-HSCT.

In patients with WAS, thrombocytopenia is the most difficult clinical complication to cure. In a patient with XLT and bleeding complications with severe thrombocytopenia, splenectomy may be the first treatment of choice. In part, the thrombocytopenia seen in XLT and WAS is due to platelet destruction in the spleen. For patients with WAS, splenectomy can be done post-HSCT for significant residual thrombocytopenia. Although splenectomy can help increase platelet counts, there is an increased risk of sepsis post-splenectomy. This risk can be reduced with vaccination prior to the procedure and prophylaxis antibiotics post procedure.

Rituximab, prednisone, and cyclophosphamide are immunosuppressive agents that are used to treat graft-versus-host disease and persistent autoimmunity post-HSCT associated autoimmune disease. However, immunosuppressive agents are not the best choice to treat refractory thrombocytopenia post-HSCT in patients with WAS. Oral antibiotics such as trimethoprim-sulfamethoxazole are helpful to prevent infections associated with WAS but does not help to treat thrombocytopenia.

2.4 Treatment of Wiskott-Aldrich Syndrome

The conventional treatment for WAS is the management of clinical symptoms. Immunoglobulin replacement is important to prevent infectious complications. Patients with recurrent infections, abnormal quantitative Igs, or poor antibody responses should be given immunoglobulin replacement therapy. In addition to immunoglobulin placement, all patients with WAS should be placed on oral trimethoprim-sulfamethoxazole or equivalent to prevent *P. jirovecii* pneumonia. Immunosuppressive agents such as corticosteroids, rituximab, and cyclophosphamide can be utilized to treat WAS associated autoimmunity. Patients should avoid live vaccines.

Splenectomy is an option and has been used to treat refractory microthrombocytopenia. It may be a desirable option to treat XLT patients with bleeding diathesis and thrombocytopenia but have an overall milder clinical disease. All patients who have undergone splenectomy will require lifelong prophylactic antibiotic therapy, most commonly with penicillin, and immunoglobulin replacement therapy. Over the years, elective splenectomy has been falling out of favor due to the increased risk of invasive pyogenic infections post-surgery despite antimicrobial prophylaxis.

HSCT is the treatment of choice for WAS, especially when a matched donor is available. The best outcomes with curative HSCT involve pre-transplant myeloablative therapy and the use of an HLA-identical sibling donor or matched unrelated donor. Patients with better clinical conditions prior to transplantation have better outcomes. Acute or chronic graft-versus-host disease can be seen in the first year following HSCT and treated with immunosuppressive agents. Persistent thrombocytopenia or the development of autoimmunity has been seen in patients with WAS who undergo HSCT. This depends on the durability of the donor cell engraftment in the recipient. This factor is termed the donor chimerism. Low donor chimerism may cause persistent thrombocytopenia, whereas highest donor chimerism may contribute to autoimmunity.

When a matched donor is not available, gene therapy is an attractive alternative for patients with WAS. Gene therapy involves infecting the patient's hematopoietic stem cells with a retrovirus that expresses WASp, ex vivo. Patients typically undergo a myelosuppressive regimen prior to reinfusion of the autologous hematopoietic stem cells. Gene therapy may be considered in a patient who has a high clinical symptom score and/or absent WASp levels and when there is no available HLA-matched donor. One of the drawbacks of gene therapy is possible insertional mutagenesis. Earlier gene therapy options for WAS utilized a gammaretroviral vector (GV) which carried a higher risk for insertional mutagenesis and the development of leukemia. However, the more recently developed recombinant lentiviral vector (LV) has lower rates of insertional mutagenesis. Gene therapy has been shown to improve bleeding diathesis, autoimmunity, eczema, and risk of severe infections. However, platelet counts remain somewhat low. Gene therapy can be beneficial to patients with severe clinical disease and higher risk of HSCT-related complications and mortality.

Question 10
What is the most common complication of HSCT in patients with WAS?

A. Autoimmune cytopenia
B. Epstein–Barr virus infection
C. Inflammatory gut granulomas
D. Leukemia
E. Large-cell lymphoma

Answer and Explanation
Answer: B

Patients with WAS who undergo HSCT can experience post-transplant viral infections, autoimmunity, and malignancy including all of the answer choices above. However, the most common post-transplant complication is viral infection/reactivation including infection with EBV, CMV, and disseminated chickenpox.

2.5 Prognosis of Wiskott-Aldrich Syndrome

Prompt diagnosis and treatment of patients with WAS are important. In a cohort of 154 patients with WAS, 36% of the patients experienced non-HSCT-related mortality at a mean age of 8 years. The major cause of death in these patients was infection-related followed by malignancy and bleeding. For patients with WAS who undergo successful HSCT, the long-term survival is greater than 80%.

However, HSCT is not without complications. HSCT can be associated with post-transplant autoimmunity, infections, and malignancy. The development of post-transplant autoimmunity usually appears within the first 2 years post-HSCT but can be delayed with a more protracted course. Autoimmune cytopenia is the most common form of post-HSCT autoimmunity. Patients who have WAS-associated autoimmunity prior to HSCT are at greater risk of severe and prolonged post-HSCT autoimmunity.

Infectious complications, including viral reactivation, are more pronounced with a cord blood donor source. Close surveillance, preventive strategies, and treatment of potential post-HSCT infections are crucial. Post-transplant malignancies reported include lymphoproliferative disorders (non-EBV and EBV associated), B cell lymphoma, myelodysplasia, myeloid leukemia, and large-cell lymphoma.

Patients with WAS are best treated and managed in centers with multidisciplinary services available including subspecialist support, psychosocial support, and genetic counseling. There has been significant strides in the management and treatment of WAS. The advances in allogeneic HSCT have provided promising improved long-term outcomes for patients with WAS. In addition, the development of gene therapy gives patients with WAS a potential alternative curative option. With timely intervention and care, patients with WAS have an excellent prognosis.

3 Summary

Combined immunodeficiency disorders (CID) refer to both T and B cell immunodeficiency that leads to a spectrum of associated abnormalities depending on the extent of the immune defects. Patients with CID are susceptible to opportunistic organisms, which lead to increased morbidity and mortality. The incidence of CID is estimated to be 1:100,000 to 1:5000 live births worldwide. Early diagnosis is crucial to prevent potentially life-threatening infections and to initiate curative therapies at an earlier age. Even though the diagnosis of CID is challenging due to the spectrum of clinical phenotypes, the increasing availability of genetic testing has helped to identify CID patients. CID should be suspected in patients with recurrent respiratory tract infections, chronic diarrhea, failure to thrive, autoimmunity, and a significant family history of recurrent infections. If suspecting a CID, a complete evaluation of humoral and cellular immunity should be initiated including measurement of immunoglobulin levels, recall antibody titers, lymphocyte subset counts, and function. Genetic testing can help to confirm the diagnosis. The treatment of choice depends on the severity of the immune defect and associated conditions, including hematologic abnormalities and autoimmunity. Immunoglobulin replacement therapy and prophylactic antimicrobials are used to prevent severe infections. Immunosuppressive agents and bone marrow stimulating agents can be helpful for associated autoimmunity and cytopenias, respectively. HSCT or possibly gene therapy can be an option for more severe clinical cases of certain CID.

The examples of combined immunodeficiency disorders are XHIGM and WAS. Primary forms of HIGM can be X-linked or autosomal recessive. XHIGM is the most common form caused by mutations in the *CD40LG* gene that encodes the CD40L protein on chromosome Xq26.3. The CD40L is expressed on T cells required for immunoglobulin CSR and activation of immune cells such as B cells, macrophages, and dendritic cells. The autosomal recessive forms of HIGM are caused by a variety of mutations in genes that are involved in CSR, somatic hypermutation, and DNA repair mechanisms, including AID and UNG. Patients with XHIGM are susceptible to *P. jirovecii* pneumonia, *Cryptosporidium* sp. induced diarrhea. Their clinical courses can be complicated by biliary and liver disease leading to sclerosing cholangitis and malignancies of the hepatobiliary system. In addition, neutropenia, osteopenia, and autoimmunity can be seen. Serum IgG and IgA levels are decreased while serum IgM levels are normal or elevated in XHIGM. Patients can have low antibody response to immunizations and impaired T cell function. The mainstay treatment of XHIGM is immunoglobulin replacement, antimicrobial prophylaxis, preventative measures to avoid *Cryptosporidium* sp. exposure, and treatment of associated complications. HSCT is the only curative therapy and has improved the well-being of XHIGM patients.

WAS is an X-linked disease caused by a defect in the *WAS* gene causing a defect in the WASp protein. WASp is present in all non-erythroid hematopoietic cells and is important in actin polymerization, cytoskeletal rearrangement, receptor engagement, and cell signaling. The loss of WASp leads to defective T cells, B cells, NK cells, monocytes, macrophages, dendritic cells, and neutrophils. The level of WASp expression leads to a range of phenotypes ranging from mild XLT to more severe WAS. Patients classically present with the triad of thrombocytopenia with small platelets, eczema, and immunodeficiency. Their clinical course can be complicated by autoimmune disease and malignancy. WAS patients have variable immunoglobulins, poor antibody response to immunizations, T cell lymphopenia, and low T cell function. Treatment relies on infection prevention with oral antimicrobials and immunoglobulin replacement therapy. In some cases, splenectomy, HSCT, and gene therapy provide more favorable results. For patients with WAS who undergo successful HSCT, the long-term survival rate is excellent.

Bibliography

1. Abolhassani H, Chou J, Bainter W, et al. Clinical, immunologic, and genetic spectrum of 696 patients with combined immunodeficiency. J Allergy Clin Immunol. 2018;141(4):1450–8. https://doi.org/10.1016/j.jaci.2017.06.049.
2. Bosticardo M, Marangoni F, Aiuti A, Villa A, Grazia RM. Recent advances in understanding the pathophysiology of Wiskott-Aldrich syndrome. Blood. 2009;113(25):6288–95. https://doi.org/10.1182/blood-2008-12-115253.
3. Buchbinder D, Nugent DJ, Fillipovich AH. Wiskott-Aldrich syndrome: diagnosis, current management, and emerging treatments. Appl Clin Genet. 2014;7:55–66. https://doi.org/10.2147/TACG.S58444.
4. Catucci M, Castiello MC, Pala F, Bosticardo M, Villa A. Autoimmunity in Wiskott-Aldrich syndrome: an unsolved enigma. Front Immunol. 2012;3:209. Published 2012 Jul 18. https://doi.org/10.3389/fimmu.2012.00209.
5. Davies EG, Thrasher AJ. Update on the hyper immunoglobulin M syndromes. Br J Haematol. 2010;149(2):167–80. https://doi.org/10.1111/j.1365-2141.2010.08077.x.
6. de la Morena MT. Clinical phenotypes of hyper-IgM syndromes. J Allergy Clin Immunol Pract. 2016;4(6):1023–36. https://doi.org/10.1016/j.jaip.2016.09.013.
7. de la Morena MT, Leonard D, Torgerson TR, et al. Long-term outcomes of 176 patients with X-linked hyper-IgM syndrome treated with or without hematopoietic cell transplantation. J Allergy Clin Immunol. 2017;139(4):1282–92. https://doi.org/10.1016/j.jaci.2016.07.039.

8. Diaz M, Daly J. Immunology: B cells break the rules. Nature. 2009;460(7252):184–6. https://doi.org/10.1038/460184a.
9. Elfeky RA, Furtado-Silva JM, Chiesa R, et al. One hundred percent survival after transplantation of 34 patients with Wiskott-Aldrich syndrome over 20 years. J Allergy Clin Immunol. 2018;142(5):1654–1656.e7. https://doi.org/10.1016/j.jaci.2018.06.042.
10. Etzioni A, Ochs HD. The hyper IgM syndrome—an evolving story. Pediatr Res. 2004;56(4):519–25. https://doi.org/10.1203/01.PDR.0000139318.65842.4A.
11. Ferrua F, Marangoni F, Aiuti A, Roncarolo MG. Gene therapy for Wiskott-Aldrich syndrome: history, new vectors, future directions. J Allergy Clin Immunol. 2020;146(2):262–5. https://doi.org/10.1016/j.jaci.2020.06.018.
12. Leven EA, Maffucci P, Ochs HD, et al. Hyper IgM syndrome: a report from the USIDNET registry. J Clin Immunol. 2016;36(5):490–501. https://doi.org/10.1007/s10875-016-0291-4.
13. Massaad MJ, Ramesh N, Geha RS. Wiskott-Aldrich syndrome: a comprehensive review. Ann N Y Acad Sci. 2013;1285(1):26–43. https://doi.org/10.1111/nyas.12049.
14. Nikolov NP, Shimizu M, Cleland S, et al. Systemic autoimmunity and defective Fas ligand secretion in the absence of the Wiskott-Aldrich syndrome protein. Blood. 2010;116(5):740–7. https://doi.org/10.1182/blood-2009-08-237560.
15. Notarangelo LD, Hayward AR. X-linked immunodeficiency with hyper-IgM (XHIM). Clin Exp Immunol. 2000;120(3):399–405. https://doi.org/10.1046/j.1365-2249.2000.01142.x.
16. Notarangelo LD, Lanzi G, Peron S, Durandy A. Defects of class-switch recombination. J Allergy Clin Immunol. 2006;117(4):855–64. https://doi.org/10.1016/j.jaci.2006.01.04.
17. Pai SY, Notarangelo LD. Hematopoietic cell transplantation for Wiskott-Aldrich syndrome: advances in biology and future directions for treatment. Immunol Allergy Clin N Am. 2010;30(2):179–94. https://doi.org/10.1016/j.iac.2010.02.001.
18. Qamar N, Fuleihan RL. The hyper IgM syndromes. Clin Rev Allergy Immunol. 2014;46(2):120–30. https://doi.org/10.1007/s12016-013-8378-7.
19. Rivers E, Thrasher AJ. Wiskott-Aldrich syndrome protein: emerging mechanisms in immunity. Eur J Immunol. 2017;47(11):1857–66. https://doi.org/10.1002/eji.201646715.
20. Sullivan KE, Mullen CA, Blaese RM, Winkelstein JA. A multiinstitutional survey of the Wiskott-Aldrich syndrome. J Pediatr. 1994;125(6 Pt 1):876–85. https://doi.org/10.1016/s0022-3476(05)82002-5.
21. Worth AJ, Thrasher AJ. Current and emerging treatment options for Wiskott-Aldrich syndrome. Expert Rev Clin Immunol. 2015;11(9):1015–32. https://doi.org/10.1586/1744666X.2015.1062366.
22. Yazdani R, Fekrvand S, Shahkarami S, et al. The hyper IgM syndromes: epidemiology, pathogenesis, clinical manifestations, diagnosis and management. Clin Immunol. 2019;198:19–30. https://doi.org/10.1016/j.clim.2018.11.007.

Human Immunodeficiency Virus (HIV) Infection and AIDS

Claudiu Georgescu

Abbreviations

AIDS	Acquired immunodeficiency syndrome
ART	Antiretroviral therapy
BAL	Bronchoalveolar lavage
CCR5	Chemokine receptor type 5
CD4	Cluster of differentiation 4
CDC	Centers for disease control and prevention
CMV	Cytomegalovirus
CSF	Cerebrospinal fluid
DHHS	The United States Department of Health and Human Services
EBV	Epstein–Barr virus
FDA	The United States Food and Drug Administration
GALT	Gut-associated lymphoid tissue
HIV	Human immunodeficiency virus
HPTN	The HIV prevention trials network
IFNγ	Interferon gamma
INSTI	Integrase strand transfer inhibitor
KS	Kaposi sarcoma
LDH	Lactate dehydrogenase
MAC	*Mycobacterium avium* complex
MSM	Men who have sex with men
NNRTI	Non-nucleoside reverse transcriptase inhibitors
NRTI	Nucleoside (nucleotide) reverse transcriptase inhibitors
PAS	Periodic acid–Schiff
PCP	*Pneumocystis jirovecii* pneumonia
PCR	Polymerase chain reaction
PPD	Purified protein derivative
PrEP	Preexposure prophylaxis
SIADH	Syndrome of inappropriate antidiuretic hormone secretion
STI	Sexually transmitted infection
TB	Tuberculosis
TMP-SMX	Trimethoprim-Sulfamethoxazole
U=U	Undetectable = Untransmissible
WHO	The World Health Organization

C. Georgescu (✉)
Division of Infectious Diseases, Department of Medicine, University of Toledo, Toledo, OH, USA
e-mail: claudiu.georgescu@utoledo.edu

Case 1

A 45-year-old woman presents to her family physician's office complaining of fever, sore throat, and a rash. She was seen 1 week ago in the emergency room where she tested negative for mononucleosis and streptococcal pharyngitis and was given a diagnosis of viral syndrome. Her last negative HIV test was during her last full-term pregnancy, 15 years ago. She has not been sexually active until very recently; 3 weeks ago she had unprotected vaginal sex with a new partner.

Question 1

What symptom is NOT part of the acute retroviral syndrome?

A. Morbilliform rash
B. Pharyngitis
C. Dyspnea
D. Cervical lymphadenopathy
E. Fever

Answer and Explanation

Answer: C

The differential diagnosis for this patient should include all etiologies of mononucleosis like syndrome, including HIV. Dyspnea is not characteristic for this syndrome, while all other listed symptoms are present in >80% of symptomatic acute HIV.

M. Mahmoudi (ed.), *Absolute Allergy and Immunology Board Review*, https://doi.org/10.1007/978-3-031-12867-7_24

1 HIV Epidemiology, Transmission, and Natural History

CDC estimates that, in 2019, approximately 1.2 million persons with HIV lived in United States, out of which circa 160,000 were undiagnosed, a segment that not only does not take advantage of treatment, but also is more likely to contribute to new infections. While the HIV incidence has been decreasing in US, the prevalence continues to increase, as the number of new infections outpaces the number of deaths. The infection disproportionately affects particular racial, sex, and age groups, with 78% of infected persons being male and 42% Black/African American (which constitute only 13% of the US population). The largest number of new diagnoses occurs in persons aged 20–34, but overall, the HIV population is aging, with almost two thirds being over 45 years old. Male to male sexual contact remains the most important method of transmission in US, as opposed to heterosexual sex and vertical transmission which are the drivers of the epidemic in sub-Saharan Africa. Vertical and postnatal mother to child transmission was not completely eliminated in the US, however universal opt-out screening during pregnancy, prenatal care, early or even pre-conception antiretroviral treatment and breastfeeding avoidance have significantly reduced the risk of HIV acquisition in infancy. Blood and blood products transmission in healthcare setting was virtually eliminated by screening in high-income countries including US but is still an ongoing threat in the developing world. Transmission through injection drug use remains an unsolved problem despite needle exchange programs and preexposure prophylaxis. Risk of HIV acquisition increases with the number of persons the index shares needles with. Saliva, sweat, tears, and urine were not shown to transmit HIV, even though the virus can be occasionally detected in low titers in these fluids.

The pathogenesis of acute HIV infection was well described in a simian model of genital mucosa acquisition, where tissue dendritic cells and macrophages act as targets for a single virion and then carriers to the lymphoid tissue. The CD4 receptor and the CCR5 coreceptor present on the surface of T cells and macrophages are essential in transmission of the virus and subsequent replication. A specific mutation to the CCR5, namely a 32-base deletion (CC5Δ32) leads to resistance to HIV infection.

Acute symptoms develop within 2–4 weeks of infection and while a number of infections can be asymptomatic, some individuals experience a nonspecific syndrome with fever, lymphadenopathy (cervical), a morbilliform rash, pharyngitis, and fatigue, rare manifestations including aseptic meningitis, Guillain-Barre syndrome or oral candidiasis are described. Viral infections including mononucleosis (acute infection with Epstein–Barr virus), acute cytomegalovirus, influenza, enterovirus, as well as other infections—toxoplasmosis, streptococcal pharyngitis, and syphilis can present in a similar manner and are part of the differential diagnosis. At this stage, a high HIV viral load would be detectable in the peripheral blood and most individuals have a positive p24 antigen. The presence of this very high viral load increases the likelihood of transmission, thus making recognition of this syndrome, linkage to care and early treatment even more important.

Early in the infection also the gastrointestinal system is an important target for the virus, one in five GALT T cells are infected in the acute HIV phase and most of them are lysed. The acute retroviral syndrome is followed by a period of viral latency (Fig. 1) associated with chronic infection, during which most individuals are asymptomatic. This period is characterized by a slow decline in the CD4+ T cell count over a period of 7–10 years. The attack on the mucosal lymphoid system continues and eventually leads to depletion of T cells in the intestinal mucosa. Rapid progression to AIDS is described to occur much earlier in individuals with a higher plasma viral load (over 100,000 copies/mL) at 6 months post-initial infection, when a setpoint is achieved. A lower T cell count at the set point is also associated with more rapid progression. Viral containment and the drop in the initial high viral load depend on anti-HIV CD8+ T cells. Throughout the chronic phase, the HIV infection elicits a significant activation of the immune response and the virus continues to evade it due to its ability to mutate frequently, related to the high rate of replication and the error prone transcription.

In late stages, the immune system collapses, which allows for very high viral loads to be detected again once a CD4+ T cell count reaches 200 cells/μL. This is a time when patients are at an increased risk for opportunistic infections and AIDS-defining malignancies.

Question 2

What is the test of choice for screening for HIV in this case?

A. HIV Western Blot
B. Combination antibody/antigen assay, followed by a differentiation assay
C. An over-the-counter home HIV test kit
D. CD4+ T cell count
E. HIV antibody ELISA

Answer and Explanation

Answer: B

All the above assays can be used for the diagnosis of HIV, however since 2014 the CDC recommends a combination antibody/antigen as the test of choice. Historically, the HIV ELISA was used for screening, and the confirmation was done with a cumbersome and time-consuming Western Blot. In 2010, FDA approved a test which included not only detection of HIV-1/2 antibodies, but also the p24 HIV-1 antigen. This test reduces the window period (between the acquisition of HIV and the time when a test would accurately detect infection) to as low as 2 weeks, although the CDC considers the window period to be 45 days, the time when the test becomes close to 100% sensitive. While the combination test does not require confirmation by Western

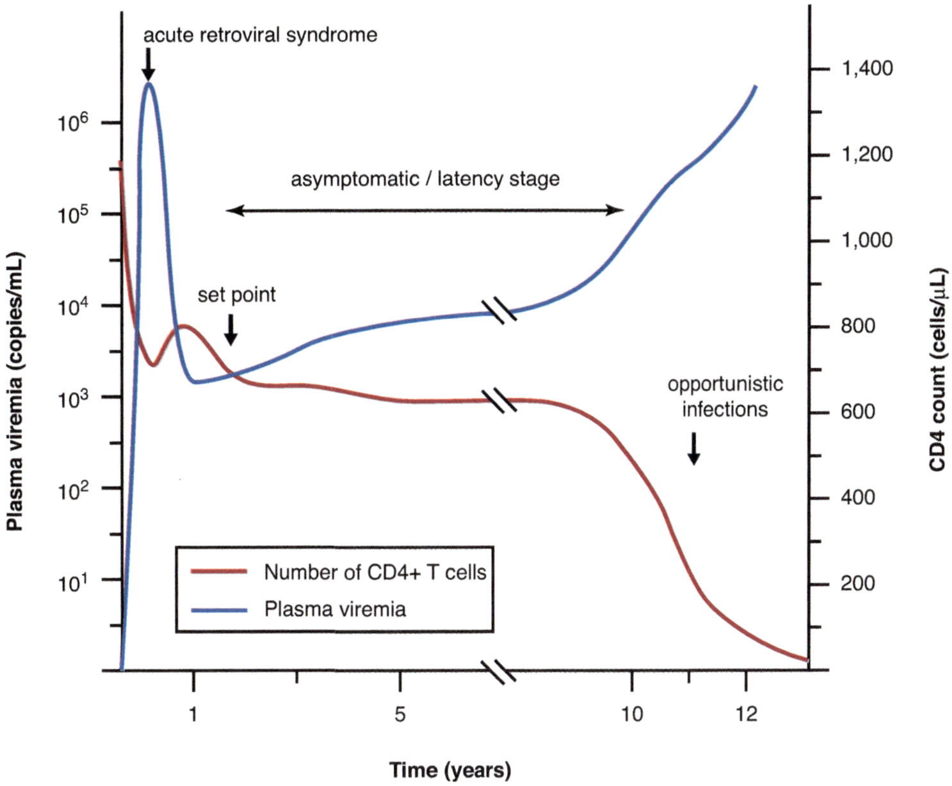

Fig. 1 Natural history of HIV infection in the absence of antiretroviral treatment. (Modified with permission from Gillespie S et al. Human Immunodeficiency Virus Infection and Acquired Immunodeficiency Syndrome, in Clinical Immunology: Principles and Practice, 5th ed., 2019, 39, 545–560.e1 [1])

Blot, it cannot distinguish between HIV-1 and HIV-2 and when positive, must be followed by an immunochromatographic differentiation assay which both confirms the HIV infection and differentiates between HIV-1 and HIV-2. A qualitative HIV-1 DNA test can also be used, but it is not as widely available, is more expensive, and does not detect HIV-2. The HIV-1 RNA quantitative tests are widely available and are occasionally used off-label for diagnosis of acute retroviral syndrome, however, they were designed to be used for monitoring treatment, rather than for diagnosis, as false positive low-level amplification can occur. Also, the quantitative molecular test can be false negative in elite controllers (persons who, despite HIV infection, have no detectable viral load in the peripheral blood in the absence of antiretroviral treatment). The over-the-counter HIV tests are convenient and offer convenient in-home testing, however, when positive, they must be confirmed with a laboratory test. Counselling and linkage to care may also not be as robust with over-the-counter kits, as patients carry the burden to actively seek care, as opposed to being offered support at the time of being informed of a positive result in the case of laboratory tests.

2 HIV Screening and Diagnosis

The diagnosis of HIV is made with serological tests but can also be done with molecular tests. A low CD4+ T cell count should not be used for screening or diagnosis, as a multitude of conditions including acute viral illnesses, malignancies, and drugs can lead to such finding.

Table 1 HIV screening

Category of individuals	Screening frequency
All individuals 13–63 of age	Once, in all health care settings
High risk individuals • Sex partners of persons with HIV • Persons who inject drugs and their sex partners • Persons who have more than one sex partner since their last HIV test • Persons who exchange money for sex or drugs	Once a year
Pregnant women	Universal opt-out screening as early as possible in pregnancy; if high risk exposure, repeat test before week 36

HIV screening is crucial in detecting asymptomatic infections and since 2006 the CDC has recommended performing HIV screening for all persons 13–64 years of age in all health care settings. Screening refers to testing in the absence of symptoms suggestive of HIV infection (Table 1).

HIV testing does not require written consent, which was demonstrated to be a barrier to HIV testing. Even though the recommendations were issued in 2006, it took until 2018 for all states to pass legislation that was consistent with this recommendation. In the case of pregnant women, the test is universally offered but pregnant women still have the right to

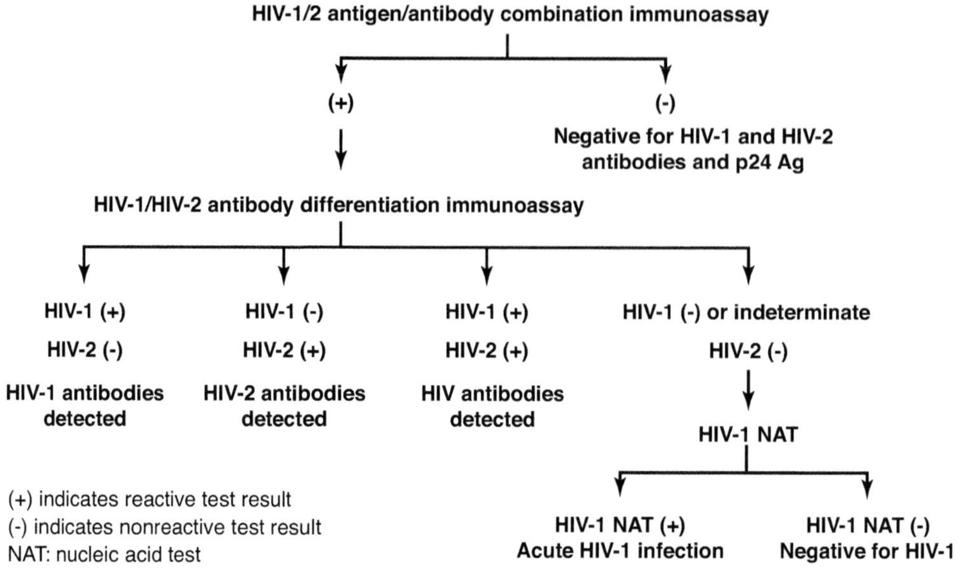

Fig. 2 The CDC recommended algorithm for HIV testing. (From the CDC: Quick Reference Guide—Laboratory Testing for the Diagnosis of HIV Infection: Updated Recommendations June 27, 2014, public domain [2])

specifically decline it; providers should discuss and address the reasons behind refusal. Also, all individuals have the right to anonymous testing.

Delivering the test results can be done without direct contact between a provider and the patient, in the case of a negative HIV test. A positive test should be delivered by a person who can provide counselling; linkage to care at the time of diagnosis is crucial in assuring proper and timely further testing and treatment.

HIV infection stigma still represents a barrier to testing among patients and providers and it may be compounded by competing priorities, language, educational, and logistical barriers (Fig. 2).

The HIV NAT is a qualitative test FDA approved for diagnosis of HIV and is used when an infected individual is tested in the first few weeks after exposure, at a time when the HIV antibodies have not developed. The test can be obtained either after a combination assay was negative in a high-risk individual with clinical suspicion of acute retroviral syndrome or when a combination assay was positive, but the differentiation test is either negative or indeterminate (in this situation only the p24 antigen is positive and antibodies have not developed yet). NAT tests are also useful in evaluating neonates for vertical transmission, where maternal antibodies are present in the newborn's bloodstream.

The combination test reliably detects 99% of infections at 45 days after exposure, however more than half of infected individuals will have a positive test as early as 2 weeks into infection. HIV NAT tests become positive at approximately 7–10 days from infection. Antibody tests start being positive around 3–4 weeks. The now obsolete Western Blot was the last to become positive, starting at 6 weeks, and in some rare situations the test became fully positive as late as 6 months. Over the counter antibody assays are less sensitive than laboratory-based tests but offer the convenience and privacy of in-home testing; while the kit includes clear instructions and access to a confidential support center, the disadvantage is that the test needs to be confirmed by a laboratory test and individuals who tested positive may not seek care immediately.

Once linked to care, initial laboratory assessment includes HIV RNA viral load, CD4+ T cell count, HIV resistance testing, basic chemistry including fasting glucose, liver function tests and fasting lipid profile, CBC with differential, chlamydia, and gonorrhea genital, rectal or urine NAT, syphilis, hepatitis A, B, and C, as well as toxoplasma serology and cryptococcal antigen, latent tuberculosis screening either by skin testing or by an interferon gamma assay; serology for dimorphic fungal infections is recommended depending on geographic location and endemicity in the area. At the initial evaluation, a physical exam and a mini-mental status examination are performed and counselling regarding natural history and transmission should be done, pneumococcal, meningococcal, and hepatitis A and B immunizations should be considered and offered.

Question 3

Patient is diagnosed with HIV-1 infection based on positive serology. Her CD4+ T cell count is 620 cells/μL. She is presenting for her first clinic visit. She is not symptomatic at this point. What is the treatment recommendation?

A. Monitor CD4+ T cell count every 6 months and initiate antiretrovirals when it drops below 500
B. Monitor CD4+ T cell count every 3 months and initiate antiretrovirals when it drops below 300
C. Initiate antiretroviral treatment only if she develops opportunistic infections

D. Initiate antiretroviral treatment immediately and continue lifelong
E. Start combination three-drug combination antiretroviral treatment but stop if HIV viral load remains undetectable for at least 6 months

Answer and Explanation

Answer: D

The treatment recommendations have changed over the years. In the 1980s when the treatment options were very limited and side effects were significant, treatment was offered only for patients with advanced disease. As more options became available, the threshold for initiating antiretrovirals and the specific regimens used have drastically changed. The current recommendation is to initiate treatment as early as possible, irrespective of the T cell count or viral load, with very few exceptions (see below opportunistic infections). In addition, multiple studies demonstrated that interruptions in treatment—"drug holidays"—lead to opportunistic infections and cardiac events, resulting in poor outcomes and should not be offered. At this point, in the absence of a viable cure, the treatment is lifelong.

3 HIV Treatment Principles

Even though treatment of HIV is beyond the scope of this chapter, it is useful to be aware of the main principles that guide HIV management. Early initiation of HIV treatment starting with the first clinic visit leads indubitably to better outcomes not only to the patient but also impacts transmission. Uncontrolled HIV replication leads to perpetual activation of the immune system leading to a constant state of inflammation. Untreated persons who live with HIV have a higher risk of not only acquiring opportunistic infections and malignancy once the CD4T cell count drops below 200 cells/μL, but HIV affects almost every organ system: HIV nephropathy (1 in 5 untreated patients have microalbuminuria), diseases of cardiovascular system (coronary artery disease is much more common than in the HIV uninfected population and HIV cardiomyopathy—while not as common and usually associated with advanced disease, leads to congestive heart failure), metabolic and endocrine disorders (hypogonadism affects up to half of infected men, subclinical hypothyroidism, and even SIADH can be seen in advanced disease).

Transmission by sexual contact is eliminated in patients with undetectable HIV viral loads as demonstrated by large cohorts of both heterosexual (HPTN 052 trial) and male-male discordant couples (PARTNER and Opposites Attract trials) followed over extended periods of time. These findings led to the WHO recommendations and to changes in the DHHS guidelines which state now that all persons with HIV should be offered antiretroviral treatment.

An effective treatment consists of combinations that usually include three active drugs from at least two different classes, although in some specific situations two drug combinations can be used (patients who switch from certain stable regimens for regimen simplification, or who have low viral loads at the time of initiating treatment). Single drug regimens are highly discouraged as they rarely lead to sustained viral suppression.

When choosing an ART regimen, one must consider a multitude of factors, including preexistent chronic kidney disease, baseline cardiovascular risk and family history of coronary artery disease, age and reproductive status in women, pill burden, interactions, lifestyle, and even weight. Choosing the right regimen for each patient can have a huge impact on the long-term outcomes, due to potential side effects and multifactorial determinants of adherence.

Interruptions in ART may lead not only to abrupt drops in CD4+ T cell counts and development of opportunistic infections, but also cardiovascular events, as proven in the SMART trial where all-cause mortality rates almost doubled and opportunistic infection related death rates almost tripled in the group that was assigned to the episodic antiretroviral therapy guided by CD4+ T cell counts (treatment was interrupted and reinstated only when CD4+ counts dropped to 250 cells/μL or if patients developed opportunistic infections). Frequent interruptions and poor adherence may also lead to development of antiviral resistance, which once acquired is perpetuated lifelong and may lead to a need for more complex regimens.

Close monitoring of viral load and basic labs is important especially in the first year of treatment, when visits and laboratory testing is more frequent. Virologic failure is the inability to achieve or maintain suppression of viral replication to HIV RNA level <200 copies/mL and should be promptly addressed with genotypic testing that can determine drugs to which the virus may have become resistant and can guide informed changes in ART that can lead to virologic suppression. This should be coupled with additional counselling regarding adherence. Lack of adherence may be related to mental health disorders and neurocognitive impairment, substance use, cost, unstable housing, adverse drug effects, and pill burden.

Huge advances in drug development led to easier to tolerate regimens with less side effects, smaller pill burden and minimal interactions. Daily drug regimens are preferred.

Integrase strand transfer inhibitors (INSTI) have become the mainstay of treatment in the past decade, as protease inhibitors (PI) have more side effects, interactions and toxicities and non-nucleoside reverse transcriptase inhibitors (NNRTI) have a lower barrier to developing resistance and undesirable side effects. Due to its efficacy and relatively high barrier to resistance, tenofovir has been one of the most used nucleotide reverse transcriptase inhibitor (NRTI) not only for treatment, but now also for preexposure prophylaxis. A newer formulation (tenofovir alafenamide) led to a smaller pill size but more importantly reduced effects on renal function and bone density loss that made the older formulation (tenofovir disoproxil) less desirable.

Newer injectable options were approved in 2021, including a long-acting combination of an INSTI (cabotegravir) and an NNRTI (rilpivirine) that can be administered monthly in a clinical setting as intramuscular injection. Newer classes of drugs that inhibit viral entry into human cells, including a CCR5 coreceptor inhibitor, a fusion inhibitor, an attachment inhibitor, and a post-attachment inhibitor monoclonal antibody offer options to individuals with multi-class resistance.

Question 4

During the first 2 years after her diagnosis, patient presented regularly to clinic appointments, viral load became undetectable on ART, but subsequently she was lost for follow up and after a gap in care of 5 years she presents to clinic complaining of weight loss and malaise and now has a CD4 of 40 cells/μL. Along with restarting ART, which of the following opportunistic infections should be considered for antimicrobial/antifungal primary prophylaxis?

A. Cryptococcal meningitis
B. Disseminated *Mycobacterium avium* complex (MAC)
C. Pneumococcal pneumonia
D. *Toxoplasma gondii* meningoencephalitis
E. Disseminated histoplasmosis

Answer and Explanation

Answer: D

Only *Pneumocystis jirovecii* and *Toxoplasma gondii* require primary prophylaxis in patients with AIDS and a CD4 count <200 cells/μL. Toxoplasma reactivation occurs more frequently at CD4+ counts under 150. Cryptococcus and MAC infections usually appear when T cell counts drop below 50, however primary prophylaxis is not indicated due to toxicity, lack of efficacy, and concerns of resistance developing with monotherapy. The lower the CD4+ counts, the higher incidence of Pneumococcal pneumonia, however prophylaxis is done with vaccination rather than antimicrobials. Prophylaxis for disseminated *Histoplasma capsulatum* infections is considered only in hyperendemic areas outside US. Primary prophylaxis presumes that the patient does not currently have a particular opportunistic infection. If this patient had symptoms consistent with any of the above infections, a directed workup would be indicated rather than simply starting primary prophylaxis.

4 Opportunistic Infections

While at the beginning of the HIV epidemic, a diagnosis of HIV was an almost certain death sentence, the current life expectancy of an infected person under ART is similar to the average lifespan of general population. Following initiation of an effective combination antiretroviral treatment (ART), a rapid decline in the viral load is expected, with a goal to achieve a viral load of less than 50 copies/μL within 6 months and somewhat slower rise in CD4+ count. Not infrequently, when ART is initiated at CD4+ counts of less than 50 cells/μL, the immune reconstitution is only partial and some patients remain lifelong at low CD4+ counts, albeit the risk for opportunistic infections remains low if the HIV viral load remains undetectable.

When considering opportunistic prophylaxis, one should take into consideration prior infections, exposures, serologies, and geography. Will discuss only a few of the most common opportunistic infections.

Pneumocystis jirovecii is a ubiquitous fungus that caused infections in 80% of AIDS patients before effective ART era and is a significant cause of mortality in untreated individuals with a CD4 countless than 200 cells/μL. Typically patients present with a dry cough, fever and dyspnea exacerbated by minimal effort. The chest radiography typically shows bilateral perihilar infiltrates, and computer tomography may show bilateral ground glass opacities and occasionally pneumatoceles or cysts that can rupture and result in pneumothorax. See more details on diagnosis below in Case 2. The preferred agent both for primary prophylaxis and treatment is trimethoprim/sulfamethoxazole (TMP-SMX), alternate prophylaxis regimens include dapsone and atovaquone. In the case of dapsone, G6PD deficiency should be ruled out to prevent hemolysis. Monthly inhaled pentamidine is used for individuals who cannot tolerate other options. Treatment options for sulfa allergic patients include clindamycin with pyrimethamine or as a last resort intravenous pentamidine, which has higher toxicity.

Toxoplasma gondii is a protozoan parasite that becomes symptomatic in immunocompromised patients through reactivation of a latent infection, however primary disseminated infections can occur. Risk factors include accidental ingestion of oocysts shed in cat feces and sporulated in the environment, eating undercooked meat or raw shellfish. Most common manifestation is encephalitis which usually presents with headaches, fever, and confusion; focal neurological abnormalities depend on location of lesions. Seizures, vision loss, and coma can develop. CT and MRI brain imaging show usually multiple ring-enhancing lesions and differential diagnosis includes lymphoma, tuberculosis, brain abscess. Treatment is usually started empirically, and brain biopsy (definitively diagnostic but carrying significant morbidity) is pursued only if response is not seen after 2 weeks of treatment. CSF toxoplasma PCR can be used for diagnosis if a lumbar puncture is not contraindicated. A negative serum IgG suggests alternative diagnosis, and a positive CSF EBV PCR suggests primary CNS lymphoma. CSF serology and serum IgM are not useful in making a diagnosis. IgG seropositive patients with a T cell count <100 cells/μL should receive prophylaxis and continue until CD4 count increases to >200 cells/μL for 3 months. The combination of choice for treatment is pyrimethamine plus sulfadiazine plus leucovorin, but high dose TMP-SMX is also effective and better tolerated. TMP-SMX is also used for prophylaxis. Pyrimethamine and leucovorin are added to dapsone in individuals with sulfa allergies (Table 2).

Mycobacterium avium complex (MAC) can cause a disseminated infection in individuals with a CD4 <50cells/μL. Organisms are ubiquitous in the environment. Patients present with fever, weight loss, fatigue, diarrhea, and abdominal pain. Laboratory abnormalities include anemia and elevated alkaline phosphatase. The organisms can be isolated from blood and bone marrow cultures. Historically, prophylaxis with a weekly macrolide was offered, however, due to lack of efficacy and concerns with development of resistance to macrolide monotherapy, the newest DHHS guideline recommends limiting prophylaxis only to individuals who cannot receive ART or do not have fully suppressive regimens available and remain viremic. Treatment consists of two or more agents, one of which should be a macrolide (clarithromycin or azithromycin). Subclinical infections may become evident in patients who initiate ART at low CD4 counts of <50 cells/μL, when patients may develop an immune reconstitution inflammatory syndrome (IRIS) manifesting as fever and painful lymphadenopathy within weeks of starting ART.

Table 2 Opportunistic infection prophylaxis

T cell counts (cells/μL)	Opportunistic infection	Primary prophylaxis (no preexistent disease)
All CD4 counts	Tuberculosis (TB)	Screen with IFNγ assay or PPD and treat latent TB
	HSV	Chronic suppression with acyclovir or valacyclovir for frequent recurrences only
<250	Coccidioidomycosis	Preemptive therapy in endemic area (Arizona, California) with fluconazole if seropositive
<200	Pneumocystis	Daily TMP-SMX double strength or single strength *or* Dapsone *or* Atovaquone *or* Aerosolized Pentamidine
<150	Histoplasmosis	Not recommended in US (lack of proven efficacy) Itraconazole is used in hyperendemic areas in South America
<100	Toxoplasma	Daily TMP-SMX double strength or single strength Dapsone + pyrimethamine + leucovorin
	Cryptococcus	Not recommended (lack of efficacy and potential development of resistance)
<50	MAC	Not recommended (monotherapy with macrolides has poor efficacy and potential development of resistance)
	CMV	Not recommended even if viremia is present (no benefit in the absence of end organ symptoms)

Cryptococcus neoformans is the main causative species of cryptococcosis in US. Exposure to dried aerosolized bird droppings may increase the risk of acquisition, but the organism is ubiquitous in environment. In people with HIV with a CD4 <50–100 cells/μL meningoencephalitis is the most common presentation, with malaise, headache, and fever developing subacutely, and meningismus being much less common than in bacterial meningitis. Altered mentation, lethargy and personality changes can also be seen. Disseminated infection, isolated lung infections, and skin involvement can be seen. Skin lesions are nodular, umbilicated, usually larger than molluscum and can ulcerate. Frequently, the serum cryptococcal antigen is positive but CSF analysis is diagnostic, with lymphocytic pleocytosis, elevated protein levels and detection of cryptococcus antigen. Opening pressure can be severely elevated and may require repeated lumbar punctures or even placement of shunts, as antifungal treatment leads to very slow improvement. Induction with amphotericin deoxycholate or with lipid formulations with the addition of oral flucytosine is usually continued for 2 weeks or until the CSF fungal cultures become negative. Cryptococcal antigen titer should not be used to guide treatment, but a high titer can be a poor prognostic factor, along with a low number of lymphocytes in CSF. Prolonged consolidation and secondary prophylaxis with oral fluconazole are usually required and cryptococcal antigen can persist for extended periods. Cryptococcal IRIS after initiation of ART in patients with unrecognized disease can be very severe and can lead to negative outcomes including death related to severe increase in intracranial pressure. ART initiation should be delayed at least 4–6 weeks after starting treatment for CNS cryptococcosis, to avoid such outcomes.

Question 5

Along with her, she brings her current male partner who tested negative recently for HIV. The couple is asking for advice on preventing transmission. You advise them that:

A. Male circumcision reduces the transmission to negligible levels
B. Using condoms was proven to be the most reliable method of preventing transmission

C. Preexposure prophylaxis for the HIV negative male partner would be highly effective
D. They should use vaginal preparations of antiretrovirals
E. The HIV negative partner should receive the HIV vaccine

Answer and Explanation
Answer: C

Unfortunately, to date, all attempts at developing an HIV vaccine have led to failure. Currently, the best strategy to reduce transmission is treatment of HIV positive individuals. When achieving undetectable HIV viral loads as a result of consistent and effective ART, sexual transmission is reduced to negligible rates, which led to the concept of "undetectable = untransmissible" (U=U) through treatment as prevention. Vaginal preparations of tenofovir have lower efficacy than systemic PrEP and while dapivirine vaginal ring is now an option for HIV negative women at risk for HIV acquisition outside of USA, it is not FDA approved and would not be indicated in this couple's situation. Systemic preexposure prophylaxis with a combination of 2 oral or injectable antiretrovirals are viable options especially if the couple is not monogamous, and the HIV positive individual does not have reliably undetectable HIV viral loads. Condoms are theoretically highly effective at prevention, have the advantage of reducing transmission of other STIs along with HIV, and they may be acceptable to many couples, but due to inconsistent use and cost, the real-life efficacy is not as high as with systemic ART. Circumcision does reduce transmission, but it is not reliable enough to be recommended as the sole method of prevention.

5 HIV Prevention

The risk of HIV acquisition by sex is highest with receptive anal intercourse (1 in 72 sex acts), followed by insertive anal intercourse (1 in 900 sex acts), receptive penile-vaginal intercourse, insertive penile-vaginal intercourse, with oral sex being the lowest but of note not completely zero risk. Concomitant STIs, especially ulcerative genital diseases (mainly HSV and syphilis) greatly increase the risk of transmission.

Male circumcision was proven to reduce female to male transmission in half in an African study, however the male to female transmission (transmission is likely related to genital secretions) is not affected and male to male transmission was not studied.

Preexposure prophylaxis (PrEP) with co-formulated tenofovir disoproxil and emtricitabine taken daily by individuals at risk for acquiring HIV can all but eliminate transmission as long as adherence is appropriate. The combination is well tolerated, but it requires frequent monitoring, and it is not devoid of side effects that may include bone density loss and kidney disease with prolonged use. Tenofovir alafenamide, a newer formulation, reduces these side effects, but is less effective in cisgender women at risk for acquiring HIV. Acquired drug resistance and transmission of virus that is resistant to the combination are rare. Long-acting injectable cabotegravir was FDA approved in December of 2021; it was shown to reduce transmission in a cohort of men who have sex with men and transgender women, when compared to tenofovir/emtricitabine. This field is rapidly evolving and additional studies are likely to provide more information in regards to other populations at risk.

For injection drug users, needle exchange programs and opioid addiction treatments have been associated with reduced transmission. On the other hand, the increase in opioid addiction rates in rural America, lack of screening, education and limited support for addiction treatment has led to an increase in HIV in these areas and even outbreaks as the one in southern Indiana in 2015.

Mother to child transmission via intrauterine and intrapartum route in the absence of treatment occurs in 24% of infected mothers but drops below 0.2% in women who start ART prior to pregnancy and maintain an undetectable viral load. The risk for transmission is incrementally higher in women who initiate ARVs during pregnancy, the later the initiation the higher the risk, reaching 2.2% in women who initiate ARVs in the last trimester. Scheduled cesarean sections are indicated only for mothers who do not achieve viral loads of <1000 copies/mL near time of delivery. In addition, intravenous zidovudine is administered intrapartum, and infants are started on presumptive HIV therapy at birth. On the other hand, for women who achieve an undetectable viral load of <50 copies/mL, the mode of delivery is determined by obstetric indications and intrapartum zidovudine is not required.

Case 2

A 33-year-old cisgender man, with history of HIV diagnosed 10 years prior to the presentation was seen in the outpatient clinic for dyspnea with minimal exertion for at least a month. He was lost for follow up, had a history of nonadherence to medications and at the time of presentation was not taking either the previously prescribed regimen of antiretrovirals nor any antibiotics for prophylaxis. He was never previously admitted to a hospital and had no known comorbidities and did not travel out of state in the past 10 years. Dyspnea was associated with a dry nonproductive cough. Patient denied hemoptysis. He endorsed 10 lb weight loss and intermittent low-grade fever. CD4+ lymphocyte count was 22 cells/µL. Creatinine was within normal limits. Upon examination, patient had extensive whitish deposits on tongue and palate consistent with oropharyngeal candidiasis, however his lungs were clear to auscultation with no crackles or rhonchi and there was no dullness on percussion. Heart exam was normal, he did not have any visible rash and there was no leg edema. He was hypoxic, with an oxygen saturation of 89%

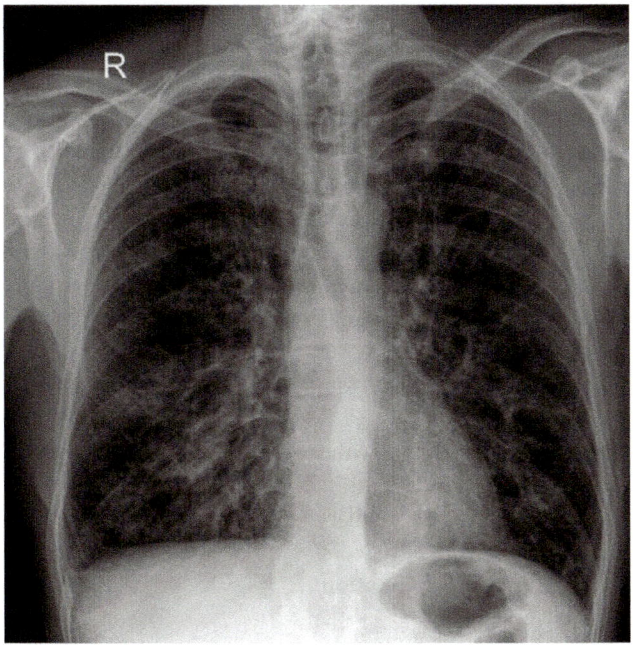

Fig. 3 Anteroposterior chest X-ray showing bilateral lung opacities. (With permission from Ramos AL et al. *International Journal of Infectious Diseases*, 2022-02-01, Vol. 115, pp. 185–188 [3])

on room air at rest and 75% after minimal effort. A chest radiograph showed bilateral infiltrates as seen in Fig. 3.

Question 1

What is the most likely etiology of the pulmonary infiltrates?

A. *Coccidioides immitis*
B. *Streptococcus pneumoniae*
C. Influenza A virus
D. Heart failure with reduced ejection fraction
E. *Pneumocystis jirovecii*

Answer and Explanation

Answer: E

Based on his history of uncontrolled HIV with a very low CD4+ count, this patient is at risk for a plethora of opportunistic infections (see Table 3), however common pathogens like influenza and noninfectious causes must be kept in the differential diagnosis. The most likely pathogen that causes dyspnea, a dry cough and bilateral infiltrates would be *Pneumocystis jirovecii*. Patients with a CD4+ of less than 200 and unsuppressed viremia are at significant risk for acquiring Pneumocystis pneumonia. The presentation is insidious, main symptoms include dyspnea, fever, sometimes cough, with no sputum production. While chest radiography can reveal infiltrates, many times bilateral, atypical presentations are frequent, from unilateral infiltrates to cavitary lesions. A significant number of cases have very vague findings on imaging, and only computer tomography may reveal the infiltrates. Pneumonia is a common presentation for patients who are newly diagnosed with HIV/AIDS, and *Streptococcus pneu-*

Table 3 Most common pulmonary complications in HIV infected individuals

Complication	Comments/"buzz words"
Infections	
Streptococcus pneumoniae	Most common bacterial infection; vaccine preventable
Haemophilus species	
Staphylococcus aureus	More common in PWID
Pseudomonas aeruginosa	Hospitalized patients with tracheostomy, ventilator exposure, Gram negative non-lactose fermenter in sputum culture
Mycobacterium tuberculosis	At any CD4+ count; person to person transmission; MDR/XDR cases more common in HIV
Mycobacterium avium complex	Rarely pulmonary, more often disseminated in AIDS
Nocardia spp.	Concomitant brain lesions, branching weakly AFB positive organism, prolonged combination drug therapy
Rhodococcus	Horse exposure; short gram positive rods
Pneumocystis jirovecii	Most common opportunistic infection at <200 T cells
Cryptococcus neoformans	More common meningoencephalitis than pneumonia; umbilicated skin lesions; serum antigen usually positive; high risk of IRIS
Histoplasma capsulatum	Ohio and Mississippi river valleys; urine antigen
Coccidioides immitis	Dry arid southwest US states; interstitial/cavitary pattern
Aspergillus species	Rare in HIV infection
Blastomyces dermatitidis	Central US; large skin lesions; broad based budding
Talaromyces (Penicillium) marneffei	Southeast Asia
Cytomegalovirus	More commonly involves colon than lung in AIDS
Toxoplasma gondii	Cat feces, raw meat; more frequent brain lesions
Malignancies	
Kaposi sarcoma	15% have isolated lung lesions, skin lesions are typical violaceous; high risk in MSM; HHV8 infection
Non-Hodgkin lymphoma	B-cell origin; EBV associated
Non-small cell lung cancer	More common than in HIV negative population possibly due to higher incidence of smoking; older than 50
Interstitial pneumonitis	Clinically indistinguishable from PCP, but CD4 >200
Other	
COPD	Likely due to higher incidence of smoking
Sarcoidosis	Usually at CD4 >200
IRIS	Infections, sarcoidosis, tumors, autoimmune conditions

moniae is likely the most common pathogen, however the presentation is much more acute, fever is more common, the lung exam is usually abnormal and reveals consolidation and rhonchi. The diagnosis suspicion is higher if the infiltrates are lobar and confirmation can be made with sputum Gram stain and culture, as well as the streptococcus urine antigen.

Question 2

Which of the following diagnostic methods can be used for Pneumocystis pneumonia?

A. Gram stain form expectorated sputum
B. PAS (Periodic acid–Schiff) stain from a transthoracic needle biopsy
C. Pneumocystis PCR from serum
D. Galactomannan antigen from serum
E. GMS (Gomori methenamine silver) stain from bronchoalveolar lavage (BAL) fluid

Answer and Explanation

Answer: E

Expectorated and induced sputum specimens have a low sensitivity for diagnosis of *Pneumocystis jirovecii* pneumonia (PCP). In addition, obtaining sputum can be very challenging, as the cough, if present, is usually nonproductive. Confirmation of diagnosis is recommended, since no clinical or radiological features are specific enough. In addition, treatment is prolonged and potentially toxic. Missing an alternative diagnosis may lead to significant morbidity and mortality. In addition, some immunocompromised patients including patients with AIDS can present with multiple opportunistic infections. While molecular methods of diagnostic are becoming broadly available and pneumocystis PCR from sputum has very high sensitivity and specificity, it may not be available in all clinical settings. Silver stain and immunofluorescence from BAL or from lung biopsy remain the gold standard. Transthoracic needle biopsy is rarely employed due to the increased risk of pneumothorax and other complication, as well as availability of less invasive methods. Nonspecific, but supportive of diagnosis are elevated 1–3-beta-D-glucan, elevated LDH. PCP can occasionally be encountered in patients with a CD4+ cell count of over 200 cells/μL especially in the absence of HIV viral load suppression. On the other hand, even in the absence of a very robust immune reconstitution, PCP is uncommon in individuals with a fully suppressed HIV viral load (undetectable).

Question 3

The patient was hospitalized due to hypoxia, was administered supplemental oxygen; TMP-SMX as well as prednisone was initiated, with a presumptive diagnosis of pneumocystis pneumonia. In addition, thrush had to be addressed. Antiretrovirals were not restarted at this point. Review of chart showed that he received multiple antiretroviral regimens since his diagnosis. He had not developed any side effects, but adherence was poor, and he never achieved complete viral suppression. He was lost for follow up for 4 years and did not take any medications during this time. Current HIV viral load is 5.2 million copies/mL. Select the correct statement regarding thrush (oropharyngeal candidiasis) in patients with HIV:

A. The diagnosis of thrush should always be confirmed with a fungal culture from a mucosal scraping
B. Oropharyngeal candidiasis is diagnosed by mucosal biopsy
C. Treatment of choice for thrush in pregnant women is fluconazole
D. Majority of patients with HIV carry fluconazole resistant strains of candida
E. The diagnosis of oropharyngeal candidiasis is clinical

Answer and Explanation

Answer: E

Oropharyngeal candidiasis occurs frequently in patients with CD4 counts less than 200 cells/μL and it may be the first manifestation in patients newly diagnosed with HIV, however thrush can be encountered in individuals with higher counts, especially if comorbid conditions including uncontrolled diabetes and incorrect use of inhaled steroids exist. The most common presentation of candidiasis is the pseudomembranous candidiasis or thrush; however, angular cheilitis and erythematous (also known as atrophic) candidiasis can be encountered. The presumptive diagnosis is clinical and only rarely confirmation with a potassium hydroxide wet mount, Gram stain or a culture are required. One should differentiate between candidiasis and other conditions encountered in patients with HIV:

- Oral hairy leukoplakia—presents as raised white lesions distributed mainly on the lateral aspect of the tongue), which do not require any specific treatment and usually resolve along with immune reconstitution in patients on antiretrovirals; these lesions are highly adherent to the tongue and cannot be removed with a tongue depressor.
- Ulcerations caused by herpes simplex and aphthous stomatitis can look similar, but usually herpetic lesions involve the vermillion border and aphthous ulcers are found on the buccal mucosa; HSV DNA PCR testing can differentiate these two conditions when in doubt (cultures are less sensitive and more time consuming). Herpes simplex responds to acyclovir derivatives and resistance is uncommon, but prolonged intermittent use can lead to selection of resistant virus. Aphthous stomatitis can be more frustrating, as the response is variable to topical steroids, anti-inflammatory mouth washes, tetracyclines and even thalidomide for severe cases. Cytomegalovirus occasionally can produce mucosal lesions, but the most common location for mucosal ulcerations would be the esophagus and lower gastrointestinal tract.

- Oral warts secondary to human papillomavirus infection—raised nonpainful lesions—diagnosis can be confirmed by biopsy which is usually indicated if suspicion for a malignant lesion exists.

Fluconazole is the drug of choice in the treatment of oropharyngeal candidiasis. Due to potential teratogenicity of fluconazole especially in the first trimester, the preferred treatment in pregnancy is topical, with miconazole, clotrimazole, or nystatin. Fluconazole resistant candidiasis is still rare, treatment is challenging and may include intravenous echinocandins. Some fluconazole resistant strains retain susceptibility to newer azoles, including posaconazole and voriconazole.

Question 4

Three days after starting fluconazole thrush has resolved. Seven days into the treatment with TMP-SMX and prednisone, the patient continues to be dyspneic, oxygen requirements fail to improve and intermittently he has low-grade fever. Alternative diagnoses are being entertained. The diagnosis of pulmonary histoplasmosis should be higher in the differential if:

A. He lives in southern Ohio and raises pigeons
B. He was recently released from being incarcerated for 5 years
C. He travelled to southeast Asia
D. He is of Asian descent and lived for the past 10 years in Arizona
E. He recently noticed several dark violaceous lesions on his chest

Answer and Explanation

Answer: A

Exposure to bat guano or bird droppings and location endemic for histoplasmosis suggests this diagnosis. Incarceration increases the risk of tuberculosis. Travel to south-Asia increases the risk of acquisition of a dimorphic fungus—*Talaromyces marneffei*. Dark violaceous lesions suggest Kaposi sarcoma.

6 Differential Diagnosis of Lung Lesions in a Patient with HIV

While Pneumocystis pneumonia is still the most common AIDS-defining opportunistic infection in US, alternative diagnoses should be considered when faced with a patient with HIV and pulmonary infiltrates.

Persons with HIV are at increased risk for developing tuberculosis and exposure to environments with higher incidence: living in densely populated areas with higher incidence of tuberculosis, visiting or migrating from endemic countries (in Africa, TB is the most common pulmonary complication in HIV).

Endemic mycoses should always be considered, depending on travel and state of residence: histoplasmosis is endemic in the Ohio and Mississippi River valleys, but also in many areas of Central and South America. Travel or residence in dry areas in Southwestern US should raise suspicion for coccidioidomycosis. *Talaromyces marneffei* can cause pulmonary and disseminated infections in HIV infected individuals from Southeast Asia. The only one endemic mycosis that requires primary prophylaxis in US is coccidioidomycosis in patients with positive serology, no symptoms and a CD4 <200 cells/μL.

Toxoplasma gondii, while well known to cause brain lesions in individuals with a CD4 count below 100cells/μL, can also cause pulmonary infections. Negative serology makes this diagnosis unlikely, however a positive IgG is not sufficient for confirmation and represents either evidence of prior exposure or active disease, the diagnosis being made by direct observation of tachyzoites in a bronchoalveolar lavage. Prophylaxis with TMP-SMX reduces the rate of reactivation of a latent infection. Alternatives for individuals with sulfa intolerance include dapsone or atovaquone plus pyrimethamine and leucovorin.

The differential of lung lesions with HIV can be extensive and more than one condition can be found in the same individual. Some opportunistic infections found in other immunocompromised hosts (i.e., transplant recipients and cancer patients), including aspergillosis, CMV infection, and nocardiosis are not as common in persons living with HIV. Noninfectious causes of lung lesions should also be considered, as interstitial lung disease and sarcoidosis are more prevalent than in HIV negative population; however, these conditions usually are found in HIV patients with higher CD4 counts (>200 cells/μL).

Question 5

Despite starting prednisone and TMP-SMX at appropriate doses as empiric treatment for PCP pneumonia, the patient fails to improve after the first week of treatment. He continues to be hypoxic and has intermittent fevers. A decision is made to pursue a bronchoscopy, which reveals no purulence, however several violaceous patches are seen throughout the bronchial tree. The bronchoalveolar lavage specimen is negative for fungal, acid-fast smears and Gram stain does not reveal any microorganisms. PCR for pneumocystis is negative (Fig. 4).

The most likely diagnosis at this point is:

A. Bronchogenic carcinoma
B. Pulmonary Kaposi sarcoma
C. Bacillary angiomatosis
D. Non-Hodgkin lymphoma
E. Histoplasmosis

Answer and Explanation

Answer: B

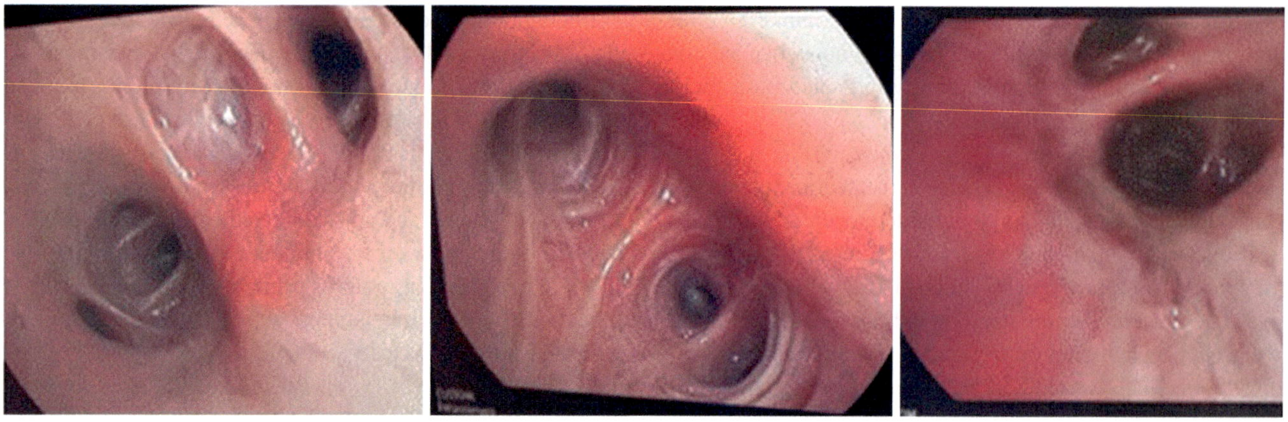

Fig. 4 Purple and erythematous lesions observed during bronchoscopy. (With permission from Ramos AL et al. Pulmonary Kaposi's sarcoma—an atypical clinical presentation. *International Journal of Infectious Diseases*, 2022-02-01, Vol. 115, pp. 185–188 [3])

The bronchial lesions seen on bronchoscopy are diagnostic for Kaposi Sarcoma (KS), a malignancy related to HHV8 infection. Usually, a biopsy is not required for diagnosis, both the skin lesions and the appearance of bronchial lesions are very typical. Pulmonary involvement can be associated with pleural effusions. While majority of patients with pulmonary KS have skin lesions, about 15% have only visceral involvement. Immune reconstitution is essential for treatment, however for extensive skin involvement as well as for visceral involvement, there is a need for addition of systemic chemotherapy. A significant number of patients with KS do have an opportunistic infection, making a comprehensive workup necessary, without early closure.

Final diagnosis: Pulmonary Kaposi Sarcoma in a patient with HIV/AIDS.

7 Conclusions

HIV continues to be mainly transmitted via sexual contact. Key changes in diagnosis and management in the past two decades include new diagnostic tools that reduce the window period to as little as 2 weeks by using the p24 antigen and treatment offered as early as immediately after diagnosis, in order to preserve the immune system, especially the gut residing lymphocytes from viral aggression and prevent transmission. New antiretrovirals are better tolerated and have lower pill burden. Options for resistant virus are being developed. PrEP, screening and treating pregnant women, as well as strategies to educate and treat intravenous opioid addiction are additional tools for preventing transmission. Opportunistic infections have decreased significantly in frequency along with the scale-up of effective ART but remain prevalent in untreated populations. Malignancies can mimic some opportunistic infections and careful evaluation may uncover multiple simultaneous opportunistic conditions.

Bibliography

1. Gillespie S, et al. Human immunodeficiency virus infection and acquired immunodeficiency syndrome. In: Clinical immunology: principles and practice, vol. 39. 5th ed. Amsterdam: Elsevier; 2019. p. 545–560.e1.
2. Centers for Disease Control and Prevention. Quick reference guide—laboratory testing for the diagnosis of HIV infection: updated recommendations, 27 Jun 2014. https://www.cdc.gov/hiv/pdf/guidelines_testing_recommendedlabtestingalgorithm.pdf. Accessed 1/10/2022.
3. Ramos AL, et al. Pulmonary Kaposi's sarcoma—an atypical clinical presentation. Int J Infect Dis. 2022;115:185–8.
4. Panel on Antiretroviral Guidelines for Adults and Adolescents. Guidelines for the use of antiretroviral agents in adults and adolescents with HIV. Department of Health and Human Services. https://clinicalinfo.hiv.gov/en/guidelines/adult-and-adolescent-arv. Accessed 1/10/2022.
5. El-Sadr WM, Lundgren J, Neaton JD, Gordin F, Abrams D, Arduino RC, et al. CD4+ count-guided interruption of antiretroviral treatment. N Engl J Med. 2006;355(22):2283–96. https://doi.org/10.1056/NEJMoa062360.
6. World Health Organization. Consolidated guidelines on the use of antiretroviral drugs for treating and preventing HIV infection: recommendations for a public health approach. 3rd ed. Geneva: World Health Organization; 2021. https://www.who.int/publications/i/item/9789240031593. Accessed 1/10/2022.
7. Sterling TR, Chaisson RE. Chapter 122: HIV. In: Mandell, Douglas, and Bennett's principles and practice of infectious diseases. 9th ed. Amsterdam: Elsevier; 2020. p. 1658–1674.e5.
8. INSIGHT START Study Group, et al. Initiation of antiretroviral therapy in early asymptomatic HIV infection. N Engl J Med. 2015;373(9):795–807. https://doi.org/10.1056/NEJMoa1506816.
9. Kitahata M, et al. Effect of early versus deferred antiretroviral therapy for HIV on survival. N Engl J Med. 2009;360:1815–26. https://doi.org/10.1056/NEJMoa0807252.
10. Chou R, et al. Preexposure prophylaxis for the prevention of HIV infection. Evidence report and systematic review for the US Preventive Services Task Force. JAMA. 2019;321(22):2214–30. https://doi.org/10.1001/jama.2019.2591.

Anaphylaxis

Jeffrey Kepes and Pavadee Poowuttikul

Abbreviations

AAAAI	American Academy of Allergy Asthma and Immunology
ACE	Angiotensin-converting enzyme
ACEi	Angiotensin-converting enzyme inhibitor
COX-1	Cyclooxygenase 1
HaT	Hereditary alpha tryptasemia
ICAM-1	Intracellular adhesion molecule 1
IM	Intramuscular
IO	Intraosseous
IV	Intravenous
NMBA	Neuromuscular blocking agent
NSAID	Non-steroidal anti-inflammatory drug
PAF	Platelet activating factor
PGE2	Prostaglandin E2
PVFM	Paradoxical vocal fold motion
SC	Subcutaneous

1 Introduction

In this chapter, we will use case-based discussion to review pathophysiology and mediators involved in anaphylaxis. We then will review the diagnostic criteria for anaphylaxis as well as the important differential diagnoses to consider in evaluation of a patient presenting with symptoms concerning for anaphylaxis. We will also discuss the etiology of anaphylaxis and the important aspects of the history and physical examination to help make the diagnosis and determine the trigger, if any, to the reaction. We will discuss that anaphylaxis can be a direct result of an allergy to an exposure like food or drug allergens or a symptom of another systemic illness. Finally, will discuss treatment for anaphylaxis at great length with a significant emphasis on the appropriate dosing and administration of epinephrine.

J. Kepes · P. Poowuttikul (✉)
Division of Allergy, Immunology and Rheumatology, Department of Pediatrics, Children's Hospital of Michigan, Detroit, MI, USA

Central Michigan University College of Medicine, Mt. Pleasant, MI, USA
e-mail: jkepes@dmc.org; ppoowutt@dmc.org

Case 1
A 3-year-old girl presents to the emergency room with diffuse urticarial lesions over her trunk and arms and in acute respiratory distress with audible wheezing. Mother is at bedside and states the symptoms started about 1 h ago, 5 min after she ate a peanut butter cracker. She has never had this reaction before, and this is her first-time eating peanuts or peanut products.

Vitals
Respiratory rate: 50
HR: 180 bpm
Blood pressure: 105/65 mmHg
Oxygen saturation: 94%

Question 1
Which mediator is most closely correlated to the severity of anaphylaxis?

A. Angiotensin-Converting Enzyme (ACE)
B. Histamine
C. Platelet Activating Factor (PAF)
D. Prostaglandin E2 (PGE2)
E. Tryptase

Answer and Explanation
Answer: C
While all these answers are correlated with anaphylaxis. Only Platelet Activating Factor (PAF) has been correlated directly with severity of anaphylaxis. In one study

M. Mahmoudi (ed.), *Absolute Allergy and Immunology Board Review*, https://doi.org/10.1007/978-3-031-12867-7_25

comparing tryptase, histamine, and PAF, only PAF was found to be elevated in all patients with severe anaphylaxis. PAF is also a direct activator of mast cells and causes the releases of histamine and PGD2 as well as increased synthesis and the release of neutrophil chemotactic chemokine CXCL8. PAF may function to amplify the feedback loop in anaphylaxis.

Tryptase is a preformed protease released by mast cells upon degranulation. It is the most specific marker for mast cell degranulation, but not severity of anaphylaxis. Tryptase is bound to heparin in these granules and dissociates upon mast cell activation. Upon degranulation, tryptase stimulates the release of IL-8 and upregulates ICAM-1 on epithelial cells. It also functions as a growth factor for epithelial cells, fibroblasts, and airway smooth muscle cells. Importantly, tryptase may also stimulate degranulation of other mast cells.

Angiotensin-Converting Enzyme (ACE) is involved in the catabolism of bradykinin. Decreased ACE causes an increase in serum bradykinin which causes increased vascular permeability leading to angioedema. While ACE is associated with severity of angioedema, ACE does not independently correlate with severity of anaphylaxis.

Histamine, a vasoactive amine, is the mediator responsible for most of the signs and symptoms of anaphylaxis. Histamine is released by preformed granules and binds to two histamine receptors (H1 and H2) on a wide variety of cells.

H1-receptors tend to be pro-inflammatory and are found on most cells in the body. Notably, H1-receptors in the gastric and respiratory mucosa cause smooth muscle contraction upon activation leading to cramping, diarrhea, and bronchospasm. H1-receptors on vascular mucosa, on the other hand, cause smooth muscle relaxation leading to vasodilation and decreased blood pressure.

H2-receptors tend to be anti-inflammatory and are found most frequently in the gastrointestinal tract but are also present in the respiratory tract as well as cardiac myocytes. Activation of H2-receptors causes increased mucous production by goblet cells as well as increased acid secretion by the gastric mucous. Importantly, H2-receptor activation increases cardiac muscle chronotropy and inotropy.

2 Pathophysiology of Anaphylaxis

Anaphylaxis is a life-threatening systemic hypersensitivity reaction which can be caused by a variety of triggers. Anaphylaxis is caused by mast cell degranulation which releases numerous vasoactive amines like histamine, enzymes, and lipid mediators that have a diverse impact on multiple tissues and organ systems. Mast cell degranulation can be caused by allergens via cross-linking of IgE on the cell surface, receptor-mediated mast cell degranulation from medications or bacterial and viral products, or from direct mast cell degranulation such as temperature, osmolarity, or physical causes.

IgE-dependent anaphylaxis is the most important mediator, especially to an allergist/immunologist. Mast cells contain FcεRI, the high-affinity IgE receptor, which is bound to allergen-specific IgE. Upon exposure to an allergen, like peanut protein, the IgE receptors cross-link and cause multiple downstream processes to take place. These downstream processes include an increase in intracellular calcium, which lead to exocytosis, or degranulation, of mast cells. Mast cell granules contain multiple preformed mediators which are responsible for the immediate effects of anaphylaxis.

Histamine, a vasoactive amine, causes increased vascular permeability, bronchospasm, and smooth muscle contraction diffusely. Histamine exists bound to tryptase and heparin in the mast cell granules prior to release. Histamine receptors are found on most cells in the body and are divided into four types, H1, H2, H3, and H4, as shown in Table 1.

Tryptase, a protease, is stored in mast cell granules and has a variety of end-organ effects including upregulation of ICAM-1 and IL-8 as well as functioning as a growth factor for multiple cell types. Notably, tryptase may also play a role in degranulation of other mast cells. Tryptase is a sensitive marker for anaphylaxis but does not correlate with disease severity.

Platelet Activating Factor (PAF) is a lipid-derived mediator created by mast cells through increased activity of phospholipase A2 after activation. Platelet activating factor has receptors which are found on neutrophils, eosinophils, and most notably mast cells. PAF directly causes increased vascu-

Table 1 Histamine receptors and their effects

Histamine receptor	H1	H2	H3	H4
Location	Smooth muscle, epithelial cells, T lymphocytes, endothelial cells, neutrophils, monocytes, mast cells	Same as H1	**Neurons**, dendritic cells	Bone marrow, mast cells
Effects	Pruritis, increase vascular permeability, bronchoconstriction, smooth muscle contraction, hypotension	Increased gastric acid production, hypotension, increased mucous production, tachycardia, **smooth muscle relaxation**	Nasal congestion, bronchoconstriction	Unknown

lar permeability and bronchoconstriction in tissues. However, it is through the potentiation of mast cell activation that PAF is believed to have the most significant impact on the severity of anaphylaxis. Multiple studies have shown that PAF level is directly correlated with the severity of anaphylaxis.

Upon activation of mast cells, there is increased synthesis of multiple cytokines, chemokines, and lipid mediators. Upregulation of phospholipase A2 leads to increased generation of multiple prostaglandins, cysteinyl leukotrienes, and PAF which all potentiate the effects of histamine as well as causing bronchoconstriction and increase vascular permeability. These mediators are not stored in mast cell granules but instead are synthesized after activation and are detectable as quickly as 15 min after activation.

Mast cells can also be activated by non-IgE mediated mechanisms and are clinically important.

Complement receptors, specifically for C3a and C5a, are plentiful on mast cells and cause degranulation and significant histamine release upon binding to their ligands. This is the most likely mechanism for anaphylaxis related to Cremophor EL which was historically prevalent in propofol and multiple chemotherapeutic agents but has since seen less use in favor of more accessible and better tolerated alternatives.

Medications, like opioids and vancomycin, cause direct mast cell degranulation via binding to opioid receptors or to other cell surface receptors like MRGPRX2 on mast cells. Historically, opioids were used as the positive control for skin testing due to this property but have fallen out of favor due to the accessibility of histamine extracts.

Multiple other physical factors like vibration, osmolarity, and temperature changes can cause direct mast cell degranulation, however the mechanism is frequently unclear or unknown.

Question 2

Which physical finding would be less supportive of anaphylaxis in this patient?

A. Hypertension
B. Incontinence
C. Syncope
D. Vomiting
E. Wheezing

Answer and Explanation

Answer: A

Anaphylaxis can present with a wide range of symptoms and severity based on numerous factors like amount and duration of exposure, genetics, concomitant medications, patient age, and comorbidities. Most signs and symptoms of anaphylaxis, as shown in Table 2, can be attributed to histamine, prostaglandins, and cysteinyl leukotrienes.

Table 2 Common manifestations and mediators of anaphylaxis

Organ system	Signs	Mediators
Upper respiratory	Sneezing	Histamine
	Rhinorrhea	Cysteinyl leukotrienes
	Congestion	
Lower respiratory	Coughing	Histamine
	Angioedema	Cysteinyl leukotrienes
	Wheezing	Prostaglandins
	Stridor	
	Hypoxia	
	Bronchoconstriction or bronchospasm	
Gastrointestinal	Nausea	Histamine
	Vomiting	
	Diarrhea	
	Abdominal pain	
Cutaneous	Angioedema	Histamine
	Hives	Cysteinyl leukotrienes
	Pruritis	Prostaglandins
	Flushing	
Cardiovascular	Hypotension	Histamine
	Tachycardia	Cysteinyl leukotrienes
	Syncope	Prostaglandins
	Incontinence	Platelet activating factor
		TNF-α

3 Diagnosis of Anaphylaxis

Anaphylaxis is defined as an acute life-threatening hypersensitivity reaction. Historically, it was believed that anaphylaxis was due to allergic triggers, but we now understand that any trigger that can cause mast cell degranulation can cause anaphylaxis. Diagnosis of anaphylaxis is clinical, although laboratory data can help to support a diagnosis. There are three major criteria proposed by the American Academy of Allergy, Asthma, and Immunology (AAAAI) to assist in the diagnosis of anaphylaxis which are listed below in Table 3.

Prompt identification of anaphylaxis is essential for quick and potentially lifesaving treatment with epinephrine and other accessory therapies.

It is important to remember that anaphylaxis can present in a multitude of forms and need not be consistent even within the same patient. A patient may present with primarily hypotension after exposure to an allergic trigger on one occasion and have urticaria and wheezing on another. Anaphylaxis may also present in varying severity depending on other factors like amount of allergen ingested, presence of other medications or illnesses, or route of administration.

Anaphylaxis usually occurs with minutes of exposure to an allergen or trigger, but various factors can affect the time

Table 3 Diagnostic criteria for anaphylaxis

Criteria 1: Acute onset of symptoms involving the skin and/or mucosa and	
1. Reduction in blood pressure or symptoms of end-organ dysfunction	
and/or	
2. Respiratory symptoms (dyspnea, cough, wheezing, stridor, hypoxemia)	
Criteria 2: Acute onset of symptoms after an exposure to a likely allergen and two or more of the following	
1. Gastrointestinal symptoms (diarrhea, nausea, vomiting, abdominal pain)	
2. Cutaneous or mucosal symptoms (angioedema, flushing, urticaria, pruritis)	
3. Reduced blood pressure or symptoms of end-organ dysfunction (incontinence, syncope or hypotonia)	
4. Respiratory symptoms	
Criteria 3: Reduced blood pressure for age after exposure to a known allergen	
Infants and children: Low systolic BP for age or decrease of 30% or more	**Adults**: Systolic BP of less than 90 mmHg or decrease in 30% from baseline

course. Route of exposure is one of the most important factors in determining time from exposure to onset of symptoms. Median onset of symptoms from exposure to oral allergens or triggers could be as long as 30–120 min, whereas subcutaneous exposures like venoms can be 15 min or less. Similarly, intravenous exposure has a median onset of symptoms of as short as 5 min or less. The timeline for the development of symptoms from exposure can help to differentiate causes and should always be assessed when evaluating a patient.

Question 3

The most common to the least common organ system involvement from anaphylaxis are

A. Respiratory > skin > cardiovascular
B. Respiratory > skin > gastrointestinal
C. Skin > nervous system > gastrointestinal
D. Skin > respiratory > gastrointestinal
E. Skin > cardiovascular > respiratory

Answer and Explanation

Answer: D

Anaphylaxis can affect most organ systems in the body, however, most cases (80–90%) will have cutaneous symptoms like urticaria, angioedema, or pruritis. Up to 70% of patients will have respiratory symptoms. It is important to remember that nasal symptoms such as rhinorrhea, congestion, and sneezing constitute respiratory symptoms and not all patients will have bronchospasm or wheezing. Finally, gastrointestinal symptoms occur in about 45% of patients and most frequently consist of nausea and vomiting, but many patients also experience severe abdominal pain or diarrhea. Cardiovascular symptoms like hypotension or syncope similarly occur in up to 45% of patients. A frequently missed sign of cardiovascular involvement is incontinence, especially in older children and adults. Importantly, loss of muscular tone, a sequela of decreased oxygenation, and blood flow can cause hypotonia, incontinence, and weakness. Cardiovascular collapse or asphyxiation from airway angioedema is the most common cause of death in anaphylaxis.

4 Presentation of Anaphylaxis

Anaphylaxis is unique in that the disease process can have numerous presentations, as shown in Table 4, and varied severity even within the same patient depending on numerous factors. However, in general, the most common presentation of anaphylaxis involves the skin or mucous membranes to some degree.

Cutaneous and mucous membrane involvement in anaphylaxis could constitute a wide variety of symptoms from generalized pruritis to lip or tongue angioedema. Cutaneous symptoms occur in 80–90% of patients with anaphylaxis and is the most recognizable symptoms by patients and providers. However, up to 20% of patients who have anaphylaxis will present with no identifiable cutaneous symptoms. This can lead to missed diagnosis and delayed treatment in patients and, in fact, the lack of cutaneous symptoms is a risk factor for fatal anaphylaxis.

Respiratory symptoms are the second most common presentation of anaphylaxis and occur in up to 70% of patients. Respiratory symptoms could be as mild as rhinorrhea and sneezing to as severe as wheezing, bronchospasm, and hypoxia. Rhinorrhea is a frequently missed or overlooked respiratory manifestation. It is important to recall that the nasal mucosa is continuous with the bronchial mucosa. Therefore, involvement of nasal mucosa could mean impending involvement of the bronchial mucosa and smooth muscle. If a provider only evaluates for involvement of the respiratory system by inquiring about wheezing or bronchospasm, they could miss a significant number of patients and lead to delayed treatment or administration of epinephrine. Delayed administration of epinephrine is another risk factor for fatal anaphylaxis.

Table 4 Involvement of organ systems during anaphylaxis

Organ system	Involvement (%)
Skin or mucous membranes	90
Respiratory	70
Cardiovascular	45
Gastrointestinal	45
Nervous system	15

Cardiovascular symptoms occur in up to 45% of patients and, along with respiratory arrest, are the most common cause of death in fatal anaphylaxis. Cardiovascular involvement in anaphylaxis can range from mild tachycardia to hypotension and cardiovascular collapse. Uniquely, both the pathology and the treatment may contribute to tachycardia in these patients. Manifestation of cardiovascular involvement may also include evidence of end-organ damage via ischemia. Incontinence, especially in previously continent children and adults, may be the only manifestation of impending cardiovascular collapse as ischemia to the bowel or urinary bladder smooth muscle causes significant bowel or bladder relaxation.

Gastrointestinal symptoms occur in up to 45% of patients and frequently involve nausea and vomiting. These symptoms can often be confused, especially in infants with gastroesophageal reflux disease (GERD). Abdominal pain can also be a frequent manifestation in anaphylaxis and must be differentiated from gastrointestinal upset. This can be increasingly difficult in children where lactose intolerance is more frequent and diarrhea and abdominal pain are common. However, the timing and reproducibility of symptoms may help to differentiate anaphylaxis from lactose intolerance or GERD. Anaphylaxis usually occurs within 30 min of exposure to an oral trigger, but only rarely occurs 2 h or more after exposure. Whereas lactose intolerance usually occurs hours after ingestion.

Finally, central nervous system (CNS) is a rare, but potentially life-threating manifestation of anaphylaxis. CNS involvement can include an impending sense of doom or anxiety, confusion or altered mental status, headache, or even potentially loss of consciousness or seizures in rare cases. Some of these symptoms can easily be dismissed by providers as being primarily psychological or psychosomatic which can further delay treatment.

As we will discuss later, early recognition of anaphylaxis and early initiation of treatment are essential to minimize morbidity and mortality.

Question 4

To support the diagnosis of anaphylaxis in this case, when is the ideal time to draw serum tryptase after symptoms onset?

A. 0–10 min
B. 15–180 min
C. 6–8 h
D. 12–18 h
E. 24–72 h

Answer and Explanation

Answer: B

Upon activation, mast cells degranulate releasing preformed mediators which mostly consist of histamine, tryptase, and heparin.

Histamine is released immediately, reaching peak serum levels 5–10 min after the onset of the reaction. Histamine has a short serum half-life and is reduced to normal levels after 30 min. Plasma histamine is rarely checked because of the infeasibility of the timeline required for collection.

N-methylhistamine is the excreted urinary metabolite for histamine and can be measured in the urine. Urine is collected for 24 h and should, ideally, be collected as soon as possible after the reaction. False positives for urinary histamine are common and frequently attributed to bacterial overgrowth or ingestion of histamine-rich foods.

Tryptase has a longer serum half-life and is detectable as soon as 15 min after the onset of the reaction. It peaks in the serum between 30 and 120 min after and remains elevated for up to 3 h. Ideally, tryptase should be drawn between 15 and 180 min after the start of a reaction to support a diagnosis of anaphylaxis. Tryptase can be normal in patients with anaphylaxis due to foods.

5 Laboratory Testing in Anaphylaxis

Laboratory testing in anaphylaxis can be a helpful tool in supporting a diagnosis or for consideration of a differential diagnosis, but it is important to note that negative laboratory testing does not rule out anaphylaxis which is a clinical diagnosis.

The most frequently used test in the workup of anaphylaxis is serum tryptase. Tryptase is a preformed mediator stored in mast cell granules and is released upon degranulation. Tryptase is detectable in the serum about 15 min after the initiation of symptoms and peaks after 30–120 min. Tryptase, when elevated, can support a diagnosis of anaphylaxis. Initiation of treatment should never be delayed to obtain a tryptase level or any other laboratory evaluation. Notably, a normal tryptase does not exclude the diagnosis of anaphylaxis especially in food-related anaphylaxis. Gastrointestinal-resident mast cells have a significantly lower concentration of tryptase in their granules than cutaneous or lung-resident mast cells. In fact, food-related anaphylaxis frequently presents with normal tryptase even in severe anaphylaxis. However, in cases where the diagnosis is unclear, tryptase can often be helpful in narrowing the differential diagnosis. Tryptase should ideally be drawn between 15 min to 3 h of the start of a reaction. However, elevated serum tryptase is sometimes seen up to 5 h after the onset of symptoms. It is also recommended that a repeat tryptase be drawn at least 24 h after the resolution of symptoms to evaluate a baseline serum tryptase level. An elevated baseline tryptase could be suggestive of a more sinister underlying process like mastocytosis or hereditary alpha tryptasemia, both of which are associated with more severe episodes of anaphylaxis.

Other mediators have been evaluated for assisting with the diagnosis of anaphylaxis, however, they are rarely useful in the immediate diagnosis. Histamine is released with tryptase in mast cell granules. It is the serum mediator which is detectable quickest in the serum before being metabolized to *N*-methylhistamine. However, histamine has an extremely low half-life which limits its utility in diagnosis of anaphylaxis. It peaks after 5–10 min and remains elevated only 30–60 min after onset of anaphylaxis. The primary metabolite, *N*-methylhistamine, is detectable in the urine as quickly as 6 h after a reaction and may be present for up to 24 h. Histamine's utility is also clouded by the fact that there are numerous other potential causes for elevated levels of histamine in the blood or urine. Scombroid poisoning causes elevation of serum histamine and its urinary metabolites while multiple foods like red wine or some vegetables or fruits like avocados can also cause very mild elevations in serum histamine. Finally, urinary histamine can be elevated in patients with certain urinary bacterial colonization or infections.

Platelet Activating Factor (PAF) has become a significant topic of research as multiple studies have shown that serum levels are correlated with severity of anaphylaxis. PAF is known to be released by mast cells after activation and has a significant impact on potentiation of mast cell degranulation. However, detection of PAF can be difficult as it appears almost immediately after the onset of reaction but is rapidly metabolized with a half-life of 5 min or less. As discussed earlier, PAF levels shortly after the onset of anaphylaxis are directly correlated with severity of anaphylaxis. Table 5 demonstrates mediators of anaphylaxis.

Question 5

What is the correct dosing and route of administration of epinephrine in this patient?

A. 0.01 mg/kg of 0.1 mg/mL intramuscularly
B. 0.01 mg/kg of 0.1 mg/mL subcutaneously
C. 0.01 mg/kg of 1 mg/mL intramuscularly
D. 0.1 mg/kg of 1 mg/mL intramuscularly
E. 0.1 mg/kg of 1 mg/mL intravenously

Table 5 Major serum mediators of anaphylaxis

Mediator	Detectable in serum (or urine) after reaction	Peak level in serum (or urine) after reaction
Histamine	0–5 min	5–10 min
Tryptase	15 min	30–120 min
Platelet activating factor	0–5 min	Unknown, half-life is <5 min
11-β-prostaglandin F2α	2 h	Unknown, obtained by 24-h urine
N-methylhistamine	6 h	Unknown, obtained by 24-h urine

Answer and Explanation

Answer: C

Epinephrine dosing for anaphylaxis is 0.01 mg/kg of the 1 mg/mL (1:1000) concentration given intramuscularly in the vastus lateralis in the lateral thigh. Epinephrine concentration for cardiac indications is 0.1 mg/mL (1:10,000) and is administered intravenously. Epinephrine in the 1 mg/mL concentration should never be administered intravenously. Intramuscular administration has multiple benefits over intravenous administration. These benefits include lower risk of adverse events like cardiac arrhythmias, ease of use, and accessibility.

6 Treatment and Prevention of Anaphylaxis

Anaphylaxis is a rapidly progressing and life-threatening disease that requires immediate and definitive treatment. The cornerstone of the treatment of anaphylaxis is epinephrine administered as quickly as possible after the initiation of symptoms. In fact, delayed administration of epinephrine is a risk factor for fatal anaphylaxis. Proper administration and dosing of epinephrine are essential. Epinephrine should be dosed at 0.01 mg/kg of the 1 mg/mL concentration (formerly called 1:1000) and administered intramuscularly in the vastus lateralis in the lateral thigh. Intramuscular administration has faster onset than subcutaneous administration. The maximum dose in adults is 0.5 mg although commercial preparations come in only three dosages (0.10, 0.15, and 0.3 mg). We recommend to use the 0.10 mg dosing for children under 10 kg, to use the 0.15 mg dosing for children between 10 and 25 kg and to use the adult 0.3 mg dose in all patients 25 kg and over. Doses may be repeated every 3–5 min. Continuous IV infusion of epinephrine should be considered in patients who have persistent symptoms despite three or more doses of IM epinephrine although there is no standardized recommendation. Intravenous administration of epinephrine should be avoided to minimize the risk of adverse events and dosing errors. Intravenous administration of epinephrine is associated with a significantly higher risk of cardiac arrhythmias and, rarely, myocardial infarctions. There is no absolute contraindication to the administration of epinephrine, although caution is advised in patients with cardiovascular disease, especially if uncontrolled. **Epinephrine is the first line therapy even in patients on beta blockers.**

Intravenous fluid should be strongly considered in all patients presenting with anaphylaxis as vasodilation and extravasation can occur suddenly. Dosing should be based off patient weight and a bolus of 20 mL/kg or 2 L in adults. Boluses can be repeated based on blood pressure and urinary output.

Oral H1 and H2 antihistamines can be supplemental therapies to assist with symptoms like pruritis, but do not have any role on attenuating mast cell degranulation or relieving bronchoconstriction. Oral antihistamines are not a substitute for epinephrine.

Albuterol or racemic epinephrine can be used to supplement IM epinephrine for patients with persistent wheezing or airway involvement but should not be used instead of epinephrine intramuscularly.

Quick recognition of symptoms and administration of epinephrine is the single most important intervention in reducing mortality in anaphylaxis, but there are multiple other risk factors for fatal anaphylaxis. As discussed earlier, cutaneous symptoms are the most common presenting symptom of anaphylaxis but up to 20% of cases present without cutaneous involvement. This can lead to misdiagnosis of anaphylaxis and delayed administration of epinephrine, which significantly increases anaphylaxis morbidity and mortality.

Certain concurrent comorbidities, especially cardiopulmonary disease and asthma also increase the risk of fatal anaphylaxis, particularly if the disease is poorly controlled. Recognition of these comorbidities is essential in triaging disposition and potential escalation of care as needed. Recognition of the risk factors for anaphylaxis is essential clinically and a frequently tested topic on exams. Table 6 demonstrates risk factors for fatal anaphylaxis.

While early recognition and treatment of anaphylaxis are vital to reducing morbidity and mortality, prevention of anaphylaxis is also key. In patients who have a known allergic trigger, avoidance of the trigger should be made of paramount importance. In patients with food allergies, they should be provided with a food allergy action plan which describes their food allergies, as well as a plan of treatment for an accidental exposure. In patients with exercise-induced symptoms, they should be advised to always exercise with a partner and to avoid any potential triggering foods prior to activity. In patients with food-dependent exercise-induced anaphylaxis, the culprit foods should be avoided for at least 4–6 h before exercise.

Patients should also be asked about comorbid conditions or medications which may potentiate the symptoms of anaphylaxis or even interfere with treatment. For example, alcohol and NSAIDs are associated with an increased risk of anaphylaxis in patients and may potentiate symptoms by diminishing the capacity for the patient to recognize their symptoms (alcohol) or synergistically amplifying the severity of symptoms though histamine-independent mechanisms (NSAIDs). Other medications may greatly diminish the effectiveness of treatment. For example, beta blockers diminish the effectiveness of epinephrine by blocking beta receptors in the heart and lung. While other medications, like opioids and medications commonly used to treat gastroesophageal reflux disease have been implicated as cofactors in anaphylaxis in multiple studies.

All patients with anaphylaxis should be referred to an allergist/immunologist for identification of potential triggers, workup for any potential underlying mast cell disorders, and continued monitoring. While routine aeroallergen skin testing is not indicated in anaphylaxis, food testing can be of utility in patients who may have a causal relationship with a food. However, routine food testing should never be performed as this is of little clinical utility in the absence of evidence of a causal relationship with a food allergen.

All patients with anaphylaxis should be discharged with an epinephrine autoinjector and educated by the provider on indications for use and proper technique. Anaphylaxis action plans are also available for distribution to families and schools to increase awareness of anaphylaxis diagnosis and treatment.

Table 6 Risk factors for fatal anaphylaxis

Risk factor
Delayed administration of epinephrine
Inability to recognize symptoms
Lack of cutaneous symptoms
Asthma
Concurrent cardiopulmonary disease
Systemic mastocytosis

Case 2

A 60-year-old man with past medical history of coronary artery disease which is well controlled with metoprolol succinate 100 mg daily and dietary changes presents to the emergency room via EMS with 30 min of facial swelling and hypotension. His blood pressure on arrival is 70/30 mmHg. He is pale and lethargic, but responsive to sternal rub and protecting his airway. His partner at bedside states that they were watching television when he began to complain of lip and eye swelling. He had not eaten anything that day and did not take any new medications. His partner notes that the patient had a similar episode 2 years ago after getting stung by a bee in their backyard, but it resolved after getting a "shot" in the emergency room.

Vitals

Respiratory rate: 25
HR: 170 bpm
Blood pressure: 70/30 mmHg
Oxygen saturation: 92%

Question 1

Which treatment should be administered first in this patient?

A. Aspirin
B. Epinephrine
C. Glucagon
D. Norepinephrine
E. Oxygen

Answer and Explanation
Answer: B

Epinephrine is the mainstay of treatment for anaphylaxis, even in patients on beta blockers or calcium channel blockers. Epinephrine increases heart rate, contractility, vasoconstriction, and bronchodilation by activity on both the alpha and beta receptors. Although beta blockers use is associated with refractory or severe anaphylaxis, studies have shown that patients on beta blockers are not more likely to require additional doses of epinephrine. However, patients who have refractory anaphylaxis despite multiple doses of epinephrine may benefit from glucagon administration.

Glucagon, like epinephrine, has both chronotropic and ionotropic effects on cardiac muscle but is independent of the beta receptors. Glucagon functions by increasing intracellular cAMP, which is also the secondary messenger for beta receptors. The most common side effects with glucagon administration are nausea and vomiting, so protecting the airway prior to administration is prudent.

Antihistamines can be used as adjuncts but should never be used as first line therapy for anaphylaxis. H1 and H2 antihistamines do not relieve bronchospasm or hypotension. However, they may be used in addition to epinephrine for symptomatic treatment of itching or urticaria during a reaction.

Norepinephrine is frequently used in intensive care units to treat severe hypotension associated with shock, especially if the patient is unresponsive to volume resuscitation. Unlike epinephrine, norepinephrine has mostly alpha-agonist activity, little beta-1 agonist activity, and no beta-2 agonist activity. Thus, norepinephrine may increase systemic vascular resistance, but has minimal impact on cardiac output or bronchospasm.

Oxygen can be used to supplement epinephrine, especially in hypoxic patients, but it is not first line therapy and does not treat underlying hypotension or bronchospasm. Care must be given, especially in children, to not cause lung injury from supratherapeutic oxygen levels.

Aspirin has no role in the treatment of anaphylaxis and is a common trigger of pseudoanaphylaxis.

Question 2

The patient is treated appropriately and clinically improves. A tryptase level is drawn and is 30 ng/mL (normal <11 ng/mL) and on repeat 48 h later is 21 ng/mL. What is the most likely diagnosis?

A. Biphasic anaphylaxis
B. Idiopathic anaphylaxis
C. Mastocytosis
D. Protracted anaphylaxis
E. Serotonin syndrome

Answer and Explanation

Answer: C

Mastocytosis is the disease caused by abnormal proliferation and survival of mast cells. Mastocytosis should be in the differential diagnosis for any patient who presents with anaphylaxis. One of the most efficacious ways to screen for mastocytosis is to check a baseline serum tryptase. After an episode of anaphylaxis, tryptase should be drawn between 15 min to 3 h and, if elevated, a repeat tryptase should be drawn at least 24 h after resolution of symptoms. Patients with anaphylaxis should have normal tryptase 24 h after resolution of their reaction, however patients with mastocytosis frequently, but not universally, have elevated baseline tryptase.

Biphasic anaphylaxis is defined as an initial anaphylaxis with resolution of symptoms for at least an hour followed by resurgence of symptoms. The second phase of anaphylaxis usually occurs within 8 h but can occur as long as 72 h after the initial reaction. Tryptase would be elevated during each phase of anaphylaxis but would not be elevated 24 h after resolution of symptoms. Risk factors for biphasic anaphylaxis are delayed administration of epinephrine and severe initial reactions. Patients who have more severe initial reactions or a history of biphasic reactions should be observed for at least 24 h for optimal identification and treatment of symptoms should they arise. While frequently used, there is no evidence that short courses of steroids after an episode of anaphylaxis prevents a biphasic reaction. In fact, the AAAAI guidelines explicitly recommend against prescribing steroids after an episode of anaphylaxis for prevention of a biphasic reaction.

Protracted anaphylaxis is anaphylaxis that lasts for hours to days without resolution of symptoms. Tryptase would likely still be elevated in cases of severe protracted anaphylaxis. However, if the patient is clinically improving, then tryptase should be decreasing with resolution of symptoms.

Serotonin syndrome is a relatively rare disorder caused by overproduction or decreased metabolism of serotonin, especially in the central nervous system. Serotonin syndrome is not associated with elevated baseline tryptase.

7 Differential Diagnosis of Anaphylaxis

Recognition of anaphylaxis is essential for minimizing morbidity and mortality, but it is also important to recognize that numerous other processes can have similar or even near-identical presentations. Being able to differentiate anaphylaxis from these other conditions is very important for clinicians and commonly tested on examinations.

Scombroid poisoning is caused by the ingestion of spoiled fish, prototypically mackerel or tuna, which have an overgrowth of bacteria that produce histidine decarboxylase which converts the amino acid histidine to the vasoactive amine histamine. This can occur in as little as 2–3 h and is not inhibited by cooking or food storage, including canning or freezing. Patients with scombroid poisoning frequently have symptoms like anaphylaxis such as pruritis, flushing, and even hypotension, but unlike anaphylaxis, they have normal tryptase levels.

Vocal cord dysfunction, also called paradoxical vocal cord motion (PVFM), is a relatively common disorder most frequently found in young athletic females. PVFM is caused by inappropriate adduction of the true vocal cords during inspiration. PVFM is typically triggered by stress, exercise, and strong or offensive odors. yPVFM is typically unresponsive to albuterol, and the treatment of choice is breathing exercises prescribed by speech pathology. Rarely, injection of botulinum toxin into the true vocal cords is needed and can be therapeutic for severe and refractory cases. PVFM does not have any skin or GI manifestations which differentiate it from anaphylaxis. Furthermore, PVFM usually has a completely normal laboratory evaluation and elevated tryptase is not expected.

Vasovagal reactions can similarly present with tachycardia, flushing, and even syncope. They can be frequently confused with allergic reactions. In adults, these reactions frequently occur with significant stressors or straining. Unlike IgE-mediate reactions, they do not have urticaria or more severe/life-threatening manifestations like hypoxia or cardiopulmonary collapse. These patients usually have quick, almost immediate, resolution of symptoms without any treatment and have completely normal laboratory studies. Treatment includes avoidance of triggers and reassurance.

Panic attacks usually present with sudden onset sensations of anxiety or impending sense of doom. Some patients may also complain of dyspnea or heart racing despite no intrinsic lung or cardiac pathology. Unlike anaphylaxis, there are usually no cutaneous symptoms and no obvious trigger. Similarly, these patients also have normal tryptase. Treatment consists of reassurance and support.

Pheochromocytomas are a rare tumor that arise from the adrenal gland that produce catecholamines, most notably epinephrine and norepinephrine, in significant amounts. Classically, these tumors present with a triad of paroxysmal (or rarely persistent) hypertension, headaches, and panic attack-like symptoms. Frequently, patients complain of periodic dyspnea, tremors, and an impending sense of doom which can mimic the symptoms of anaphylaxis, especially during their initial onset. Pheochromocytomas are detected by measuring serum or urinary catecholamine metabolites, ideally following an episode. Treatment involves surgical resection of tumor, if applicable, and alpha and beta receptor blockade. In contrast to anaphylaxis, pheochromocytomas do not typically present with urticaria, episodes are not attributable to any potential exposure, and tryptase is within normal limits.

Symptoms of carcinoid syndrome are due to neuroendocrine tumors which can secrete a variety of mediators including histamine, prostaglandins, and most notably, serotonin. These tumors usually, but not universally, originate from the gastrointestinal tract. Rarely, they can originate from the lung mucosa. Symptoms can imitate anaphylaxis, especially in histamine-predominant tumors. In these patients, paroxysmal diarrhea, flushing, tachycardia, and even rarely bronchospasm can be the major presenting symptoms. Diagnosis involves evaluation of urinary 5-hydroxyindoleacetic acid (5-HIAA) which has over 90% sensitivity and specificity for carcinoid syndrome. Some differentiating factors from anaphylaxis are the lack of any inciting trigger or exposure, normal tryptase levels, and presence of an underlying malignancy. Treatment of carcinoid syndrome involves treatment of underlying malignancy, if present, and suppression of serotonin secretion. Somatostatin is a peptide that bound to somatostatin receptors on neuroendocrine tumors. Binding of somatostatin receptors inhibits mediator secretion including serotonin. A synthetic analog of somatostatin known as octreotide is the most prescribed therapeutic used in carcinoid syndrome.

Hereditary angioedema (HAE) is a relatively rare disorder, caused by either the quantitative or qualitative deficiency in C1-inhibitor, or rarely Factor XII deficiency. The disease pathology is caused by over production of bradykinin causing increased vascular permeability leading to tissue edema. Clinically, HAE presents as recurrent angioedema without urticaria or pruritis. HAE can affect any part of the body and thus symptoms can be as diverse as dyspnea if the angioedema involves the airway or abdominal cramping if the angioedema involves the gut. It is not uncommon for HAE to present with a prodromal rash called erythema marginatum which frequently appears on the stomach just prior to the onset of angioedema. Erythema marginatum is nonpruritic, appears as a confluent erythematous rash, and should be differentiated from urticaria. Diagnosis of HAE involves assessing complement levels and the function, and quantity of C1-inhibitor in the blood. Normal complement levels do not rule out HAE. Treatment involves replacement of C1-inhibitor either acutely or prophylactically. Plasma kallikrein inhibitors and bradykinin receptor blockade are also used in select patients. HAE is differentiated from anaphylaxis by the absence of urticaria and pruritis. HAE also has normal levels of tryptase and urinary histamine metabolites, but frequently has abnormal complement studies, especially during acute attacks.

Mastocytosis is the disease caused by abnormal or inappropriate mast cell accumulation in tissues. Mastocytosis can involve any organ system, but most frequently involves the skin. Skin lesions in mastocytosis, called urticaria pigmentosa, are reddish-brown papules that can present anywhere on the body. Darier's sign is the induction of an

Table 7 Differential diagnosis and differentiating features of anaphylaxis

Differential diagnosis	Key features	Tryptase	Identifying studies
Vasovagal reaction	Flushing, tachycardia, associated with stress or straining	Normal	None
Panic attack	Anxiety, fear, tachycardia, diaphoresis	Normal	None
Vocal cord dysfunction	Isolated shortness of breath usually associated with anxiety or stress	Normal	Flattened inspiratory loop in spirometry
Scombroid poisoning	Ingestion of scombroid fish (mackerel, tuna, etc.), flushing, tachycardia, rarely urticaria	Normal	Elevated plasma histamine
Pheochromocytoma	Headache, tachycardia, **hypertension**	Normal	Elevated plasma and urinary catecholamines
Carcinoid syndrome	Flushing, diarrhea, tachycardia. Associated with malignancy, especially GI	Normal	Elevated 5-HIAA
Hereditary angioedema	Isolated angioedema without urticaria. Unresponsive to epinephrine or antihistamines. Usually familial	Normal	Potentially abnormal complement studies (C1q function/level, C3)
Mastocytosis	Frequently associated with persistent cutaneous lesions (urticaria pigmentosa), other organ involvement (osteopenia, liver or kidney dysfunction)	Elevated baseline	KIT D816V mutation

urticarial lesion after stroking of an urticaria pigmentosa lesion. It is pathognomonic for urticaria pigmentosa. Symptoms of mastocytosis can be either from direct localized overpopulation of mast cells in the tissue or from excessive release of mast cell mediators systemically. Gastrointestinal complaints are among the most common symptoms, with diarrhea and abdominal cramping being frequent complaints. The symptoms severity of mastocytosis can vary widely. Rarely excessive mast cell degranulation can cause signs and symptoms identical to anaphylaxis from an allergic trigger. These episodes can be potentially life threatening and should be treated identically to anaphylaxis from any cause with prompt administration of intramuscular epinephrine. Treatment of mastocytosis is complex and dependent upon the degree of local and systemic involvement and symptoms. While anaphylaxis has elevated tryptase after an acute episode, levels of tryptase should normalize within 24 h after resolution of symptoms. However, patients with mastocytosis frequently have persistently elevated levels of tryptase even when asymptomatic. Tryptase levels ≥20 ng/mL at ≥24 h after resolution of anaphylaxis are highly suggestive, but not diagnostic, of mastocytosis. Table 7 lists the differential diagnosis of anaphylaxis and their important features.

Question 3

Which type of tryptase level is most likely to be elevated in anaphylaxis?

A. Mature α tryptase
B. Mature β tryptase
C. Pro-α tryptase
D. Pro-β tryptase
E. Meta-tryptase

Answer and Explanation
Answer: B

Mature β tryptase is released by mast cells upon activation and degranulation. It reaches the highest level in the blood after anaphylaxis and decreases rapidly as symptoms resolve. Mature α tryptase is secreted by mast cells constitutively and is elevated in the blood of patients with hereditary α tryptasemia (HaT). HaT occurs in 5% of the population and is an autosomal dominant disease. However, total tryptase is usually normal or only mildly elevated in patients with HaT. Commercial tryptase does not differentiate between pro- or mature tryptase or between α and β tryptase. HaT should be considered in anaphylactic patients with normal tryptase levels, especially in cases with severe or recurrent anaphylaxis, and those attributable to insect venoms.

Tryptase is synthesized constitutively by mast cells in both the immature (pro-tryptase) form and a mature form. Immature tryptase does not have any enzymatic activity.

There is no such entity as meta-tryptase.

Question 4

Which statement is true regarding perioperative anaphylaxis?

A. Cutaneous symptoms are the most common presenting symptom
B. Mortality is higher than in other forms of anaphylaxis
C. Neuromuscular blocking agents are the most common cause in the United States
D. There is a significant male predominance
E. Tryptase is often normal

Answer and Explanation
Answer: B

8 Perioperative Anaphylaxis

Perioperative anaphylaxis is defined as anaphylaxis that occurs within a short window surrounding general anesthesia or surgical intervention. It can occur at any age. The exact incidence is unknown, but likely 1 in 10,000–20,000 cases. It is to be noted that this number is complicated by difficulties in obtaining the true number of surgical interventions and use of general anesthesia. Furthermore, there are significant regional differences in the incidence which may be due to differences in general anesthesia, antibiotics, and perioperative medications administered. Here, we will only discuss perioperative anaphylaxis related to cases in the United States. Perioperative anaphylaxis is more common in women than in men, and the incidence tends to increase with age. However, in children there is no such gender disparity.

Perioperative anaphylaxis, much like other forms of anaphylaxis, can be IgE-mediated or non-IgE-mediated. Around 60% of cases of perioperative anaphylaxis are attributable to an IgE-mediated mechanism. The etiological agent responsible varies based on the region, but the most common cause in the US are antibiotics. In Europe, the most common cause is neuromuscular blocking agents (NMBA). However, nearly any medication or exposure in the perioperative environment could be causal. Latex, topical sterilization agents like chlorohexidine, opioids, and even reversal agents have all been implicated in causing perioperative anaphylaxis.

Antibiotics are the single most common cause of perioperative anaphylaxis in the US. They account for almost 50% of cases in the US, but less than 20% in Europe. Penicillin or cephalosporins administered just prior to the initiation of surgery are the most likely culprits. Establishing a causal agent can be difficult or, in some situations, not possible to determine a clear etiology. This is because many medications and interventions are administered at the same time or within minutes of each other in the perioperative setting. These cases often require a thorough review of the history, anesthesia record, flowsheets, and looking for a clear temporal relationship between administration of the medication and the onset of symptoms. Patients who have received the antibiotic multiple times in the recent past are less likely to have an allergy to the antibiotic, whereas a new exposure to an inducing agent or NMBA could lead towards a potential cause.

Skin testing for penicillin has high utility in ruling out a penicillin allergy and has significant utility in perioperative anaphylaxis patients. Skin testing to other medications, including NMBAs, are not standardized; however, skin testing has significant sensitivity in detecting an IgE-mediated allergy in evaluation of these agents. These diagnostic evaluations should not take the place of a thorough history and physical exam or review of the anesthesia flowcharts.

Perioperative anaphylaxis has a significantly higher mortality than anaphylaxis due to other causes. Some studies postulate that it may be as high as 1.4%, compared to 1% mortality rate from anaphylaxis of all causes in hospitalized patients. Perioperative anaphylaxis also has significantly higher morbidity. The causes of the elevated morbidity and mortality in perioperative anaphylaxis are likely multifactorial. First, perioperative anaphylaxis which are due to IgE-mediated mechanisms are usually due to medications administered intravenously. This means that the median onset of anaphylaxis is significantly faster than orally administered medications. Secondly, many symptoms of anaphylaxis may be missed or attributed to a normal reaction from anesthesia or other medications administered in the perioperative setting, or even because of the procedure itself. As discussed earlier, the most common manifestations, and often the first presenting signs, in anaphylaxis are cutaneous symptoms like urticaria or pruritis. However, in surgical patients they are frequently sedated and unable to express that they are experiencing pruritis. Furthermore, patients are also draped and gowned so skin changes can go unnoticed by the surgical team. These factors can cause significant delay in the diagnosis and treatment of anaphylaxis which subsequently leads to higher morbidity and mortality.

As with other forms of anaphylaxis, ideally, a tryptase should be drawn between 15 min to 3 h from the onset of symptoms but should not delay therapy. This can help differentiate true anaphylaxis from adverse drug reactions or from surgical complication in hindsight.

Early recognition of symptoms and low threshold for the initiation of treatment are essential in minimizing morbidity and mortality. Early treatment of symptoms with epinephrine and adjunctive therapies like fluid resuscitation and vasopressors if required has been shown to decrease mortality. All interventions to treat anaphylaxis should be immediately available in a surgical suite. The perioperative setting is one of the few, if not the only, setting in which intravenous epinephrine is more commonly used than intramuscular epinephrine because of the ease of availability.

Question 5
What is the most common cause of fatal anaphylaxis?

A. Aeroallergens
B. Drugs
C. Foods
D. Idiopathic
E. Venoms

Answer and Explanation
Answer: B

The most common cause of anaphylaxis in the United States is attributed to foods in children with peanuts, milk, and eggs being the most common trigger in children. In adults, the most common cause of anaphylaxis is medications with antibiotics being the most common trigger. While food related anaphylaxis is common trigger overall, it is not the most common cause of fatal anaphylaxis. Drug-induced anaphylaxis is much more common in adults and has a significantly higher morbidity than foods or venoms. In fact, medications account for almost 60% of fatal anaphylaxis.

9 Incidence, Etiology, and Prognosis of Anaphylaxis

Obtaining the true incidence of anaphylaxis is difficult due to differences in coding of the diagnosis, delayed recognition, and the inability to find a causative agent. Studies have reported wide ranges in the incidence of anaphylaxis, but it is believed that the true incidence is anywhere from 40 to 200 per 100,000 person-years. In the United States, there are believed to be up to 1500 deaths a year attributed to anaphylaxis with most of these deaths related to medications. This translates into a rate of approximately 0.21–0.76 per million.

In children, the most common cause of anaphylaxis is foods, occurring in up to 4% of children. The most common triggers being peanuts, milk, and egg. Milk and peanut are the most common causes of fatal food-related anaphylaxis in children, while shellfish is a common cause of fatal food anaphylaxis in adults.

Foods are the single most common cause of anaphylaxis in both children and adults. Medications are the most common cause of fatal anaphylaxis. The most common etiological agent for fatal anaphylaxis is antimicrobials, especially beta-lactam antibiotics. Non-steroidal anti-inflammatory (NSAID) agents are another increasingly common cause of anaphylaxis, especially in adults. NSAIDs are unique in that they can also cause anaphylaxis through non-IgE-mediated mechanisms. NSAIDs inhibit COX-2 which causes shifting of the arachidonic acid pathway to leukotriene and prostaglandin production.

Venoms are a less common cause of anaphylaxis with frequency steadily decreasing over the past few decades for unclear reasons. Anaphylaxis due to venoms occurs in up to 0.8% of children and almost 3% of adults, however fatal anaphylaxis is becoming increasingly uncommon.

Radiocontrast media (RCM) is another rare, but clinically important, cause of anaphylaxis. The pathophysiology in RCM-mediated anaphylaxis is believed to be due to the hyperosmolarity of the contrast mediated and not believed to be caused by an IgE-dependent mechanism. These reactions usually occur quickly after administration of IV contrast. There are numerous pre-treatment protocols available that may reduce the risk of reactions. These protocols include using high dose corticosteroids and antihistamines up to 24 h prior to administration of contrast.

Immunotherapy is a clinical important cause of anaphylaxis for the practicing allergist/immunologist. The incidence of anaphylaxis after conventional schedule allergen immunotherapy is less than 1%, but still can be potentially life threatening, especially in patients with comorbid asthma or heart disease. Alternative schedules for allergen immunotherapy, like cluster or rush immunotherapy, have a significantly higher incidence of anaphylaxis. This incidence can be as high as 30% depending on schedule, comorbidities, and allergen. Fatal anaphylaxis, while rare, has happened with allergen immunotherapy and occurs in roughly 1 in 2–2.8 million injections.

Exercise-induced anaphylaxis is a rare disorder present in 0.03% of the population. In these patients, they have systemic histamine release shortly after exercise necessitating the use of an epinephrine autoinjector. Foods have rarely been associated with exercise-induced anaphylaxis but it is much less common (0.017% of the population). Patients with either of these conditions should be prescribed an epinephrine autoinjector and educated on proper avoidance of food triggers. All patients with exercise-induced anaphylaxis should also be advised to exercise with a partner should they become symptomatic.

Multiple potential risk factors for anaphylaxis have been evaluated to assist in risk stratifying patients. To date, there is no evidence that race or ethnicity or socioeconomic status has any impact on incidence or severity of anaphylaxis. Atopy has only been associated as a risk factor for food-induced anaphylaxis, but its role in other forms of anaphylaxis is highly controversial. Poorly controlled asthma and other primary mast-cell disorders like mastocytosis are clear risk factors for anaphylaxis in all age groups. Finally, there are age-dependent differences in the prevalence of anaphylaxis based off sex. Males have an increased incidence of anaphylaxis below the age of 15, however in older teenagers and adults, there is a clear female-predominance. One exception to this is a persistent male-predominance for venom-induced anaphylaxis across all ages.

Another risk factor for anaphylaxis including route of administration with IV medications has an increased risk for anaphylaxis and increased severity over oral medications.

Finally, in a large percentage of cases, the cause of anaphylaxis is unavailable or unable to be determined (idiopathic). The incidence of this is unknown, largely due to difficulties in coding the episodes using ICD-10 or inability to obtain a clear history. Many of these cases may not truly be idiopathic. However, these cases account for just under 20% of fatal episodes of anaphylaxis. Table 8 demonstrates the incidence of anaphylaxis and fetal anaphylaxis by etiology.

Table 8 Incidence and percentage of fatal anaphylaxis by etiology

Mechanism	Incidence of anaphylaxis	Percentage of fatal anaphylaxis cases
Medications	9.1 per 100,000 person-years	59%
Idiopathic or unknown	Unknown	19.30%
Venoms	10.2 per 100,000 person-years	15.20%
Food	15.2 per 100,000 person-years	6.70%
All-cause	42 per 100,000 person-years	N/A

10 Summary

Anaphylaxis is a serious and potentially life-threatening disorder that can affect multiple organ systems. Anaphylaxis is most often characterized by the presence of cutaneous symptoms, however, the absence of cutaneous symptoms does not rule out the diagnosis. Prompt diagnosis and initiation of treatment are essential to minimizing morbidity and mortality. Treatment of anaphylaxis consists of 0.01 mg/kg of intramuscular epinephrine. Multiple laboratory tests may assist in the diagnosis of anaphylaxis if unclear, however, they are not required. Tryptase has the most utility in aiding in the diagnosis if the clinical picture is unclear, however, it must be drawn quickly after the onset of reaction and has a high false negative rate depending on the underlying etiology.

Bibliography

1. Reber LL, Hernandez JD, Galli SJ. The pathophysiology of anaphylaxis. J Allergy Clin Immunol. 2017;140(2):335–48.
2. Zilberstein J, et al. Anaphylaxis. J Emerg Med. 2014;47(2):182–7.
3. Vadas P, Perelman B, Liss G. Platelet-activating factor, histamine, and tryptase levels in human anaphylaxis. J Allergy Clin Immunol. 2013;131(1):144–9.
4. Payne V, Kam PCA. Mast cell tryptase: a review of its physiology and clinical significance. Anaesthesia. 2004;59(7):695–703.
5. Zilberstein J, McCurdy MT, Winters ME. Anaphylaxis. J Emerg Med. 2014;47(2):182–7. https://doi.org/10.1016/j.jemermed.2014.04.018. Epub 2014 Jun 2.
6. White JL, et al. Patients taking β-blockers do not require increased doses of epinephrine for anaphylaxis. J Allergy Clin Immunol Pract. 2018;6(5):1553–8.
7. Shaker MS, et al. Anaphylaxis—a 2020 practice parameter update, systematic review, and grading of recommendations, assessment, development and evaluation (GRADE) analysis. J Allergy Clin Immunol. 2020;145(4):1082–123.
8. Lyons JJ, et al. Heritable risk for severe anaphylaxis associated with increased α-tryptase–encoding germline copy number at TPSAB1. J Allergy Clin Immunol. 2020; https://doi.org/10.1016/j.jaci.2020.06.035.
9. Poowuttikul P, Seth D. Anaphylaxis in children and adolescents. Pediatr Clin N Am. 2019;66(5):995–1005.
10. Hungerford JM. Scombroid poisoning: a review. Toxicon. 2010;56(2):231–43.
11. Sjöblom S-M. Clinical presentation and prognosis of gastrointestinal carcinoid tumours. Scand J Gastroenterol. 1988;23(7):779–87.
12. Reubi JC, et al. Detection of somatostatin receptors in surgical and percutaneous needle biopsy samples of carcinoids and islet cell carcinomas. Cancer Res. 1990;50(18):5969–77.
13. Bork K, et al. Treatment of acute edema attacks in hereditary angioedema with a bradykinin receptor-2 antagonist (Icatibant). J Allergy Clin Immunol. 2007;119(6):1497–503.
14. Mertes PM, et al. Anaphylaxis during anesthesia in France: an 8-year national survey. J Allergy Clin Immunol. 2011;128(2):366–73.
15. Mertes PM, et al. Reducing the risk of anaphylaxis during anaesthesia: guidelines for clinical practice. J Investig Allergol Clin Immunol. 2005;15(2):91–101.
16. Gibbs NM, et al. Survival from perioperative anaphylaxis in Western Australia 2000–2009. Br J Anaesth. 2013;111(4):589–93.
17. Umasunthar T, et al. Incidence of fatal food anaphylaxis in people with food allergy: a systematic review and meta-analysis. Clin Exp Allergy. 2013;43(12):1333–41.
18. Turner PJ, et al. Fatal anaphylaxis: mortality rate and risk factors. J Allergy Clin Immunol Pract. 2017;5(5):1169–78.
19. Jerschow E, et al. Fatal anaphylaxis in the United States, 1999–2010: temporal patterns and demographic associations. J Allergy Clin Immunol. 2014;134(6):1318–28.
20. Lee S, et al. Trends, characteristics, and incidence of anaphylaxis in 2001-2010: a population-based study. J Allergy Clin Immunol. 2017;139(1):182–8.

Diagnostic Testing in Allergic Diseases

Massoud Mahmoudi

1 Introduction

An allergic patient presents with multiple complaints. An allergist, before being a treating provider, is a diagnostician. Although various diagnostic testing are available, choosing a right test for the right patient is the first attempt to diagnose an allergic condition in order to formulate a treatment plan. In this short chapter, two cases are presented which helps to exercise the approaches to diagnosis of an allergic patient.

Case 1

A 42-year-old Caucasian male is referred to your clinic from the primary provider with multiple complaints. He has had allergies "all life" and has been using over the counter antihistamines with some relief. He has also experienced off and on hives for 6 years but unable to identify the trigger. A week ago started using a body deodorant for his armpits and noted itching and redness. Two weeks ago when touching a balloon in his son's birthday experienced itchy hands and lips. He was frustrated with all these problems and is seeking your help. He is an information technology support coordinator, lives with his spouse, two young sons, and a cat. He never smoked but consumes occasional alcohol. There was no history of tobacco use or recreational drugs. The family history is significant for mother with allergic rhinitis and a younger brother with asthma.

In the first visit, he appeared as a tall, well nourished male in no acute distress. Vitals were all within normal limits. On examination, head, ears, eyes, and nose (with the exception of edematous nostrils and pinkish mucous membranes), lungs abdomen, and extremities were all within normal limits. The skin was dry but no hives were present at the time of examination.

M. Mahmoudi (✉)
Department of Medicine, University of California San Francisco,
San Francisco, CA, USA

Question 1

Which one the following diagnostic test is reasonable to include in your work up?

A. Prick skin test to aeroallergens
B. Patch testing
C. Serum IgE to latex
D. Serum thyroid hormone, levels, anti-thymoglobulin and thyroid peroxidase antibodies, and complete metabolic panel
E. All of the above

Answer: E

The patient presented with multiple complaints. He is atopic, and an investigation of his current condition should include a work up for allergic rhinitis by prick skin testing to aerollergens, patch testing to assess contact dermatitis, and serum IgE to latex due to his skin reaction to balloon (natural rubber latex). For evaluation of hives, underlying etiology such as hepatitis should be ruled out. And since the hives is chronic, presence of thyroid antibodies should also be investigated. The decision regarding the priority of the test ultimately depends on severity and labor disabling of the symptoms.

Question 2

You decide to start with aeroallergens prick skin testing. In which one of the following situation you are still able to perform the test?

A. The hive have covered the lower extremities
B. The patient is unable to stop antihistamines
C. The entire back is covered with a large and dark tattoo
D. Continuous pruritus
E. The patient took 10 mg dose of cetirizine 10 h prior to the visit

M. Mahmoudi (ed.), *Absolute Allergy and Immunology Board Review*, https://doi.org/10.1007/978-3-031-12867-7_26

Answer: C

General pruritus and hives are due to release of histamine. Since the allergen skin testing is based on release of histamine from mast cells after exposure to specific allergen, the reaction to the tested allergen would show inaccurate results; i.e., the test result would read as false positives.

When the patient is unable to stop antihistamine before prick skin testing, the H1 receptor is blocked by antihistamine and therefore, the result of testing would be inaccurate; i.e., there would be false negative as the allergens would not elicit histamine release to react on skin.

The allergen prick skin testing is performed on the upper and mid back. This is because of larger surface area and sensitivity. Ideally, the test area should be clear of rash and dark pigmentation. Testing on a darker skin color is a challenge but is possible; however, general dark tattoos make the testing impossible. In such a situation, the volar aspect of the arms is used for the testing.

2 Indication of Allergy Testing

The indication of the allergy testing is summarized in Table 1:

3 Types of Allergy Testing

The most utilized test by allergists is the **prick or epicutaneous testing**. The test is performed simply by taking the allergen by a test device and pricking the skin. Various devices are available for the test application. Choosing the test device depends on various factors such as effectiveness, the facility of administration, cost, and the preference of the allergist (Fig. 1).

Intradermal allergy testing is another in vivo testing which has limited utility. When the prick test is negative, you like to make sure there is no underlying allergies by performing the intradermal testing. The advantage of the test is that it is more sensitive compared to the prick test but less specific.

Radioallergosorbent test (known as RAST) is an in vitro test where the specific IgE antibody in the patient' serum is detected by radiolabeled conjugated anti-IgE (a traditional method) and recently by an enzyme-linked anti-IgE antibody (also known as ImmunoCAP) (Fig. 2).

Table 1 Indication for prick skin testing

Allergic rhinitis
Allergic rhinoconjunctivitis
Allergic asthma
Atopic dermatitis
Food allergy
Venom allergy
Drug allergy

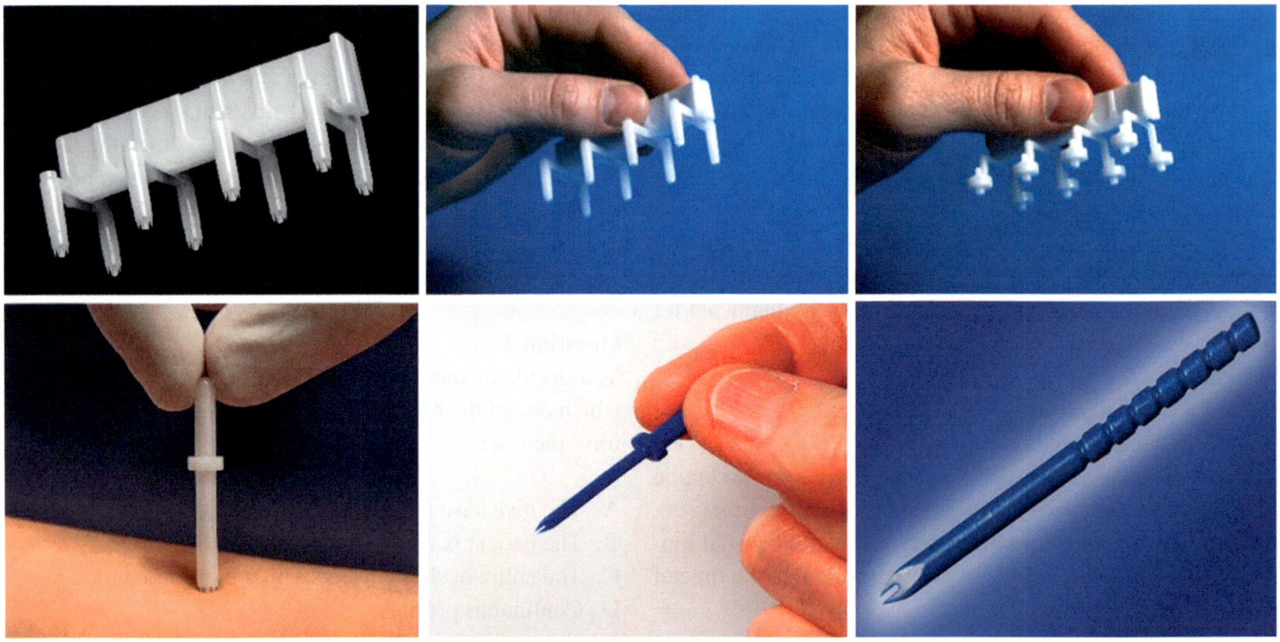

Fig. 1 Examples of devices used for prick sin testing: Multi-Test devices (from left to right) Multi-Test® PC (pain control), Multi-Test® II, Multi-Test®. Examples of single test devices (from left to right) UniTest® PC, Duotip-Test® II, Duotip-Test®. (Courtesy of Lincoln Diagnostics)

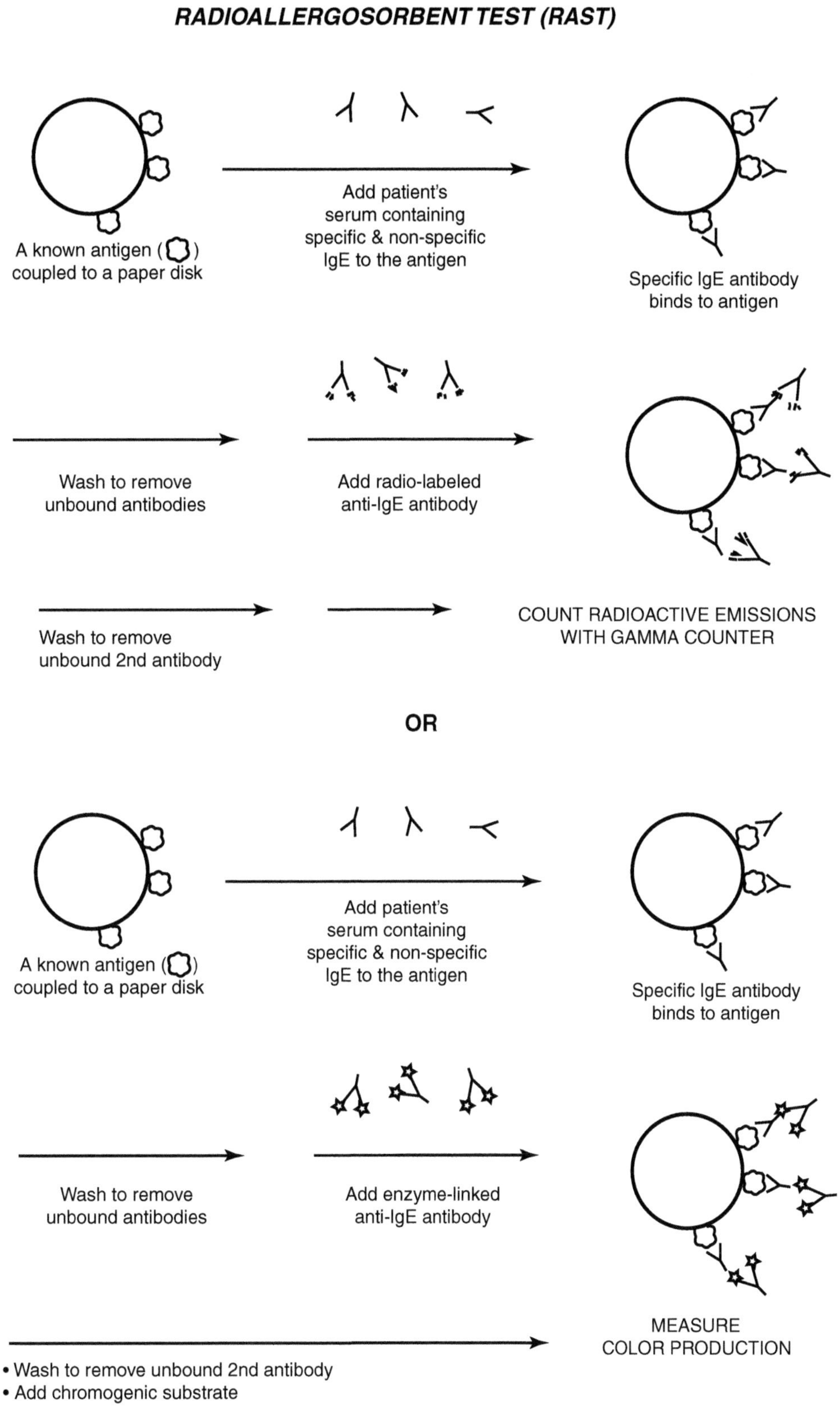

Fig. 2 Radioallergosorbent test. (From Immunology made ridiculously simple; M. Mahmoudi; MedMaster, 2009, with permission)

4 Mast Cells and Histamine

The mast cells are small, 10–15 μm immune cells with bilobed and multilobed nucleus which originate from bone marrow and are distributed in skin and connective tissue. They are the main players in hypersensitivity reaction. The mast cells have two unique features: they carry a high affinity receptor on their surface known as FcεRI which binds to the IgE and contain granules carrying various mediators, some of which are common with basophils (Fig. 3).

Release of histamine from mast cells, also known as degranulation is a multistep process which include the following (Fig. 4):

- Binding of allergen to the B cells
- Binding of B cells with TH2 cells leading to activation of TH2 cells
- B cells undergo class switching and become antibody producing cells and produce IgE
- Binding of IgE to FcεRI on surface of mast cells
- Re-exposure of the host to the same allergen
- Mast cell degranulation and release of various mediators including histamine, the key to formation of wheal (tissue edema) and flare (redness) in the area

After application of the allergen in skin (prick or intradermal) wait 15–20 min. The specific allergen causes degranulation of mast cell, releasing histamine and other mediators. The positive reaction is then compared with two controls (positive control, histamine, and a negative control, usually saline). The reaction is measured and compared with the positive control to assess the ratio of the reaction.

Common Mediators in Mast Cells and Basophils

Secretory protein	Type of protein	Functions
Histamine	Vasoactive amines	• Vasodilates, increases vascular permeability • Bronchospasm
Neutral proteases (Tryptase, chymase carboxypeptidase, cathepsin)	Enzyme	• Degrades tissues • Damages microbial structures
Chondroitin Sulfate	Proteoglycan	Participates in structural matrix of granules
Platelet-activating factor	Lipid mediator	• Attracts leukocytes • Increases vascular permeability • Activates neutrophils, eosinophils, and platelets • Bronchocoostricts
Leukotriene C4 (LTC4) , D4 (LTD4), and E4 (LTE4)	Lipid mediator	• Bronchospasm • Increases vascular permeability and constriction of arterial, arteriolar, and intestinal smooth muscle
Prostaglandin D2 (PGD2)	Lipid mediator	• Vasodilates • Increases vascular permeability • Bronchoconstricts • Inhibits platelet aggregation • Stimulates neutrophil chemotaxis (attracts neutrophils)
TNF-α	Cytokines	Activates neutrophils
IL-4	Cytokines	• Promotes TH2 differentiation, isotype switching to IgE • B cell proliferation • Eosinophil and mast cell growth and function
IL-5	Cytokines	• Stimulates eosinophil production, activation, growth, and differentiation • Stirrulates B lymphocyte proliferation
IL-6	Cytokines	Induces fever, acute phase response (liver)
IL-13	Cytokines	Similar to IL-4

Fig. 3 Common mediators in mast cells and basophils. (From Immunology made ridiculously simple; M. Mahmoudi; MedMaster, 2009, with permission)

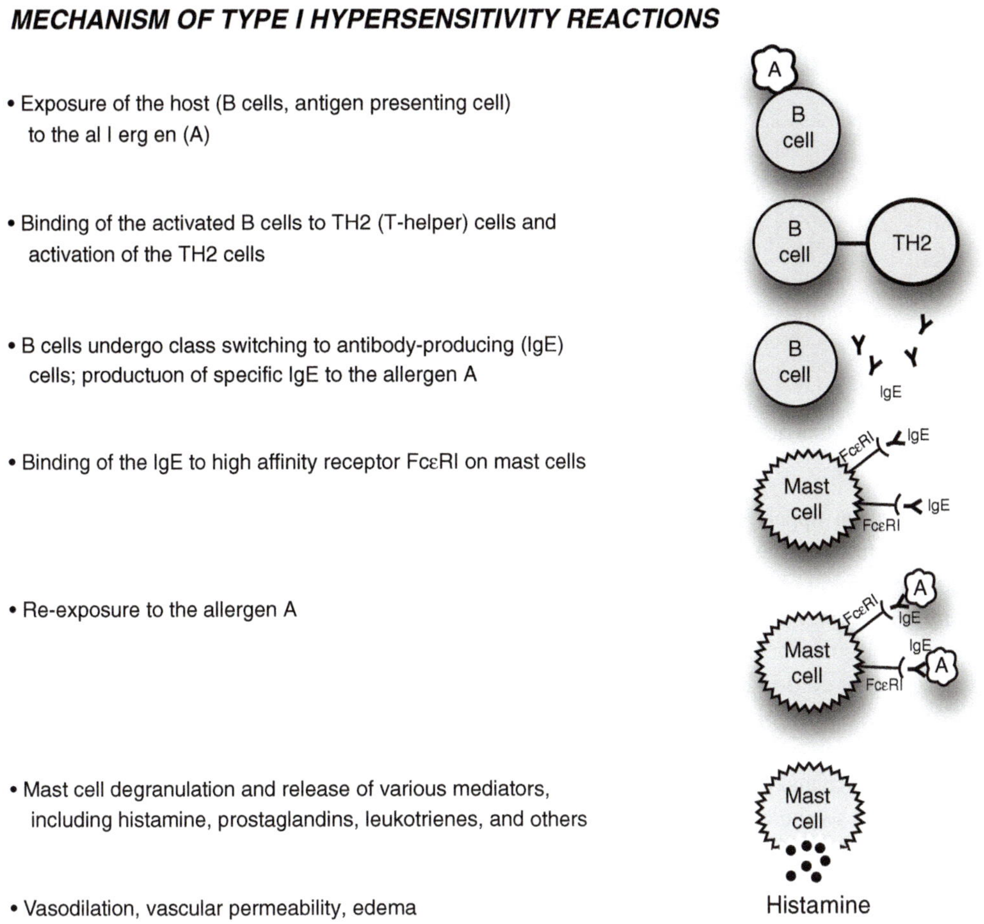

Fig. 4 Release of histamine from mast cells (degranulation), a type I hypersensitivity reaction. (From Immunology made ridiculously simple; M. Mahmoudi; MedMaster, 2009, with permission)

5 Utility of Allergy Testing in Allergic Diseases

Based on the type of allergic disease, the appropriate test is selected. Table 2 summarizes the choices of allergy testing.

Table 2 Choice of allergy testing

Condition	Prick skin test	Cap	Patch testing	Oral challenge
Allergic rhinitis	+	+		
Atopic dermatitis	+	+	+	
Allergic contact dermatitis			+	
Food allergy	+	+	+[a]	+
Venom allergy	+	+		
Drug allergy	+			

[a] Although not standardized, some allergists perform the patch for food allergens

Question 3

In one of the follow up visits you plan to address the rash,redness, and itching of the axilla. Which of the following test is appropriate?

A. Skin biopsy of the rash
B. Food allergen prick skin testing
C. Serum IgE to latex
D. Patch testing
E. Ice cube test

Answer: D

Skin biopsy of rash is non-specific and less likely results in a diagnosis. Food allergy causes a widespread reaction such as urticaria and angioedema, or localized area of contact such as laryngeal edema, which eventually spread to be a systemic presentation. In this case, we need to look for a contactant. The patient reported the reaction after using of a deodorant and therefore patch testing is the right test for evaluation.

The mechanism of patch testing is cell-mediated type 4 Gell and Coombs reaction. The allergen in question embedded

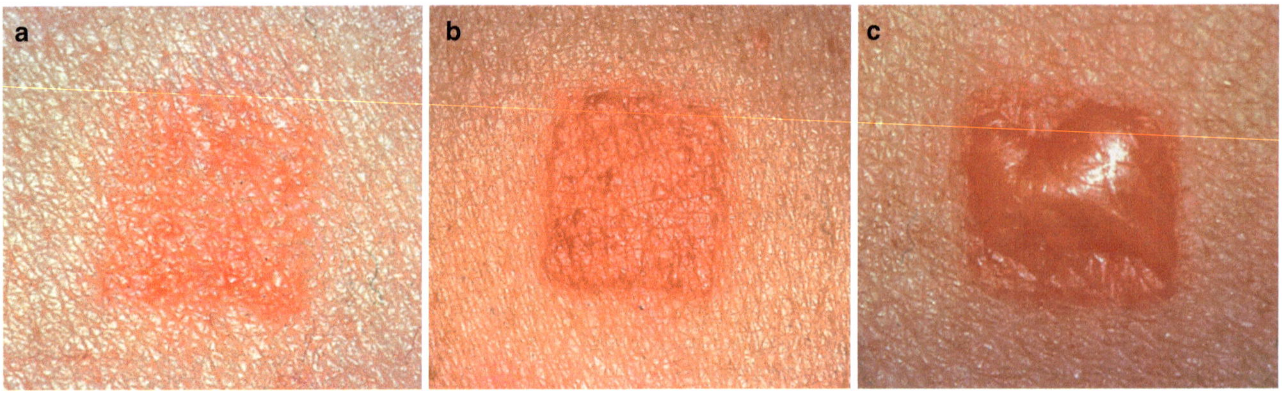

Fig. 5 Patch test reactions: (**a**) mild, (**b**) moderate, and (**c**) severe. (Courtesy of SmartPractice)

in the test patch is placed on upper back. The patches are removed after 48 h for the first reading of the result and again in 48 h for the final interpretation of the results. A typical test result ranges from mild to severe, a blister-like like reaction, Fig. 5.

The ice cube test is test for cold-induced urticaria.

Question 4

The patient follows your advice and pays attention to his diet to identify the possible food as the culprit of itching. He is still unsure of the food that triggers his reaction. He presets to the clinic for the testing and reports that he is not itching, has been off of antihistamine for 5 days, and has been free of hives for the past 8 days. Which one of the following is reasonable first diagnostic test?

A. Prick-prick test with the suspected food
B. Open oral Food challenge test
C. Prick skin test to food allergen
D. Single blind food challenge test
E. Double blind food challenge test

Answer: C

The first diagnostic test is prick skin test to suspected food allergen. If the test is negative but the patient continues to be symptomatic with the suspected food, you may use the "Prick-Prick test." Simply the patient brings the suspected food for the testing. First you prick the food and then prick the skin. The ultimate testing is when the patient ingests the food in question, known as oral food challenge. The oral food challenge is either, an open label, single blind, or a double blind challenge:

The open label oral challenge is when the patient and the allergist know what is being tested. The patient ingests a minute amount of the food and is observed for possible reaction.

Single blind food challenge—In this test, the allergist is aware of the test food but the patient is not.

Double blind placebo-controlled test—The suspected food is prepared without knowledge of the allergist and the patient. Then the patient ingests the placebo and later the food prep and is observed for an allergic reaction. This test is considered the gold standard of the food allergy testing.

The food challenge should be closely observed by the allergist and resuscitative tools must be available.

Question 5

Per your instruction, the patient started logging the events that resulted in hives. After reviewing the list you note that he develops hive with exposure to cold air and water. Which one of the following is the appropriate diagnostic test?

A. Emerge both hands in cold ice water for 15 min
B. Stay in a cold room for 15 min
C. Ice cube test
D. Serum cryoglobulinemia
E. Swimming in a cold water for 10 min and check for hives

Answer: C

The patient's symptom points to diagnosis of cold-induced urticaria. The best test is an ice cube test.

Cold-induced urticaria is a type of physical urticaria which results after exposure to a cold environment. Some examples of such exposures are holding a cold object, swimming in cold water, or staying in a cold room. Although the common reaction is urticaria, other skin reaction such as pruritus, burning, or angioedema may develop with such exposures. At times, the reactions are systemic such as angioedema, hypotension, arthralgia, faintness or shock, vertigo, anaphylaxis, or death. Cold-induced urticaria is either familial or acquired. The familial form is a rare autosomal dominant trait disorder which does not respond to the ice cube test. The acquired form is either primary (idiopathic) or secondary and responds immediately to the cold stimulation test. The secondary type is associated with various other conditions such as human deficiency virus (HIV), hypothyroidism, and some medications.

The ice cube test is performed by simply placing an ice cube on the volar aspect of the forearm for 5 min. Then after 5 min, the ice cube is removed. The positive test result is a wheal and flare reaction in the area of ice contact. Although the test may be negative in patients with cold-induced urticaria, the ice cube test is still simplest and safest test for screening such condition. Placing hand in cold water or staying in a cold room for 15 min may result in positive reaction and is not necessary as a screening test. Cryoglobulins are immunoglobulins that precipitate in temperatures below body temperature of 37 °C. Secondary cold-induced urticaria may be associated with serum cryoglobulinemia (presence of cryoglobulins in patient's serum). In presented case, this is not the first screening test. Swimming in cold water is not safe as it may cause anaphylaxis.

Case 2

A 13-year-old male student presents to your clinic with multiple complaints. He and his 15-year-old brother moved to town 3 months ago to attend school and have been living with friend of family. He reports 6 weeks history of intermittent dry cough, mild shortness of breath, nasal congestion, nausea, and headaches. Family history is significant for allergic rhinitis in mother and atopic dermatitis in sister. In his new residence, he has no pet. He does not smoke or drink alcoholic beverages. His only medication is a multivitamin. Except intermittent skin infections which usually resolves with antibiotics, he has been healthy. The physical examination, with the exception of minor nasal hypertrophic turbinates and mild anterior expiratory wheezing, is within normal limits.

Question 6

Which one of the following test does best evaluate his nasal symptoms?

A. Nasal smear evaluation for eosinophils
B. Laryngoscopy to assess nasal polyps
C. Prick skin testing to aeroallergens
D. Limited views sinus CT
E. Intradermal testing for dust mites and mold allergens

Answer: C

The logical initial test is prick skin test to aeroallergens. If the test is negative, then limited intradermal testing to dust mites and mold allergens is the next step in situation where the prick and intradermal testing are negative, limited views CT of sinuses are recommended. Then nasal smear for eosinophiles and laryngoscope may be considered at later stage of the diagnosis.

Question 7

Which one the following test is the first diagnostic test for the patient's respiratory symptoms?

A. Chest X-ray, PA and lateral views
B. Spirometry
C. Complete pulmonary function test with DLCO
D. Methacholine challenge test
E. High resolution CT of the chest

Answer: B

The spirometry is the first diagnostic test of choice. The Forced Expiratory Volume in one second (FEV1) is measured pre and post bronchodilator. An increase of 12% and 200 mL FEV1 post bronchodilator is considered a positive response in an asthmatic patient (Fig. 6). A complete pulmonary function test comprises of more parameters such as Diffusing Capacity of Carbon Monoxide (DLCO) which is helpful in distinguishing asthma from other respiratory pathology.

If the spirometry test is negative and you still are suspicious of underlying asthma, the next test is a gold standard methacholine challenge test. A 20% reduction in FEV1 after the methacholine test is considered a positive test for the diagnosis (Fig. 7).

If you suspect pneumonia or heart failure, a chest X-ray is a logical choice. A high-resolution CT of chest is reserved for assessment of interstitial disease.

You don't hear from him for a year and one day he presents to your clinic with new complaint. He reported that after he left his old residence, all his symptoms resolved. He believes the basement he and his bother used to live in was likely the cause of his symptoms. This time he asks your help to evaluate his skin infections. He has noted that his skin infections are more frequent now and requests some test to find out the cause of his condition.

Question 8

Which one of the following test helps to diagnose the skin reaction.

A. C-reactive protein
B. CBC with differentials
C. CBC with culture of the abscess
D. Stool culture for ova and parasite
E. Assay for CH50

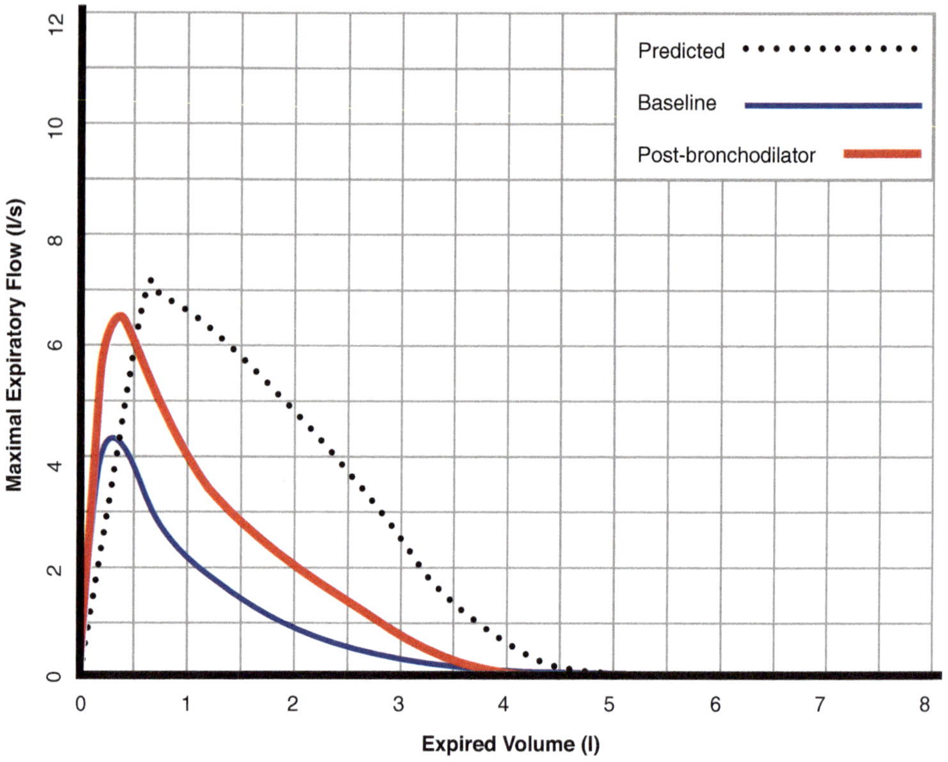

Fig. 6 Positive bronchodilator response in an asthmatic patient. Note shift upward and rightward of the red curve. Significant is 12% response in and at least 200 mL. (Daniel E. Maddox, Adult Asthma in Allergy and Asthma: Practical Diagnosis and Management; M. Mahmoudi, Editor; Springer publisher, 2016 with permission [1])

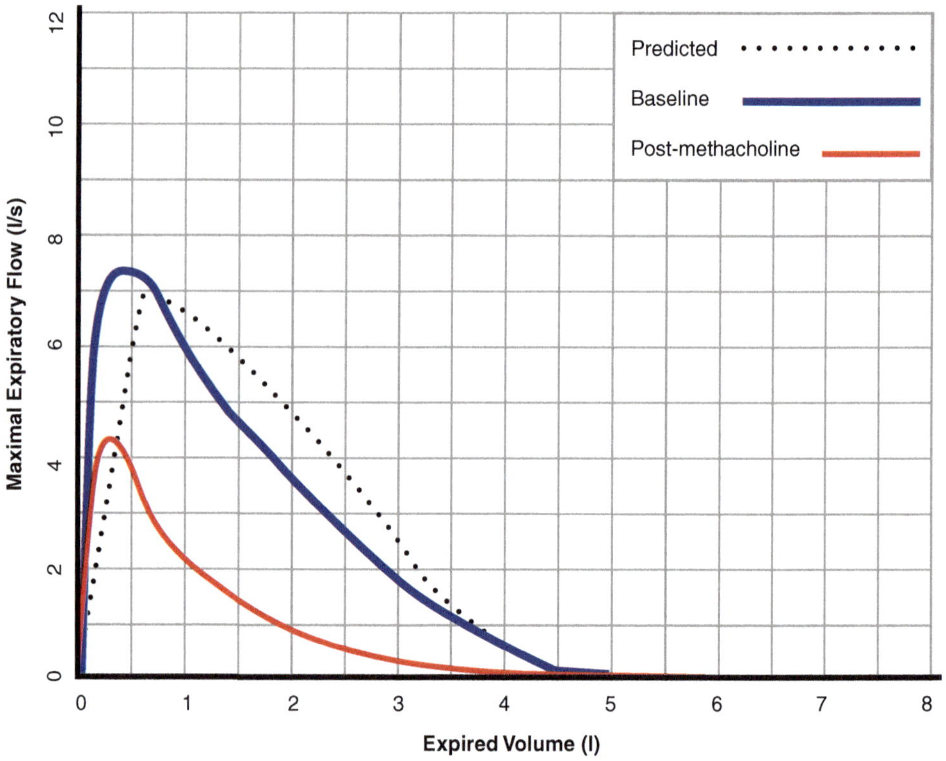

Fig. 7 Positive methacholine challenge in an asthmatic patient. A 20% reduction in FEV1 is a positive response. (Daniel E. Maddox, Adult Asthma in Allergy and Asthma: Practical Diagnosis and Management; M. Mahmoudi, Editor; Springer publisher, 2016 with permission [1])

Answer: C

It is not clear whether the skin rashes are due to allergic contact dermatitis. However, past history of infection and use of antibiotics points to direction of infection etiology and not allergy. Therefore, the CBC with culture of abscess is the logical choice. The C-reactive protein is one of the markers of inflammation and is a non-specific diagnostic test in this case. There are no associated gastrointestinal conditions, and stool culture for ova and parasite is not indicated. The CH50 is reasonable test for investigating the complement deficiency; however, at this stage, there are no indications of such deficiency.

Question 9

Which one of the following tests is considered an assessment of the innate immunity?

A. Quantitative serum immunoglobulins assay
B. CD40 ligand-CD40 interactions assay
C. Lymphocyte proliferative responses to mitogens test
D. Nitroblue tetrazolium test

Answer: D

The first line of defense of a host against microbial and nonmicrobial substances is innate or natural immunity. The host defends itself by various mechanisms that include physical barrier such as skin and mucosal lining of gastrointestinal and respiratory tracts; secretions such as lysozyme-containing bodily secretions, phagocytosis, and by help of natural killer cells. The nitroblue tetrazolium is a neutrophil function assay.

The quantitative serum immunoglobulin analysis tests for B cell immune function.

The CD40 ligand-CD40 interactions assay is a test for assessment of B cell immune function (adaptive immunity). The lymphocyte proliferative responses to mitogens is a screening test for T cells immunity (adaptive immunity).

Question 10

Due to frequent skin infection and growth of *Staphylococcus aureus*. You suspect phagocyte cell defect.

All of the following tests are considered screening test for phagocyte cell defect **except**?

A. CD62 L shedding
B. DHR flow cytometry
C. Nitroblue tetrazolium tests (NBT)
D. STAT1/STAT4
E. Flow cytometry for leukocyte adhesion molecules (CD11/CD18 and CD15a)

Answer: A

The CD62 L shedding is a test for toll like receptor (TLR) defect.

Neutrophils, the most abundant leukocytes cells are phagocytic cells which originate from bone marrow and after maturation release into circulation. The steps involved to reach and destroy the invading organism include adhesion to the vascular wall, chemotaxis into the tissue, engulfing the organism, and finally killing the engulfed organism. A defect in any stage of normal neutrophil function can lead to inability of the neutrophils to kill the microorganisms. The four stages are adhesion, chemotaxis, phagocytosis, and destruction of the engulfed bacteria (Fig. 8).

The neutrophils' nicotinamide adenosine dinucleotide phosphate (NAPDH) oxidase by producing superoxide kills the engulfed microorganisms. In chronic granulomatous disease (CGD), there is a defect in NAPDH oxidase and that makes the neutrophil incapable of killing the microorganism. One of the traditional tests to identify such a deficiency is known as nitroblue tetrazolium tests (NBT). Normally, when the cells are stimulated they are able to reduce NBT. Such a reduction results in dark blue precipitate which can be observed microscopically. In a disease sate, the defective NAPH oxidase enzyme is unable to reduce NBT and therefore there is a lack of activity and color change.

The other test is dihydrorhodamine 123 (DHR) flow cytometry. The normal neutrophils are able to reduce DHR resulting in the fluorescent form. Such a transaction is detected by flow cytometry.

STAT1/STAT4 is an advance test for assessment of phagocytic defect. The flow cytometry for leukocyte adhesion molecules (CD11/CD18 and CD15a) is for assessment of phagocytic defect (see Fig. 8).

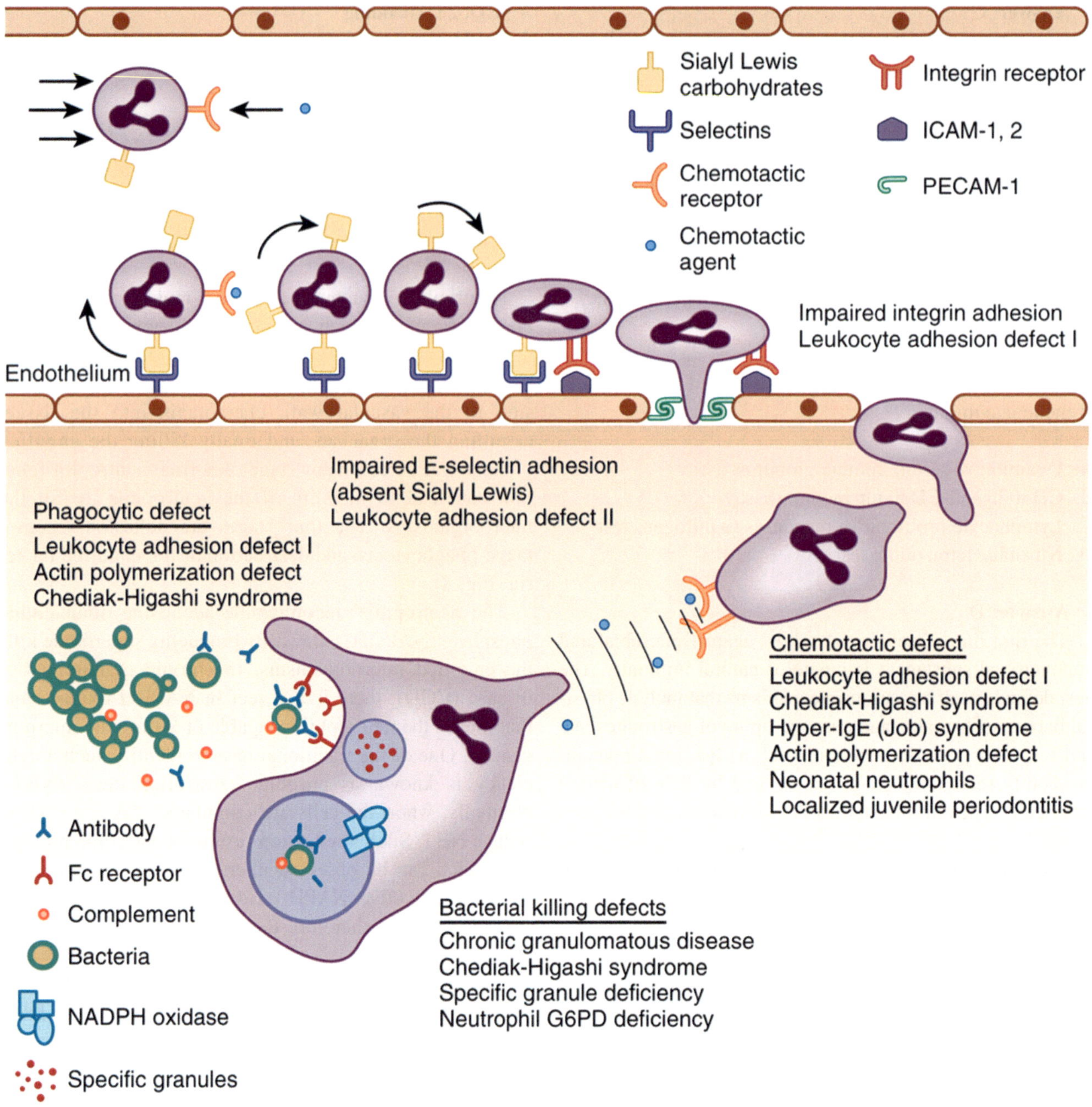

Fig. 8 Defects of neurtophils in : adhesion chemotaxis, phagocytosis, and destruction of tue engulfed bacteria. Hematology: Basic Principles and Practice, sixth edition, Mary C. Dinauer and Thomas D. Coates, Disorders of Phagocyte Function, Chapter 48, Pages 655–673, Copyright Elsevier, 2012, with permission [2]

Bibliography

1. Maddox DE. Adult asthma. In: Mahmoudi M, editor. Allergy and asthma: practical diagnosis and management. Cham: Springer; 2016. p. 223–39.
2. Dinauer MC, Coates TD. Disorders of phagocyte function. In: Hoffman R, Benz Jr EJ, Silberstein LE, Heslop H, Weitz J, Anastasi J, editors. Hematology basic principles and practice. 6th ed. Philadelphia: Elsevier; 2012. p. 655–73.
3. Coates TD. Laboratory evaluation of neutrophil disorders. 2021. Up to date. http://uptodate.com. Accessed 29 May 2021.
4. Mahmoudi M. Other important components of the immune system. In: Mahmoudi M, editor. Immunology made ridiculously simple. Miami: MedMaster; 2009. p. 15–21.
5. Mahmoudi M. Hypersensitivity. In: Mahmoudi M, editor. Immunology made ridiculously simple. Miami: MedMaster; 2009. p. 22–31.
6. Mahmoudi M. Common diagnostic tests. In: Mahmoudi M, editor. Immunology made ridiculously simple. Miami: MedMaster; 2009. p. 41–5.
7. Mahmoudi M. Cold-induced urticaria. JAOA. 2001;101(5 Suppl):1–4.
8. Rosenzweig SD, Kobrybski L, Fleisher TA. Laboratory evaluation of primary immunodeficiency disorders. In: Sullivan KE, Stiehm ER, editors. Stiehm's immune deficiencies inborn errors of immunity. 2nd ed. Cambridge: Academic Press; 2020. p. 115–31.
9. Zerbe CS, Marciano BE, Holland SM. Chronic granulomatous disease: pathogenesis, clinical manifestations, and diagnosis. 2021. Up to date. http://uptodate.com. Accessed 21 May 2021.

Antihistamines and Corticosteroids

Tara Carr

Case 1
Mrs. Smith is 46 years old and came to your clinic for evaluation of persistent, bothersome rhinitis symptoms. She notes a lifelong history of seasonal symptoms, particularly in the Spring and Fall seasons, during which time she suffers from nasal congestion, sneezing, rhinorrhea, and itching of the nose, throat, and eyes. Her family recently adopted a pet cat, and since that time her symptoms have been significantly worse, and persistent despite lack of pollen exposure. She has taken over-the-counter diphenhydramine as-needed for her symptoms, which gives partial, temporary relief but is impractical for her to use every day because it makes her very sleepy. Skin prick testing performed on Mrs. Smith confirms immediate hypersensitivity to multiple regionally prevalent tree pollens, grass pollens, weed pollens, and cat dander.

Question 1
The wheal and flare reaction seen as part of the positive skin prick test was due to:

A. Preformed histamine released from basophils
B. Preformed leukotriene released from mast cells
C. Immediately synthesized prostaglandin released from mast cells
D. Preformed histamine released from mast cells

Answer: D
Histamine is a low-molecular weight amine synthesized from l-histidine exclusively by the enzyme histidine decarboxylase. Histamine is stored in preformed granules within mast cells and basophils. Histamine can therefore be released immediately upon mast cell or basophil activation, which, in the setting of allergy, includes the antigen-induced cross-linking of preformed specific IgE attached to the high affinity IgE receptor on those cells. While histamine is largely thought to be released from mast cells and basophils, in fact histamine can also be produced and released by gastric parietal cells, lymphocytes, and some central nervous system neurons, in addition to some commensal bacteria (*E. coli*, *Lactobacillus vaginalis*).

The major effector function of histamine in allergic reactions relates to its ability to cause itch, erythema, and vascular permeability, and smooth muscle contraction (particularly noticeable in the gut, lung, and cardiovascular system). During skin testing, the released histamine causes endothelial cell contraction and dilation of cutaneous vasculature, leading to local erythema, induration, and pruritus. In fact, histamine also is involved in the regulation of cell proliferation and differentiation, hematopoiesis, embryonic development, regeneration, and wound healing. Further, histamine can affect cognition, memory, sleep-wake cycle, and endocrine homeostasis. Most of histamine (>97%) is metabolized before excretion in the urine, through N-methylhistamine and diamine oxidase pathways.

Antihistamine medications are indicated for the treatment and symptom control of multiple allergic diseases, including allergic rhinitis, allergic conjunctivitis, and urticaria.

Question 2
What histamine receptor was most involved in the positive skin prick test response?

A. H1R
B. H2R
C. H3R
D. H4R

Answer: A
Four histamine receptors have been described, named H1R, H2R, H3R, and H4R. All signal through g-protein coupled receptors. Active and inactive functional states exist in equilibrium, and are stabilized by agonists and inverse ago-

T. Carr (✉)
Asthma and Airway Disease Research Center, University of Arizona, Tucson, AZ, USA
e-mail: taracarr@arizona.edu

M. Mahmoudi (ed.), *Absolute Allergy and Immunology Board Review*, https://doi.org/10.1007/978-3-031-12867-7_27

nists, respectively. Therefore, all can have spontaneous receptor activity.

H1R is expressed on smooth muscle cells (vascular, respiratory, GI), immune cells (neutrophils, eosinophils, mono/macrophages, dendritic cells, T/B cells, endothelial and epithelial cells), and mediates pro-inflammatory responses in response to histamine release. H2R is expressed on gastric parietal cells, as well as CNS and immune cells, and contributes to increased gastric acid secretion. The H3R is a presynaptic autoreceptor on neurons in the CNS and peripheral nervous system. H3R regulates levels of a variety of neurotransmitters, including norepinephrine, acetylcholine, serotonin, and dopamine. H4R is expressed on bone marrow and peripheral immune cells and may contribute to autoimmunity and inflammation.

Question 3
Antihistamines have anti-allergic and anti-inflammatory effects.

A. True
B. False

Answer: A
Histamine may contribute to allergic inflammation in a variety of ways. Located near histamine-releasing cells, dendritic cells, and other antigen presenting cells can respond to histamine to increase antigen presentation and cytokine production toward Th1 cells. On T cells, activation of H1R enhances IFN-g production and Th1 proliferation. Medications that inhibit H1R suppress these pathways through inhibiting cell adhesion molecules and release of mast cell and basophil mediators.

Question 4
Potential adverse effects of H1 antihistamines include all of the following except:

A. Sedation
B. Urinary retention
C. Sinus tachycardia
D. Decreased appetite

Answer: D
H1 antihistamines are generally classified as older generation or newer generation, with the newer generation drugs notable for improved receptor specificity and side effect profile (Fig. 1, Table 1). First-generation H1 antihistamines can cross the blood–brain barrier more readily than the second-generation drugs, and therefore cause increased sedation, decreased cognitive and psychomotor performance, and increased appetite. Additional side effects of H1 antihistamines reflect function at other receptors. Antimuscarinic and anticholinergic effects include dry mouth, urinary retention, and sinus tachycardia. Additionally, the first generation H1 antihistamines with anti-alpha-adrenergic effect may contribute to dose-related prolonged QT interval and may cause symptomatic hypotension.

Question 5
Mrs. Smith has been following allergen avoidance, including mitigation of cat exposure, and is using a long acting, second-generation antihistamine regularly with benefit. However, she continues to have significant pruritus of her eyes and nasal congestion. She is interested in additional topical therapy. Which of the following is not an expected side effect from intranasal or intraocular antihistamine treatment?

A. Reduced degranulation of mast cells
B. Reduced cell signaling through H1R
C. Increased risk of cataracts
D. Increased mucosal dryness

Answer: C
Topical applications of antihistamine are available for both intraocular and intranasal administration. With a relatively fast onset of action, these medications may be appropriate for both regular and intermittent use. Many require dosing more frequently than once daily for maximum benefit. All of the therapeutics in Table 2 have both mast cell-stabilizing properties and anti-H1R properties, potentially increasing the clinical benefit beyond that of the systemic antihistamines. As these do not contain corticosteroids, the side effect profile is mostly limited to increased dryness of the ocular or nasal mucosa, and some formulations have a bitter taste.

Case 2
You are seeing at a 48-year-old woman for treatment of asthma, allergic rhinoconjunctivitis. Her symptoms are poorly controlled, but she refuses to use medication other than antihistamines and "vitamin supplements" because steroids are "going to suppress my immune system" and doesn't want "to take hormones."

Question 6
Cortisol is released by the:

A. Hypothalamus
B. Anterior pituitary
C. Posterior pituitary
D. Adrenal cortex

Answer: D
Two general classes of corticosteroids are produced by the adrenal cortex: mineralocorticoids (MCs) and glucocorti-

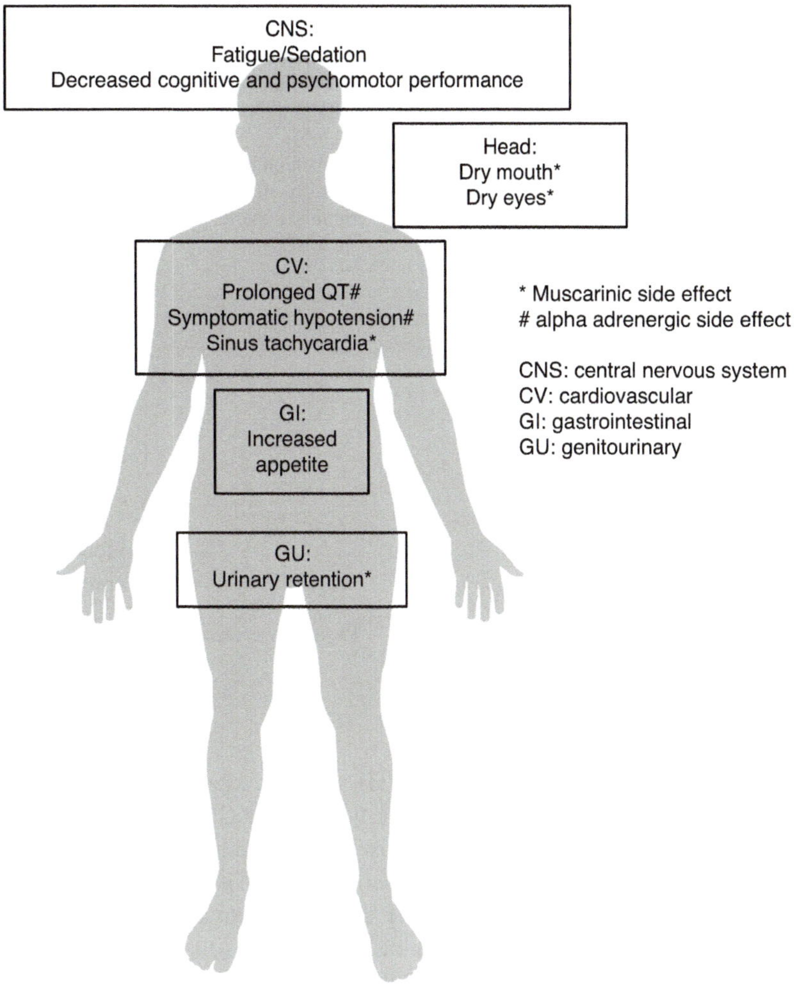

Fig. 1 Side effects of antihistamines are due to function at histamine receptor, muscarinic receptors, and alpha adrenergic receptors. *Muscarinic side effect, #alpha adrenergic side effect. *CNS* central nervous system, *CV* cardiovascular, *GI* gastrointestinal, *GU* genitourinary

Table 1 Systemic antihistamines, clinical comparison

Generic name	Brand name	Sedation, side effects	Onset of action	Duration of action	Daily dose (pediatric)	Daily dose (adult)
First generation						
Diphenhydramine	Benadryl	+++	30–60 min	+		
Chlorpheniramine	Chlor-Trimeton	++	3 h	+		
Hydroxyzine	Atarax	+++	2 h	+		
Doxepin		+++		+++		
Second generation						
Loratadine	Claritin	+	2 h	+++	5–10 mg	10 mg
Desloratadine—metabolite of loratadine	Clarinex	–	2 h	+++	1–5 mg	5 mg
Cetirizine—metabolite of hydroxyzine	Zyrtec	++	30–60 min	+++	2.5–10 mg	10 mg
Levocetirizine—L-enantiomer of cetirizine	Xyzal	+	30–60 min	+++	2.5 mg	5 mg
Fexofenadine	Allegra	–	2 h	+++	15–30 mg bid	60 bid–180 qd

coids (GCs; also called glucocorticosteroids). GCs are essential for life and support various body functions including fetal development, stress responses, gluconeogenesis, arousal, and immune regulation. MCs principally affect the regulation of fluid and electrolyte balance. MC activity in corticosteroid medications intended for GC effect may produce fluid and electrolyte side effects, and therefore they are not entirely without relevance.

Circulating cortisol levels follow a diurnal pattern, under control of the circadian clock. Hypothalamic release of the

Table 2 Topical antihistamines

	Brand name	Dosage	Age range (years)	Additional information
Intranasal				
Azelastine	Astelin	2 sprays per nostril BID	6+	For perennial and seasonal allergic rhinitis; nonallergic rhinitis
Olopatadine	Patanase	2 sprays per nostril BID	6+	
Intraocular				
Ketotifen	Zaditor, Alaway	1 drop bid	3+	
Azelastine	Optivar	1 drop bid	3+	
Olopatadine	Patanol 0.1%; Pataday 0.2%; Pazeo 0.7%	1 drop bid 1 drop qd 1 drop qd	3+ 2+ 2+	
Epinastine	Elestat	1 drop bid	2+	
Alcaftadine	Lastacaft	1 drop qd	2+	Pregnancy category B
Bepotastine	Bepreve	1 drop bid	2+	

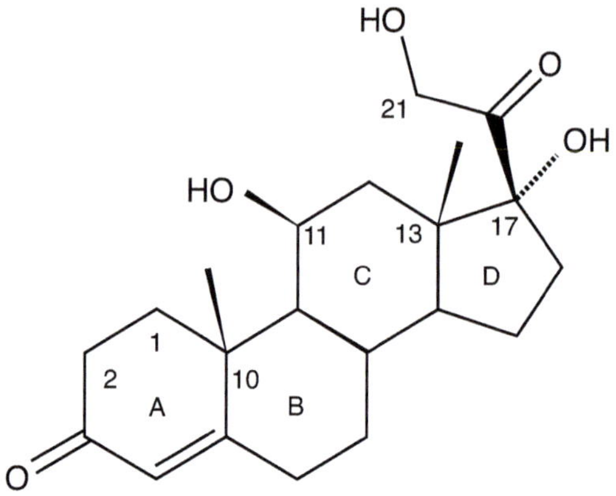

Fig. 2 Chemical structure of cortisol

corticotropin releasing hormone (CRH) induces anterior pituitary release of adrenocorticotropic hormone (ACTH), which in turn stimulates release of cortisol from the adrenal cortex. Cortisol has negative feedback effect on ACTH and CRH secretion. Cortisol and ACTH secretion normally reaches peak levels in the early morning around 8 a.m., then declines throughout the day to a nadir around midnight. Timing of exogenous GC dosing to mimic physiologic cortisol release, first thing in the morning, may reduce side effects and improve anti-inflammatory activity. Daily endogenous secretion of cortisol is about 10–20 mg in healthy individuals, but environmental stress or increased circulating levels of cytokines in the setting of illness can raise levels to as high as 400–500 mg/day.

Hydrocortisone (the synthetic compound structurally identical to endogenous cortisol) is the parent molecule from which other natural and synthetic GCs derive. Essential features of the anti-inflammatory GC consist of the following (Fig. 2): (a) a two-carbon chain at the 17th position, (b) methyl groups at carbons 10 and 13, (c) a ketone oxygen at C3, (d) a double bond between C-4 and C-5, (e) a ketone oxygen at C-20, and (f) a hydroxyl group at C-11. Modifications of either the nucleus or the side chains produce different GC agents. For example, alterations at the C-17 and C-21 positions result in corticosteroids with enhanced binding affinity to the glucocorticoid receptor and reduced systemic effects and thus are commonly seen in topical GC preparations.

Free GCs readily diffuse across cell membrane to affect cell function. First, the GC rapidly binds to the cytoplasmic glucocorticoid receptor (GRalpha; GRbeta), which is widely expressed in tissue and cells, after which heat shock proteins dissociate. The receptor is phosphorylated, and the GR-GC complex is rapidly transported to the nucleus. In the nucleus, it has multiple effects on gene regulation. Genes can be activated directly through GR binding to positive glucocorticoid responsive elements in nucleus, genes can be inhibited through GR binding to inhibitory glucocorticoid responsive elements, and genes can be indirectly repressed through interaction with transcription factors.

Question 7

Glucocorticoids reduce the numbers and/or function of all of the following cells except:

A. Eosinophils
B. Basophils
C. Th1 cells
D. Th2 cells
E. Innate lymphoid cells

Answer: E

Within 4–6 h of systemic GC administration, apoptosis leads to marked decrease in the number of circulating eosinophils, basophils, and monocytes. Increased circulating neutrophils are a result of demargination and increased production and survival. The adaptive immune system is also influenced with reduction in B lymphocytes, impaired CD8+, CD4+ Th1 and Th2 cells, and less cytokine expression from these cells. Steroids do not significantly influence the innate immune system.

Question 8

Glucocorticoids will cause more side effects if they have

A. High first pass hepatic metabolism
B. High volume of distribution

C. High protein binding
D. High lipophilicity

Answer: B

Cortisol is 95% protein-bound in circulation, binding to albumin and the cortisol binding globulin. The unbound form is the active form. Dexamethasone does not bind protein, which contributes to its potency. Interconversion of active and inactive forms of GC can contribute to the regulation of GC activity at the tissue level. For example, cortisol is partially converted to the inactive cortisone by subtype 2 of the enzyme 11-beta-dehydrogenase. GCs are lipophilic and readily absorbed from the GI tract and other tissues. Synthetic GCs for inhalation or other topical use are modified for increased lipophilicity and higher receptor affinity, which improves tissue retention, duration of action, and efficacy. Finally, the volume of distribution of a GC reflects how much is circulating systemically, which is increased with increasing water solubility of a drug. GCs are rapidly metabolized in the liver through first pass hepatic metabolism. While this is highly relevant to oral GC preparations, a portion of a dose of inhaled or intranasal GC is also swallowed—estimated up to 90% with some devices—and subsequently absorbed from the gastrointestinal tract.

Question 9

Which of the following is NOT true re: steroid side effects?

A. Reduce dosing to every other day to minimize side effects
B. Side effects are cumulative over time
C. Topical administration allows for reduced systemic dose
D. Steroids have no effect on growth in children

Answer: D

Adverse effects of GCs are influenced by duration of therapy, dosing regimen, and mode of administration (systemic, topical). Emerging evidence suggests that the adverse effects are cumulative over time. The factors determine the occurrence and severity of adverse effects. Side effects can affect most organ systems to a variable severity and frequency (Fig. 3). To reduce the degree of side effects from fluid retention, treatment agents should have little or no MC activity. To prevent need for systemic steroids or to reduce dose, patients should receive optimized topical steroids and concomitant steroid-sparing medications (i.e., nasal steroids with nasal antihistamines). Single-dose oral GCs should be administered in the morning to minimize disruption of the HPA axis. Alternate-day systemic GC therapy can minimize side effects

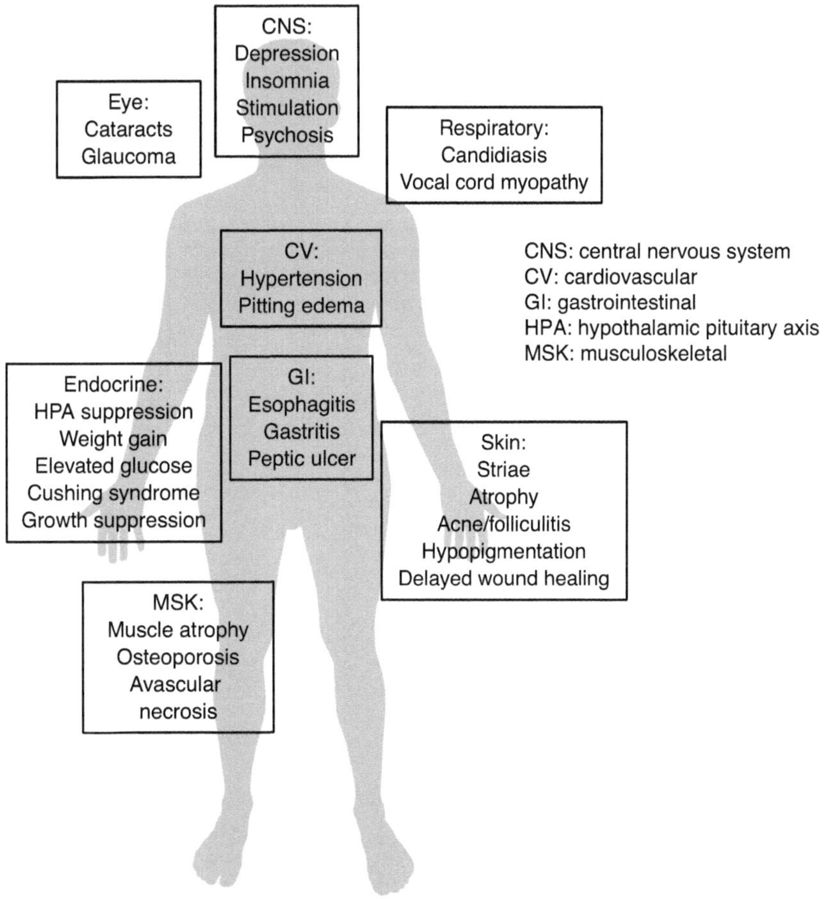

Fig. 3 Side effects of corticosteroids. *CNS* central nervous system, *CV* cardiovascular, *GI* gastrointestinal, *HPA* hypothalamic pituitary axis, *MSK* musculoskeletal

and cumulative dose; oral agents with tissue half-lives in the 12–36-h (intermediate) range, such as prednisone, prednisolone, and methylprednisolone are appropriate for this use. Children receiving GC therapy should be regularly evaluated for growth, and adults using chronic GC therapy should receive vitamin D and calcium supplementation, with monitoring of bone mineral density periodically.

Question 10

To reduce side effects from inhaled steroids, all of the following interventions are effective EXCEPT:

A. Mouth rinsing
B. Use of spacer device
C. Use of MDI preparation instead of DPI
D. Use of higher MMAD drug

Answer: D

Inhaled steroids are the mainstay of therapy for asthma, as well as other lung diseases. The steroid delivery device—the type of inhaler—determines the amount of drug delivered to the lungs and subsequently the clinical benefit. The common devices are metered-dose inhaler (MDI), dry powder inhaler (DPI), and the nebulizer. Hydrofluoroalkane (HFA) propellants have replaced chlorofluorocarbon (CFC) propellants in MDI inhalers. Smaller particle size, i.e., lower mass mean aerodynamic diameter (MMAD), may have improved deposition profile into the lung. Using a pMDI with extra-fine particles can also improve lung deposition.

Side effects of oral candidiasis and dysphonia appear to be dose-dependent but are also related to the physical use of inhalers. For example, dysphonia is reported in up to 58% of patients using ICS due to deposition of the active steroid in the oropharynx leading to local myopathy. A spacer and/or a change to an MDI preparation may alleviate oral candidiasis and dysphonia due to reduction in drug deposition to the posterior oropharynx. Mouth rinsing is useful to reduce local deposition.

Bibliography

1. O'Hehir RE, Hershey GKK, Bacharier LB, Peebles S, Holgate ST, Burks AW, Broide DH. Middleton's allergy principles and practice. 9th ed. Edinburgh: Elsevier; 2019.
2. Castillo M, Scott NW, Mustafa MZ, Mustafa MS, Azuara-Blanco A. Topical antihistamines and mast cell stabilisers for treating seasonal and perennial allergic conjunctivitis. Cochrane Database Syst Rev. 2015;(6):CD009566.
3. Grammer LC, Greenberger PA. Patterson's allergic diseases. 8th ed. Philadelphia: Lippincott Williams & Wilkins; 2018.
4. Sullivan PW, Ghushchyan VH, Globe G, Schatz M. Oral corticosteroid exposure and adverse effects in asthmatic patients. J Allergy Clin Immunol. 2018;141(1):110–6 e7.

Bronchodilators

Sumeet Sandhu and Maria-Anna Vastardi

1 Introduction

There are different categories of bronchodilators including beta-agonists, such as short-acting beta2 agonists and long-acting beta2 agonists and anticholinergics. Beta-agonists utilize G protein-coupled receptors or GPCR to increase cAMP with subsequent smooth muscle relaxation. Anticholinergics bind to the muscarinic receptors on smooth muscle to block acetylcholine at the neuromuscular junction, resulting in bronchodilation. The EPR Guidelines on Asthma by the National Asthma Education and Prevention Program (NAEPP) Coordinating Committee guidelines and the GOLD Guidelines from the Global Strategy for the Diagnosis, Management and Prevention of Chronic Obstructive Pulmonary Disease provide up to date evidence-based recommendations on how to manage asthma and COPD.

Case 1

A 10-year-old male with no significant past medical history presents with shortness of breath and cough for the last 2 years. The mother reports that the cough mostly occurs a few minutes after his gym class, and he cannot continue playing. His symptoms resolve about 30–40 min after resting. Mother denies any fever or nighttime cough. He has been healthy and growing well according to his pediatrician. No hospitalizations since birth and no history of recurrent infections.

Question 1

Which category of bronchodilators, if any, would be the most appropriate for this patient?

A. Leukotriene receptor antagonist
B. Short-acting beta2 agonist (SABA)
C. Long-acting beta2 agonist (LABA)
D. Anticholinergic

Answer: B

This patient's presentation is consistent with exercise-induced bronchospasm. According to the EPR Guidelines on Asthma by the National Asthma Education and Prevention Program (NAEPP) Coordinating Committee guidelines, the most appropriate management for exercise-induced bronchospasm is SABAs. All SABAs act rapidly within 5 min and have a duration of action lasting 3–4 h. SABAs are not recommended for regular use. They are primarily used as "rescue medications" during an asthma exacerbation to provide acute relief of airflow obstruction and before exercise to pretreat exercise-induced bronchoconstriction. For mild to moderate asthma exacerbations, SABA via pMDI can be given up to 4–10 puffs every 20 min during the first hour and then 4–10 puffs every 3–4 h, up to 6–10 puffs every 1–2 h for the following hours of treatment. Most common side effects to monitor include tachycardia and tremors. Using SABA alone as treatment is no longer recommended as patients who frequently use SABA are found to have worse outcomes, lower lung function, and a higher risk of asthma-related death even if symptoms are well controlled on SABA alone. On a physiological level, overuse of SABAs causes down regulation of the beta-receptor, resulting in a decreased response and leading to more use of the medication.

Question 2

Which medication would be the most appropriate for this patient?

A. Albuterol sulfate HFA
B. Albuterol sulfate nebulizer solution
C. Ipratropium
D. Formoterol fumarate

Answer: A

S. Sandhu (✉) · M.-A. Vastardi
Division of Allergy and Immunology, Department of Medicine, SUNY Downstate Medical Center, Center for Allergy and Asthma Research, SUNY Downstate Health Sciences University, Brooklyn, NY, USA
e-mail: Maria-anna.vastardi@downstate.edu

M. Mahmoudi (ed.), *Absolute Allergy and Immunology Board Review*, https://doi.org/10.1007/978-3-031-12867-7_28

For pre-treatment of exercise-induced bronchospasm in a patient of this age, albuterol sulfate HFA w/ pMDI is most appropriate. As per EPR guidelines, for patients 5 years and older, MDI is the preferred device. Common SABA brands and doses are listed below (Table 1). Ipratropium is an anticholinergic. Using an anticholinergic with a SABA has been found to be associated with fewer hospital admissions and improvement in spirometry FEV1 in patients with moderate to severe exacerbations. Formoterol fumarate is a LABA. For exercise-induced bronchospasm pre-treatment, a beta2 agonist of short duration is sufficient and most appropriate.

Question 3

Where do beta2 agonists bind?

A. M3 muscarinic receptor
B. IgE
C. B2 receptor
D. Leukotriene receptor

Answer: C

Beta2 agonists bind to B2 receptors which are found in the smooth muscle, epithelium, and alveoli of the lung and inflammatory cells including mast cells, eosinophils, lymphocytes, and neutrophils. Beta-agonists bind to the G protein-coupled receptors or GPCR, which activates adenylyl cyclase, resulting in increased cAMP. CAMP activates protein kinases A and G. PKA phosphorylates myosin light-chain kinase which results in muscle relaxation. Leukotriene receptor antagonists like montelukast bind to leukotriene receptors and are currently recommended as an add-on therapy. Of note, leukotriene receptor antagonists have an FDA black box warning regarding the risk of mental health adverse effects with this medication. Omalizumab is an IgE biologic that binds to free IgE in the blood and is another add-on therapy that is approved for patients with moderate or severe allergic asthma. Muscarinic receptor antagonists are add-on medications for asthma that is uncontrolled with medium or high dose combination inhaled corticosteroid with LABA. Adding this medication helps to improve lung function.

Question 4

A year later, the mother reports that the patient has been using his SABA daily including times when he is not exercising. You see him in clinic and on exam, he is now obese, is wheezing throughout and has significant retractions. After his exacerbation, what would be the preferred step-up therapy for this patient?

Table 1 Short-acting beta2-agonists (SABAs) names and doses

Generic name	Brand name	Use	Usual dosage
Albuterol sulfate	ProAir RespiClick	Relief of asthma symptoms and prevention of exercise-induced bronchospasm ages 4 and above	2 puffs every 4–6 h as needed
	ProAir Digihaler	ProAir Digihaler connects with a companion mobile application.	
Albuterol sulfate HFA	ProAir HFA Proventil HFA Ventolin HFA Generic albuterol HFA	For quick relief of asthma symptoms. Prevention of exercise-induced bronchospasm. Age 4 and above.	1 or 2 inhalations every 4–6 h for quick relief of asthma symptoms. 2 puffs 15 min before exercise to prevent exercise-induced bronchospasm.
Albuterol sulfate inhalation solution	Unit-dose vials of albuterol 0.083% comes as: Generic	For quick relief of asthma. Age 2 and above.	One unit-dose vial. Every 4–6 h.
Albuterol sulfate nebulizer solution	AccuNeb inhalation solution	For quick relief of asthma. Age 2–12.	0.63 and 1.25 mg in 3 mL unit-dose vials.
	Albuterol sulfate 0.5%	For quick relief of asthma symptoms. Age 2 and above.	Dose will vary.
Levalbuterol HCl	Xopenex inhalation solution Generic	For the treatment and prevention of bronchospasm. Approved: Ages 6–11 (0.31 mg) and older. Ages 12 and older, adjust from 0.63 to 1.25 mg.	0.31 mg (green); 0.63 mg (green); 1.25 mg (red) in 3 mL unit-dose vials; one vial every 6–8 h. Also concentrate of 1.25 mg in 0.5 mL (need to dilute). For use in nebulizers only. Not studied in combination with other medication.
Levalbuterol HFA	Xopenex HFA Generic	For treatment or prevention of bronchospasm. Adults and children older than 4 years of age.	2 inhalations repeated every 4–6 h.

A. No changes, continue to use albuterol prn
B. Add low dose inhaled corticosteroids-LABA combination
C. Add low dose inhaled corticosteroids
D. Add leukotriene receptor antagonist alone

Answer: C

The preferred step-up therapy for this patient is adding a low dose inhaled corticosteroid. Alternative options would be the addition of a leukotriene receptor antagonist alone. A low dose inhaled steroid with LABA would be the preferred next step-up therapy if the patient was still symptomatic on the current regimen. This patient has high SABA use, and his previous exercise-induced bronchospasm asthma has now worsened so he requires an adjustment to his asthma medication. Since the patient is experiencing symptoms on most days, a low dose inhaled corticosteroids-LABA combination or a low dose inhaled corticosteroid and leukotriene receptor antagonist can be added.

Question 5

If this patient was 12 years or older, with uncontrolled daily daytime asthma and FEV <60%, how could his asthma regimen be adjusted?

A. Change to medium dose inhaled corticosteroid and LABA combination
B. Consider adding long-acting muscarinic antagonist (LAMA)

Answer: A or B

For ages 12 years and older, for step 4 care, either option can be considered.

Case 2

A 55-year-old female presents with dry cough and dyspnea that started a few years ago. She reports that she has issues doing her daily activities on most days, like going up and down the stairs. She has tried albuterol, which has helped minimally. She reports a 15-year pack-year smoking history and is wheezing on exam. Her pulmonary function test shows that FEV1/FVC is <60% predicted post-bronchodilator and low DLCO.

Question 1

What is an appropriate treatment regimen for her?

A. Start a LABA
B. Start a LAMA
C. Steroid course
D. Continue albuterol as needed

Answer: A or B

The patient's presentation is consistent with COPD. According to GOLD 2021 recommendations, she can be started on a LABA or a long-acting muscarinic antagonist (LAMA). For COPD, it is recommended to use LABAs alone without inhaled corticosteroids. For asthma, using a LABA without an inhaled corticosteroid is not recommended as it can worsen asthma exacerbations. LABAs have a duration of action lasting 12–24 h. The only LABA that has a rapid onset is formoterol. Common LABA brands and doses are listed below (Table 2).

Table 2 Long-acting beta2-agonists (LABAs) names and doses

Generic name	Brand name	Usual dosage
Albuterol sulfate	VoSpireER extended-release tablets	For relief of bronchospasm. Age 6–12: 4 mg every 12 h Age 12 and older: 4 or 8 mg every 12 h
Formoterol fumarate (inhalation powder)	Foradil aerolizer (Fradel is the capsule. Aerolizer is the inhaler)	Prevent asthma symptoms (age 5 and older) Prevent exercise-induced bronchospasm (age 5 and older). Occasional use once a day. One capsule is 12 mcg; taken every 12 h (age 5 and older)
Salmeterol Xinafoate	Serevent Diskus (Serevent is the medication. Diskus is the inhaler)	Prevent asthma symptoms in patients 4 and older One inhalation every 12 h Prevent exercise-induced bronchospasm in patients 4 and older
Combination		
Budesonide in combination with formoterol (bronchodilator)	Symbicort 80/4.5 Symbicort 160/4.5 Generic	Maintenance of asthma. Age 12 and older Symbicort 80/4.5: 2 puffs twice a day Symbicort 160/4.5: 2 puffs twice a day Prescriber should designate the exact strength 80/4.5 or 160/4.5
Fluticasone furoate, umeclidinium, and vilanterol inhalation powder	Trelegy Ellipta 100/62.5/25 mcg Trelegy Ellipta 200/62.5/25 mcg	Maintenance treatment of asthma ages 18 and older 1 puff per day
Fluticasone propionate and salmeterol	AirDuo Respiclick 55/14, 113/14 and 232/14 and generic of each.	12 and over maintenance treatment of asthma1 puff twice a day AirDuo Digihaler (connects with a companion mobile application)
Fluticasone in combination with salmeterol (bronchodilator)	Advair Diskus 100/50 Advair Diskus 250/50 Advair Diskus 500/50 Advair HFA 45/21 Advair HFA 115/21 Advair HFA 230/21	Control/prevent asthma Advair Diskus 100/50 is for age 4–11 Advair Diskus 100/50, 250/50 and 500/50 are for age 12 and older One inhalation twice daily Advair HFA (all 3 strengths) Age 12 and older 2 inhalations twice a day

(continued)

Table 2 (continued)

Generic name	Brand name	Usual dosage
Fluticasone furoate 100 mcg and vilanterol 25 mcg	Breo Ellipta	Maintenance of asthma age 18 and above 1 puff once a day
Fluticasone furoate 200 mcg	Breo Ellipta	Maintenance of asthma age 18 and above 1 puff once a day
Mometasone in combination with formoterol (bronchodilator)	Dulera 100/5 and 200/5	Control/prevent asthma For ages 12 and above two puffs two times a day

Reprinted with permission from the American Academy of Allergy, Asthma and Immunology (AAAAI). Visit AAAAI.org for additional information and updates

Question 2

Her CBC reveals eosinophil level of 450 μL. How does this affect your management of the patient?

A. No changes
B. An inhaled corticosteroid can be added
C. Steroid course
D. Consider adding mepolizumab

Answer: B

According to GOLD guidelines, an inhaled corticosteroid can be added for COPD patients who have been hospitalized, have had two exacerbations in a year that have required oral corticosteroids or COPD patients who have an eosinophil level more than 300 μL. Studies are still ongoing regarding mepolizumab and it is not FDA approved for COPD.

Question 3

Where do anticholinergics bind?

A. M3 muscarinic receptor
B. IgE
C. B2 receptor
D. Leukotriene receptor

Answer: A

Anticholinergics bind to muscarinic receptors expressed on airway smooth muscle and block the neurotransmitter acetylcholine at the neuromuscular junction. This decreases intracellular levels of cyclic guanosine monophosphate (cGMP). This results in bronchodilation. They can be used for exercise-induced bronchospasm for airflow obstruction relief but SABAs are first line due to better protective effects and faster onset. Muscarinic receptor antagonists are add-on medications for asthma that is uncontrolled with medium or high dose combination inhaled corticosteroid with LABA. Adding this medication helps to improve lung function. Anticholinergics are approved for COPD patients, and the most common side effect is mouth dryness.

Question 4

A month later, you see the patient in your clinic for follow up. Are there any changes to her COPD regimen that can be made to improve lung function and reduce exacerbations?

A. No changes, there is no data yet
B. Add a LABA
C. Add a LAMA
D. Switch to combination therapy with a LABA and LAMA

Answer: D

LABAs and LAMAs improve lung function and reduce exacerbations. LAMAs reduce hospitalizations. Combination therapy with LABA and LAMA reduces exacerbations compared to monotherapy.

Question 5

Are there any medications that can be added to improve her FEV1 on her PFTs?

A. No
B. Add a SABA
C. Add a LAMA
D. Consider adding mepolizumab

Answer: B or C

As per GOLD guidelines, regular and prn use of a SABA or short-acting antimuscarinic antagonist improves FEV1. Combination therapy with LABA and LAMA increases FEV1 compared to monotherapy. Studies are still ongoing regarding mepolizumab and it is not FDA approved for COPD.

Bibliography

1. Cloutier MM, Baptist AP, Blake KV, Brooks EG, Bryant-Stephens T, DiMango E, Dixon AE, Elward KS, Hartert T, Krishnan JA, Lemanske RF, Ouellette DR, Pace WD, Schatz M, Skolnik NS, Stout JW, Teach SJ, Umscheid CA, Walsh CG. 2020 Focused updates to the asthma management guidelines: a report from the National Asthma Education and Prevention Program Coordinating Committee Expert Panel Working Group. J Allergy Clin Immunol. 2020;146:1217–70.
2. Burks A, Holgate S, O'Hehir R, Broide D, Bacharier L, Hershey G, Peebles R, Middleton E. Middleton's allergy: principles and practice. Amsterdam: Elsevier; 2020. p. 2020.
3. Global strategy for the diagnosis, management and prevention of chronic obstructive pulmonary disease. 2021. Available from: https://goldcopd.org.

New Asthma Therapeutics

Katheryn Birch, Maaz Jalil, Craig Sewell, Marija Rowane, Haig Tcheurekdjian, Devi Jhaveri, Theodore Sher, and Robert Hostoffer

Abbreviations

A/I	Allergy/immunology
AR	Allergic rhinitis
CRSwNP	Chronic rhinosinusitis with recurrent nasal polyposis
FeNO	Fractional exhaled nitric oxide
ICS	Inhaled corticosteroid
IDT	Intradermal testing
IL	Interleukin
LABA	Long-acting beta$_2$-adrenergic agonist
LAMA	Long-acting muscarinic agonist
SPT	Skin prick testing
Ig	Immunoglobulin

1 Introduction

The clinical components of asthma are composed of three characteristics which include edema, mucous production, and airway contraction. These characteristics have been treated with various modalities which include beta-agonists, inhaled corticosteroids, muscarinic agonists, and leukotriene modifiers. Advances in the phenotyping and endotyping of asthma have resulted in a more precise approach to treatment. New therapeutic options such as biological medications that block IL4, IL5, and IL13 are now available, with more in development. We aim to introduce the reader to key decision-making points for asthma phenotypes and the appropriate use of biologic treatments.

K. Birch · M. Jalil · M. Rowane · H. Tcheurekdjian · D. Jhaveri
T. Sher · R. Hostoffer
University Hospital's Pediatric and Adult Allergy/Immunology Fellowship, Division of Pulmonary, Critical Care and Sleep Medicine, UH Cleveland Medical Center, Cleveland, OH, USA

C. Sewell (✉)
University Hospital's Pediatric and Adult Allergy/Immunology Fellowship, Division of Pulmonary, Critical Care and Sleep Medicine, UH Cleveland Medical Center, Cleveland, OH, USA

Allergy/Immunology Associates, Inc., Mayfield Heights, OH, USA

Case 1

A 54-year-old female, with a history of moderate persistent asthma, chronic rhinosinusitis with recurrent nasal polyposis (CRSwNP) status post (s/p) polypectomy, presented for the control of her asthma. The patient was diagnosed with asthma 6 years prior, after recurrent episodes of coughing and dyspnea on exertion, attributed to new mold exposure and recurrent sinusitis. She reported medication compliance with her maintenance combined high-dose inhaled corticosteroid (ICS) mometasone furoate/long-acting beta$_2$-adrenergic agonist (LABA) formoterol fumarate dihydrate (MF/F) inhalation aerosol (200 μg/5 μg 2 inh bid), in addition to the short-acting beta$_2$-agonist bronchodilator albuterol sulfate 0.083% inhalation nebulization solution (2.5 mg/3 mL q4h prn), unchanged since initial diagnosis. She discontinued a brief trial of montelukast sodium (10 mg tablet po qd) due to lack of symptom improvement. The patient denied hospital admissions and intubation. She admitted to four emergency department (ED) visits for severe asthma exacerbations since established specialty care, and she required prednisone (10 mg oral tablet, 12-day taper) once in the past 3 months with temporary improvement. Her persistent coughing, wheezing, and nasal congestion with nocturnal asthma symptoms and dyspnea required daily albuterol over the past 3 months. Prolonged cold, dry air exposure, chronic sinusitis flare-ups and, rarely, physical exertion triggered her acute exacerbations. The patient confirmed an active, productive cough, wheezing, and shortness of breath (SOB) with multiple nighttime awakenings weekly. Albuterol helped relieve her symptoms temporarily. She denied fever, rash, or weight loss. Her self-reported Asthma Control Test (ACT) totaled 15.

Further subjective history was negative for previous diagnoses of perennial/seasonal allergies, food allergies, eczema, and eosinophilic esophagitis. The family history was significant for asthma (father), perennial allergic rhinitis (sister),

M. Mahmoudi (ed.), *Absolute Allergy and Immunology Board Review*, https://doi.org/10.1007/978-3-031-12867-7_29

and psoriatic arthritis (maternal niece). She worked as an occupational therapist, resides in a 1950s townhouse without domestic animals, and denied tobacco use or significant smoke exposure.

Her vitals identified an overweight body mass index (BMI) (27.8, normal range [NR] 18.5–24.9 kg/m^2). The physical examination revealed bilateral nasal turbinate edema (2+) with polyps, mild expiratory wheezing on lung auscultation bilaterally with productive cough and clear sputum. No integumentary findings, including eczema and rashes, were identified. The ocular examination was unremarkable.

The spirometry revealed a pre-bronchodilator forced expiratory volume in one second (FEV_1) of 66%, with 18% reversibility to 77% and 73 ppb forced expiratory nitric oxide (FeNO) (Table 1, Fig. 1). Skin prick testing (SPT) of aeroallergen extracts was controlled with saline (negative, 0 mm) and histamine (positive, 8 mm) and revealed clinically insignificant wheal-and-flare reactions to cockroach (2 mm) and unreactive to *Aspergillus fumigatus*. Serum IgE was within reference range (32, NR <115 kU/L). Her complete blood count (CBC) revealed eosinophilia [absolute eosinophils 510 (NR 15–500 cells/μL), eosinophils 8% (NR 0–4%)]. Her urinary leukotriene levels were unremarkable. Her chest X-ray was unremarkable.

Question 1

The patient described in the above scenario best fits which diagnosis:

A. Allergic bronchopulmonary aspergillosis (ABPA)
B. Aspirin-exacerbated respiratory disease (AERD)
C. Churg-Strauss disease
D. Chronic eosinophilic asthma
E. Sarcoidosis

Answer: D

The patient was diagnosed with severe persistent asthma with eosinophilic phenotype. The remainder of the workup failed to support other potential causes for this constellation of symptoms. The patient had adult-onset asthma, diagnosed at 48 years old, with a history of CRSwNP in the absence of allergic sensitization/atopy. She experienced recurrent asthma flare-ups, despite high-dose ICS/LABA therapy, a leukotriene inhibitor, and recurrent prednisone use. Laboratory evaluation further supported the diagnosis with significant eosinophilia (absolute eosinophils 510, eosinophils 8%) and spirometry demonstrating reversible defect (18%) and elevated FeNO (73 ppb).

The differential diagnosis included ABPA, supported by the patient's productive cough, uncontrolled asthma, eosino-

Table 1 Case 1 Pulmonary function test

	Bronchodilator					
Measurement (unit)	Predicted[a]	Pre-	Predicted (%)[a]	Post-	Predicted (%)[a]	Percent change (%)
FVC (L)	3.45	2.86	83	2.87	83	0
FEV_1 (L)	2.71	1.78	66	2.09	77	18
FVC/FEV_1	0.79	0.62	78	0.73	92	11
PEF (L/s)	6.58	3.67	56	3.41	52	−7

FVC forced vital capacity, *FEV_1* forced expiratory volume in one second, *PEF* peak expiratory flow
[a] Standard predicted values derived from Hankinson et al.

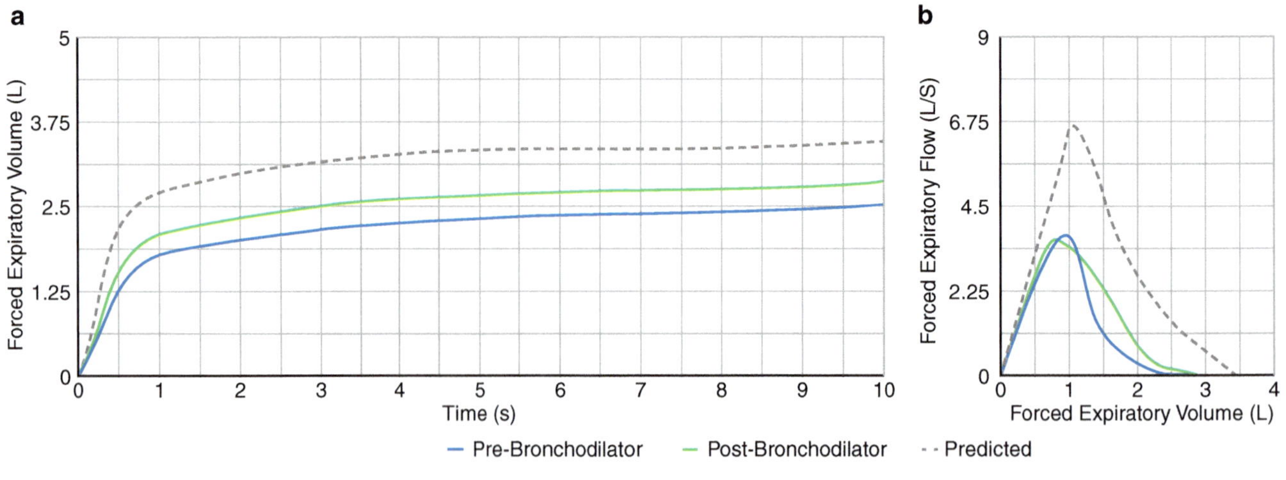

Fig. 1 Case 1 Spirometry. (**a**) Volume-time curve. (**b**) Flow-volume loop

philia, and improvement with prednisone. However, her clinically negative chest X-ray (CXR, for central infiltrates), aeroallergens SPT, aspergillus specific SPT did not support this further consideration of ABPA.

AERD was also on the differential for elevated eosinophils and CRSwNP, but her unremarkable IgE and lack of nonsteroidal anti-inflammatory drug (NSAID) use excluded this prospective diagnosis. In addition, urinary leukotrienes were absent.

Churg-Strauss was suggested by history of asthma, peripheral eosinophilia, and sinus and pulmonary symptoms but ultimately excluded by absence of neurologic (i.e., neuropathy), systemic, or skin manifestations. There were no obvious sites for biopsies to determine granuloma formation.

Sarcoidosis also involves eosinophilia. However, this diagnosis was less probable on the differential due to the patient's unremarkable chest X-ray (CXR) (i.e., absent hilar adenopathy, nodules, ground-glass opacities, upper lobe infiltrates for sarcoidosis).

Subacute evolution of asthma additionally proposed bronchiolitis obliterans on the differential. This diagnosis was unsubstantiated by lack of notable environmental or occupational exposures, severe respiratory illness preceding uncontrolled asthma status, and pulmonary function test (PFT) reversibility.

Question 2

Which is not an indication for use of biologic therapy in the treatment of asthma?

A. Asthma not controlled despite treatment with high-dose ICS/LABA
B. Allergic, eosinophilic asthma phenotype
C. Requiring maintenance oral corticosteroids for asthma control
D. Diagnosis of asthma-COPD overlap syndrome (ACOS)
E. Recurrent emergency room visits for asthma exacerbations

Answer: D

Severe persistent asthma is characterized by daily symptoms and nighttime awakenings requiring continuous daily/nightly short-acting beta-agonist (SABA) use despite high-dose ICS/LABA maintenance therapy or other advanced treatment modalities (Table 2). These patients may also have frequent emergency room visits with an inability to wean oral steroids for prolonged periods of time. Biologic therapy has recently been shown to decrease oral steroid use and emergency room visits especially in patients with T2-high/Type 2 eosinophilic asthma. Characteristics of eosinophilic asthma warranting biologic therapy include elevated eosinophils, IgE, FeNO, and Periostin. Additionally, history and physical examination remain a significant key as signs of atopy and allergic sensitization in the setting of early, severe asthma appears to be responsive to biologic therapy such as omalizumab. The five monoclonal antibodies presently approved for add-on biologic treatment of severe asthma target IgE (omalizumab), interleukin-5 (IL-5) (mepolizumab and reslizumab), IL-5 receptor (benralizumab), and IL-4/

Table 2 Recommended maintenance and rescue controllers for asthma

Symptom severity score	Maintenance		Rescue	
	First line	Alternative	First line	Alternative
1	LD ICS-formoterol PRN[a]	SABA PRN with low-dose ICS[b]	LD ICS-formoterol PRN[a]	SABA PRN
2	LD ICS daily or LD ICS-formoterol PRN[a]	LTRA qd or SABA PRN with low-dose ICS[b]		
3	LD ICS-LABA	MD ICS or low-dose ICS + LTRA[c]	LD ICS-formoterol PRN for patient prescribed maintenance and rescue therapy[d]	
4	MD ICS-LABA	HD ICS + tiotropium/LTRA[c]		
5	HD ICS-LABA, refer for phenotypic assessment and/or add therapy, e.g.,LAMA[e], tiotropium, ICS-LABA-LAMA, biological therapy	LD OCS[f]		

Adapted from Global Initiative for Asthma (GINA), Lee et al., Schend et al.
ICS inhaled corticosteroid, *PRN* as needed, *SABA* short-acting β2-agonist, *LTRA* leukotriene receptor antagonist, *LABA* long-acting β2-agonist, *OCS* oral corticosteroids, *IL* interleukin, *Ig* immunoglobulin, *HD* high-dose, *MD* middle-dose, *LD* low-dose
[a] Off-label
[b] Separate or combination ICS and SABA inhalers
[c] Consider + sublingual immunotherapy only if diagnosed allergic rhinitis + specific dust mite allergy capable of FEV1 >70% predicted
[d] Only applicable if prescribed bud-form or beclomethasone dipropionate form maintenance and reliever therapy
[e] ICS/LABA/LAMA (fluticasone furoate/umeclidinium/vilanterol) triple therapy recently FDA-approved for asthma and chronic obstructive pulmonary disease overlap (2020)
[f] Caution side effects

Table 3 Th2 targeting of biologics in asthma

Generic names	Mechanism of action	Biomarkers
Omalizumab	Blocks IgE interaction with FcεRI	Total IgE 30–700 kU/L (12-yo+) or 30–1300 kU/L (6–11-yo)
Mepolizumab	Inhibits IL-5 binding to IL-Rα	Blood eosinophil >400 cells/μL
Reslizumab	Inhibits IL-5 binding to IL-Rα	Blood eosinophil >150 cells/μL
Benralizumab	Inhibits IL-5 Rα ADCC-induced eosinophil apoptosis	Blood eosinophil >150 cells/μL
Dupilumab	Receptor antagonism of IL-4/-13 Targets IL-4Rα	Blood eosinophil >150 cells/μL

Adapted from Tabatabaian, Ray et al., Schend et al., Pelaia et al.
T_h T-helper cell, *Ig* immunoglobulin, *FcεRI* high-affinity receptor for the fragmented crystallizable region of immunoglobulin, *IL* interleukin

IL-13 receptor complex (dupilumab) (Table 3). Although new triple inhaler therapy has been advocated for ACOS more recently, biologic therapy has not been yet approved.

Question 3

Which cytokine is more prominent in the pathophysiology of this case's asthmatic phenotype, and which biological would be most appropriate?

A. IL-33, reslizumab
B. IL-4, omalizumab
C. IL-5, mepolizumab
D. IL-13, dupilumab
E. TSLP, tezepelumab

Answer: C

IL5 is a major contributor responsible for activation of eosinophil production and basophil differentiation. It is also involved in B cell growth stimulation. It is secreted by Th2 cells and mast cells. As eosinophils mature, they express the IL5 receptor (CD125). Increased IL5 has been associated with eosinophilia, asthma, and allergic rhinitis. IL5 also prolongs the life of eosinophils by delaying apoptosis. Mepolizumab is an anti-IL-5 monoclonal antibody administered subcutaneously every 4 weeks that is primarily used for patients with severe, persistent, allergic (T2-high, eosinophilic phenotype) asthma who remain uncontrolled, despite recurrent oral steroids and high-dose ICS/LABA use.

IL33 activates group 2 innate lymphoid cells (ILC2s) to promote allergic asthmatic responses. It is also involved in airway remodeling. Reslizumab is also an anti-IL5 monoclonal antibody used for severe asthmatics with eosinophilia, however it is administered intravenously. Omalizumab binds to free IgE and is best considered for severe asthmatics that are highly allergenic. IL13 is involved with IgE production from B cells and increases mucus production from goblet cells in the airway epithelia. It also recruits eosinophils and helps to retain them in the airway mucosa. Dupilumab binds to the IL4 receptor, thus blocking the actions of IL4 and IL13 and is most often considered for asthmatics with concomitant atopic dermatitis or nasal polyps. Tezepelumab is a monoclonal antibody against thymic stromal lymphopoietin (TSLP), released from epithelial cells, and is involved in allergic inflammation.

Question 4

Which of the following is characteristic of T2-high or eosinophilic asthma?

A. Elevated fractional exhaled nitric oxide (>25 ppb) and serum IgE (>150 kU/L)
B. History of allergic rhinitis with positive skin prick testing to numerous trees, grasses, and ragweed
C. Elevated peripheral blood eosinophils (>300–400 cells/μL)
D. Positive chest X-ray denoting infiltrates
E. All of the above

Answer: C

T2 high asthma is also known as eosinophilic asthma due to the abundance of eosinophils found in these patients. These eosinophils stimulate chronic inflammation by release of their granule contents which include a myriad of inflammatory mediators including cytokines, chemokines, cytotoxic proteins, and cysteinyl leukotrienes.

T2 low asthma is relatively less understood than T2 high asthma. It is characterized by the absence of T2 high biomarkers including eosinophilia, IgE, FeNO. The physiology behind this endotype is due to Th1 and Th17 T-cell activation. Instead of eosinophils or neutrophils, paucigranulocytic cells predominate. Clinically, these patients are less responsive to corticosteroids. Severe neutrophilic asthma has been associated with certain clinical features including chronic infection with atypical bacteria, smooth muscle abnormalities, smoking, and obesity.

Several biomarkers have been assessed to help identify patients with T2 high asthma. Sputum eosinophilia was shown to correlate with T2 gene expression, however clinically it is difficult to analyze and obtain. Serum eosinophilia has become a surrogate marker for T2 high asthma, and it is the most accepted biomarker for T2 inflammation. Serum IgE is the classic marker for atopy in individuals. Its usefulness in differentiating T2 high asthma is limited, but clinically it is still used to identify asthmatics with the allergic phenotype, as well as to screen for ABPA, which can manifest as difficult to treat asthma with elevated IgE.

Question 5

How does mepolizumab work mechanistically for its observed pharmacologic response?

A. Decreases basophil FC epsilon receptor 1 (FCεR1) expression
B. Binds to IL-5Rα present on eosinophils and basophils, inducing antibody-dependent cell-mediated cytotoxicity of eosinophils and basophils
C. Binds to and blocks IL-5
D. Binds to IL-4Rα and inhibits signaling of IL-4 and IL-13

Answer: C

Advanced pathophysiology investigation over the past several decades has advanced traditional step management to phenotype/endotype-specific treatments, targeted to individual aberrant components of asthma pathobiology. Biological therapies for severe asthma refractory to conventional treatment allow successful target immune-inflammatory responses underlying uncontrolled allergic or non-allergic eosinophilic T2-predominant asthma. The five monoclonal antibodies presently approved for add-on biologic treatment of severe asthma target IgE (omalizumab), IL-5 (mepolizumab and reslizumab), IL-5 receptor (benralizumab), and IL-4/IL-13 receptor complex (dupilumab) (Table 3). Other biologics for severe T2-predominant asthma under advanced clinical development include tezepelumab, REGN3500, and fevipiprant, which target thymic stromal lymphopoietin (TSLP) binding to its receptor complex; IL-33 binding to its ST2 receptor; and the prostaglandin D_2 (PGD_2) chemoattractant receptor-homologous molecule expressed on T2 cells (CRTH2) receptor. Severe neutrophilic or paucigranulocytic asthma are distinct phenotypes to further investigate for novel biological experimental development (Fig. 2).

Case 2

A 35-year-old female, with the past medical history of asthma, allergic rhinitis (AR), obstructive sleep apnea (OSA), and supraventricular tachycardia, presented at the start of fall to establish care for asthma management and to evaluate her acute productive cough. The patient reported a well-controlled asthma, with rare albuterol use since her diagnosis at 11-years-old. Last winter she contracted influ-

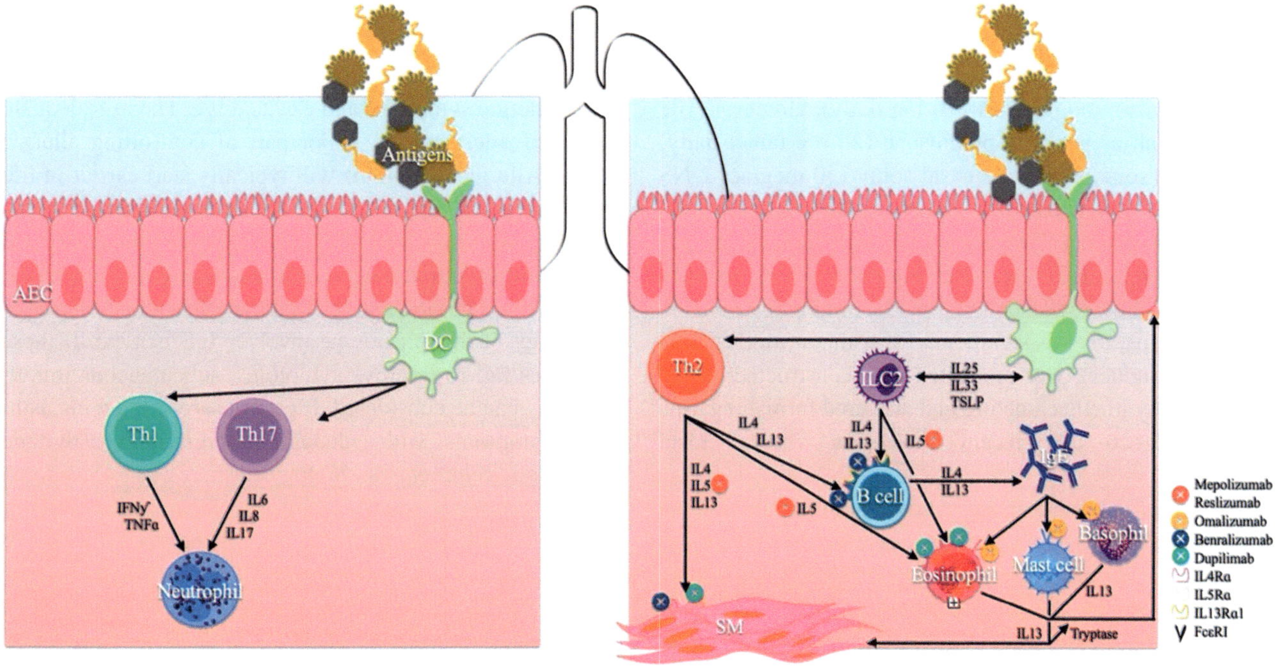

Fig. 2 T-helper 2 (Th2)-low (*L*) versus-high (*R*) inflammatory pathways mediating asthma pathophysiology. Type 2 (T2), or Th2-high, inflammatory pathway activates cytokines, including interleukins (ILs)-4, -5, and -13, derived from Th2 and innate lymphoid cells (ILC2). Chronic activation of eosinophils, basophils, and mast cells in the inflammatory cascade modifies airway smooth muscle (SM) and epithelial cells (AECs) into asthma pathology. T2-low inflammatory pathway stimulates neutrophilic inflammation through IL-6, IL-17, IL-8 from Th17 and interferon-gamma (INFγ) and tumor necrosis factor-alpha (TNFα) from Th1 lymphocytes. Present asthma biologics target interleukins, immunoglobulin E (IgE), and their receptor interactions. *DC* dendritic cell, *FcεRI* high-affinity receptor for the fragmented crystallizable region of immunoglobulin, *IL4Ra* IL-4 receptor-alpha, *IL5Ra* IL-5 receptor-alpha, *IL13Ra1* IL-13 receptor alpha-1. (Adapted from Manka et al., Pelaia et al., and Tabatabaian et al.)

enza that required a prolonged, high-dose prednisone taper, and has had multiple asthma exacerbations requiring prednisone since then with minimal symptom alleviation. She was also started on fluticasone propionate inhalation powder (100 mcg/act 1 inh daily), montelukast (10 mg daily). After a brief summer hiatus, she relapsed the week prior, necessitating 3–4 albuterol activations per day, resumed prednisone use, and work absence. She admitted to intermittent wheezing, productive cough, nocturnal symptoms, and SOB with exertion exacerbated by cold air exposure. Her self-reported ACT totaled 6.

Her AR was well-managed with subcutaneous immunotherapy, until she discontinued a decade of maintenance dosing during her early 20s. She recently began continuous positive airway pressure ventilation (CPAP) per pulmonary sleep study identification of mild OSA. Other pulmonary workup, including spirometry and chest X-rays, were ordered for suspected interstitial lung disease-induced complications of influenza and were unremarkable. She also attributed ongoing rhinorrhea, post-nasal drip, nasal congestion, and mild periorbital edema to perennial allergic rhinitis, reportedly heightened when outdoors, during winter, and in the past 2 years.

The patient denied prior chronic diagnoses of eczema or autoimmune disease. She reported an allergy to sulfa antibiotics, as well as indoor and outdoor environmental allergies. She asserted compliance with her current medication regimen of fluticasone propionate inhalation powder (100 mcg/act 1 inh daily), montelukast (10 mg daily), albuterol (108 mcg/act 2 inh as needed), prednisone (20 mg tablet daily) taper, fluticasone propionate nasal spray (50 mcg/act 2 NS twice per day as needed), azelastine hydrochloride (HCl) 0.10% nasal spray (137 mcg/act 1 NS daily as needed), and diphenhydramine HCl tablets (25 mg every 6 h as needed). She admits these collectively provide mild improvement. Her family history was significant for asthma (mother) and eczema (daughter). Her residence was constructed in the 1970s with synthetic carpeting and cluttered furnishing. She denied tobacco use, recent travel, or known novel exposures.

Her initial physical examination findings included an obese habitus (BMI 42.4 kg/m^2), boggy nasal turbinates (2+) bilaterally, posterior oropharynx cobble stoning, dry cough, and mild end-expiratory wheezes bilaterally. Pulmonary function testing demonstrated a FEV_1 of 45.91% with a post-bronchodilator improvement (Table 4, Fig. 3). Her FeNO was 26 ppb (on prednisone taper day 5). Skin prick testing for aeroallergens demonstrated multiple clinically positive hypersensitivities: Hickory, June, Meadow fescue, Orchard, Red top, Sweet vernal, timothy, *Alternaria alternata*, *Dermatophagoides farinae*, mixed grasses (4+ mm diameter wheel), *D. pteronyssinus*, and mixed molds (3+ mm). Her CBC with differential revealed absolute eosinophils within normal limits (100 cells/μL) and mildly elevated eosinophils (1.1%). Her serum IgE was significantly elevated (311 kU/L) with moderately elevated levels of specific IgE antibodies to *A. alternata*, Bermuda grass, June grass, Meadow fescue, Timothy grass, common ragweed, Birch tree, Elm tree, Pecan tree, house dust, *D. farinae*, and *D. pteronyssinus*.

Question 1

Based on the above what is your diagnosis:

A. Eosinophilic asthma
B. Neutrophilic asthma
C. Allergic asthma
D. Occupational asthma
E. Obesity-related asthma

Answer: C

Patients with allergic asthma will frequently have uncontrolled allergic symptoms and elevated IgE. Having adequate control of allergies is a major part of controlling allergic asthma. Allergic symptoms will typically start earlier in life when allergic symptoms start to manifest. Allergic symptoms are frequently controlled with allergen avoidance, oral antihistamines, intranasal steroids, intranasal antihistamines, ocular antihistamines, and leukotriene inhibitors. The pathophysiology of this phenotype involves IgE isotype. If these modalities fail to improve symptoms, subcutaneous immunotherapy can be considered. If patients are still having asthmatic symptoms with adequate allergic and asthmatic medication regimen, a biological may be considered (Table 5).

Patients with eosinophilic asthma phenotype typically do not respond well to corticosteroids. The onset of the disease

Table 4 Case 2 Pulmonary function test

	Bronchodilator					
Measurement (unit)	Predicted[a]	Pre-	Predicted (%)[a]	Post-	Predicted (%)[a]	Percent change (%)
FVC (L)	4.92	2.24	45.59	3.36	45.91	68.38
FEV_1 (L)	4.05	1.24	30.65	1.86	68.38	45.91
FVC/FEV_1	0.83	0.55	66.97	0.55	66.88	66.88
PEF (L/s)	–	3.47	–	4.90		–

FVC forced vital capacity, *FEV_1* forced expiratory volume in one second (FEV_1), *PEF* peak expiratory flow
[a] Standard predicted values derived from Hankinson et al.

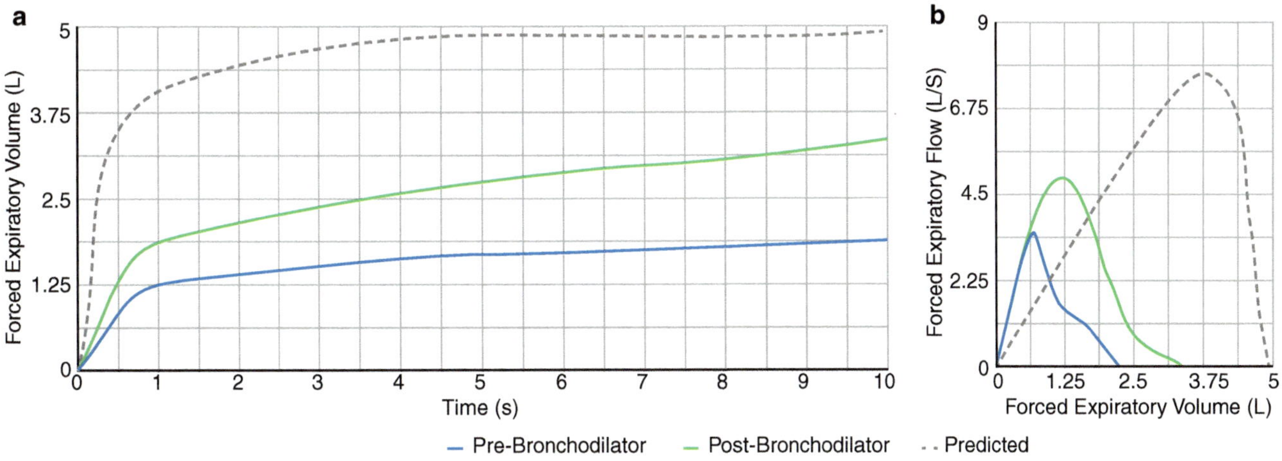

Fig. 3 Case 2 Spirometry. (**a**) Volume-time curve. (**b**) Flow-volume loop

Table 5 Asthma phenotypes

Phenotypes	Clinical nature	Pathophysiology	Therapeutic options
Early-onset allergic	Early onset; associated with atopic comorbidities Mild-to-severe asthmatic symptoms	Elevated specific IgE T_h2 cytokine secretion Thickened SBM	Anti-IgE biologic (omalizumab) Subcutaneous immunotherapy T_h2-targeted CS
Late-onset eosinophilic	Adult onset; less atopic Severe asthmatic symptoms, including AERD	CS-resistant sputum eosinophilia Excessive T_h2-mediated IL-5 production	Anti-IL-5/IL-5R, IL4R biologics (dupilumab, mepolizumab, reslizumab, benralizumab) Cysteinyl leukotriene receptor antagonists
Exercise-induced	Exacerbations triggered by exercise Milder asthmatic symptoms	Mast-cell activation T_h2-mediated IL-9 and cysteinyl leukotriene secretion	Cysteinyl leukotriene receptor antagonists Beta-agonists deleted
Obesity-related	Adult onset with female predominance; non-atopic Frequent OCS use	Paucity of T_h2 biomarkers Oxidative stress	Weight reduction
Neutrophilic	Air trapping causing low FEV_1 Frequent OCS use	Sputum neutrophilia T_h17-induced IL-8 secretion	Azithromycin, erythromycin

Adapted from Tabatabaian, Ray et al., Schend et al.
Ig immunoglobulin, T_h T-helper, *IL* interleukin, *SBM* subepithelial basement membrane, *OCS* oral corticosteroids, *AERD* aspirin-exacerbated respiratory disease, *IL* interleukin, FEV_1 forced expiratory volume in one second, *C-X-C* motif chemokine ligand

usually is later in life. Eosinophilic asthma is associated with nasal polyps. Their presentation does not have a demonstrable allergic trigger. Their eosinophilic counts are substantially higher than other asthmatic phenotypes. These patients require leukotriene modifiers, high-dose inhaled corticosteroids, and/or oral steroids. Their pathophysiology may involve increased secretion of IL5, and/or IL4 and IL13 (Table 5).

Neutrophilic asthma is characterized by normal eosinophils, lack of atopy, and increased neutrophils in the sputum. Neutrophilic asthma has been associated with smoking, obesity, and atypical bacterial infections. IL6 is elevated in neutrophilic asthma and is associated with inflammatory reactions and metabolic dysfunction. These patients do not respond to corticosteroids and have more difficulty with typical asthmatic treatments. Some patients have shown improvement with macrolide antibiotics (Table 5).

Occupational asthma is a type of reactive airway that is triggered by antigens/allergens found in workplace venues. Initially and historically, the asthma episodes will occur in and around the work environment. As the pathology progresses, the asthma symptoms may be more difficult to attribute to the environment. The asthma treatment process results in symptom resolution in these patients, although dependency on allergen/antigen exposure and clinical symptoms, occupational change may be required (Table 5).

Obesity asthma is more common in females. They may have severe symptoms with normal lung function. These patients often do not respond to corticosteroids. The mechanism is thought to be from the excess oxidative stress. IL6 is associated with increased inflammation and is thought to contribute to the asthmatic symptoms and has been shown to decrease with weight loss. Weight loss is extremely important to stress to these patients (Table 5).

Question 2

Based on your evaluation of Case 2, which biological would be most appropriate?

A. Benralizumab
B. Omalizumab
C. Mepolizumab
D. Dupilumab
E. Reslizumab

Answer: B

The recombinant, humanized monoclonal, anti-IgE antibody omalizumab was the first biologic approved for severe allergic asthma in 2003 in the United States. Anti-IgE immune complex formation on the two IgE Cε3 constant domains prevents its interaction with both high-affinity FcεRI and low-affinity FcεRII/CD23 membrane receptors, abrogating all IgE-dependent cellular and molecular events involved in the immune/inflammatory cascade underlying allergic asthma. Omalizumab as adjunctive asthma therapy has conducted randomized controlled trials up to 60 weeks of treatment, as well as real-world studies, to substantiate its well-established safety, tolerability, and sustainability profile. Omalizumab therapy resulted in significant improvement in symptom control, pulmonary functions, quality of life, school or work absences, and cost effectiveness, as well as reduced asthma exacerbation rate. Among numerous trials with supportive evidence of its efficacy, the longitudinal Patient-Reported Outcomes and Xolair® In the Management of Asthma (PROXIMA) study observing patients requiring Global initiative for asthma (GINA) Step 4 therapy identified the comorbidities chronic rhinosinusitis without nasal polyps (CRSsNP) and CRSwNP as the most frequent comorbidities (17.9%, 13.8%) mitigated by omalizumab. Post hoc analysis of PROXIMA compared omalizumab's advantageous efficacy in controlling severe asthma with CRSwNP versus sNP through Asthma Control Questionnaire-5 (ACQ-5) scores (40% vs. 28.1% ACQ-5 <1), FEV_1 (>400 mL, >15% from baseline/Minimal Clinically Important Difference), and exacerbation rates (100% vs. 85.4% reduced).

Question 3

Allergic asthma is driven predominately by which type of immune cells?

A. T-helper 1
B. T-helper 2
C. Eosinophils
D. T regulatory cells

Answer: B

Allergic asthma is driven primarily by T-helper 2 or Th2 cells that are activated by specific allergens. This asthmatic phenotype features an abundance of IL-4, IL-5, and IL-13. ILC2s also play a critical role in propagating the T2 high pathway. The pathophysiology of T2 high asthma starts with a dysregulated epithelial barrier, where allergens, pollutants, and microbes can gain access to stromal tissue. Airway epithelial cells respond to this stimulus by releasing specific cytokines or alarmins such as TSLP, IL- 25, and IL-33. These alarmins promote signaling of the type 2 allergic response. IL-25 and IL-33 activate ILC2s, and TSLP readies antigen presenting cells (APC) to promote the T2 allergic pathway by interacting with certain T and B cells. Activated ILC2s produce IL-5 and IL-13. The main interleukins, IL-4, IL-5, and IL-13 are responsible for the effects and propagation of T2 immunity, which results in high IgE antibodies and eosinophilia. IL-4 is responsible for the differentiation of Th2 cells, B-cell activation, and IgE class-switching. IL-13 is involved in B-cell differentiation, and causes goblet hypersecretion, airway hyperresponsiveness, and eosinophil trafficking. IL-5 is crucial in the differentiation, maturation, and survival of eosinophils. T2 high asthma is also known as eosinophilic asthma due to the abundance of eosinophils found in these patients. These eosinophils stimulate chronic inflammation by the release of their granule contents which include a myriad of inflammatory mediators including cytokines, chemokines, cytotoxic proteins, and cysteinyl leukotrienes.

Question 4

What is the most serious side effect associated with omalizumab?

A. Conjunctivitis
B. Headaches
C. Aseptic meningitis
D. Anaphylaxis
E. Emesis

Answer: D

All biological medications used for asthma generally have a very good safety profile. For omalizumab, the most reported adverse events include injection site reactions, viral infections, and upper respiratory tract infections. There is a United States (U.S.) boxed warning for anaphylaxis associated with omalizumab, however, the incidence of this was found to be low at 0.1–0.2%. Due to these findings, physicians take precautions including implementing a post-injection wait period, since anaphylaxis occurs within 2 h of the injection, and also prescribing epinephrine autoinjectors to patients.

There has been speculation of increased risk of malignancy with omalizumab use, however pooled analysis of phase I to IV clinical trials showed no clear association. One role of IgE in the body is to protect against parasitic infections. Omalizumab's effect on IgE can increase the risk of parasitic infection, especially in parasitic endemic regions. There have been reports of Churg-Strauss with omalizumab treatment; however, causality has not been proven.

Other adverse events associated with biologics include conjunctivitis which has been found to occur in approximately 10% of all users of dupilumab, and headaches which occur more commonly with mepolizumab, occurring in 19% of patients.

Question 5

What additional ICS therapy may have been considered as an add-on therapy?

A. Fluticasone furoate plus umeclidinium plus vilanterol (FF/UMEC/VI)
B. Tiotropium bromide
C. Fluticasone furoate
D. Vilanterol
E. Budesonide plus formoterol fumarate dihydrate

Answer: A

Approximately, 30–50% of symptomatic patients with uncontrolled, moderate, or severe asthma have exhausted standard high-dose ICS/LABAs and do not meet the T_h2-predominant phenotype and/or insurance eligibility for presently approved biological therapies. Recent approval of a once-daily single-inhaler fluticasone furoate plus umeclidinium plus vilanterol (FF/UMEC/VI) offers the first single-inhaler triple therapy for asthma and meets the need for further step-up management. Single-inhaler triple therapy is a well-established COPD treatment, but the Clinical Study in Asthma Patients Receiving Triple Therapy in a Single Inhaler (CAPTAIN) double-blind, randomized, phase 3A trial was the first study to demonstrate its efficacy in asthma.

Bibliography

1. Hankinson JL, Odencrantz JR, Fedan KB. Spirometric reference values from a sample of the general U.S. population. Am J Respir Crit Care Med. 1999;159(1):179–87.
2. Global Initiative for Asthma Board of Directors and Committees. Management of asthma. In: 2020 GINA main report: asthma management and prevention for adults and children older than 5 years: a pocket guide for health professionals. Fontana-on-Geneva Lake: Global Initiative for Asthma; 2020. p. 18–39.
3. Lee LA, Bailes Z, Barnes N, Boulet LP, Edwards D, Fowler A, et al. Efficacy and safety of once-daily single-inhaler triple therapy (FF/UMEC/VI) versus FF/VI in patients with inadequately controlled asthma (CAPTAIN): a double-blind, randomised, phase 3A trial. Lancet Respir Med. 2020; https://doi.org/10.1016/S2213-2600(20)30389-1.
4. Schend J, Rowane M, Sanan N, Hostoffer R. An osteopathic modular approach to asthma: a narrative review. J Am Osteopath Assoc. 2020;120:774–82.
5. Tabatabaian F. Asthma phenotypes and biomarkers. In: Mahmoudi M, editor. Allergy and asthma: the basics to best practices. Cham: Springer Nature; 2019. p. 275–88.
6. Ray A, Camiolo M, Fitzpatrick A, Gauthier M, Wenzel SE. Are we meeting the promise of endotypes and precision medicine in asthma? Physiol Rev. 2020;100(3):983–1017.
7. Pelaia C, Crimi C, Vatrella A, Tinello C, Terracciano R, Pelaia G. Molecular targets for biological therapies of severe asthma. Front Immunol. 2020;11:603312.

Vaccination

Dhami Jaspreet, Wang Vivian, Wang Ziwei, Pham Brittney, Yabuno Jamie, and Joseph Yusin

1 Introduction

The importance of global vaccines for various infectious agents has been emphasized through our experience with the current COVID-19 vaccine. Physicians trained in allergy and immunology are considered experts in vaccines and adverse reactions from vaccines. This chapter will emphasize how to approach vaccines for various populations, including immunodeficient patients, how to determine best paths for patients who miss vaccine doses, and how to address adverse reactions to vaccines, as noted in Case 2, where the patient experienced an anaphylactic reaction to the measles, mumps, and rubella (MMR) vaccine. In addition, mechanisms behind adverse vaccine reactions are discussed.

Case 1

A 24-year-old female diagnosed at age 22 with common variable immune deficiency (CVID) currently on monthly gamma globulin presents to the clinic. She lives with her husband who is healthy. She is planning to become pregnant and wants to make sure she is up to date with vaccines. She currently works as an RN in the local hospital.

Question 1

Review of the medical records dating back to the age of 18 years old do not show any vaccines given to our patient. What vaccinations do you recommend for this patient?

D. Jaspreet · W. Vivian · J. Yusin (✉)
Division Allergy Immunology, VA Greater Los Angeles Healthcare System, Los Angeles, CA, USA
e-mail: Jaspreet.Dhami@va.gov; Vivian.Wang@va.gov; Joseph.yusin2@va.gov

W. Ziwei · P. Brittney · Y. Jamie
David Geffen School of Medicine at UCLA, Los Angeles, CA, USA
e-mail: ZiweiWang@mednet.ucla.edu; BQPham@mednet.ucla.edu; JYabuno@mednet.ucla.edu

A. Administer all recommended inactivated and live age-appropriate vaccines per the CDC immunization schedule
B. Administer all recommended inactivated age-appropriate vaccines and select inactivated vaccines outside recommended age groups per the CDC immunization schedule
C. Administer all recommended live age-appropriate vaccines per the CDC immunization schedule
D. Do not administer any further vaccines

Answer: B

Discussion

CVID is a primary immune deficiency characterized by compromised antibody production, resulting in low levels of serum immunoglobulins G, A, and/or M. This immune defect predisposes to a variety of clinical presentations including recurrent infections, inflammatory diseases, autoimmune disorders, and increased incidence of lymphoma. Unlike combined immunodeficiencies that have impaired cellular (T-cells) and humoral (B-cells) immunity, CVID is primarily a dysfunction in B cell differentiation. Although patients with CVID generally have poor antibody immune response to immunizations, vaccines are still recommended due to the variable spectrum of disease and responses with antibody production.

All age-appropriate inactivated vaccines are recommended for patients with CVID (see Tables 1 and 2), with the understanding that patients may not mount an appropriate immune response. Pneumococcal and Haemophilus influenzae type b (Hib) are risk-specific recommended vaccines, even outside of recommended age groups. IgG levels can be measured following immunization to evaluate for protective immunity. For example, the response to protein or polysaccharide antigens of respective pneumococcal vaccines. Conversely, live vaccines are not recommended in patients with CVID. Contraindicated vaccines include oral poliovirus (OPV), smallpox, live attenuated influenza

M. Mahmoudi (ed.), *Absolute Allergy and Immunology Board Review*, https://doi.org/10.1007/978-3-031-12867-7_30

Table 1 Recommended adult immunization schedule by age group, United States, 2021 (U.S. Department of Health and Human Services, Centers for Disease Control and Prevention)

Vaccine	19–26 years	27–49 years	50–64 years	≥65 years
Influenza inactivated (IIV) or Influenza recombinant (RIV4)	1 dose annually			
or	or			
Influenza live, attenuated (LAIV4)	1 dose annually			
Tetanus, diphtheria, pertussis (Tdap or Td)	1 dose Tdap each pregnancy; 1 dose Td/Tdap for wound mangement (see notes)			
	1 dose Tdap, then Td or Tdap booster every 10 years			
Measles, mumps, rubella (MMR)	1 or 2 doses depending on indication (if born in 1957 or later)			
Varicella (VAR)	2 doses (if born in 1980 or later)		2 doses	
Zoster recombinant (RZV)			2 doses	
Human papillomavirus (HPV)	2 or 3 doses depending on age at initial vaccination or condition	27 through 45 years		
Pneumococcal conjugate (PCV13)	1 dose			1 dose
Pneumococcal polysaccharide (PPSV23)	1 or 2 doses depending on indication			1 dose
Hepatitis A (HepA)	2 or 3 doses depending on vaccine			
Hepatitis B (HepB)	2 or 3 doses depending on vaccine			
Meningococcal A, C, W, Y (MenACWY)	1 or 2 doses depending on indication, see notes for booster recommendations			
Meningococcal B (MenB)	2 or 3 doses depending on vaccine and indication, see notes for booster recommendations; 19 through 23 years			
Haemophilus influenzae type b (Hib)	1 or 3 doses depending on indication			

Recommended vaccination for adults who meet age requirement, lack documentation of vaccination, or lack evidence of past infection

Recommended vaccination for adults with an additional risk factor or another indication

Recommended vaccination based on shared clinical decision-making

No recommendation/ Not applicable

(LAIV), Bacillus Calmette–Guérin (BCG), Ty21a (live typhoid), yellow fever, MMR, Measles-Mumps-Rubella-Varicella (MMRV), and most live bacteria vaccines. This is due to concern for inadvertent live vaccine-related pathologic disease.

Question 2

Review of further records indicates she received a Hepatitis B vaccine (HepB) 2 years ago, though did not receive any repeat HepB vaccines since. Regarding keeping her protected against Hepatitis B virus, what is the next step?

A. Continue the remainder of the HepB vaccine series as soon as possible
B. Nothing further since she is adequately protected against Hepatitis B virus
C. Restart the entire HepB series
D. Check her Hepatitis B core (anti-Hbc) titer to see if she is fully protected against Hepatitis B virus

Answer: A

Discussion

HepB dosing schedules may vary based on renal function and vaccine formulations but are generally administered as a three-dose series over the course of 6 months (at 0, 1, and 6 months). Interruption of the recommended vaccine schedule does not require restarting the series or adding on more doses, but the remainder of the series should be administered as soon as possible.

In this case, the patient received her first dose of the series, but none since. It would be advised for her to obtain the second dose as soon as she is able and, if receiving a three-dose series, to separate the second and third doses by at least 2 months. If a patient was only missing the third dose, they should have it administered when convenient.

Receiving vaccines at longer than advised intervals do not alter final antibody concentrations, but there is concern that protection may not be reached until receiving the recommended number of doses. IgG Hepatitis B surface antibody

Table 2 Recommended adult immunization schedule by medical condition and other indications, United States, 2021 (U.S. Department of Health and Human Services, Centers for Disease Control and Prevention)

Vaccine	Pregnancy	Immuno-compromised (excluding HIV Infection)	HIV Infection CD4 count <200 mm³	HIV Infection CD4 count ≥200 mm³	Asplenia, complement deficiencies	End-stage renal disease; or on hemodialysis	Heart or lung disease, alcoholism[1]	Chronic liver disease	Diabetes	Health care personnel[2]	Men who have sex with men
IIV or RIV4 (or)	1 dose annually										
LAIV4	Not Recommended					Precaution				1 dose annually (or)	
Tdap or Td	1 dose Tdap each pregnancy	1 dose Tdap, then Td or Tdap booster every 10 years									
MMR	Not Recommended*	Not Recommended		1 or 2 doses depending on indication							
VAR	Not Recommended*	Not Recommended			2 doses						
RZV					2 doses at age ≥50 years						
HPV	Not Recommended*	3 doses through age 26 years			2 or 3 doses through age 26 years depending on age at initial vaccination or condition						
PCV13		1 dose									
PPSV23		1, 2, or 3 doses depending on age and indication									
HepA					2 or 3 doses depending on vaccine						
HepB					2, 3, or 4 doses depending on vaccine or condition				<60 years ≥60 years		
MenACWY	1 or 2 doses depending on indication, see notes for booster recommendations										
MenB	Percaution	2 or 3 doses depending on vaccine and indication, see notes for booster recommendations									
Hib		3 doses HSCT[3] recipients only			1 dose						

Recommended vaccination for adults who meet age requirement, lack documentation of vaccination, a lack evidence of past infection

Recommended vaccination for adults with an additional risk factor or another indication

Precaution–vaccination might be indicated if benefit of protection outweighs risk of adverse reaction

Recommended vaccination based on shared clinical decision-making

Not recommended/ contraindicated–vaccine should not be administered.

*Vaccinate after pregnancy.

No recommendation/ Not applicable

1. Precaution for LAIV4 does not apply to alcoholism. 2. See notes for influenza; hepatitis B; measles, mumps, and rubella; and varicella vaccinations. 3. Hematopoietic stem cell transplant

(anti-HBs) titer levels in an immunized patient are considered to be protective when >10 mIU/mL. The duration that titers remain above protective level is often proportional to the peak titer reached after completing the vaccination series. In immunocompromised patients, post-vaccination testing may be done 1–2 months after the last dose. If below protective levels, immunocompromised patients at risk of exposure may be given a booster or second vaccine series. If patients still fail to respond after undergoing repeat vaccinations, they are unlikely to benefit from further vaccination.

Question 3

What vaccines should you counsel your patient with immunodeficiency to receive prior to and during pregnancy?

A. Avoid all vaccines
B. Proceed with only inactivated vaccines prior to and during pregnancy
C. Proceed with all inactivated and some live vaccines prior to and during pregnancy
D. Proceed with all inactivated and all live vaccine required prior to and during pregnancy for fetal protection

Answer: B

Discussion

Non-immunocompromised individuals should receive tetanus, diphtheria toxoids, and acellular pertussis (Tdap) and MMR vaccines, if necessary, prior to pregnancy. During pregnancy, non-immunocompromised individuals should receive influenza, with consideration of Hepatitis A and B vaccines if indicated. In general, inactivated vaccines are generally safe to administer in immunocompromised patients, while live vaccines need to be carefully considered given the risk of causing infection depending on the severity of the patient's immunocompromised state. In this patient's case, her history of CVID on monthly IVIG would make live vaccines (such as MMR) contraindicated given the risk of developing infection from vaccine administration and the unlikely benefit of humoral response given her ongoing monthly gamma globulin treatments. Pregnant patients with a history of CVID should generally receive Tdap prior to pregnancy and inactivated influenza during pregnancy as immunoglobulin preparations may not contain antibiotics to the circulating strain and there may be some cellular immunity benefits.

Question 4
Which class of immunoglobulin is predominantly transferred across the placenta from the mother to the fetus during pregnancy?

A. IgA
B. IgG
C. IgM
D. IgD
E. IgE

Answer: B
Discussion
Humoral immunity can be subdivided into either active or passive immunity. Active immunity occurs when an individual's endogenous immune response produces antibodies after exposure to infection or vaccination. This response may take days or weeks to develop but provides long-lasting immunity. Conversely, passive immunity refers to the transfer of preformed antibodies to an individual and confers immediate though short-lived immune protection. Passive immunity is essential not only in newborn immunity, but also has important therapeutic applications as well.

Maternal passive immunity allows for the transfer of the mother's preformed IgG antibodies to the fetus prior to birth. The transmission of these antibodies occurs predominantly during the third trimester of pregnancy and are mediated by the fetal IgG-specific Fc receptor across the placenta. This process allows for immediate protection of newborns as the neonatal immune system requires several months to fully mature. Newborn titers of IgG antibodies are largely dependent on maternal concentrations of specific antibodies during pregnancy. Therefore, maternal vaccination during pregnancy is essential to conferring adequate immune protection to the newborn. Notably, live attenuated vaccines are contraindicated in pregnancy due to the potential risk of transplacental viral transmission to the fetus. Breast milk and colostrum are another example of maternal passive immunity and result in the transfer of secretory IgA antibodies for mucosal protection in the infant's gastrointestinal and respiratory tract.

The efficacy of passive immunity has resulted in the development of several therapeutic options for both immunocompetent and immunodeficient patients. Immune globulin, a pooled plasma preparation from several thousand donors containing IgG antibodies against an array of antigens, can be administered to immunodeficient individuals for the treatment of various infections or autoimmune and inflammatory diseases. Similarly, hyperimmune globulin, plasma derived from an individual with high titers of a specific antibody against specific pathogens, has important applications in post-exposure prophylaxis for several infectious diseases.

Question 5
The patient's husband has avoided getting vaccines, since he has been told that the patient could get exposure to infections in what he has been vaccinated for. What vaccinations do you recommend for the patient's husband?

A. Continue to avoid all vaccinations
B. Administer only recommended live vaccines per the CDC annual schedule
C. Administer all recommended inactivated age-appropriate vaccines and select live vaccines (MMR and varicella) per the CDC annual schedule
D. Administer only recommended inactivated vaccines and avoid live vaccines per the CDC annual schedule

Answer: C
Discussion
Healthcare providers often receive questions regarding the safety of vaccinations from immunocompetent household contacts of immunocompromised individuals. Both primary care providers and specialists thus serve an important role in ensuring these patients receive the appropriate recommended vaccinations. Per IDSA and CDC guidelines, immunocompetent household contacts of immunocompromised patients may safely receive all recommended inactivated vaccines per the CDC-ACIP annual schedule. The following live vaccines can also be safely administered based on the CDC annual schedule or travel guidelines:

- Recommended per CDC annual schedule: Combined MMR, rotavirus in infants 2–7 months old, varicella, zoster
- Recommended per CDC travel guidelines: Yellow fever, oral typhoid

The annual flu vaccine is recommended for all patients who are 6 months of age or older and share a household with immunocompromised individuals. The inactivated influenza vaccine may be safely administered to all patients. The live attenuated influenza vaccine can be administered to healthy, non-pregnant patients between 2 and49 years old unless they live in a household with an immunocompromised patient who has severe combined immune deficiency (SCID), received a hematopoietic stem cell transplant within the last 2 months, or developed graft vs host disease.

In this case, the patient's husband should receive all recommended inactivated vaccines, the MMR and varicella live vaccines if he has not previously received them, and either the inactivated or live attenuated flu vaccine.

Case 2
A 60-year-old male presents to your office after experiencing urticaria/angioedema within 30 min of receiving the MMR

vaccine. Since no records of vaccination were available, he was advised to receive MMR vaccine, since he lives in an area where high prevalence of measle outbreaks has occurred over the past few years due to vaccine hesitancy within his community. A blood test failed to reveal protective titers to MMR.

Question 1

What is the most likely allergen responsible for the anaphylaxis he experienced following the MMR vaccine?

A. The actual vaccine itself is most likely responsible for the reaction and not from egg, gelatin, or latex.
B. Ovalbumin elicited the reaction since the MMR vaccine is grown in fibroblast from chick embryos.
C. Patient most likely had a reaction to thimerosal the ovalbumin amount within the vaccine is not enough to elicit a reaction and he has tolerated ingestion of Jell-O.
D. Gelatin within the MMR vaccine is the most likely culprit of the reaction he experienced given the high content of gelatin within the vaccine.

Answer: D

Discussion

Common post-vaccine reactions include local and constitutional symptoms, such as fever. These reactions do not contraindicate future vaccination. Local injection-site reactions are generally mild, and include local swelling, redness, and soreness. Delayed-type hypersensitivity reaction to a vaccine component (such as neomycin, thimerosal, and aluminum) can result in a transient small nodule at the injection site. Patients who have allergic contact dermatitis to neomycin are not contraindicated to receive vaccines that contain neomycin, though a small transient papule at the site of injection can be expected. Similarly, those with thimerosal contact allergy can have large local reactions to vaccines that contain thimerosal, though most patients are asymptomatic. Aluminum containing vaccines can infrequently lead to persistent injection-site nodules. Severe allergic reactions to vaccines are very rare. Out of 25,173,965 vaccinations that were administered between January 1, 2009 through December 31, 2011, 33 cases of anaphylaxis were confirmed. The estimated risk of an allergic reaction is 1.31 (95% CI, 0.90–1.84) cases per million vaccine doses. Allergic reactions to vaccines are rarely due to the actual vaccine itself. Instead, these reactions are generally caused by components and excipients. Vaccine components include inactive ingredients such as gelatin, egg protein, formaldehyde, thimerosal, or neomycin, which can cause specific IgE-mediated reactions. Vaccine excipients are added to vaccines to improve the stability, solubility, absorption, and prevent microbial growth. Polyethylene glycol (PEG) and polysorbate are common excipients in vaccines and injectable medications to improve water-solubility (Fig. 1). Excipients represent a significant cause of specific IgE-mediated and immediate reactions in vaccines.

Gelatin is present in many vaccines as a stabilizing agent. Examples of vaccines that contain gelatin include MMR, varicella, and Japanese encephalitis vaccines. Gelatin has been frequently shown to be the culprit for anaphylactic reactions from these vaccines. If a patient has an allergy to foods with gelatin (e.g., gummy bears, jelly), further evaluation should be obtained prior to administering any vaccines that contain gelatin. Having a negative history of gelatin ingestion does not rule out having a potential allergy to gelatin-containing vaccines as they are administered via different routes. Gelatin used in vaccines are derived from either bovine or porcine sources which are extremely cross-reactive. Individuals with beef or pork sensitivities can also be sensitized to gelatin which potentially places them at risk for reactions to these vaccines. Egg protein is present in MMR and influenza vaccines, though having an egg allergy is not a contraindication to receiving these vaccines. HepB and quadrivalent human papilloma vaccine (HPV4) contain residual yeast protein, though adverse reactions are very infrequent. Patients with latex allergy are at risk when receiving vaccines with vial stoppers made from dry natural rubber (DNR) latex, though the theoretical risk is very small. Although these reactions are rare, patients who report a history of immediate hypersensitivity reactions to gelatin, yeast, latex, thimerosal, or neomycin must be evaluated with skin testing before immunization.

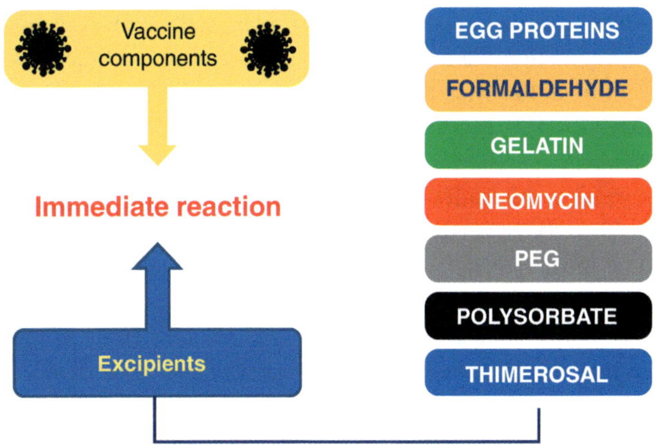

Fig. 1 Vaccine compound allergens. (Kounis NG, Koniari I, de Gregorio C, Velissaris D, Petalas K, Brinia A, Assimakopoulos SF, Gogos C, Kouni SN, Kounis GN, Calogiuri G, Hung MY. Allergic Reactions to Current Available COVID-19 Vaccinations: Pathophysiology, Causality, and Therapeutic Considerations. Vaccines (Basel). 2021 Mar 5;9(3):221. https://doi.org/10.3390/vaccines9030221. PMID: 33807579; PMCID: PMC7999280, with permission)

Question 2

The most likely mechanism behind what appeared to be anaphylaxis following the MMR vaccine could be explained by:

A. Release of mast cell contents following direct interaction with the mast cell membrane
B. Release of mast cell mediators following cross-linking of IgE on the surface of the mast cell
C. A T-cell mediated reaction following mast cell release of mediators
D. An anaphylactoid reaction rather than true anaphylaxis, considered nonimmunologic anaphylaxis

Answer: B

Discussion

Vaccine-induced allergic reactions can cause immediate and delayed-typed presentations. Mast cell activation and degranulation can be mediated by different pathways. The most well-understood mechanism is via type I hypersensitivity reaction which involves IgE and antigen cross-linking of FcεRI on mast cells. These reactions are mediated via the interaction between specific preformed IgE antibodies that are targeted to their counterpart vaccine components acting as antigens. Symptoms occur within minutes up to 4 h after exposure. Confirmation of the reaction is via detection of specific IgE antibodies. Serum tryptase can confirm an anaphylactic reaction though is only elevated within the first 3 h. Vaccine excipients and/or components are usually the cause of these reactions. Non-IgE-mediated mast cell activation and degranulation, also referred to as nonimmunologic anaphylaxis also play a significant role in vaccine reactions. Direct activation of the complement system leads to the generation of anaphylatoxins C1q, C3a, C4, C5a, and Factor B which are potent stimulators of mast cell activation and degranulation via a non-IgE-mediated pathway. Direct activation of the Mas-related G protein-coupled receptor X2 (MRGPRX2) can also result in direct mast cell activation, however specific IgE and tryptase may be undetectable and normal in this pathway.

Type IV hypersensitivity (delayed) reactions are the second most common type of allergic reactions. Symptoms appear around 48 h and peaks between 72 and 86 h after receiving the vaccine. These reactions are mediated via activation of T -cells and monocytes/macrophages and release of cytokines that lead to inflammation. Antimicrobial agents in vaccines such as thimerosal and aluminum are common causes for these types of reactions.

Initial studies estimated allergic reactions to mRNA COVID-19 vaccines to range from 2.5 to 11.1 cases per million doses, though recent data showed 4.7 and 2.5 cases per million doses administered with 9,943,247 doses of Pfizer-BioNTech and 7,581,428 doses of Moderna vaccines reportedly in the US from December 14, 2020 to January 18, 2021. Another study showed that the rate was 2.47 cases per 10,000 vaccines out of 64,900 employees who received their vaccines at Mass General Brigham. The Pfizer-BioNTech and Moderna mRNA vaccines use a lipid-based nanoparticle carrier system that is stabilized by polyethylene glycol (PEG) which have been implicated in anaphylactic reactions. PEG is also used a laxative. It is commonly added as an excipient in medications and previously has been shown to be rare causes of anaphylaxis. Delayed large local reactions have also been reported to present between 1 and 2 weeks after the first dose and a few days after the second dose. Case reports of shingles have also been reported following vaccination, though there is no clear association between the vaccine and shingles at this time.

Question 3

What are the appropriate next steps in diagnosis and management given this patient's history?

A. Administer skin testing for vaccine and vaccine excipients.
B. Obtain further history regarding ability to tolerate egg ingestion
C. Proceed with vaccination under observation since benefits of vaccination outweigh risk associated with vaccine administration given his age/risk factors
D. Prescribe premedication regimen prior to vaccine administration and proceed with administration under observation
E. Obtain further history regarding latex allergy

Answer: A

Discussion

Patients who have anaphylactic reaction after immunization should undergo immediate-type allergy skin testing with vaccine and vaccine excipients to confirm the reaction was IgE-mediated and determine the responsible excipient of the vaccine. Lack of further investigation can lead to patients avoiding future vaccines unnecessarily or increased risk of other vaccines that contain similar excipients. Egg, gelatin, latex, and yeast are vaccine excipients most often responsible for anaphylactic reactions, with egg and gelatin being the most common culprits for type I hypersensitivity reactions. Gelatin from either bovine or porcine origin is added to many vaccines as a stabilizer and has been the culprit for anaphylactic reactions to MMR. In this case, it will be important to focus on gelatin skin testing as the MMR vaccine contains a high gelatin content. It is prudent to exclude a prior history of allergy to ingestion of gelatin, however a negative history does not exclude potential for an allergic reaction to vaccine containing gelatin. Other notable gelatin-containing vaccines include influenza, rabies, typhoid, varicella, yellow fever, and zoster. MMR vaccines are created in cell cultures using chicken embryo fibroblasts, but the vaccine contains minimal to negligible amounts of egg protein, making anaphylaxis risk low even in children with mild or severe egg allergies.

The patient should undergo both vaccine skin test and vaccine component skin testing. To start, a prick test with the full-strength vaccine and vaccine excipients should be performed. If epicutaneous skin test is negative, intradermal testing with 0.02 cc of vaccine with 1:100 dilution should be pursued. Vaccine ingredient skin tests can be done concomitantly with prick tests with egg, Saccharomyces cerevisiae yeast, gelatin, and IgE antibody testing to latex depending on the vaccine. In our patient's case, gelatin prick testing should be pursued given his history of reaction to the MMR vaccine and the significant content of gelatin within the vaccine that has been associated with IgE-mediated MMR vaccine reactions. Gelatin can be prepared by dissolving 1 teaspoon (5 g) of any sugared gelatin power (i.e., Jell-O) in 5 mL of normal saline to create a skin prick test solution. If skin testing is negative and additional doses of the vaccine are requested, the vaccine can be administered under observation for at least 30 min.

Other vaccine component testing that can be performed for other vaccines include: DTaP, Td, Tdap: Milk, Hepatitis B: Yeast, Influenza: Egg, gelatin, Varicella, or zoster: Gelatin, and Yellow Fever: Egg, gelatin4. PEG is the major excipient in both the Pfizer and Moderna mRNA COVID-19 vaccines, whereas polysorbate 80 is the excipient in the Johnson and Johnson COVID-19 vaccine.

Question 4

The patient undergoes gelatin skin testing and the results are positive. You discuss these results with your patient. What is the most appropriate advice for other vaccines containing gelatin as excipient?

A. Advise the patient to avoid proceeding with vaccination given his contraindication of prior reaction and positive gelatin skin testing result
B. Premedicate and administer the full-strength vaccine under observation
C. Receive vaccine in graded fashion
D. Repeat the gelatin skin test for confirmation testing given the unreliable nature of vaccine component testing standardization

Answer: C

Discussion

Because skin testing to gelatin is positive, the vaccine will need to be administered in a graded fashion with 30 min of observation. If the full normal dose volume of vaccine is 0.5 mL, give vaccine at 15 min intervals as tolerated: 0.05 mL 1:10 dilution, 0.05 mL full-strength, 0.1 mL full-strength, 0.15 mL full-strength and 0.2 mL full-strength, and staff must be prepared to treat anaphylaxis. If skin testing had been negative, the vaccine could have been administered under 30 min of observation. For patients with egg allergy with a history suggestive of only hives, the vaccine can be administered in the primary care physician's office with 30 min of observation. Patients with a history of anaphylaxis to egg should receive their vaccine in an allergist's office. Patients who have a history of an allergic reaction the influenza vaccine require further assessment with skin testing to the vaccine and its components. For patients who receive a vaccine containing latex or Saccharomyces cerevisiae yeast (HepB and HPV4) and have an IgE-mediated reaction the vaccine, skin testing should be done for these allergens. If skin testing is positive and further doses of the vaccine are indicated, the vaccine will again need to be administered in a graded fashion in a monitored setting.

There are few contraindications to vaccine administration. Patients with history of Guillain-Barre syndrome within 6 weeks of influenza vaccine should avoid subsequent immunizations. Development of severe encephalopathy after pertussis vaccine administration is a contraindication to future pertussis vaccination. Development of encephalitis after yellow fever vaccine is a contraindication to further yellow fever vaccines. Even with these notable exceptions, all patients with vaccine reactions after administration should undergo appropriate investigation to both minimize future risks to patients but also to support public health measures and prevent unnecessary avoidance of vaccinations.

Question 5

In addition to the need for MMR vaccine, the patient will require vaccination against COVID-19 prior to work up for evaluation of his reaction from the MMR vaccine. What is the next appropriate step?

A. Avoidance of the COVID-19 vaccine until workup for MMR vaccine has been undertaken.
B. Administer the COVID-19 vaccine since excipients within currently approved COVID-19 vaccines differ from the excipients found in the MMR vaccine.
C. Administer only vaccines that currently are approved for a one time dose and avoid those that require 2 doses to limit the possibility of developing sensitization.
D. Avoidance of the COVID-19 vaccine until work up for MMR vaccine has been undertaken and include testing for COVID-19 and its excipients.

Answer: B

Discussion

The incidence of an anaphylactic reaction to a COVID-19 vaccine is very rare. Possible culprits that could explain such reactions include the excipients associated with the currently

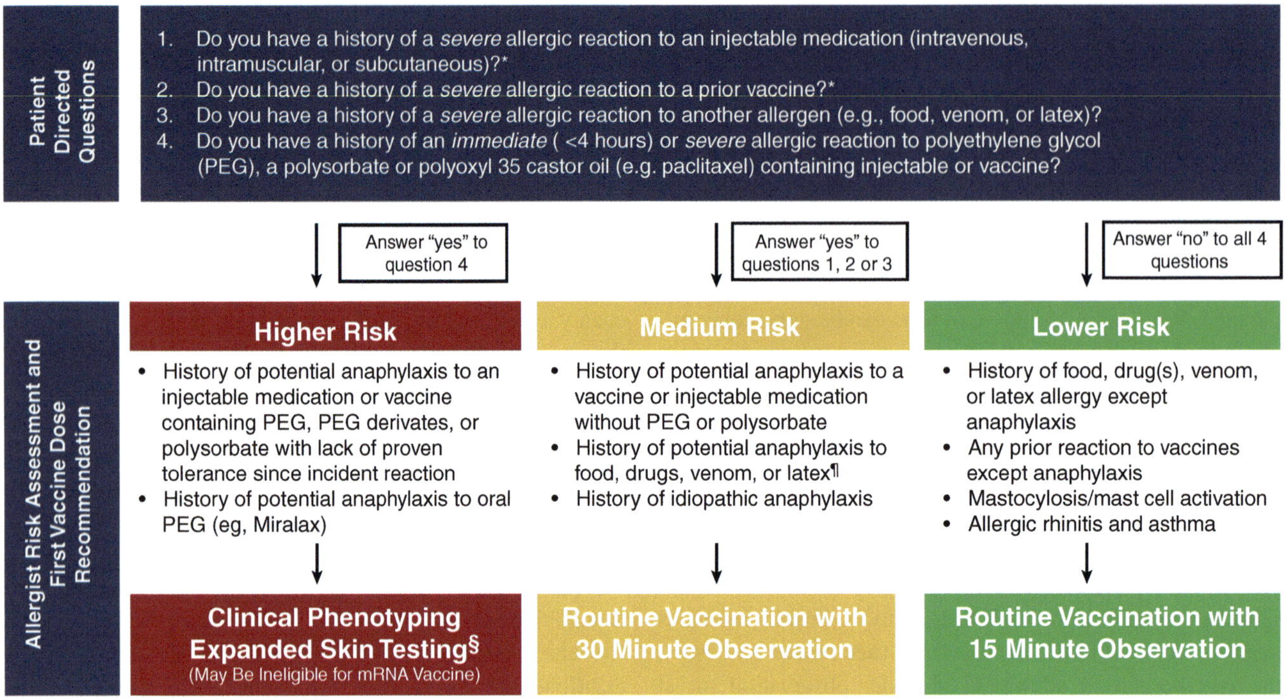

Fig. 2 Risk stratification pathways with categories based initial mRNA COVID-19 vaccination. (Banerji A, Wickner PG, Saff R, Stone CA Jr, Robinson LB, Long AA, Wolfson AR, Williams P, Khan DA, Phillips E, Blumenthal KG. mRNA Vaccines to Prevent COVID-19 Disease and Reported Allergic Reactions: Current Evidence and Suggested Approach. J Allergy Clin Immunol Pract. 2021 Apr;9(4):1423–1437. https://doi.org/10.1016/j.jaip.2020.12.047. Epub 2020 Dec 31. PMID: 33388478; PMCID: PMC7948517, with permission)

approved vaccines. PEG has been thought of as the main culprit though prior to the current pandemic, allergic reactions to PEG were very rare. A recent paper suggests the actual mechanism behind the reaction could be more consistent with direct mast cell activation vs. true IgE, since patients were able to tolerate the second dose with only premedication with an antihistamine, in addition tryptase was not elevated and PEG was negative. In addition, the MMR vaccine excipients are different from excipients currently found in approved COVID-19 vaccines. It has been advised that anyone with a history of immediate onset hypersensitivity reactions (urticaria, wheezing, angioedema, anaphylaxis) to any prior vaccine or injected medication be monitored for 30 min following the vaccination (Fig. 2).

Bibliography

1. Kounis NG, Koniari I, de Gregorio C, Velissaris D, Petalas K, Brinia A, et al. Allergic reactions to current available covid-19 vaccinations: pathophysiology, causality, and therapeutic considerations. Vaccines. 2021;9:221.
2. Banerji A, Wickner PG, Saff R, Stone CA, Robinson LB, Long AA, et al. mRNA vaccines to prevent COVID-19 disease and reported allergic reactions: current evidence and suggested approach. J Allergy Clin Immunol Pract. 2021;9:1423–37.
3. Albrecht M, Arck PC. Vertically transferred immunity in neonates: mothers. Mechanisms and mediators. Front Immunol. 2020;11:555.
4. Cunningham-Rundles C. How I treat common variable immune deficiency. Blood. 2010;116:7–15.
5. Rubin LG, Levin MJ, Per Ljungman E, Davies G, Avery R, Tomblyn M, et al. 2013 IDSA clinical practice guideline for vaccination of the immunocompromised host. Clin Infect Dis. 2014;58:309–18.
6. Sobh A, Bonilla FA. Vaccination in primary immunodeficiency disorders. J Allergy Clin Immunol Pract. 2016;4:1066–75.
7. McNeil MM, Weintraub ES, Duffy J, Sukumaran L, Jacobsen SJ, Klein NP, et al. Risk of anaphylaxis after vaccination in children and adults. J Allergy Clin Immunol. 2016;137:868–78.
8. Sakaguchi M, Nakayama T, Inouye S. Food allergy to gelatin in children with systemic immediate-type reactions, including anaphylaxis, to vaccines. J Allergy Clin Immunol. 1996;98:1058–61.
9. Kelso JM, Jones RT, Yunginger JW. Anaphylaxis to measles, mumps, and rubella vaccine mediated by IgE to gelatin. J Allergy Clin Immunol. 1993;91:867–72.
10. Kelso JM, Greenhawt MJ, Li JT, Nicklas RA, Bernstein DI, Blessing-Moore J, et al. Adverse reactions to vaccines practice parameter 2012 update. J Allergy Clin Immunol. 2012;130:25–43.
11. Banerji A, Wolfson AR, Wickner PG, Cogan AS, McMahon AE, Saff R, et al. COVID-19 vaccination in patients with reported allergic reactions: updated evidence and suggested approach. J Allergy Clin Immunol Pract. 2021;9:2135–8.
12. Shimabukuro TT, Cole M, Su JR. Reports of anaphylaxis after receipt of mRNA COVID-19 vaccines in the US-December 14, 2020-January 18, 2021. JAMA—J Am Med Assoc. 2021;325:1101–2. https://doi.org/10.1001/jama.2021.1967
13. Blumenthal KG, Freeman EE, Saff RR, Robinson LB, Wolfson AR, Foreman RK, et al. Delayed large local reactions to mRNA-1273 vaccine against SARS-CoV-2. N Engl J Med. 2021;384:1273–7.
14. McMahon DE, Amerson E, Rosenbach M, Lipoff JB, Moustafa D, Tyagi A, et al. Cutaneous reactions reported after Moderna and

Pfizer COVID-19 vaccination: a registry-based study of 414 cases. J Am Acad Dermatol. 2021;85:46–55.
15. Furer V, Zisman D, Kibari A, Rimar D, Paran Y, Elkayam O. Herpes zoster following BNT162b2 mRNA Covid-19 vaccination in patients with autoimmune inflammatory rheumatic diseases: a case series. Rheumatology. 2021;60:SI90–5.
16. Chung EH. Vaccine allergies. Clin Exp. Vaccine Res. 2014;3:50–7.
17. Kazatchkine MD, Kaveri SV. Immunomodulation of autoimmune and inflammatory diseases with intravenous immune globulin. N Engl J Med. 2001;345:747–55.
18. Hamborsky J, Kroger A, Wolfe SE, Centers for Disease Control and Prevention. Epidemiology and prevention of vaccine-preventable diseases. Atlanta, GA: Centers for Disease Control and Prevention; 2015.
19. Dreskin SC, Halsey NA, Kelso JM, Wood RA, Hummell DS, Edwards KM, et al. International Consensus (ICON): allergic reactions to vaccines. World Allergy Organ J. 2016;9:32.
20. Furer V, Zisman D, Kibari A, Rimar D, Paran Y, Elkayam O. Herpes zoster following BNT162b2 mRNA Covid-19 vaccination in patients with autoimmune inflammatory rheumatic diseases: a case series. Rheumatology. 2021. https://doi.org/10.1093/rheumatology/keab345
21. Greenhawt M, Turner PJ, Kelso JM. Administration of influenza vaccines to egg allergic recipients: A practice parameter update 2017. Ann Allergy, Asthma Immunol. 2018. https://doi.org/10.1016/j.anai.2017.10.020
22. Stone CA, Rukasin CRF, Beachkofsky TM, Phillips EJ. Immune-mediated adverse reactions to vaccines. Br J Clin Pharmacol. 2019. https://doi.org/10.1111/bcp.14112

Allergen Immunotherapy

Karla E. Adams and James M. Quinn

Case 1

A 28-year-old female with a history of asthma and chronic rhinitis has been referred to allergy and immunology for consideration of immunotherapy in her hometown of Durham, North Carolina. Her visit is in February. She has had asthma and rhinitis since childhood. She has never been hospitalized for her asthma and it has been 2 years since she has required an urgent visit or systemic steroids for her asthma. She reports using albuterol once daily on most but not every day for the past 2 months, no nighttime awakenings, and no activity limitations. Her asthma symptoms are triggered during peak allergy seasons and with upper respiratory infections. Her rhinitis symptoms include sneezing, clear rhinorrhea, nasal congestion, and postnasal drip. She has been unsatisfied with the control of her nasal symptoms despite multiple different medical regimens over many years. The peak seasons for her symptoms are the spring and the fall but she has mild-moderate symptoms perennially. She lives in a single family newly constructed home with her husband and no children or pets. The home has carpeting throughout. There is no rodent presence/infestation. She has dust mite bedding encasements. She is in year 1 of a 3-year master's program with plans to remain in the area and start a family after completion of her master's program.

Her current medications are fluticasone propionate 44 mcg MDI 2 inhalations twice daily, albuterol MDI 2 inhalations as needed, fluticasone propionate 50 mcg 2 sprays each nostril daily, cetirizine 10 mg daily, montelukast 10 mg daily, and azelastine 1 spray each nostril twice daily. Refill patterns and self-report suggest full adherence and proper technique for her medications.

Her pulmonary function tests can be seen in Fig. 1.

Her prick skin testing results can be seen in Fig. 2.

K. E. Adams (✉) · J. M. Quinn
Allergy and Immunology Department, San Antonio Uniformed Services Health Education Consortium—Wilford Hall, San Antonio, TX, USA
e-mail: karla.e.adams2.mil@mail.mil

She would like to discuss both subcutaneous immunotherapy (SCIT) and sublingual immunotherapy (SLIT).

Question 1

The patient has some basic questions regarding SCIT and SLIT. She would like to know if she is currently a candidate to initiate immunotherapy. What is the best answer?

A. She is currently a candidate to initiate either SCIT or SLIT
B. She is currently a candidate to initiate SCIT but patients with asthma are not candidates for SLIT
C. She is currently a candidate to initiate SLIT and SLIT is preferred over SCIT in all patients with asthma because SLIT is safer
D. She is not currently a candidate to initiate SCIT or SLIT

Answer: D

Discussion

The 2007 National Heart, Lung, and Blood Institute's Expert Panel Report 3 (EPR-3) defined asthma as not well controlled when the short-acting beta agonists (SABA) such as albuterol are used more than 2 times per week. The 2020 National Heart, Lung, and Blood Institute's Expert Panel Report 4 (EPR-4) recommended that asthma should be well controlled before initiating immunotherapy. The 2011 Allergen Immunotherapy Practice Parameter Third Update (2011 AIT PP) recommended against initiating immunotherapy unless a patient's asthma is stable because patients with uncontrolled asthma are at increased risk for systemic reactions. The 2020 Clinical Practice of Allergen Immunotherapy for Allergic Rhinoconjunctivitis and Asthma: An Expert Panel Report (2020 AIT EPR) recommended that SCIT should be postponed in uncontrolled asthma. The package inserts for all the FDA approved SLIT tablets identified uncontrolled asthma as a contraindication to therapy. The 2017 Sublingual Immunotherapy Focused Allergen Immunotherapy Practice Parameter Update (2017 SLIT Update) discussed the lack of evidence regarding SLIT in

M. Mahmoudi (ed.), *Absolute Allergy and Immunology Board Review*, https://doi.org/10.1007/978-3-031-12867-7_31

Plethysmograph Report

Age: 28 Height(in): 65 Weight(lb): 135 Gender: Female
Race: Black Diagnosis: Medication:
Dyspnea Rest: No Dyspnea Exercise: No
Cough: No **Persistent: No** Productive (cc):
Smoker: No How Long(pk/yrs): Stopped(yrs):
Cigarettes: No Cigars: No
Technician: Temp: 23 PBar: 764

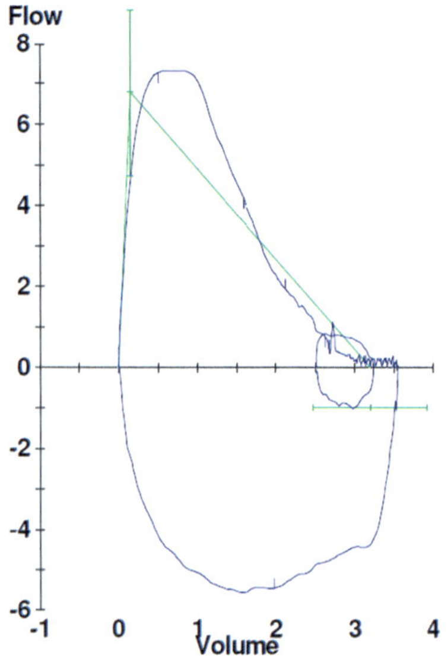

Spirometry

		Ref	Pre Meas	Pre % Ref	Post Meas	Post % Ref	Post % Chg
FVC	Liters	3.19	3.55	111			
FEV1	Liters	2.63	2.69	102			
FEV1/FVC	%	83	76				
FEF25-75%	L/sec	2.90	2.24	77			
PEF	L/sec	6.76	7.30	108			
FET100%	Sec		7.45				
FIVC	Liters		3.54				
FIF50%	L/sec		5.43				
FVL ECode			000010				
MVV	L/min	105					

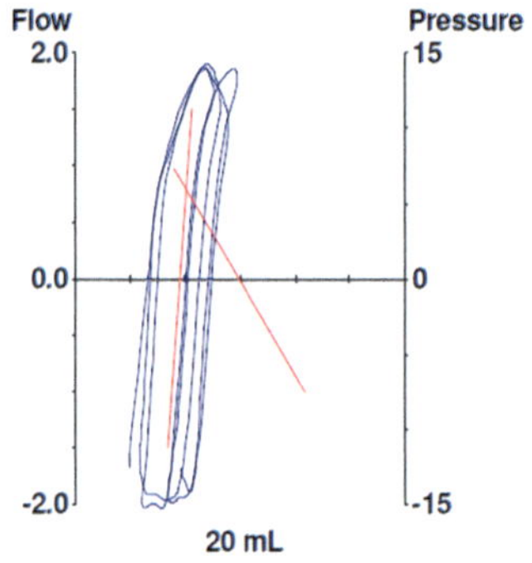

Lung Volumes

		Ref	Pre Meas	Pre % Ref	Post Meas	Post % Ref	Post % Chg
VC	Liters	3.38	3.62	107			
TLC	Liters	5.10	5.71	112			
RV	Liters	1.63	2.09	128			
FRC PL	Liters	2.74	2.36	86			
ERV	Liters		0.27				
IC	Liters		3.30				
Raw	cmH2O/L/sec		0.90				
Raw f	BPM		95				
Vtg (Raw)	Liters		2.78				
Vtg	Liters		2.78				

Single Breath Lung Volumes

TLC Sb	Liters	5.10	4.72	93
RV Sb	Liters	1.63	1.09	67
RV/TLC Sb%		33	23	

Fig. 1 Pulmonary function testing for Case 1

uncontrolled asthma. Nevertheless, the 2017 SLIT Update considered uncontrolled asthma a relative contraindication to starting SLIT. This patient is not currently a candidate for either SCIT or SLIT due to uncontrolled asthma.

The diagnosis of asthma itself is not a contraindication to SLIT. SLIT is not contraindicated in patients with well controlled asthma that is not severe.

There is evidence that SLIT has a lower risk than SCIT of systemic reactions, anaphylaxis, and fatal anaphylaxis. The 2020 AIT EPR concluded SLIT is generally viewed as safer than SCIT. However, while safety is an important consideration, there are multiple factors in the risk and benefit assessment that would not always favor SLIT over SCIT in patients with asthma.

Question 2

Which of the following approaches to immunotherapy would be the most appropriate and expected to have the greatest efficacy for the treatment of her AR and asthma?

A. SCIT with Bermuda grass, Timothy grass, and dust mite (df and dp) extracts
B. SCIT with Bermuda grass, Timothy grass, dust mite (df and dp), and mouse extracts
C. SLIT aqueous drops for Bermuda grass, Timothy grass, and dust mite (df and dp) extracts
D. SLIT grass tablets and dust mite tablets

Answer: A

ALLERGY PRICK SKIN TEST RECORD - INHALANTS, POLLENS, INSECTS

CRITERIA FOR PRICK EVALUATION

All skin test reactions are measured in mm and the largest diameter of wheal and flare are recorded.

A wheal of 3 mm or greater with accompanying erythema is considered a positive unless commented

Hollister Stier QUINTIP Devices are used for Prick Testing and tests are read at 15 minutes.

TESTING MATERIAL

PRICK TESTING EXTRACT CONCENTRATIONS:

Trees, and Weeds: 1:10 w/v or 1:20 w/v
Grasses (Bermuda 10,000 BAU)
Grasses (Rye, Timothy): standardized 100,000 BAU
Grasses (Bahia and Johnson): 1:20 w/v or 1:10 w/v
Molds, Dog, Cockroach: 1:10 w/v
Dp and Df: 10,000 AU   Cat: 10,000 BAU

	Wheal	Flare	Comment
1 Diluent Control	0	0	
Trees			
2 Alder, Red	0	0	
3 Ash, White	0	0	
4 Birch Mix	0	0	
5 Box Elder	0	0	
6 Maple, Red	0	0	
7 Cottonwood, Common	0	0	
8 Cypress, Bald	0	0	
9 Elm, Siberian	0	0	
10 Hackberry	2	2	
11 Mesquite	0	0	
12 Mountain Cedar	0	0	
13 Mulberry Red	0	0	
14 Oak Mix	2	2	
15 Olive, European	0	0	
16 Palm, Queen	0	0	
17 Pecan	2	2	
18 Sycamore, East	0	0	
19 Walnut, Black	2	2	
Grasses			
20 Bahia	0	2	
21 Bermuda	7	10	
22 Johnson	2	2	
23 Rye	0	2	
24 Timothy	5	8	
Environmentals			
25 D. farinae	5	9	
26 D. pteronyssinus	6	12	
27 Cat pelt (AP)	0	2	
28 Dog (AP)	0	0	
29 Mouse	4	6	

	Wheal	Flare	Comment
50 Histamine Control	6	12	
Weeds			
30 Nettle (Urtica)	0	2	
31 Cocklebur	0	0	
32 Dock/Sorrel Mix	0	0	
33 Lambs Quarters	0	0	
34 Marshelder, Rough	0	2	
35 Burning Bush (Kochia)	0	0	
36 Pigweed Mx (CPR)	2	2	
37 Plantain, English	0	0	
38 Ragweed Mx	7	11	
39 Russian Thistle	0	0	
40 Sage/Mugwort	0	0	
Molds			
41 Alternaria	0	0	
42 Aspergillus fum	0	0	
43 Chaetomium	0	0	
44 Curvularia	0	0	
45 Epicoccum	0	0	
46 Fusarium Mx	0	0	
47 Helminthosporium	0	0	
48 Cladosporium	0	0	
49 Penicillium Not	0	0	
Extra/Optional			
51			
52			

Name of Patient XXXXXXXXXXXXXXXXXX   **DOB** XXXXXXXX   **Last 4**

Doctor/Technician XXXXXXXXXXXXXXXXXXXXXXX   **Date** XXXXXXXX

☐ Follow up scheduled on:

☐ Patient was sent to front desk to schedule f/u

Fig. 2 Skin testing for Case 1

Discussion

There is robust evidence for the efficacy SCIT for properly selected patients using appropriately selected allergens. The 2020 Rhinitis Practice Parameter Update of the Joint Task Force on Practice Parameters (2020 Rhinitis Update) recommended SCIT as a safe and effective therapy for allergic rhinitis. In addition, the EPR-4 supported the efficacy of SCIT for the treatment of asthma in people with asthma and allergic sensitization. The body of evidence of the efficacy of SCIT for allergic rhinitis and asthma is concentrated on dust mite, grass, ragweed, and cat antigens. There is also efficacy for several other aeroallergens represented in the literature. However, there is scant evidence for the efficacy of mouse SCIT. The evidence is limited to uncontrolled case series and individual cases. The 2020 AIT EPR did not include mouse among the antigens approved for use by any regulatory agency in the world. In addition, for this patient there is no apparent exposure to mouse; it does not represent a clinically relevant allergen. Mouse extract should not be included in her immunotherapy.

There is some evidence supporting the efficacy of SLIT aqueous drops and SLIT tablets for both allergic rhinitis and asthma. However, the evidence for the efficacy of SLIT aqueous drops suffers from significant limitations. Efficacy endpoints for SLIT aqueous drops demonstrate varied and sometimes non-standardized endpoints that limit the quality of comparisons and systematic reviews. Furthermore, efficacy data is based on studies using varied products and varied doses resulting in limited quality and strength of the data. Most of the literature for SLIT aqueous drops is based on European products in European populations. Not surprisingly, SLIT aqueous drops are widely used in Europe for allergic rhinitis. Although "off label" use of SLIT aqueous drops occurs in the United States, there are no FDA products approved for SLIT aqueous drops. The 2017 SLIT Update did not endorse the use of non-FDA approved SLIT formulations.

In the United States, there are two FDA approved SLIT tablets for grass, one tablet product for ragweed, and one tablet product for dust mite. These tablets have demonstrated safety and efficacy for the treatment of allergic rhinitis. They have not been approved for the treatment of asthma. The 2017 SLIT Update recommended the use of FDA approved SLIT products for the treatment of allergic rhinitis. The EPR-4 recommended against the use of SLIT in the treatment of asthma until further information is available.

The two FDA approved grass tablets' major allergy is limited to the group-5 grass antigen. One product uses only Timothy grass as the source of the group-5 antigen and the second product uses a mix of 5 different northern pasture grasses for the group-5 antigen. There is limited cross-reactivity between the northern pasture grasses and the common southern grass—Bermuda grass. Northern pasture grass immunotherapy is generally considered insufficient for treatment of Bermuda grass allergic rhinitis.

The 2017 SLIT Update found the data insufficient to make recommendations regarding the safety and efficacy of multiple tablet SLIT for multiple allergens and recommended further study.

Regarding the comparative efficacy of SLIT and SCIT, there is limited data allowing head-to-head comparison. Nevertheless, indirect comparison of endpoints has led to the widely accepted conclusion that SCIT has an incremental benefit over SLIT.

Question 3

The patient returns in 3 months after changing her asthma controller to fluticasone 45 mcg/salmeterol 21 mcg MDI 2 inhalations twice daily. She has done well and only had asthma symptoms and used her albuterol MDI 2 times in the last month; her spirometry remains normal. Increasing her azelastine to 2 sprays each nostril twice daily has not significantly improved her rhinitis. She is following up to discuss more about immunotherapy. In conjunction with shared medical decision-making, which of the following is an indication for immunotherapy for this patient?

A. Prevent the sensitization to new allergens
B. Reduced the long-term use of medications
C. Reduce the risk of developing food pollen allergy syndrome
D. Reduce the risk of fatal asthma outcomes

Answer: B

Discussion

The 2020 Rhinitis Update identified the importance of shared medical decision-making in approaching immunotherapy. The 2020 Rhinitis Update made the recommendation to offer immunotherapy (SCIT or SLIT) to people who preferentially choose immunotherapy to avoid cost, side effects, or long-term use of medications. The recommendation was also to offer SCIT or SLIT to people with moderate/severe AR when symptoms are not controlled by avoidance and/or pharmacotherapy. The 2020 Rhinitis Update also identified the desire to prevent or reduce the severity of comorbid conditions (such as asthma) as an indication for immunotherapy (SCIT or SLIT). Separately, the 2020 Rhinitis Update gave a conditional recommendation with moderate evidence to offer immunotherapy in patients with controlled mild to moderate asthma and coexisting AR.

While there has been data suggesting that immunotherapy may prevent sensitization to new allergens, especially in children, the evidence has been inconsistent and unconvincing and it was not recommended as an indication in the 2020 Rhinitis Update. The 2020 AIT EPR reached similar conclusions.

Similarly, there are case series, case reports, and observational studies of immunotherapy reducing the symptoms of pollen food allergy syndrome (PFAS). However, the 2011 AIT PP and the 2017 SLIT Update concluded that more evidence was needed prior to any recommendation regarding SCIT and SLIT in PFAS treatment and neither parameter addressed prevention. There are no FDA approved products for the prevention or treatment of PFAS.

In aggregate, there is broad evidence supporting that immunotherapy (SCIT and/or SLIT) results in several beneficial outcomes in the treatment of asthma. Evidence of variable consistency and strength has identified beneficial outcomes of SCIT and/or SLIT in reduced asthma symptoms, reduced medication use, reduced allergen specific bronchial hyperresponsiveness, reduced non-specific bronchial hyperresponsiveness, improved FEV_1, and improved quality of life, among others. However, there has not been significant evidence supporting a reduction in asthma mortality with SCIT or SLIT.

Question 4

Which of the following is the best answer regarding risks associated with immunotherapy for this patient?

A. It is safe to receive her SCIT if she has an upper respiratory infection and she has needed albuterol 1–2 times daily for 3 days but her peak flows are normal
B. Planning a pregnancy before the completion of a full course (minimum 3 years) of SCIT is a contraindication for immunotherapy
C. Premedication with antihistamines should be recommended for all aeroallergen SCIT and SLIT patients to reduce the risk of a systemic reaction
D. Waiting for 30 min in the office after SCIT should be observed, but there is still a 15% or greater risk of a delayed reaction

Answer: D

Discussion

The 2011 AIT PP, 2020 AIT EPR, EPR-4, and the 2018 European Guidelines on Allergen Immunotherapy (2018 European AIT) recommend a minimum 30-min wait after SCIT. The selection of 30 min is based on multiple studies and survey data that report the large majority of systemic reactions occur within 30 min of SCIT and all but a few of the most severe reactions occur within 30 min. Nevertheless, reactions beginning after 30 min are uncommon but consistently found. Rates of delayed reactions vary among reports often approximating 15% but sometimes higher. Fortunately, most of the delayed reactions are mild and not life-threatening.

The 2011 AIT PP concluded that SCIT in a pregnant patient is considered to be low risk and with shared medical decision-making it is reasonable to continue maintenance injections. However, the risks and benefits of aeroallergen SCIT are generally not viewed favorably for the build-up phase or increasing doses of immunotherapy in pregnancy. Aeroallergen SCIT should not be initiated during pregnancy. If a patient becomes pregnant while on aeroallergen SCIT build-up, her dose should be kept the same and if the dose is unlikely to be effective, then her immunotherapy should be discontinued until after pregnancy. The 2018 European AIT is in keeping with these same recommendations. The 2017 SLIT Update recommended very cautious use of FDA approved SLIT products in pregnant or breastfeeding women acknowledging insufficient data on their safety during initiation or maintenance therapy. This patient does not plan on pregnancy for another 2 years or more. As a result, it would be expected that she would be on maintenance therapy and possible nearing completion of a 3-year course when she plans to consider pregnancy. In this scenario, a recommendation to consider SCIT with shared medical decision-making would be indicated.

The 2011 AIT PP recommended screening to assess the patient's current health before each injection. Screening of patients with asthma before each SCIT injection is advised because uncontrolled asthma is a risk factor for severe systemic reactions to immunotherapy including fatal and near-fatal reactions. Screening was recommended to assess asthma symptoms. Additional screening with objective lung function in the form of peak flows or pulmonary function testing could be considered. The EPR-4 similarly recommended against initiating, increasing, or administering SCIT while individuals have asthma symptoms. Utilizing validated screening questionnaires such as the ACT may be acceptable as well.

The 2020 AIT EPR supported the recommendation to postpone SCIT in the setting of uncontrolled asthma. It also recommended against administering SCIT in the setting of a fever or a respiratory infection.

Some data has suggested that premedication may reduce the risk of a systemic reaction with SCIT. However, when specifically looking at aeroallergen SCIT, there have been inconsistent and contradictory findings leaving no clear firm recommendations for or against premedication. Much of the data is focused on accelerated cluster or rush schedules and less on conventional schedules. The 2020 AIT EPR reinforced the uncertain benefit of premedication in either conventional SCIT or SLIT.

Question 5

As noted above, SCIT is indicated as treatment for select patients with AR and asthma. For what other conditions is SCIT recommended?

A. Anaphylaxis to peanut when peanut specific IgE is identified

B. Chronic urticaria when aeroallergen specific IgE is identified
C. Moderate atopic dermatitis when house dust mite IgE is identified
D. Severe asthma refractory to standard medications and biologics (Step 6 in EPR-4 guideline)

Answer: C
Discussion
The 2011 AIT PP recommended SCIT as an indication for potential benefit in patients with atopic dermatitis associated with aeroallergen sensitivity. Atopic Dermatitis: A Practice Parameter Update 2012 similarly recommended consideration of allergen immunotherapy in select patients with atopic dermatitis with aeroallergen sensitivity. In both of these practice parameters, the strongest evidence was with dust mite antigens.

Conversely, the 2011 AIT PP specifically identified chronic urticaria as a condition for which SCIT is not indicated. Echoing this recommendation, the 2014 update to the practice parameter for chronic urticaria considered immunotherapy as an unproven treatment and recommended against its use.

The 2011 AIT PP concluded clinical trials do not support the use of SCIT for food allergy. The 2014 Food Allergy Practice Parameter Update concluded that there was inadequate evidence to recommend the benefit over the risk of therapy with SCIT for food allergy.

The 2017 SLIT Update summarized that there are no FDA approved SLIT products for PFAS, food allergy, latex allergy, atopic dermatitis, or venom allergy. The update summarized some of the limited data and acknowledged the ongoing investigations for some of these indications.

The 2011 AIT PP, 2020 AIT EPR, and EPR-4 all recommend against initiating or administering SCIT in severe asthma. In EPR-4, SCIT is limited in recommendations to mild and moderate persistent asthma and specifically limits SCIT as an adjunctive option to therapy steps 2–4, not step 6.

Case 2

A 26-year-old male is being evaluated in the allergy clinic due to a history of rhinitis symptoms since early childhood. He has tried multiple medications to include intranasal corticosteroids and oral antihistamines without significant relief of his symptoms. He reports watery, red and itchy eyes as well as intermittent clear rhinitis and congestion. His symptoms are most prominent in the spring, summer, and fall; however, he has noted immediate rhinitis and sneezing whenever he visits a family member that owns cats. He denies any history of respiratory symptoms with his rhinitis symptoms. He has never used an inhaler and has no history of asthma.

He also reports a history of recurrent bee stings since childhood. Several of the stings have led to whole body hives and difficulty breathing. The most recent sting occurred at the age of 25 for which he was seen in the emergency department. Within minutes of the sting, he broke out in an urticarial rash over his torso, had swelling of his face, cough, and lightheadedness. He does not recall a retained stinger. In the emergency room, he was treated with intramuscular epinephrine and intravenous antihistamines with quick resolution of his symptoms. He was discharged with an epinephrine auto injector which he carries but has never had to use. He has not been stung by bees or any other insects since then. He has never been evaluated for venom allergy.

He denies any other medical problems. His only current medications are an intranasal corticosteroid, daily oral antihistamine, and an epinephrine auto injector for as needed use. Patient consents to allergy testing and would like to pursue immunotherapy for both aeroallergens and venom if indicated.

Allergy testing is ordered and completed. The patient showed skin test sensitization to trees, grasses, weeds, and cat. Venom testing shows positive IgE to yellow jacket, white faced hornet, and yellow hornet. Based on the testing results, two prescriptions for the patient (aeroallergen and venom subcutaneous immunotherapy) are provided in Table 1.

Question 1

For the aeroallergen prescription shown in Table 1, which of the following statements best describes the dosing for the allergens included?

A. All allergens are within the projected effective dosing range
B. Standardized extracts should be added at highest tolerated dose
C. Cat allergen is under dosed based on minimum projected effective dose range
D. Only Bermuda grass, weeds and cat allergens are within the recommended projected effective dose range

Answer: D
Discussion
Dosing for allergen SCIT is vital as it will determine the effectiveness of the treatment. The goal is to achieve effective dosing while minimizing side effects such as local or systemic reactions. A maintenance vial includes the highest concentration of each individual allergen in order to deliver the full therapeutic dose for each allergen. To determine the volume of extract needed to achieve a target SCIT dose, use the following equation: $V1 \times C1 = V2 \times C2$ where $V1$ is the final volume prepared, $C1$ is the target concentration of the allergen extract to be prepared, $V2$ is the volume of the extract that will be added to the dilution, and $C2$ is the con-

Table 1 Immunotherapy prescription for Case 2

Allergen extract prescription: VIAL #1				Venom extract prescription: VIAL #2		
Allergen contents	mL of extract	Manufacturer concentration	Final allergen dose[a]	Allergen contents	mL of extract	Manufacturer concentration
Trees				**Venoms**		
Box elder	0.5	1:20 wt/vol	1:400 wt/vol	Mixed vespid	Std	300 mcg
American elm	0.5	1:20 wt/vol	1:400 wt/vol			
Grasses						
Bermuda	1	10,000 BAU/mL	500 BAU			
Kentucky blue grass	1	100,000 BAU/mL	5000 BAU			
Orchard	1	100,000 BAU/mL	5000 BAU			
Weeds						
Pigweed	1	1:20 wt/vol	1:200 wt/vol			
Russian thistle	1	1:20 wt/vol	1:200 wt/vol			
Environmental						
Cat hair	4	10,000 BAU/mL	2000 BAU			
Total mL of extract	**10**					
Total mL of diluent	**0**					
Total mL of vial	**10**					

Key: *BAU* bioequivalent allergy unit, *wt/vol* weight/volume, *Std* standard
[a] Calculated allergen dose assumes a maintenance dose of 0.5 mL of 1:1 vol/vol concentration

centration of the extract to be added. Using this case as an example, the target concentration for cat allergen ($C1$) is 2000 BAU per 0.5 mL dose (where $V2 = 4$ mL; $C2 = 10{,}000$ BAU/mL; $V1 = 10$ mL; maintenance dose = 0.5 mL). A list of available extracts in the United States, concentrations and recommended probable therapeutic effective dose range are provided in Table 2.

Standardized extracts have consistent allergen content between lots and should be used when possible as their use decreases the risk of adverse reactions due to variable extract potency that may be found in non-standardized extracts. Standardization of extracts in the U.S. is based on potency determination by identifying the intradermal concentration of an allergen extract that produces a sum orthogonal diameter erythema of 50 mm or the $ID_{50}EAL$. Current U.S. licensed aeroallergen extracts include cat hair and pelt, dust mite (*Dermatophagoides pteronyssinus* and *D. farinae*), short ragweed, and grasses (Bermuda, Kentucky blue grass, perennial rye, orchard, timothy, meadow fescue, red top and sweet vernal). Venom extracts for flying Hymenoptera (FH) are standardized and available for testing and treatment (Table 2). Each of the five flying Hymenoptera venoms (honey bee, yellow jacket, wasp, white faced hornet, yellow hornet) is available in their own individual lyophilized extracts at a concentration of 100 µg/mL. A mixed vespid extract which has three closely related members of the Vespidae family (yellow jacket, white faced hornet, and yellow hornet) is available at a concentration of 300 µg/mL. Imported fire ant (IFA) whole body extract (WBE) of *Solenopsis invicta* and *S. richteri* is available for the treatment of IFA hypersensitivity. Unlike FH venom extracts, IFA WBE is a non-standardized glycerinated extract.

The selection of allergens to be included in each SCIT prescription depends on several factors. Each prescription must be individualized and include only the relevant allergens for which the patient is sensitized (i.e., evidence of specific IgE to that allergen has been determined via skin testing or serologic testing) and that are clinically relevant. Knowledge of the aerobiology patterns for the location the patient lives is also important as it can assist with determining which allergens are relevant. Finally, determining the exact composition of each individual prescription requires careful attention be paid to the compatibility of extracts. For instance, extracts that are high in proteases such as mold extracts should not be mixed with aeroallergens due to concern for degradation of the pollens.

Question 2

You decide to rewrite the aeroallergen prescription in Table 1 in order to ensure adequate projected therapeutic doses for all the allergens is achieved. Which of the following describes steps you could take to ensure all allergens are within the acceptable effective ranges?

A. Decrease cat allergen to 3 mL and increase each tree to 1 mL. Leave all other allergens the same
B. Decrease all the grasses to 0.25 mL of each grass. Leave all other allergens the same

Table 2 Example of available allergen extracts in the U.S., probable effective doses and typical extract volume range for immunotherapy prescriptions

Allergen	Manufacturer concentration	Effective dose range	Extract volume range for 10 mL vial	Extract volume range for 5 mL vial
Standardized extracts				
D. farinae/D. pteronyssinus	10,000 AU/mL	500–2000 AU	1–4 mL	0.5–2 mL
Cat hair/cat pelt	10,000 BAU/mL	1000–4000 BAU	2–8 mL	1–4 mL
Bermuda	10,000 BAU/mL	300–1500 BAU	0.6–3 mL	0.3–1.5 mL
Other grasses[a]	100,000 BAU/mL	1000–4000 BAU	0.2–0.8 mL	0.1–0.4 mL
Short ragweed	100 μg/mL	6–12 μg Amb a 1	1.2–2.4 mL	0.6–1.2 mL
	100,000 AU/mL	1000–4000 AU	0.2–0.8 mL	0.1–0.4 mL
Hymenoptera venom	100 μg/mL for single venom 300 μg/mL for mixed vespid	50–200 μg of each venom	NA	NA
Non-standardized extracts				
AP dog	1:100 wt/vol	15 μg Can f 1	2 mL	1 mL
Pollen	1:20 wt/vol	0.5 mL of 1:100–1:200 wt/vol	1–2 mL	0.5–1 mL
Mold/CR	1:20 wt/vol	Highest tolerated dose	NA	NA
Imported fire ant	1:20 wt/vol WBE	0.5 mL of 1:100 wt/vol[b]	2 mL	1 mL

Key: *AP* acetone precipitated, *AU* allergy unit, *Amb a 1* major allergen for ragweed, Antigen E, *BAU* bioequivalent allergy unit, *CR* cockroach, *NA* not applicable, *WBE* whole body extract, *wt/vol* weight/volume

[a] Standardized grasses include Kentucky blue grass, perennial rye, orchard, timothy, meadow fescue, red top and sweet vernal

[b] Recommended maintenance dose for IFA is 0.5 mL of 1:100 wt/vol WBE though maintenance doses as high as 0.5 mL of 1:10 wt/vol WBE have been used

C. Remove Kentucky blue grass and orchard allergens from the prescription since these are both standardized allergens that cross-react with Bermuda grass
D. Remove orchard grass from the prescription since inclusion of Kentucky blue grass will suffice due to cross-reactivity patterns; decrease Kentucky blue grass volume to 0.8 mL, increase tree allergen volume to 1 mL each, add 0.2 mL diluent to complete 10 mL vial

Answer: D

Discussion

Table 3 Allergen cross-reactivity patterns

Taxonomic relationship	Cross-reacting members	Taxonomic relationship	Cross-reacting members
Betulaceae family	Birch, alder, hazel, hornbeam	*Artemisia* genus	Sages, mugwort, wormwood
Cupressaceae family	Juniper, cypress, cedar	*Ambrosia* genus	Short ragweed, giant ragweed Western ragweed, false ragweed
Oleaceae family	Ash, European olive, privet	*Atriplex* genus	Saltbush, wingscale
Fagaceae family	Beech, oak, chestnut	Pooideae subfamily	Kentucky blue grass, meadow fescue Perennial rye, orchard, red top Sweet vernal, Timothy
Populus genus	Cottonwood, poplar, Aspen	Pyroglyphidae family	*D. pteronyssinus*, *D. farinae*

The initial aeroallergen prescription for the patient shown in Table 1 is notable for under dosing of tree allergens (as written the concentration for each tree allergen is at 1:400 wt/vol per 0.5 mL maintenance dose) and overdosing Kentucky blue grass and orchard allergens (as written the concentration for each of these allergens is 5000 BAU per 0.5 mL dose). See Table 2 for recommended effective dosing. Answer A is incorrect since Kentucky blue grass and orchard grass would remain at doses outside the recommended effective range. Answer B is incorrect as this answer does not resolve the under dosing of tree allergens and under doses Bermuda allergen. Answer C is incorrect because northern grasses do not cross-react with Bermuda grass (see Table 3). While there are several options to rewrite the aeroallergen prescription to ensure that all allergens are at projected effective doses, the best option of the answers provided is answer D. Answer D utilizes an appropriate cross-reactivity pattern (northern grasses), maximizes the dose of the utilized northern grass in the extract, and increases tree allergens to projected effective therapeutic doses.

Knowledge of cross-reactivity patterns allows for limiting the number of allergens in a SCIT prescription to ensure coverage of all clinically relevant allergens. Cross-reactivity amongst members of the same family or genus has been described for several allergens. Table 3 includes a list of cross-reacting allergens based on taxonomic relationships. If a patient has evidence of sensitization to multiple cross-reacting allergens, inclusion of a single allergen in an immunotherapy prescription is acceptable. For example, there is extensive cross-reactivity amongst northern grasses (subfamily Pooideae; e.g., Kentucky blue grass, orchard, perennial rye, timothy) so choosing one representative member to include in a SCIT prescription will provide coverage for other members of this subfamily. Knowledge of local aerobiology patterns can also help in that locally prominent species can be selected.

Question 3
The patient is offered immunotherapy to both aeroallergens and venom. He is concerned about recurrent venom stings and is wondering about rapid build-up schedules. Select the most accurate statement regarding build-up immunotherapy schedules:

A. Aeroallergen immunotherapy build-up can occur via conventional, cluster, or rush protocols without an associated increased risk of reactions
B. Compared to aeroallergen rush schedules, flying Hymenoptera rush immunotherapy schedules have not been associated with a similarly increased risk of systemic reactions
C. Premedication therapy such as H_1 receptor antagonists and H_2 receptor antagonists should only be considered for accelerated venom immunotherapy schedules and not aeroallergen immunotherapy accelerated schedules
D. Pretreatment with omalizumab has been associated with increased safety of cluster and rush protocols in patients with asthma and venom hypersensitivity but not in allergic rhinitis patients

Answer: B
Discussion
Immunotherapy consists of a build-up phase where increasing doses and concentration of an allergen extract are delivered until the maintenance phase is reached. The maintenance phase consists of the delivery of the therapeutic concentration of the allergen extract to the individual patient. Starting concentrations for immunotherapy depend on the allergen extract being used. For instance, aeroallergen immunotherapy (AIT) is typically started at dilutions of 1:10,000 volume/volume (vol/vol) or 1:1000 vol/vol. Flying Hymenoptera venom immunotherapy (VIT) is typically started at an initial dose of 1 μg for single venoms and 3 μg for a mixed vespid extract. IFA whole body extract (WBE) immunotherapy follows the AIT pattern in that the starting dilution is 1:10,000 vol/vol or 1:1000 vol/vol. In general, each SCIT kit contains three to four 10-fold volume dilutions from the maintenance concentrate. A conventional build-up schedule delivers increasing doses of allergen extract every 1–3 days over a 3–6-month window. Cluster immunotherapy delivers 2–3 doses of an allergen extract per visit, is typically associated with fewer scheduled doses, and allows for build-up to occur over a period of weeks. Rush schedules are designed to deliver multiple doses of an allergen extract over the course of days in order to reach the maintenance dose in that timeframe. Ultra-rush protocols have also been utilized, in particular for VIT and IFA WBE, where the maintenance dose can be achieved over the course of hours. Accelerated schedules are beneficial in that they allow the patient to achieve the maintenance dose and therapeutic effectiveness in a faster time frame compared to conventional schedules.

The benefit of cluster and rush schedule does come at a price of increased reactions depending on the allergen being used. For example, rush AIT schedules are associated with an increased risk of local and systemic side effects. Systemic reaction rates with aeroallergen rush protocols range from 15% to 100% in a non-premedicated cohort. Rush and ultra-rush VIT protocols, however, have not been associated with increased systemic side effects. In one study of 57 patients undergoing ultra-rush VIT, only 7% of patients experienced a mild systemic reaction. Premedication with an H_1 receptor antagonists prior to accelerated SCIT schedules has been shown to decrease the risk of systemic reactions associated with both aeroallergen and VIT rush protocols. Combination premedication regimens (e.g., H_1 receptor antagonists, H_2 receptor antagonists, and oral corticosteroids) similarly have a favorable profile in reducing the risk of systemic reactions with accelerated SCIT schedules. Finally, omalizumab has been shown to improve the safety of accelerated schedules for patients with asthma, allergic rhinitis, and venom hypersensitivity when used prior to or in conjunction with SCIT build-up.

Question 4
Immunologic changes with immunotherapy include all of the following except:

A. Increased levels of allergen specific IgG_1, IgG_4, and IgA have consistently been shown to correlate with clinical improvement
B. Increased IL-10 and TGF-β secretion by allergen specific T cells
C. Inhibition of IgE-allergen binding mediated by IgG_4
D. Increased IFNγ/IL 4 ratio

Answer: A
Discussion
Table 4 has a summary of the expected immunologic changes with immunotherapy. While IT is associated with increased IgG_1, IgG_4, and IgA, measurement of these immunoglobulins has not consistently been shown to be predictive of improved clinical symptoms. Additional immunoglobulin changes associated with IT includes an early increase in allergen specific IgE followed by a subsequent late decrease in allergen specific IgE. One of the functions of IgG_4 is that it serves as a blocking factor by inhibiting binding of IgE to an allergen. Increased production of allergen specific IgA is also seen in immunotherapy and provides a blocking function at mucosal sites. Cellular changes associated with IT include suppression of T lymphocyte proliferation, upregula-

Table 4 Expected immunologic changes with immunotherapy

	Humoral response	Innate response
Increased	Specific IgE (early)	Regulatory T lymphocytes
	Specific IgG_1 and IgG_4	Expression of Foxp3
	IgA at mucosal surfaces	IL-10 and TGF-β secretion
		IFNɣ/IL 4 ratio
		Th1 cytokines (IL 12, IFNɣ)
Decreased	Specific IgE (late)	Th2 cytokines (IL 4, IL 5, IL 13)
	Seasonal rise in specific IgE	Basophil reactivity
		Allergen specific lymphocyte proliferation

tion of $CD4^+CD25^+$ T regulatory cells (Tregs), increased production of IL-10 and TGF-β mediated by Tregs, and a shift from Th2 cytokines to Th1 cytokines (e.g., increased IFNɣ/IL 4 ratio). IL-10 itself is an inhibitor of T cell proliferation and is associated with increased IgG_4 and decreased IgE production. Finally, clinical changes expected with immunotherapy include decreased end-organ effect seen with decreased nasal or ocular challenge response as well as decreased respiratory response to non-specific bronchial challenge with histamine.

Question 5

The patient asks about duration of immunotherapy. Which of the following statements best describes the length of immunotherapy treatment that is recommended for this patient?

A. The recommended length of duration for a treatment course with aeroallergen and venom immunotherapy is 3–5 years for all patients
B. While a standard duration for venom immunotherapy is recommended for all patients, risk factors for relapse should be taken into account with consideration for prolonged treatment duration in those deemed to be at higher risk for relapse
C. Aeroallergen immunotherapy should always be continued indefinitely if the patient is experiencing clinical improvement regardless of length of duration of immunotherapy treatment course
D. Aeroallergen immunotherapy should be stopped only when decreased sensitivity is noted in a repeat skin test

Answer: B

Discussion

Venom immunotherapy, if indicated, should be continued for a minimum of 5 years. Risk factors for relapse include history of a severe initial sting reaction (e.g., syncope or hypotension), a systemic reaction while on VIT, honey bee allergy or in patients with an elevated basal serum tryptase measurement. Other factors that may prompt a longer duration of VIT include frequent exposures (e.g., beekeepers), underlying medical problems that may lead to an inability to tolerate anaphylaxis (e.g., cardiovascular disease), use of antihypertensive medications (e.g., beta blockers which may be associated with increased risk for or severity of anaphylaxis) and finally in patients that are anxious about recurrent stings that may impair quality of life. Mast cell disease has also been identified as a risk factor for increased risk for venom allergy, increased risk for reaction severity, and increased risk for relapse after stopping venom immunotherapy. The decision to continue VIT in these patients should be individualized, however, indefinite VIT should be strongly considered. Less is known about the optimal duration for IFA WBE immunotherapy. Therefore, consideration for treatment courses of longer than 5 years is recommended.

Aeroallergen SCIT efficacy should be monitored with routine follow-up appointments. Clinical response to treatment and tolerability should all be taken into account when determining prolonged SCIT length beyond the initial 3–5-year treatment course. Clinical remission may be noted in some patients after only a 3-year course of immunotherapy which may indicate that discontinuation of AIT after 3 years is reasonable. There are no available tests to evaluate which patients may be at risk for relapse after AIT discontinuation. For patients on AIT, skin test reactivity or serologic evidence of specific IgE do not correlate with clinical response or risk for relapse. While some experts have recommended repeat testing in venom allergic patients to evaluate for a decrease in specific IgE to undetectable or a negative skin test response, the current recommendations do not require repeat serologic or skin testing prior to the discontinuation of VIT.

Bibliography

1. National Asthma Education and Prevention Program. Third Expert Panel on the Diagnosis and Management of Asthma. Expert panel report 3: guidelines for the diagnosis and management of asthma. Bethesda, MD: National Heart, Lung, and Blood Institute (US); 2007. 440 pp. https://www.ncbi.nlm.nih.gov/books/NBK7232/.
2. National Asthma Education and Prevention Program. 2020 Focused updates to the ashtma management guidelines. Bethesda, MD: National Heart, Lung, and Blood Institute (US); 2020. https://www.nhlbi.nih.gov/health-topics/all-publications-and-resources/2020-focused-updates-asthma-management-guidelines.
3. Cox L, Nelson H, Lockey R, Calabria C, Chacko T, Finegold I, et al. Allergen immunotherapy: a practice parameter third update. J Allergy Clin Immunol. 2011;127(1 Suppl):S1–S55.
4. Calderon MA, Waserman S, Bernstein DI, Demoly P, Douglass J, Gagnon R, et al. Clinical practice of allergen immunotherapy for allergic rhinoconjunctivitis and asthma: an expert panel report. J Allergy Clin Immunol Pract. 2020;8:2920–36.
5. Greenhawt M, Oppenheimer J, Nelson M, Nelson H, Lockey R, Lieberman P, et al. Sublingual immunotherapy: a focused allergen immunotherapy practice parameter update. Ann Allergy Asthma Immunol. 2017;118:276–82.

6. Dykewicz MS, Wallace DV, Amrol DJ, Baroody FM, Bernstein JA, et al. Rhinitis 2020: a practice parameter update. J Allergy Clin Immunol. 2020;146:721–67.
7. Roberts G, Pfaar O, Akdis CA, Ansotegui IJ, Durham SR, Gerth van Wijk R, et al. EAACI guidelines on allergen immunotherapy: allergic rhinoconjunctivitis. Allergy. 2018;73:765–98.
8. Schneider L, Tilles S, Lio P, Boguniewicz M, Beck L, LeBovidge J, et al. Atopic dermatitis: a practice parameter update 2012. Allergy Clin Immunol. 2013;131:295–9.
9. Sampson HA, Aceves S, Bock SA, James J, Jones S, Lang D, et al. Food allergy: a practice parameter update—2014. J Allergy Clin Immunol. 2014;134:1016–25.
10. Bernstein JA, Lang DM, Khan DA, Craig T, Dreyfus D, Hsieh F, et al. The diagnosis and management of acute and chronic urticaria: 2014 update. J Allergy Clin Immunol. 2014;133:1270–7.
11. Turkeltaub PC, Rastogi SC, Baer H, Anderson MC, Norman PS. A standardized quantitative skin-test assay of allergen potency and stability: studies on the allergen dose-response curve and effect of wheal, erythema, and patient selection on assay results. J Allergy Clin Immunol. 1982;70(5):343–52.
12. Golden DB, Demain J, Freeman T, Graft D, Tankersley M, Tracy J, et al. Stinging insect hypersensitivity: a practice parameter update 2016. Ann Allergy Asthma Immunol. 2017;118(1):28–54.
13. Cox L. Advantages and disadvantages of accelerated immunotherapy schedules. J Allergy Clin Immunol. 2008;122(2):432–4.
14. Schiavino D, Nucera E, Pollastrini E, De Pasquale T, Buonomo A, Bartolozzi F, et al. Specific ultrarush desensitization in Hymenoptera venom-allergic patients. Ann Allergy Asthma Immunol. 2004;92(4):409–13.
15. Wesley Burks M, Holgate ST, O'Hehir RE, Bacharier LB, Broide DH, Khurana Hershey GK, Peebles S. Middleton's allergy: principles and practice, 2 vols. Philadelphia: Elsevier; 2019. 1840 p.

Index

M. Mahmoudi (ed.), *Absolute Allergy and Immunology Board Review*, https://doi.org/10.1007/978-3-031-12867-7

D

GPSR Compliance

The European Union's (EU) General Product Safety Regulation (GPSR) is a set of rules that requires consumer products to be safe and our obligations to ensure this.

If you have any concerns about our products, you can contact us on ProductSafety@springernature.com

In case Publisher is established outside the EU, the EU authorized representative is:

Springer Nature Customer Service Center GmbH
Europaplatz 3
69115 Heidelberg, Germany

Batch number: 10731401

Printed by Printforce, the Netherlands